CONSERVATION BIOLOGY

CONSERVATION BIOLOGY
Evolution in Action

Edited by
Scott P. Carroll
Charles W. Fox

OXFORD
UNIVERSITY PRESS
2008

OXFORD
UNIVERSITY PRESS

Oxford University Press, Inc., publishes works that further Oxford University's objective of excellence in research, scholarship, and education.

Oxford New York

Auckland Cape Town Dar es Salaam Hong Kong Karachi
Kuala Lumpur Madrid Melbourne Mexico City Nairobi
New Delhi Shanghai Taipei Toronto

With offices in
Argentina Austria Brazil Chile Czech Republic France Greece
Guatemala Hungary Italy Japan Poland Portugal Singapore
South Korea Switzerland Thailand Turkey Ukraine Vietnam

Published by Oxford University Press, Inc.
198 Madison Avenue, New York, New York 10016
www.oup.com

Library of Congress Cataloging-in-Publication Data

Conservation biology: evolution in action / edited by Scott P. Carroll and Charles W. Fox.
p. cm.
Includes bibliographical references and index.
ISBN: 978-0-19-530679-8; 978-0-19-530678-1 (pbk.)
1. Conservation biology. I. Carroll, Scott P. II. Fox, Charles W.
QH75.C6615 2008
576—dc22 2007044664

9 8 7 6 5 4 3 2 1

Printed in the United States of America
on acid-free paper

Preface

At no time in the nearly four billion years since the origin of life on earth has our planet seen such tremendous environmental change. Even the major mass extinctions in prehuman earth history (Raup & Sepkoski, 1982) are mere blips in comparison with the current biodiversity crisis. Human actions and impacts such as the elimination, fragmentation, and conversion of habitats; mass poisoning; overharvesting; species introductions; and climate change dramatically alter the local and global carrying capacities of other species. But they also do more. By modifying the challenges organisms face, and the resources they have to address those challenges, we are altering the conditions under which behavioral and physiological traits are expressed and in which ecological interactions occur. These changes affect the selective environments encountered by organisms, influencing evolutionary dynamics, which in turn feed back to affect ecological dynamics. Conservation problems are thus eco-evolutionary in nature, rather than just ecological, demographic, or genetic (Kinnison & Hairston, 2007). This ecoevolutionary nature of responses to environmental change is the focus of this book.

There is thus a clear and present need to develop practical approaches to managing our biodiversity problems that consider a role of evolution occurring during the time frame of the conservation program. Evolutionary theory is the predictive core of the biological sciences, and it provides the foundation for designing new and integrative strategies. Central to our perspective is the discovery that a great many organisms, from microbes to trees, are rapidly evolving in response to their changing environments. As risks and resources change in form, distribution, and abundance, they create new niches, affect competition, add or subtract enemies, and generally recast the landscape for surviving taxa. Selection is now operating in new directions and at new intensities, and the degree to which populations respond adaptively can determine their capacity to persist. Moreover, adaptive evolution, emerging from the demographic and genetic chaos suffered by "refugee taxa," may prove to be of foremost importance in altering the form and structure of species, interspecific interactions, and communities in the coming years, decades, and millennia.

If heretofore unanticipated, widespread evolution is itself a major component of global change. Static models—those that treat the ecological players as passive bystanders in the ecological play—are now obsolete. Understanding and managing ongoing adaptation to global change requires new perspectives to accommodate, exploit, and manage evolutionary processes of conservation concern, which include population structuring and the pace and extent of gene flow, the maintenance and expression of phenotypic polymorphisms and plasticity, niche specialization versus generalism, costs versus benefits of harvesting and of genetic engineering, and diversity management decisions above and below the species level. If we succeed in protecting species and biota demographically, a chief outcome may be to provide raw material for both targeted and unmanaged evolution. Accordingly, the principal challenge of evolutionary conservation biology is to predict and then manage evolutionary dynamics, and make conservation (and preservation) plans that maximize evolutionary potential—for example, by protecting communities that have unprecedented assemblages of juxtaposed, and rapidly evolving, remnant taxa. It is our hope that the authors of this volume provide insights that ultimately contribute to the success of such efforts.

This volume is intended to introduce, explore, and elaborate evolutionary approaches to conservation biology. The volume is divided into five parts, each of which is preceded by a brief introduction and commentary. The chapters in Part I, "Population Structure and Genetics of Threatened Taxa," present the history and general concepts of conservation genetics, and examine the interaction of genetic and demographic factors. Part II, "Conserving Biodiversity within and among Species," focuses on evolutionary processes, their relationship to biodiversity at different taxonomic levels, and how they influence practical conservation issues, including the reintroduction of threatened taxa and the loss of distinctive populations to hybridization. The chapters in Part III, "Evolutionary Responses to Environmental Change," examine both genetic and phenotypic modes of adaptation to the stresses and opportunities associated with global change phenomena. Part IV, "Conservation of the Co-evolving Web of Life," examines the evolutionary and co-evolutionary causes and consequences of changing interspecific dynamics, including species invasions, extinctions, and host parasite dynamics. The fifth and concluding portion of the volume, "Evolutionary Management," presents evolutionary analyses of three critically important areas: reserve design, management of transgene flow into the wilds, and the sustainable harvest of wild populations.

All chapters in this book were reviewed by peers, usually two or three scientists with expertise in the topics covered by the chapter. These reviewers offered insightful commentary on the chapters and have made this a much better volume. We thank Karina Acevedo-Whitehouse, Paul Agapow, Fred Allendorf, Suzanne Alonzo, Mike Angilletta, Tristan Armstrong, Leslie Blancas, Janette Boughman, Juan Bouzat, Linda Broadhurst, Jeremy Burdon, Mar Cabeza, Christina Caruso, Denis Couvet, Richard Cowling, George Gilchrist, John Kelly, Holly Kindsvater, Mike Kinnison, Mike Loeb, Arne Mooers, Patrick Nosil, Stephen O'Brien, Julian Olden, Otso Ovaskainen, William Perry, David Reed, Gerald Rehfeldt, Kevin Rice, Kim Scribner, Mike Singer, David Tallmon, John Thompson, Peter Thrall, Andrew Weeks, Alastair Wilson, and a few reviewers who asked to remain anonymous, for their constructive comments on individual chapters. We especially thank Mike Loeb for copy-editing chapters and compiling the final version of the book.

Last, and most important, we thank the authors for their dedication to this project. The success of this volume, and its influence on the conservation community, ultimately depends on the quality of the chapters and thus on the hard work, creativity, and insight of the contributing authors. Thanks to all of you!

Scott P. Carroll
Charles W. Fox

REFERENCES

Kinnison, M. T., & N. G. Hairston Jr. 2007. Eco-evolutionary conservation biology: Contemporary evolution and the dynamics of persistence. Funct Ecol. 21: 444–454.
Raup, D., & J. Sepkoski. 1982. Mass extinctions in the marine fossil record. Science 215:1501–1503.

Contents

List of Contributors

Allendorf, Fred
Division of Biological Sciences
University of Montana
Missoula, Montana 59812, USA

Altizer, Sonia
Odum School of Ecology
University of Georgia
Athens, Georgia, 30602, USA

Avise, John C.
Department of Ecology and Evolutionary Biology
University of California, Irvine
Irvine, California, 92697, USA

Benkman, Craig W.
Department of Zoology and Physiology
University of Wyoming
Laramie, Wyoming, 82071, USA

Boulding, Elizabeth Grace
Department of Integrative Biology
University of Guelph
Guelph, Ontario, N1G 2W1, Canada

Brook, Barry W.
Research Institute for Climate Change and Sustainability
School of Earth & Environmental Sciences
The University of Adelaide
South Australia 5005, Australia

Carroll, Scott P.
Department of Entomology and Center for Population Biology
University of California, Davis
Davis, California, 95616, USA

Dieckmann, Ulf
Evolution and Ecology Program
International Institute for Applied Systems Analysis (IIASA)
A-2361 Laxenberg, Austria

Dudash, Michele R.
Department of Biology
University of Maryland
College Park, Maryland, 20742, USA

Etterson, Julie R.
Department of Biology
University of Minnesota, Duluth
Duluth, Minnesota, 55812, USA

Faith, Daniel P.
The Australian Museum
Sydney, NSW 2010, Australia

Folk, Donna G.
Department of Biology
College of William & Mary
Williamsburg, Virginia, 23187, USA

Fox, Charles W.
Department of Entomology
University of Kentucky
Lexington, Kentucky, 40546, USA

Gilchrist, George W.
Department of Biology
College of William & Mary
Williamsburg, Virginia, 23187, USA

Grether, Gregory F.
Department of Ecology and Evolutionary Biology and Center for Tropical Research, Institute of the Environment
University of California, Los Angeles
Los Angeles, California, 90095, USA

Harrison, Susan
Division of Environmental Studies
University of California, Davis
Davis, California, 95616, USA

Heino, Mikko
Institute of Marine Research
N-5817 Bergen, Norway;
Department of Biology, University of Bergen
N-5020 Bergen, Norway;
Evolution and Ecology Program
International Institute for Applied Systems Analysis (IIASA)
A-2361 Laxenburg, Austria

Hendry, Andrew P.
Redpath Museum and Department of Biology
McGill University
Montreal, Quebec, H3A 2K6, Canada

Kinnison, Michael T.
School of Biology and Ecology
University of Maine
Orono, Maine, 04469, USA

Marvier, Michelle
Department of Biology
Santa Clara University
Santa Clara, California, 95053, USA

Murren, Courtney J.
Department of Biology
College of Charleston
Charleston, South Carolina, 29424, USA

Neel, Maile C.
Department of Natural Resource Sciences and Landscape Architecture, and Department of Entomology
University of Maryland
College Park, Maryland, 20742, USA

Ouborg, N. Joop
Department of Molecular Ecology
University of Nijmegen
6525 Ed Nijmegen, The Netherlands

Parchman, Thomas L.
Department of Biology
New Mexico State University
Las Cruces, New Mexico, 88003, USA

Pedersen, Amy B.
Department of Animal and Plant Sciences
University of Sheffield
Sheffield, S10 2TN, UK

Reed, David H.
Department of Biology
University of Mississippi
University, Mississippi, 38677, USA

Rhymer, Judith M.
Department of Wildlife Ecology
University of Maine
Orono, Maine, 04469, USA

Sambatti, Julianno B. M.
Department of Botany
University of British Columbia
British Columbia, V6T 1Z4, Canada

Siepielski, Adam M.
Department of Zoology and Physiology
University of Wyoming
Laramie, Wyoming, 82071, USA

Smith, Thomas B.
Department of Ecology and Evolutionary Biology and Center for Tropical Research, Institute of the Environment
University of California, Los Angeles
Los Angeles, California, 90095, USA

Stahl, Eli
Department of Biology
University of Massachusetts Dartmouth
North Dartmouth, Massachusetts, 02747, USA

Thompson, John N.
Department of Ecology and Evolutionary Biology
University of California, Santa Cruz
Santa Cruz, California, 95060, USA

Vergeer, Philippine
School of Biological Sciences
University of Leeds, UK

Vermeij, Geerat J.
Department of Geology and Center for Population Biology
University of California, Davis
Davis, California, 95616, USA

Watters, Jason V.
Chicago Zoological Society—Brookfield Zoo
Brookfield IL 60513

I

POPULATION STRUCTURE AND GENETICS OF THREATENED TAXA

The purpose of this book is to present the field of conservation biology in an evolutionary–genetic framework. In particular, our aim is to illustrate where evolutionary thinking has much to offer the field of conservation biology, but for which the importance of evolution and genetics is underappreciated (such as the conservation importance of natural selection). However, it is misleading to imply that evolutionary insight does not already play an important role in conservation biology. The risks of deleterious genetic changes through inbreeding in captive breeding programs, for example, have long been recognized, as has the loss of genetic variation in small natural populations. More recently, the field has begun to appreciate that human actions are changing the genetics of populations by altering selective environments, with consequences for both population viability and management strategies.

It is thus fitting that a book dedicated to the importance of evolutionary processes in conservation biology begins with chapters presenting the general concepts of conservation genetics. John Avise sets the stage for the rest of the book with a brief history of conservation genetics, sketching the maturation of this field to its modern state (chapter 1). Both Darwin and Mendel noted the consequences of inbreeding for populations, and many of the concepts underlying modern conservation genetics have been advancing since the origin of population genetics. However, before the 1970s, researchers only infrequently applied these concepts to species and problems of conservation significance (for example, in reserve design and the captive breeding of plants and animals [reviewed in Frankel & Soulé, 1981]). It was only after the introduction of early molecular techniques (electrophoresis) to the study of wild organisms (after 1966), and publication of a landmark paper by Frankel (1974), that conservation genetics coalesced as a tractable topic of empirical study. Subsequent publication of conservation biology textbooks and edited volumes that had substantial genetic and evolutionary components, including *Conservation Biology: An Ecological–Evolutionary Perspective* by Soulé and Wilcox (1980), *Conservation and Evolution* by Frankel and Soulé (1981), and the symposium volume *Genetics and Conservation* edited by Schonewald-Cox and colleagues (1983), solidified conservation genetics as a field of study. Recognition that evolution and genetics are relevant to conservation biology is now widespread (but see the later discussion).

Although evolutionary biology as a discipline originated with studies of natural selection (Darwin, 1859), conservation genetics has primarily been concerned with genetic variation within and among populations, and especially the genetic consequences of small population size. As David Reed notes in chapter 2, as populations decline in size they become more susceptible to stochastic processes (processes influenced by chance events). Although the outcome of a stochastic process can be predicted, it cannot, by definition, be known with certainty. This contrasts with natural selection, which is a deterministic process, meaning that specific outcomes are inevitable given a defined set of conditions. Both deterministic and stochastic processes act simultaneously in real populations, such that few "outcomes" of relevance to conservation biologists can ever be predicted with 100% certainty.

Reed explores ways in which stochastic processes affect population growth and viability. Casting his net broadly, he considers the influence of random variation in population age structure, reproduction, mortality, sex ratios, and dispersal (collectively, demographic stochasticity); random variation in extrinsic environmental variables that affect demography (environmental stochasticity); and random changes in allelic frequencies, such as accumulation of mutations or loss of alleles resulting from genetic drift (genetic stochasticity). We learn, for example, how loss of genetic variation and fixation of deleterious mutations reduces the mean fitness and adaptive potential of populations, contributing to population decline and, often, population extinction. Even when populations recover from bottlenecks, stochastic genetic changes that occurred when a population was small can have long-term consequences that include increased susceptibility of populations to future threats. Stochastic genetic events can also leave genetic "footprints" in populations from which we may infer a diversity of unobservable demographic and biogeographical characteristics of a lineage's history (Frankham et al., 2002). Such retrospective analyses may reveal both key past events and potential future vulnerabilities.

Even though biologists recognize the influence of genetic stochasticity on population growth rate, and thus population viability, the relative importance of genetic "problems" in comparison with demographic "problems" affecting the viability of small populations is widely disputed. This is a debate that Barry Brook addresses in chapter 3. Although the population sizes required to buffer against genetic threats are generally smaller than those required to buffer against nongenetic threats, this does not imply that genetic problems experienced by small or fragmented populations are unimportant. Empirical and theoretical studies of real extinction events demonstrate that no single cause, either genetic or demographic, is solely responsible for most extinctions. Instead, at small population sizes, a variety of stochastic hazards interact synergistically. For example, small populations tend to show increased genetic drift and inbreeding, both of which lead to a loss of genetic variation. Loss of genetic variation in turn affects mortality and reproductive rates, often reducing population growth rates and making populations more susceptible to stochastic events (environmental or demographic) that are ultimately the "cause" of extinction.

In nature, populations of most species exhibit considerable spatial structure, with regions of greater abundance separated by regions of low abundance or absence. Each subpopulation may face local extinction, so that species persistence becomes defined by the balance between local extinctions and the recolonization of empty patches. Metapopulation theory, developed to analyze these natural dynamics, has a sensitivity to loss and rescue that makes it well suited to addressing species conservation problems. Because of its emphasis on spatial structure, a chief application of metapopulation theory is to predict better the biotic consequences of anthropogenic habitat fragmentation.

As Julianno Sambatti and colleagues argue in chapter 4, in metapopulations, demography, genetics, and selection interact in ways that are often not considered in demographically stable species. Fragmentation not only reduces local population size, it also alters patterns of adaptation and gene flow among demes. For example, selection within highly fragmented metapopulations may reinforce adverse circumstances conducive to species collapse. Nondispersing (or selfing) genotypes may become favored within patches as a result of reduced reproductive success of dispersers or outcrossers. Moreover, even if dispersers successfully establish new populations, high local extinction rates may nonetheless eliminate dispersing genotypes altogether. Yet gene flow among populations created by such formerly adaptive individuals may be crucial to counterbalance inbreeding depression within the remaining populations. Metapopulation theory provides a framework for understanding how such ecological and evolutionary consequences of population fragmentation are inextricably linked, and points us toward clues about altered selection that may permit biologists to identify and monitor traits promoting species survival in fragmented landscapes.

We conclude this first section of the book by exploring the consequences of breeding and mating systems for the management of threatened and invasive species, using plants as a guide (Michele Dudash and Courtney Murren, chapter 5). In most species, not all individuals contribute offspring to the next generation. Species vary in the proportion of individuals and in which types of individuals reproduce, as well as in the amount of variation among individuals in reproduction. The most obvious consequence of such variation in breeding systems is that the relationship between census size and the number of individuals contributing genes to the next generation varies substantially among species, such that the breeding system generally must be known before basing management decisions on census data. Even among species that have similar breeding systems (and thus a similar proportion of reproductive individuals), how mates are chosen (the mating system, such as the frequency of selfing vs. outcrossing phenotypes) is often highly variable, and will affect the genetic structure of populations. Understanding mating and breeding systems will help to determine the genetic consequences of declining population size and population fragmentation.

The chapters of this section focus primarily on the conservation significance of genetic variability that occurs within populations, including metapopulations. However, a second major theme of early conservation genetics was the importance of genetic variation among conspecific populations. Current perspectives on the value of recognizing, conserving, and promoting the evolution of variation among conspecific populations are the focus of Part II of this volume.

Charles W. Fox
Lexington, Kentucky

Scott P. Carroll
Davis, California

1

The History, Purview, and Future of Conservation Genetics

JOHN C. AVISE

Direct and indirect effects of human population growth are precipitating sharp declines of biodiversity worldwide. The field of conservation biology has been defined as "a response by the scientific community to the biodiversity crisis" (Meffe & Carroll, 1997, p. 4). Biodiversity *is*, ultimately, genetic diversity, a product of evolutionary processes. Thus, the field of conservation genetics could be defined as "a response by the scientific community to the genetic diversity crisis." However, any definition this broad is unduly vague and fails to convey what practicing conservation geneticists actually do. At the other end of the spectrum, conservation genetics has sometimes been portrayed as a discipline devoted mostly to problems associated with inbreeding and the loss of adaptive genetic variation in small populations. However, any definition this narrow is unduly restrictive.

Perhaps a more useful approach is to define conservation genetics as the study of genetic patterns or processes in any context that informs conservation efforts. Theoretical population genetics and phylogenetics, as well as molecular and other empirical studies of genetic patterns and processes in captive and natural populations, have all played key roles in the emergence of conservation genetics as a recognizable subdiscipline of conservation biology. My goals in this review are the following: survey the extensive scientific literature that self-describes as being in the realm of conservation genetics; categorize major research themes within this field; comment on past accomplishments and future prospects with regard to each of those primary themes; and, lastly, outline a panorama of the burgeoning field of conservation genetics in the broader framework of conservation biology.

A BRIEF HISTORY OF CONSERVATION GENETICS

Before the 1960s, genetic properties of most species could be inferred only indirectly (and rather insecurely) via descriptions of organismal phenotypes such as morphologies and behaviors. Since then, a succession of powerful molecular technologies has given researchers direct access to voluminous genetic information stored naturally in nucleic acids and proteins. Patterns of genetic diversity within and among individuals, kinship groups, populations, species, and supraspecific taxa can now be investigated using molecular genetic data in addition to phenotypic traits. As this book attests, conservation biologists now incorporate genetic appraisals routinely in studies of plant and animal mating systems; behavior and natural history; population structure resulting from past and present demographic factors; gene flow, genetic drift, and selection; speciation, hybridization, introgression, phylogeny, systematics, and taxonomy; forensic identification of wildlife and wildlife products; and many additional topics relevant to conservation

TABLE 1.1 Some Historical Milestones in Conservation Genetics*.

1966	Lewontin and Hubby introduce allozyme methods to population biology.
1973	The U.S. Endangered Species Act sets a legal precedent to save rare taxa.
1974	Frankel publishes an article on genetic conservation as an evolutionary responsibility.
1975	Frankel and Hawkes edit a volume on genetic resources in crops. Martin edits a book on the captive breeding of endangered species.
1979	Ralls, Brugger, and Ballou draw attention to inbreeding depression in captive demes. Avise and colleagues as well as Brown and Wright introduce mtDNA methods to population biology.
1980	Soulé and Wilcox publish the first of several conservation books with an evolutionary genetic as well as ecological orientation.
1982	Laerm and colleagues publish perhaps the first multifaceted genetic appraisal of a wild, endangered species.
1983	Schonewald-Cox and colleagues edit the first major book on conservation and genetics. O'Brien and colleagues initiate studies on inbreeding effects in wild felids. Mullis invents polymerase chain reaction (PCR) for in vitro amplification of DNA.
1985	The Society for Conservation Biology is formed. Jeffreys and colleagues introduce multilocus DNA fingerprinting methods.
1986	Ryder brings the phrase *evolutionarily significant unit* to wide attention.
1987	Ryman and Utter edit a book on population genetics in fishery management. The journal *Conservation Biology* is launched. Avise and colleagues coin the term *phylogeography* and outline the field.
1988	Lande distinguishes genetic from demographic issues in small populations.
1989	The Captive Breeding Specialist Group initiates "population viability analyses" for endangered taxa. The U.S. Fish and Wildlife Service opens a wildlife forensics lab in Ashford, Oregon. Tautz (among others) introduces microsatellites as a source of polymorphic nuclear markers.
1990	Hillis and Moritz edit a book on molecular approaches to systematics.
1991	Vane-Wright and colleagues raise issues about phylogenetic diversity and conservation worth. Falk and Holsinger edit a volume on conservation genetics in rare plants.
1992	Avise introduces a regional phylogeographic perspective to conservation. Hedrick and Miller discuss genetic diversity and disease susceptibility in conservation. Groombridge edits a taxonomic and genetic inventory of global biodiversity.
1993	Thornhill edits a book on the natural history of inbreeding and outbreeding.
1994	Avise publishes the first major textbook on molecular genetic approaches in ecology, evolution, and conservation. Loeschcke, Tomiuk, and Jain edit a volume on conservation genetics. Burke edits an issue of *Molecular Ecology* on conservation genetics.
1995	Ballou, Gilpin, and Foose edit a book on genetic and demographic management of small populations.
1996	Avise and Hamrick as well as Smith and Wayne edit books on molecular approaches to conservation genetics. O'Brien initiates a biannual conservation genetics course, sponsored by the American Genetic Association and Smithsonian, that has trained many conservation geneticists.
1997	Hanski and Gilpin edit an important volume on metapopulations.
1998	Allendorf edits an issue of *Journal of Heredity* on conservation genetics in the sea.
1999	Landweber and Dobson edit a volume on genetics and species extinction. Wildt and Wemmer address reproductive technologies in conservation biology.
2000	The journal *Conservation Genetics* is launched. Avise publishes the first textbook on phylogeography.
2002	Frankham, Ballou, and Briscoe publish the first "teaching textbook" on conservation genetics.
2005	Purvis, Brooks, and Gittleman edit a book on phylogeny and conservation.

*With all due apologies to numerous additional authors whose works could justifiably have been cited as well.
See Meffe and Carroll (1997) for a history of the broader field of conservation biology.

efforts. Conservation genetic studies are often targeted on particular populations or species that are imperiled, but they can also be aimed at composite biotas or comparative themes, or toward uncovering conservation lessons from species that are not currently in danger of extinction.

The formal birth of conservation genetics occurred with the publication in 1983 of *Genetics and Conservation,* edited by Schonewald-Cox and colleagues. This nascent discipline had emerged from the well-established conceptual frameworks of population genetics, ecological genetics, quantitative genetics, evolutionary genetics, and phylogenetics, as now applied to biodiversity issues. Conservation genetics in 1983 was not a tightly knit field, but rather an ensemble of genetic approaches loosely united by a shared relevance to conservation efforts. I think that the same can be said of this eclectic field today.

Table 1.1 summarizes many of the milestone events, both before and after 1983, in the history of conservation genetics. Several breakthroughs

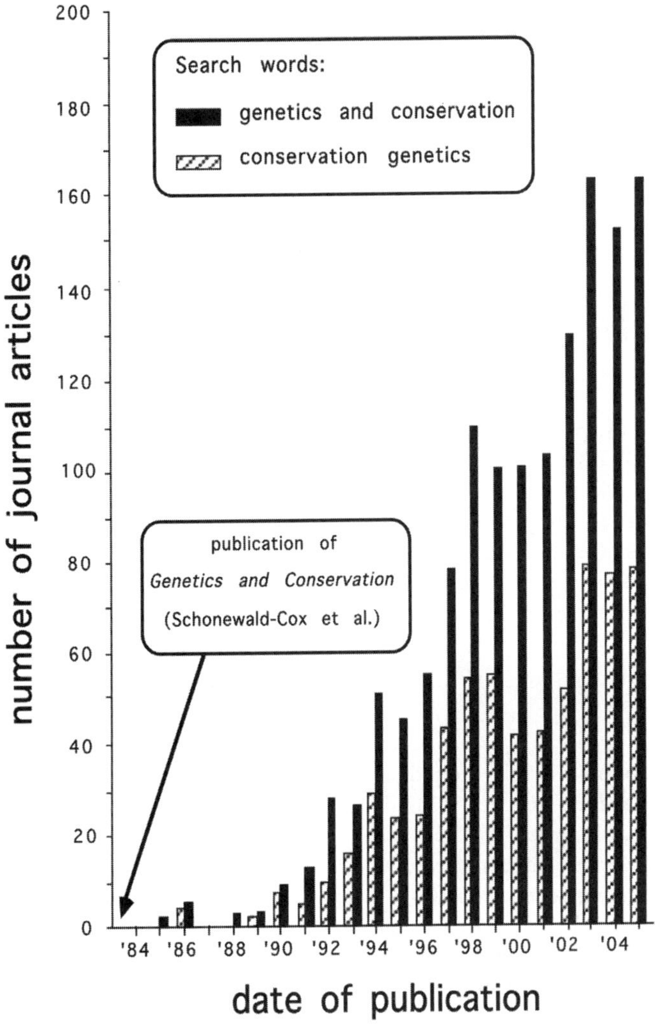

FIGURE 1.1 Number of conservation genetics articles (as identified in computer searches using the indicated key words) published from 1983 to 2006.

involved introductions of laboratory methods for revealing molecular genetic variation, followed by pioneering applications of each technique to conservation or management. Other landmark publications involved the introduction of key concepts to conservation genetics, journal reviews, edited volumes, and authored textbooks.

To quantify how the field of conservation genetics has grown since 1983, and to assess what its practitioners mean when they use the words conservation and genetics jointly, late in the year 2006 I conducted a search of the scientific literature. I used the computer database available at http://isiwebofknowledge.com and searched the terms genetics and conservation joined by the Boolean operator *and,* along with the operator-free search term conservation genetics. Figure 1.1 shows a temporal breakdown of the nearly 2,000 articles identified. These papers represent only a fraction of the genetics literature relevant to conservation, but they nonetheless provide a useful guide to the historical trajectory and to the traditionally perceived scope of the discipline.

After a lag in the 1980s, the number of conservation genetic publications per year has grown dramatically and consistently. The impact of molecular biology on the field is evidenced by the fact that approximately two thirds of all publications centered on analyses of protein or DNA data. The remaining studies that emerged from the computer searches involved conservation-relevant genetic theory, or empirical conservation assessments based on other data such as species lists, phenotypes, or biogeographic patterns.

MAJOR CONSERVATION GENETIC THEMES

The literature searches also provided an opportunity to identify and quantify research foci in conservation genetics. Many topical breakdowns are possible; I arbitrarily chose to categorize publications into the five primary subject areas pictured in the top half of Figure 1.2 What follows are some brief comments about each of these topics.

Variation within Populations

History and Purview

About 25% of journal articles found with the search term conservation genetics focused on genetic issues within small captive or natural populations. Loss of genetic variation under inbreeding—the result of mating among genetic relatives—was the most common theme in these articles, but a smaller number of studies addressed three related areas of research: the longer term demographic history of a single population as deduced by, for example, coalescent theory; parentage, kinship, or gender identification of relevance to captive breeding programs; or the microspatial dispersal of organisms in the context of natural history, reproductive modes, and mating systems in nature.

Deleterious effects of inbreeding have been understood for centuries (Darwin, 1868), and the genetic causes and consequences of inbreeding depression remain important research topics today (Brook, this volume; Hedrick & Kalinowski, 2000). For example, susceptibility to inbreeding depression has been quantified in many empirical studies, and the avoidance of inbreeding depression has been a major goal in the design of captive breeding programs and the management of small or isolated natural populations.

Many of the earliest studies identified in my literature searches addressed theoretical and empirical effects of inbreeding (and outbreeding) on fitness components such as viability and fertility. And following the introduction of molecular tools for population biology, the consequences of inbreeding could also be evaluated in terms of diminished heterozygosity at specific loci. This technological advance in turn led to the widespread use of multilocus molecular data (for example, from allozymes or microsatellites) to quantify genomic variation, which was sometimes used as a measure of population genetic "health," or adaptive potential.

These interpretations were prompted by reports of a positive correlation between heterozygosity and traits associated with enhanced reproduction, such as growth rate or disease resistance (Mitton, 1997). For several reasons, however, caution is indicated in concluding that observed levels of molecular variation predict population viability (Hedrick, 2001). One problem is that studies that show positive correlation between heterozygosity and fitness are more likely to be published than those showing a non-significant correlation, resulting in a publication bias. A second problem is that many molecular studies have been based on too few loci to rank order individuals (or even populations) reliably by heterozygosity. And lastly, magnitudes of variation in molecular markers often correlate poorly with quantitative genetic variation that is more likely the target of natural selection and thus the product of adaptive evolution (Reed & Frankham, 2001).

Representative Examples from the Literature Search

Laikre (1999) demonstrated high genetic load and severe inbreeding depression in zoo-maintained populations of brown bears, gray wolves, and lynx. A similar result was found by van Oosterhout and colleagues (2000) for captive populations of a butterfly species that is normally outbred in the wild. Authors of both studies discussed the ramifications of such findings for population management.

Frankham and colleagues (2000) revisited a longstanding theoretical prediction from quantitative genetics. Theory suggests that equalization of family sizes in controlled breeding programs should reduce genetic adaptation to captivity and thereby enhance prospects for successful reintroductions

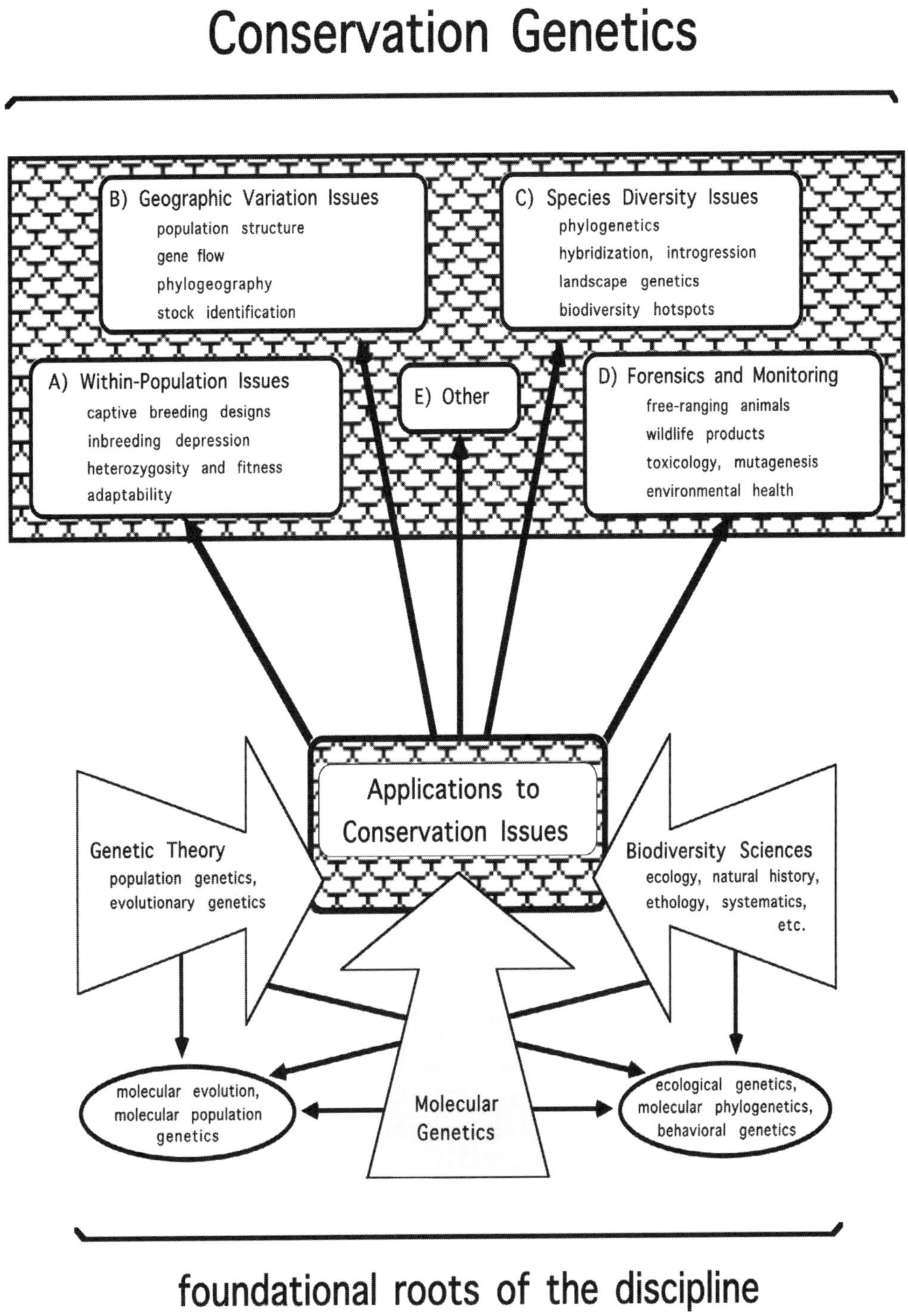

FIGURE 1.2 The purview of conservation genetics (upper half of figure), and the field's empirical and conceptual foundations within the time-honored life sciences (lower half of figure).

to nature. The authors cast doubt on the universality of this prediction when they showed that fruit flies raised for 25 generations under either equal or variable family sizes had similar reductions in reproductive success when returned to the wild. Armbruster and Reed (2005) reviewed the literature to test another outstanding prediction: Deleterious effects of inbreeding should be more evident in harsh environments than in benign environments. However, what they found was that inbreeding depression increased significantly under stress in only about 50% of the 34 studies they reviewed.

Other types of within-population assessments relevant to conservation are illustrated by the following studies. Spong and colleagues (2000) observed high heterozygosity at microsatellite loci in Tanzanian leopards. Using statistical inference, they concluded that the evolutionary effective population size was about 40,000 individuals, and that the number of leopards in this geographic region had been large and stable for several thousand years. In a different kind of intrapopulation conservation application, Sacchi and coworkers (2004) used sex-linked genetic markers to reveal the sex (otherwise unknown) of particular individuals in the endangered short-toed eagle.

Geographic Variation

History and Purview

Nearly 50% of conservation genetic articles in my literature searches involved comparisons among conspecific populations. These studies often addressed geographic population structure in particular species; spatial dispersal and gene flow, including both ongoing and historical patterns of genetic transfer among populations; genetic drift; or the delimitation of genetic and demographic stocks. Each analysis was typically conducted in the context of recognizing ecological and evolutionary sources of intraspecific genetic variation for conservation purposes.

Traditional population genetics and phylogeography were the two main areas of focus within this broader category of research. Papers in population genetics usually addressed geographic variation in allelic frequencies at allozyme or microsatellite loci, whereas studies of phylogeography often analyzed intraspecific gene trees of mitochondrial (mt) DNA. Although both these research programs seek to illuminate the causes and consequences of spatial genetic patterns in nature, population genetics is primarily concerned with contemporary forces molding populations. In contrast, phylogeography evaluates historical genealogical processes. The most powerful empirical studies to emerge from my literature search incorporated elements of both methodological tool boxes, and utilized data from multiple nuclear and cytoplasmic genes.

An important distinction in conservation biology is between management units (MUs) and evolutionarily significant units (ESUs). By definition, MUs are populations that currently exchange so few individuals as to be, in effect, demographically independent from one another at the present time (regardless of how recent or extensive their prior historical genetic connections); ESUs, by contrast, are populations with long histories of genetic separation. All ESUs are potential MUs, but not all MUs are ESUs. Population genetic and phylogeographic analyses have both been extremely useful in identifying otherwise cryptic MUs and ESUs in hundreds of imperiled and other species. For detailed examples from this vast literature, readers are directed to a recent extensive review by Avise (2000).

Species Diversity

History and Purview

This broad category includes genetic studies dealing with issues at and above the species level of taxonomy. About 12% of the author-identified "conservation genetics" papers were of this type. Given that a much broader literature exists on speciation, hybridization, introgression, and molecular phylogenetics, the articles identified in my database search clearly represent only a small fraction of evolutionary studies with potential relevance to biodiversity assessments and conservation.

Representative Examples from the Literature Search

Demarais and colleagues (1992) used cytonuclear markers to document that an endangered fish in the American Southwest (*Gila seminuda*) is the product of introgression between *G. robusta* and *G. elegans*. Similarly, Aparicio and coworkers (2000) confirmed by molecular analyses that a Spanish population of endangered plant, *Phlomis × margaritae,* arose via interspecific crosses. In an allied sort of conservation-relevant application, Ellstrand (2003b) reviewed evidence that most of the world's major food crops occasionally hybridize with wild relatives, a process that may compromise the genetic integrity of native progenitor species and perhaps even cause their introgressive extinction (in other words, genetic swamping [Rhymer, this volume; Rhymer & Simberloff, 1996]).

Outstanding issues regarding taxonomy and systematics have also been resolved. For example, Friesen and colleagues (1996) provided molecular evidence for elevating the taxonomic status of an endangered seabird to a full species. And in a similar vein, Hickson and coworkers (1992) used molecular genetics to justify splitting into separate species each of several morphologically cryptic forms of New Zealand skinks. With regard to phylogenetic patterns originating farther back in time, Bowen and colleagues (1993) discussed conservation ramifications of a molecular phylogeny for extant species of marine turtles, all of which are listed as threatened or endangered.

Considerable discussion has centered on phylogenetic distinctiveness as a measure of taxon "worth" when priority decisions are made regarding investment of finite time and conservation resources (Faith, this volume; Purvis et al., 2005). A basic notion is that unique (in other words, long-separated) evolutionary lineages contribute disproportionately to the planet's overall genetic diversity, such that their extinction would constitute a far greater loss of biodiversity than would the extinction of species that have extant close relatives. Although phylogenetic considerations can be important in particular instances, my own guess is that they will seldom override more traditional criteria used by societies to decide which species or biotas merit greatest protection (Avise, 2005). These conventional ranking criteria (explicit or implicit) often include a species' inherent charismatic appeal to humans, its rarity or restricted distribution per se, or its ecological or economic significance.

Wildlife Forensics

History and Purview

About 5% of the studies uncovered in my literature searches used molecular genetics in forensic identification. Two contexts were paramount: censuses of free-ranging animals using genetic samples collected from hair, feathers, skin, or feces; and ascertainment of the geographic source of confiscated wildlife products such as elephant tusks or rhinoceros horns. A great boon to forensic applications was the invention of the polymerase chain reaction (PCR), a relatively noninvasive molecular technique that permits in vitro amplification of specific DNA sequences from very small samples of tissue like that found in hair or a drop of blood.

Representative Examples from the Literature Search

In one study, Taberlet and colleagues (1997) used PCR-based analyses of microsatellite loci to determine the genotype of hair and feces from a wild population of endangered Pyrenean brown bears. Unfortunately, they concluded that the bear population consisted of just a few individuals. And Palsbøll (1999) reviewed applications and potential pitfalls to the use of molecular markers to "genetically tag" wild animals.

Illicit trade in animal products was investigated by Baker and colleagues (2000). These researchers used mtDNA sequences to identify protected cetaceans in commercially available stocks of whale meat. Similarly, Roman and Bowen (2000) showed with mtDNA markers that about 25% of "turtle meat" stocks in markets of Louisiana and Florida were incorrectly identified to source species.

Other Topics

Other studies captured in my database searches covered a miscellany of topics and genetic approaches that I could not easily pigeonhole into any of the aforementioned subject headings. For example, several papers examined in vitro maintenance of DNA banks, germplasms, or tissues, such as cryopreserved cells of endangered species (Ryder et al., 2000; Wildt et al., 1997). Another handful of distinctive papers explored the use of genetic markers as tools for monitoring mutagenic or other biological impacts of chemical toxins and pollutants. As examples, Street and colleagues (1998) assessed the genetic consequences of hydrocarbon spills on copepods, and Bickham and coworkers (2000) reviewed potential applications for population genetic data and models in the fields of biomonitoring and ecotoxicology.

To my surprise, the computer searches identified only a few articles involving genetic engineering in a conservation context. This gap in the literature is remarkable given the prominence of genetic engineering in arenas such as agriculture, animal husbandry, and environmental bioremediation (Avise, 2004b). Assessing the ecological or genetic costs and benefits of releasing genetically modified organisms into the environment (Wolfenbarger & Phifer, 2000) falls within the purview of conservation genetics. Similarly, a conversation-oriented perspective could be applied to engineering plant or animal species to carry, for example, transgenes that protect against pests or diseases (Adams et al., 2002). Although numerous such papers exist, they were seldom identified under the search term genetics *and* conservation.

Likewise, my targeted searches uncovered few "conservation genetics" articles involving methods of reproductive manipulation. These methods would include, for example, in vitro fertilization, artificial insemination, embryonic transfer, or organismal cloning. Again, this gap in the literature is surprising because many of these methods already are (and others may soon become) standard practice in captive breeding programs for endangered species (see, for example, Cohen, 1997).

Reproductive and genetic efforts often grade into one another. For example, more than 15 years of basic research into the refractory reproductive biology of cheetahs finally yielded successful techniques for artificial insemination using fresh or frozen–thawed sperm (Wildt & Wemmer, 1999). These methods in turn can be used in breeding programs to facilitate reproduction and to minimize inbreeding depression in this endangered cat.

FUTURE DIRECTIONS

Paradoxically, the field of conservation genetics will have ample room for growth as the world's biotas become increasingly stressed from the relentless pressures of human overpopulation. No crystal ball is needed to predict that as evolutionary lineages become increasingly threatened by human activities, conservation genetic studies will be expanded to many more populations, species, and higher taxa, as well as to additional ecological settings.

However, newly emerging molecular genetic technologies will offer unprecedented opportunities for exploring and understanding organismal genomes in ways that are germane to the theory and practice of conservation biology. In the following sections (which mirror the five topical areas in Fig. 1.2), I briefly highlight what I believe are excellent opportunities for empirical and conceptual advances as biologists further enter the genomics era.

Local Kinship

Many standard molecular genetics assays can be brought to bear on a host of within-population assessments. For example, offspring parentage, mating systems, determination of individual sex, and identification of clonal lineages are the types of problems that are now routinely addressed with molecular genetic assays. However, other within-population applications are only poorly addressed by the kinds of methods currently in wide use. One area of much-needed improvement is in estimates of coefficients of relatedness for other than full-sibs or parent–offspring pairs. Most studies to date lack the statistical power to infer with confidence the relationships among nondescendent kin such as half-siblings, cousins, grandparents, and so forth. In the near future, far more comprehensive genomic scans might be accomplished with markers such as single-nucleotide polymorphisms (SNPs), which permit individuals to be genotyped for a very large number of independent loci. Hundreds or even thousands of independent molecular polymorphisms will soon be available for many model, and perhaps nonmodel, taxa. These markers should help considerably in refining empirical estimates of genetic relatedness and assist molecular ecologists in their understanding of behavior and other phenotypes.

Conspecific Populations

A major current challenge in the field of phylogeography (especially for proper ESU identification) is to increase the number of independent gene genealogies examined within particular species. For biological and technical reasons, most phylogeographic studies to date have focused on mtDNA, but this maternally transmitted molecule carries only a minuscule fraction of any population's total hereditary history, most of which is collectively recorded at nuclear loci. Thus, although the biological hurdles (for example, intragenic recombination) and technical hurdles (haplotype isolation) remain high in many cases, in principle much stands to be gained by extending powerful genealogical appraisals to multiple nuclear loci.

Another challenge for the field will be to develop more realistic models that link current population genetic patterns to historical demographic processes. For reasons of mathematical tractability, much of traditional population genetic theory was built on equilibrium outcomes, but of course most natural populations are in continual or episodic flux in relevant parameters like population size and gene flow (see Reed, this volume). Phylogeographic perspectives (including coalescent theory and branching process models) have already moved population genetics toward greater realism by addressing some of the idiosyncratic, nonequilibrium demographic histories of particular populations and species. However, much remains to be accomplished, especially in developing a multilocus coalescent theory that tackles the expected variances across gene genealogies under alternative historical demographic scenarios. Hopefully, results of such theory could then be used to interpret genealogical data (analogous to those from mtDNA) that in the not-too-distant future might be gathered routinely in molecular surveys of large numbers of unlinked nuclear loci.

Supraspecific Issues

The genomics era will likewise offer—indeed, demand—information from many more loci in conservation applications that involve species and higher order taxa. For example, careful genomic study of large numbers of unlinked markers can provide important insights into variation in introgression patterns of natural populations, as well as illuminate the nature and frequency of horizontal gene transfer. Barriers to the hybridization-mediated exchange of genetic material between biological species are often semipermeable rather than absolute, but this phenomenon can only be revealed in multilocus assessments. Interspecific gene flow can affect organismal fitness, phylogenetic reconstructions, and species identifications, and in general be relevant to conservation efforts in many other ways.

Functional genomics is a vibrant branch of genetics that seeks to identify direct mechanistic links between genes and particular adaptations. Genes of the major histocompatibility complex, which influence resistance to pathogens and infectious disease, are just one example of loci in threatened or endangered species with patterns of variation that can provide functional information relevant to conservation efforts.

As science further enters the genomics era, another exciting opportunity will be afforded: to reconstruct, once and for all, the Tree of Life. Ideally, this collective scientific effort should include robust estimates not only of branching topologies, but also of ancestral nodal dates estimated with a molecular genetic clock that has been integrated with traditional fossil and biogeographic evidence. At least for many major taxa and clades, this grand phylogenetic synthesis should be completed within the next two decades, and it will stand as one of the grand achievements in the history of biology. Conservation biology will benefit from this massive tree reconstruction effort because improved evolutionary road maps of biodiversity will assist efforts to protect threatened taxa.

Forensic Applications

The goal of DNA bar coding methodologies is to use large-scale taxonomic screening of one or a few reference loci to assign individuals to species and to tease apart cryptically varying taxa (Hebert et al., 2003a). The most frequently used gene for this task is cytochrome *c* oxidase I (COI) from the mitochondrion. Forensic identifications—often of relevance to conservation—will undoubtedly be enhanced by COI sequencing, and any standardization of genetic methods and data has some inherent advantages. However, it should also be appreciated that basing forensic identifications on only one or a few genes has several potential pitfalls (Moritz & Cicero, 2004), and that ultimately much richer genomic characterizations will be desirable, especially in problematic situations.

Additional Topics

Evolutionary response to environmental changes (such as climate alterations, introductions of invasive species, and so on) is a primary theme of this current volume. Conservation genetics has close links to this topic, too, if for no other reason than that evolution is, by definition, genetic change across time. How will organisms and their genomes respond to current and near-future environmental challenges? The general answer is clear: They will respond exactly as populations have responded across the millennia—by adapting or by going extinct. The only differences between current ecological changes (such as global warming and habitat fragmentation fueled by human actions) and those of bygone eras are, arguably, the faster pace of many current ecological shifts, and, less arguably, the global pervasiveness of the environmental alterations.

There are two empirical generalities from quantitative genetics that are relevant to conservation biology. The first is that most species have the genetic capacity to adapt rapidly to environmental challenges; the second is that the capacity for adaptation has limits. Similarly, two relevant empirical generalities from population genetics are that most species are spatially structured, but at the same time, populations are historically connected at various temporal depths. Two generalities from evolutionary genetics are that adaptive evolution is pervasive, but so, too, is genetic extinction. Two empirical generalities from phylogenetics are that biodiversity can be exuberant and tenacious but, paradoxically, that it can also be fragile. Thus, the overarching question for conservation biology is what balance, if any, will be achieved in the coming decades between each of these counterposing genetic forces? How many populations will adapt in place to the environmental challenges and how many will succumb? How many populations will shift their ranges to track the new environmental alterations and how many will have no migration corridors or other means of dispersal to find suitable habitats? In general, what fraction of populations and species will make it through this critical "bottleneck century?" These are the kinds of questions that will increasingly occupy the attention of conservation biologists.

CONCLUSIONS

My attempt to characterize conservation genetics based on the field's self-described literature has identified several major aspects of the discipline beyond its more traditional focus on inbreeding challenges in small populations. Many of these extensions were made possible by the fact that molecular markers have opened the entire biological world for genetic scrutiny at many levels in life's genealogical hierarchy, ranging from parentage and kinship in local populations to deep phylogeny in the Tree of Life. However, data from these new molecular technologies would have been of little use in conservation efforts or elsewhere had it not been for the interpretive frameworks already provided (and since elaborated) by such time-honored fields as population and quantitative genetics, phylogenetics, and systematics. Defining the boundaries of conservation genetics will always be arbitrary to some extent because the field is intimately wedded to many of the other evolutionary and biodiversity disciplines.

Despite having compiled this overview, I do not wish to be interpreted as claiming any undue priority for genetic perspectives per se within the broader field of conservation biology. Genetic data and theory can be empirically and conceptually illuminating in many conservation efforts, but these endeavors are only a part of a larger mission. The truly pressing issue for the 21st century is the degree to which standing biodiversity, and the ecological and evolutionary processes that foster its maintenance, can be preserved at least quasi-intact for future generations. The ongoing biodiversity crisis is fundamentally a problem of environmental alteration and habitat loss caused by the collective weight of burgeoning human numbers. We have already destroyed a noticeable fraction of the planet's evolutionary genetic heritage. No genetic efforts, however valiant, can make more than a modest dent in solving the greater conservation problem—a challenge that will require full engagement of the life sciences as well as enlightened societal attitudes and steadfast political will.

SUGGESTIONS FOR FURTHER READING

Two edited volumes on molecular approaches to conservation genetics—one by Smith and Wayne and another by Avise and Hamrick—both appeared in 1996. An excellent authored textbook on conservation genetics is by Frankham and coworkers (2002), and a broader treatment of conservation biology is by Groom and colleagues (2005). Recommended source books that introduce various disciplines closely allied to conservation genetics are as follows: phylogeography (Avise, 2000), molecular phylogenetics and systematics (Hillis et al., 1996), molecular markers in ecology and evolution (Avise, 2004), inbreeding and outbreeding (Thornhill, 1993), and metapopulation biology (Hanski & Gilpin, 1997).

Avise, J. C. 2000. Phylogeography: The history and formation of species. Harvard University Press, Cambridge, Mass.
Avise, J. C. 2004. The hope, hype, and reality of genetic engineering. Oxford University Press, New York.
Avise, J. C., & J. L. Hamrick (eds.). 1996. Conservation genetics: Case histories from nature. Chapman & Hall, New York.
Frankham, R., J. D. Ballou, & D. A. Briscoe. 2002. Introduction to conservation genetics. Cambridge University Press, Cambridge, UK.
Groom, M. J., G. K. Meffe, & C. R. Carroll (eds.). 2005. Principles of conservation biology. 3rd ed. Sinauer Associates, Sunderland, Mass.
Hanski, I. A., & M. E. Gilpin. 1997. Metapopulation biology: Ecology, genetics, and evolution. Academic Press, New York.
Hillis D. M., C. Moritz, & B. K. Mable (eds.). 1996. Molecular systematics. 2nd ed. Sinauer Associates, Sunderland, Mass.
Smith, T. B., & R. K. Wayne (eds.). 1996. Molecular genetic approaches in conservation. Oxford University Press, New York.
Thornhill, N. W. (ed.). 1993. The natural history of inbreeding and outbreeding. University of Chicago Press, Chicago, Ill.

2

Effects of Population Size on Population Viability: From Mutation to Environmental Catastrophes

DAVID H. REED

During the past several centuries, an increasing rate of anthropogenic impacts on the global environment has caused a dramatic loss, degradation, and fragmentation of wilderness habitats, and has initiated a concomitant increase in extinction rates (see, for example, Lawton & May, 1995). Thus, one of the highest priorities in conservation biology is to understand how environmental quality, patterns of environmental stochasticity, genetic quality of individuals, the evolutionary potential of a population, and demographic stochasticity interact to determine persistence of populations affected by decreasing population size and gene flow.

Population size has very strong impacts on the viability of populations. This is expected from both ecological and evolutionary theory, and is strongly supported by observations of natural populations (reviewed by Reed et al., 2003c), by analysis of extinction rates calibrated against the fossil record, and by experimental results (see, for example, Belovsky et al., 1999; Reed & Bryant, 2000). Smaller populations are more vulnerable to extinction for three fundamental reasons: (1) they generally have less evolutionary potential and therefore are less capable of tracking a changing environment (see, for example, Reed et al., 2003a; Swindell & Bouzat, 2005), (2) they have decreased fitness and lower mean population growth rates (see, for example, Reed, 2005; Reed & Frankham, 2003), and (3) because their population growth rate (and therefore population size) is temporally more variable (see, for example, Reed & Hobbs, 2004; Thomas 1990).

This chapter outlines the effects of population size on the persistence of populations and explores the importance of evolutionary theory in guiding decision making in biological conservation. I focus primarily on stochastic threats to biodiversity and the interactions between stochastic and deterministic factors (Box 2.1). Stochastic threats facing populations are traditionally categorized as being demographic, environmental, or genetic in nature (Shaffer, 1981). I define and elaborate on these three forms of stochasticity later in this chapter. There is a tendency in the literature to use the terms *anthropogenic threat* and *deterministic threat* interchangeably. Yet, anthropogenic effects such as habitat conversion, harvesting, and so forth, can be stochastic (for example, tied to fluctuations in the local economy), and stochastic threats (for example, inbreeding depression) can lead to deterministic declines in population size. Furthermore, categories such as stochastic or deterministic are seldom as dichotomous as often portrayed. There is inevitably a mix of deterministic and stochastic factors interacting simultaneously and synergistically within populations. Observation and modeling of extinction events suggests that most extinction occurs after population size is reduced, either through natural or anthropogenic perturbations, to the point where stochastic factors may deliver the *coup de*

BOX 2.1 Deterministic and Stochastic Processes

A deterministic process is one that is predictable. Given certain starting conditions, the process proceeds to a fixed end point that does not vary among replicates (for example, populations). A stochastic process is inherently unpredictable and the end points do vary among replicates. Stochastic processes involve one or more variables that are probabilistic in nature. That is, the parameters vary, usually over time, and each time point represents a random draw from some probability distribution. Models of population dynamics usually include a mix of deterministic and stochastic factors. For example, population growth rates are typically both deterministic (for example, growth rate decreases on average with increasing density), but also stochastic (for example, changes in environmental quality and random changes in the demographic constitution of the population affect the population growth rate at each time step). Thus, population growth could be modeled stochastically using a random draw from a distribution of possible growth rates, with the mean growth rate changing in a predictable way with changes in density. Similarly, inbreeding leads to a decrease in fitness on average, but the specific effects of increases in the inbreeding coefficient will be unpredictable and dependent on the nature of the deleterious genetic load in the population prior to inbreeding (Armbruster & Reed, 2005).

grace (Fagan & Holmes, 2006; Lande et al., 2003; O'Grady et al., 2004; Reed et al., 2003a).

It is generally appreciated that demographic and genetic stochasticity increase as population size decreases. I will attempt to convince you that this is often true for environmental stochasticity as well. I devote the majority of this chapter to presenting an evolutionary framework for thinking about environmental stochasticity, genetic stochasticity, the interaction between these forces, and how persistence of populations depends on how population size affects population dynamics. It is important to keep in mind that it is not population size per se that leads us to believe that smaller populations are at greater risk of extinction relative to larger populations; but, rather, our expectations concerning the causal relationships among population parameters, such as temporal variation in population size, population fitness, and evolutionary potential.

AN EVOLUTIONARY FRAMEWORK

For an established population, the probability of extinction over a given period of time is determined by the long-term stochastic growth rate of the population (Box 2.2) and the carrying capacity of the environment. This holds true as long as environmental perturbations are not temporally autocorrelated and the temporal series of population growth rates is approximately normally distributed. The latter seems to be true for the majority of cases (Inchausti & Halley, 2001; Reed & Hobbs, 2004). The former is generally not true, but its impact on the relationship between extinction risk and the stochastic growth rate in natural populations is mostly unexplored.

The power of the stochastic growth rate to predict the likelihood of extinction is demonstrated in Figure 2.1, where the probability of extinction is given for three different species when time (25 generations) and carrying capacity (250 individuals) are fixed, but the stochastic growth rate is allowed to vary. The three graphs are nearly identical despite differences among species in life history and ecology. More important, the results are from computer programs that use very different types of model structure to provide estimates of extinction. The first species was simulated with a simple count-based model using the procedures described in Heering and Reed (2005), the second species was simulated using the individual-based Vortex software program (Lacy, 2000; version 9.50, Brookfield, Illinois, USA03.Carr; www.vortex9.org/vortex.html), and the third species was simulated using a cohort

BOX 2.2 Population Growth Rates and Stochasticity

The net replacement rate, R_0, is a measure of population growth. It can be defined as the mean number of female offspring produced per female in the population. If R_0 is more than one, the population is growing; if it is less than one, the population is decreasing, and a replacement rate of one means the population is stable. The net replacement rate is the sum of age-specific birth rates (m_x) multiplied by age-specific survivorship (l_x).

$$R_0 = \sum_{x-1}^{\infty} l_x m_x$$

Because populations are age structured and environments vary temporally, population growth rates are not constant. The magnitude of fluctuations in the population growth rate are just as important to the risk of extinction as the average growth rate. The net reproductive rate can be made stochastic by using the geometric mean of a time series of population growth rates, computed by taking the arithmetic mean of the natural log-transformed values of the population growth rates and then using the exponential function to transform this mean back to its original scale. The geometric mean is always less than or equal to the arithmetic mean, with the deviation between the two decreasing as the variance among the observed R_0 values decreases.

model (Reed, unpublished) with a stochastic food supply and indirect genetic effects.

One may be skeptical about the claim that the probability of extinction over a period of time is determined solely by the long-term stochastic growth rate. What about density-dependent reproduction and mortality? What about variation in life history? Do these details matter? These factors do indeed affect extinction risk. For example, the strength and form of density-dependent effects on reproduction or mortality affects temporal variation in growth rates (usually reducing variation) and therefore increases the stochastic growth rate relative to a model without density dependence. All three results represented in Figure 2.1 model density dependence in a different way: linear, a ceiling carrying capacity, and nonlinear, respectively. Changing the carrying capacity of the model does not change the shape of the curve, but does, however, shift the curve left (increasing K) or right (decreasing K). Changing the period of time over which extinction is to be measured also shifts the curve either left (decreasing time) or right (increasing time), but does not change the shape of the curve.

Given enough time, natural variation in population growth rates condemns to extinction even those populations with growth rates that are, on average, positive. Thus, we might predict that *organisms should evolve traits that maximize potential growth rate and minimize temporal variance in growth rates.* Because populations with low stochastic growth rates are more likely to go extinct, the mean potential growth rate among populations will increase over evolutionary time and the temporal variance in population growth rates will also decrease over time.

The notion that populations evolve to maximize potential growth rate is generally appreciated. However, the expectation that populations evolve to minimize temporal variance in growth rates—a form of fitness homeostasis—is not usually made explicit. Any process or mechanism that (1) decreases the maximum potential growth rate of the population, (2) increases temporal variation in population growth rate, or (3) decreases carrying capacity of the environment will increase the probability of extinction.

It has long been recognized that there is continuous variation in life history strategies such as

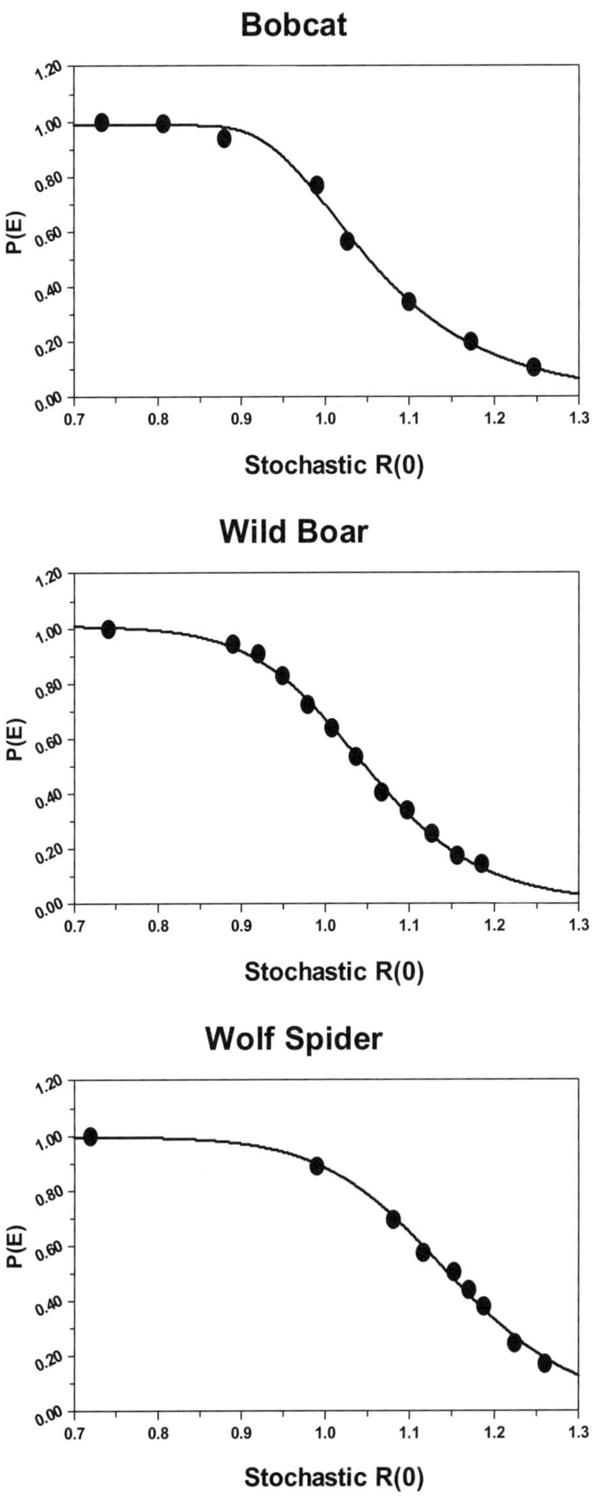

FIGURE 2.1 The relationship between probability of extinction, P(E), and the stochastic net replacement rate, R_0. All models use a different model structure and are for different species.

body size at reproductive maturity, life span, and number and size of offspring. Indeed, patterns of variation in life histories often reveal a fundamental evolutionary trade-off. Selection may favor large body size as a physiological buffer against environmental stochasticity, reducing temporal variation in population growth rates (even the frequency of environmental catastrophes scale to generation length [Reed et al., 2003b]), or a species can evolve small body size and increase maximum population growth rate by shortening the time until sexual maturity. Depending on how rapidly the environment is changing, one could even suggest that periods of mass extinction that are often (but not always) biased against larger organisms may in fact reflect a change in the cost–benefit ratio of the two strategies.

Life history trade-offs help to explain the observation that the net replacement rate (R_0) and the temporal coefficient of variation (CV) in population growth rates per generation are quantitatively similar for a broad array of taxa. Any taxonomic group of organisms that was able to evolve a life history strategy that avoided this trade-off (in other words, larger R_0 values and less temporal variation) would have a significant evolutionary advantage over competitors. This might also aid us in understanding some macroevolutionary patterns. Longer persistence times with increasing carrying capacity would suggest that being a carnivore might be a poor evolutionary strategy, because sitting on top of the food chain potentially limits the number of individuals the environment can support. In fact, mammalian carnivores have wide geographic ranges compared with other mammalian taxa and are rarely dietary specialists (Pagel et al., 1991). We would also expect maximum body size to scale with the number of individuals of that size that can form a long-term viable population. Consistent with this expectation, body mass of the largest carnivores and herbivores scales with available land area, and those numbers differ for ectotherms versus endotherms (Burness et al., 2001).

This summary of theory and observation suggests several important lessons. First, all else being equal, increasing temporal variation in population growth rates decreases persistence times. Second, decreasing the mean population growth rate decreases persistence times. Third, decreasing the carrying capacity of the environment decreases persistence times. It is important to note that these relationships are probably not linear. For example, depending on the initial mean population fitness, the effects of decreasing fitness on the probability of extinction could range from negligible to devastating (Fig. 2.1). This also implies that habitat degradation (environmental stress) and inbreeding depression have very similar effects on population viability, although their root causes and management options for amelioration are very different.

FORMS OF STOCHASTICITY

The dynamics of a population determine its extinction risk. Population dynamics are determined by a mix of deterministic and stochastic forces, and the ability to predict population dynamics is fundamental to conservation biology and population ecology (Lande et al., 2003). What follows is a detailed description of what is considered to be the three major stochastic forces influencing patterns of population fluctuations.

Demographic Stochasticity

Demography is a description of population age structure as it relates to fecundity rates, mortality rates, sex ratios, and sex-specific dispersal rates. One way to define demographic stochasticity is: the variation in population growth rates that arises through the sampling variance in demographic rates. Demographic stochasticity is inversely proportional to population size; as long as the population size is relatively large, effects of stochasticity are small and thus the average for various demographic rates provides an accurate description of population dynamics.

Demographic stochasticity has important consequences for population dynamics. In particular, individual variation in birth and mortality rates can cause the rate of population growth to fluctuate randomly even in a constant environment. As noted, an increase in temporal variation of population growth rates decreases the stochastic population growth rate and persistence times. The relative importance of demographic stochasticity to extinction is somewhat controversial (Brook, this volume). Lande (1988) argues that demography may usually be of more immediate importance than population genetics in determining the persistence of wild populations. However, Lande (1993) suggests that demographic stochasticity is probably a significant factor only in populations of 25 individuals or less, which is well below the threshold where genetics becomes a concern (see, for example, Lande, 1994; Reed & Bryant, 2000). Others suggest population size must exceed 50 or even 100 individuals. Fox and Kendall (2002) argue that the importance of demographic stochasticity is overestimated in most models of extinction because individual heterogeneity in demographic rates that are constant throughout an individual's life are ignored (for example, an individual of higher genetic quality might have a lower than average probability of mortality at each stage of life).

The relative contributions of demographic and environmental stochasticity to variation in population growth rate can rarely be directly measured (but see Vucetich & Peterson, 2004). However, if we have a time series of population sizes and we assume that the number of births and deaths is approximately Poisson distributed, the demographic variance is equal to $(1 + r)/N$, where r is the mean intrinsic rate of growth of the population over the census period and N is the mean population size during the same time period. We can then divide this predicted variance in r as a result of demographic stochasticity alone by the total variance in r. Figure 2.2 illustrates the relationship between the percentage of variance in r resulting from demographic stochasticity and the mean population size. The graph suggests that demographic and environmental stochasticity are equal when $N \sim 10$, that environmental variation is 10 times as large at $N \sim 150$, and is 100 times as large at $N \sim 10,000$. These time series were selected from the Population Dynamics Database (Natural Environment Research Council, 1999) on the basis of their quality and the length of the census period (see Reed & Hobbs, 2004), but the observations still contain error variance that will inflate total temporal variation slightly and decrease the relative importance of demographic stochasticity.

Environmental Stochasticity

Populations are subject to constant variation in environmental influences such as food supply, density of parasites and competitors, rainfall, and temperature. Lande and colleagues (2003) defined environmental stochasticity as temporal fluctuations in mortality and reproductive rates that affect all individuals within a population in the same or a similar fashion. They suggest that the impact of environmental stochasticity is roughly the same for small and large populations. Under this view, separation of the temporal variation in population growth rates into demographic and environmental causes depends on being able to ascertain the proportion of variation that explicitly depends on population size.

Most of us have an intuitive sense of what environmental stochasticity is: changes in the environment that directly change the mean birth and/or death rates in a population. Defining demographic and environmental stochasticity in a satisfying way is difficult. Moreover, partitioning the total temporal variation among different forms of stochasticity is, in practice, almost impossible, and often leads to very crude estimates such as the ones I present in Figure 2.2.

A major problem with defining environmental variation is a result of the fact that individuals and genotypes often do not respond to changes in the environment in a homogeneous fashion. Genotype-by-environment interactions for fitness are widespread, and a single response to an environmental perturbation is unlikely to exist. Individuals differ in their resistance to starvation, desiccation, and disease because of genotype or body condition before exposure to these stresses. Another problem is that it can often be difficult to separate demographic and environmental forms of stochasticity, not just mechanistically but philosophically. For example, a taxon with environmental sex determination such as in snapping turtles (*Chelydra serpentina*) could experience an unusually warm season and thus produce broods of hatchlings that consist entirely of females. Would such a striking variation in sex ratio, and the accompanying variation in population growth rates, be attributed to demographic or environmental stochasticity?

Furthermore, despite the definition provided by Lande and colleagues (2003), I would suggest that smaller populations may often experience greater amounts of environmental stochasticity than larger ones. For example, a fire that destroys all the forest habitat occupied by a squirrel species in a small patch of old-growth forest may destroy only a small proportion of a larger habitat patch. Similarly, a hail storm that destroys all the eggs of the Nile crocodile (*Crocodilus niloticus*) along the banks of particular lake will not have nearly as detrimental consequences for the species across a much larger geographic area. Evidence from the fossil record supports this intuitive notion. A positive

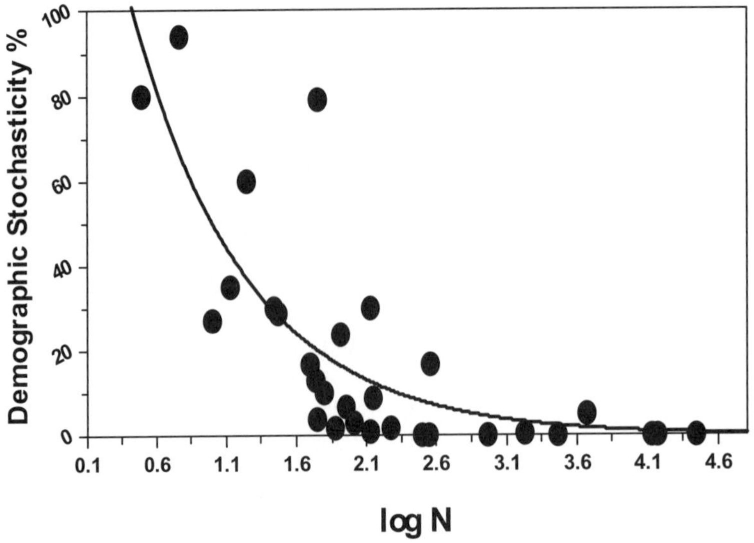

FIGURE 2.2 Relationship between population size (log N) and the proportion of the total temporal variance in population size attributable to demographic stochasticity. (Data are from the Global Population Dynamics Database [Natural Environment Research Council, 1999].)

correlation exists between geographic distribution and persistence in species of Miocene carnivores (Viranta, 2003) and marine invertebrates (Jablonski et al., 1985). However, there is a confounding positive correlation between local abundance and geographic range. Rosenzweig (1995) presents an analysis of unpublished data on 66 species of ants from Barro Colorado Island (Panama). The analysis suggests that species persistence through an El Niño drought was significantly affected by both population size and by ubiquity. An effect of geographic range is predicted by theory even without an increase in absolute numbers. This is the "don't-put-all-your-eggs-in-one-basket" argument that frequently arises in discussions of metapopulation structure or reserve design. However, it is not as well appreciated that even continuous blocks of habitat capable of supporting viable populations of vertebrates will contain considerable habitat heterogeneity and provide a buffer against most forms of environmental stochasticity (Reed, 2004).

Genetic Stochasticity

I will define genetic stochasticity as the degree to which the fate of new mutations, or preexisting genetic variation, is determined by stochastic rather than deterministic processes. The two major detrimental effects of genetic stochasticity are (1) a lowering of population fitness through fixation of alleles other than the one that provides the highest fitness and (2) a loss of potentially adaptive genetic variation (evolutionary potential). Others define genetic stochasticity as simply random genetic drift; sometimes it is defined as inbreeding depression, loss of genetic diversity, and the accumulation of deleterious mutations. The latter two in the group of three might better be described as the *results* of random genetic drift.

Genetic stochasticity differs from demographic stochasticity in that it affects primarily the mean population growth rate rather than the temporal variability of the growth rate (environmental stochasticity generally affects both). Nonetheless, it is likely that inbreeding also increases variability in population growth rates by, for example, amplifying demographic stochasticity (inbreeding can lead to skewed sex ratios) or, more important, by magnifying the effects of environmental perturbations through genotype–environment interactions, leading to inbred populations being more sensitive to environmental stress (Armbruster & Reed, 2005; Reed et al., 2007a, b).

Loss of Fitness

Inbreeding depression is a decrease in fitness that results from increasing homozygosity across the genome. Increased homozygosity can be the result of drift removing genetic variation faster than mutation can replace it. Alternatively, inbreeding depression can be the result of increased homozygosity within individuals without the loss of genetic diversity at the population level (for example, through consanguineous mating). Loss of fitness seems to be primarily caused by the increased expression of (partially) recessive deleterious alleles, but is also due in some part to increased homozygosity at overdominant loci. Inbreeding is frequently the result of habitat fragmentation or other factors that limit gene flow.

Although mechanistically distinct, the population genetic consequences of inbreeding and drift are similar and can be difficult to separate in natural populations. Authors often use the term *inbreeding* for effects of "small population size," whether this effect is the result of random genetic drift, nonrandom mating, or a combination of both. It is important that authors are clear about which mechanisms are being lumped under the umbrella term of *inbreeding,* because different mechanisms may lead to different outcomes or may require different management solutions.

Despite their deleterious nature, alleles can become fixed through random genetic drift (Box 2.3). Moreover, as population size declines, the fixation of deleterious, but effectively neutral, alleles can cause significant declines in fitness. The accumulation of deleterious alleles is sometimes listed as a separate phenomenon from random genetic drift. However, the accumulation and fixation of slightly deleterious mutations is the eventual outcome of drift, provided this process continues long enough.

Extinction through the accumulation of deleterious alleles has been called *mutational meltdown* (Lande, 1994; Lynch et al., 1995b). Mutational meltdown is a feedback process by which a decline in fitness resulting from the increasing frequency of deleterious alleles reduces population size. Reduction in population size in turn causes accelerated fixation of deleterious alleles, further suppressing population size and eventually causing growth rates to become negative—leading to extinction. The risk of extinction to small ($N_e < 50$) populations of sexually reproducing species, through the accumulation of mildly deleterious alleles, has been supported empirically in the laboratory and in natural populations (see, for example, Fry, 2001; Higgins & Lynch, 2001; Reed & Bryant, 2000; Rowe & Beebee, 2003; Zeyl et al., 2001).

Loss of Adaptive Potential

Besides loss of fitness, random genetic drift causes a reduction in the amount of genetic variation contained within a population. Although the loss of fitness through inbreeding depression is often viewed as an immediate threat to population persistence, the loss of genetic variation is often viewed as a long-term threat that limits the evolutionary potential of populations. Although this dichotomy has some truth to it, it ignores the complex interactions among the environment, fitness, and evolutionary potential.

It is well established that the amount of genetic diversity a population contains is strongly and positively correlated with population size. Both theoretical treatments and empirical results have strongly concluded that genetic variation, whether measured as heritability, heterozygosity, or allelic diversity, correlates positively with evolutionary potential (see, for example, Eisen, 1975; Jones et al., 1968; Reed et al., 2003a; Swindell & Bouzat, 2005). Some counterexamples exist from laboratory studies with experimentally bottlenecked populations. However, these bottlenecked populations are proving to be exceptions rather than the rule. Most bottlenecked populations experience deep declines in fitness accompanying their rather modest increases in heritability, and the increases in evolutionary potential are often environment specific and may be transient. The greater adaptability provided by increased genetic diversity is illustrated by the experimental hybridization of two fruit fly species from the genus *Dacus.* Although hybridization lowered fitness of the earliest generations, hybrids nevertheless had greater evolutionary potential than either species alone over the course of 16 generations of selection for thermal stress. Indeed, hybrids eventually outperformed both parental species (Lewontin & Birch, 1966).

A rapidly growing body of literature links increased genetic variation with increased resistance to infectious diseases and parasites (see, for example, Acevedo-Whitehouse et al., 2003; Altizer & Pederson, this volume; Baer & Schmid-Hempel, 1999; Hale & Briskie, 2007; Reid et al., 2003;

BOX 2.3 The Evolutionary Fate of Alleles in Natural Populations

An allele can be either deleterious or beneficial. However, despite an allele's effects on fitness, it may be effectively neutral because its eventual loss or fixation is determined solely or primarily by random genetic drift, rather than by selection. This occurs when $s < 1/2N_e$, where s is the selection coefficient and N_e is the effective population size (Kimura, 1983). If $s \ll 1/2N_e$, then drift is much stronger than selection and the fate of the allele is essentially stochastic.

There are very little data on the distribution of selection coefficients against deleterious alleles in natural populations, and what is available is extremely heterogeneous. The distribution is undoubtedly different for new mutations as opposed to standing genetic variation that has already been filtered by natural selection. Reasonable estimates for the median selection coefficient might be in the neighborhood of 0.01, with the distribution of selection coefficients following a log-normal distribution. If this estimate is accurate, it can be see from the equation presented earlier that a population with an effective population size of 50 individuals would be effectively neutral at half its polymorphic protein coding loci. An effective population size of 2,500 might be needed for selection to prevail at 95% of the polymorphic loci (all loci with $s \geq 0.0002$). Observed effective population sizes seem to be in the neighborhood of 10% to 20% of census size.

Spielman et al., 2004a). This is particularly noteworthy because disease is an extremely important factor in determining the geographic range of a species; and, moreover, epidemics can strongly affect population persistence.

Mutation

Given its importance, it is somewhat astonishing that discussions of genetic stochasticity rarely mention mutation (but see Houle & Kondrashov, 2006). Mutation is probably the poster child for a stochastic process in genetics because (1) both the generation of mutational variation and the immediate fate of those mutations is stochastic, (2) the generation and eventual fixation or loss of new mutations is heavily dependent on effective population size, and (3) mutation is the source of all heritable phenotypic novelty. In fact, the entire evolutionary process is dependent on the mutation rate, the distribution of mutational effects on fitness, and the norm of reaction for those mutations across relevant environmental conditions. Theory shows that large populations have a twofold advantage when it comes to beneficial mutations—specifically, large populations produce a larger number of beneficial mutations and those beneficial mutations that arise are less likely to be lost through drift.

In the following list I outline some important points about mutation:

1. Larger replicates of genetically identical starting populations increase in fitness faster than smaller replicates (see, for example, Estes et al., 2004). This pattern may be mostly the result of the increased number of beneficial mutations rather than the accumulation of deleterious mutations.
2. The rate of fitness increase is much faster in genetically identical replicate populations that are adapting to a novel environment than in an environment to which these populations are already adapted (see, for example, Giraud et al., 2001).
3. Larger mutation rates are more beneficial in a novel environment (see, for example, Giraud et al., 2001).
4. The majority of mutations are deleterious. However, the exact proportion of mutations that is beneficial is controversial. Reasonable estimates range anywhere from 0.01% to 10% of all mutations (see, for example, Lynch & Walsh, 1998). Heterogeneity in reported estimates is not surprising, because the proportion of mutations that are beneficial depends on the relative novelty of the environment, the fitness measure used, and

the ability of the researcher to detect mutations of small effect.

5. Deleterious mutations are probably dominated by mutations of relatively large effect. Average selection coefficients against new deleterious mutations are in the neighborhood of 2% to 5% (see, for example, Lynch & Walsh, 1998). It should be kept in mind that these are averages and the distribution is certainly positively skewed.
6. There are very few estimates of selection coefficients for new beneficial mutations, but one study estimates the mean selection coefficient at 0.02 (Imhof & Schlötterer, 2001).
7. The probability of fixation for an advantageous mutation can be affected by the initial frequency of a mutation, such as with premeiotic clustering of mutations (Woodruff et al., 2004).

Which Form of Stochasticity Is Most Important?

Other authors in this volume have weighed in on the debate regarding the form of stochasticity that most significantly influences population persistence (Boulding, this volume; Brook, this volume). Table 2.1 shows the suggested scaling of mean time to extinction with population size under demographic, environmental, and genetic stochasticity separately. Discussion of the relative importance of different forms of stochasticity is similar to the oftentimes acrimonious debate over the role of nature versus nurture in human behavior—another question that has no definitive answer. For instance, if one has a (hypothetical) genetically homogenous experimental population, one might conclude that the environment is necessarily responsible for all behavioral variation among individuals. However, controlling for genetic variation does not mean that genes are unimportant and do not influence the behavioral trait in question. Or to take a counterexample, dwarf versions of the Nile crocodile (*C. niloticus*) exist in very marginal habitats. Differences in the size of the crocodiles might be entirely the result of the fixation of different alleles in the two populations. However, this would not render environment unimportant. The reason these different alleles were fixed in these two populations probably reflects differences in selection resulting from the relative abundance of resources at each location. Laboratory experiments or computer simulations of extinction will suffer the same types of flaws. As soon as the parameters and variables (initial population size, carrying capacity, time, and so forth) in the experiment or model are set, the conclusions are limited to that set of parameter space and may or may not reflect natural populations. Models also are limited regarding what they can tell us because they can only reflect reality to the extent to which we have inputs that reflect reality.

You would probably never speak of half of your eye being the result of genes and half of your eye being the result of the environment. You also cannot partition an extinction event so that half of it is the result of environmental stochasticity and half is the result of genetic stochasticity. It

TABLE 2.1 Scaling of Mean Time to Extinction as a Function of Carrying Capacity or Effective Population Size (N_e) under Demographic, Environmental, and Genetic Stochasticity Independently.

Risk Factor	Mean Time to Extinction
Demographic stochasticity	$(1/K)\varepsilon^{2Kr/V}$
Environmental stochasticity	$K^{2r*/V(e)-1}$
Fixation of new mutations (variable s)	$N_e^{1+1/cv}$

K is the carrying capacity of the environment, r is the population growth rate, r^* is the mean value of r, V is the variance among individuals in their Malthusian fitness, V(e) is the temporal variance in the population growth rate resulting from changes in the environment, and cv is the coefficient of variation of selection coefficients (s) for new mutations squared.

Adapted from Table 3 of Lande (1995).

turns out that for ecologically important traits, the environment and genes are both important, and the interaction between them is where the really interesting stuff occurs. I suspect that the really interesting stuff about extinction is in the interactions among the various factors contributing to population persistence as well.

Part of the debate over identifying the form of stochasticity that is most important is actually an argument over the relevance of inbreeding depression, and genetics generally, to conservation. Those arguing against the importance of genetics in conservation point to populations with low genetic variation (as measured by molecular markers) that have recovered admirably from population crashes, populations that have been thriving for many generations at small population sizes (although some of these populations are heavily managed or are domesticated species released to islands that lack predators), and point to the fact that extinctions on islands seem to be mostly the result of the introduction of mammalian carnivores (Blackburn et al., 2004) and habitat destruction rather than genetic causes.

Those arguing for the importance of genetics point to the dramatic recovery of populations after the introduction of new genetic material (Madsen et al., 1999; Pimm et al., 2006; Vilà et al. 2003; Westemeier et al., 1998), sometimes after attempts at ecological restoration failed. They will also refer to studies that have clearly demonstrated the contribution of increased inbreeding levels to elevating the probability of extinction in natural populations of plants and butterflies (Newman & Pilson, 1997; Saccheri et al., 1998), albeit only in highly inbred populations.

There are two points I want to make about the role of genetics in population persistence. First, one of the important take-home points from Figure 2.1 is the sharp transition from a viable population, to an at-risk population, to a doomed population. It is easy to translate this result to realistic contexts in conservation. For example, if a population of birds is inoculated against disease and then supplemented with food in times of shortage, the population will probably have stochastic growth rates large enough that all but the most severe fitness loss resulting from inbreeding will have little or no effect on the short-term persistence of the population. For many natural populations, growth rates will be high enough that ecological considerations will trump genetic concerns, but for others such will not be the case.

Second, it is not surprising that most populations that are reduced to a few tens of individuals, or less, are able to rebound to much larger population sizes and a small proportion of others become fixed for deleterious alleles that permanently reduce fitness to the point at which the population declines despite improvements in the habitat. One can never know which replicate of a population will contain the allele for resistance to a key disease or, conversely, the population that contains a segregating load of mutations that cannot be purged and eventually causes extinction. That is why it is called *genetic stochasticity!*

The genetic factors affecting extinction probability are so entangled with environmental and demographic stochasticity that it is impossible to disentangle them. The relative importance of each form of stochasticity will often depend on the specifics of the situation and on the point in time in the population's history at which you start your accounting.

A related question is: Does it matter which form is most important? O'Grady and colleagues (2006) make the case that, if we do not prioritize these risks, money might be spent ineffectually and poor management decisions made. I agree up to a point. Incorrect management decisions will certainly be more likely if ecological or genetic concerns are ignored. Changes in the environment provide constant challenges to the persistence of populations and determine their geographic distributions whereas changes in allelic frequencies (via selection on genetic variation maintained through mutational inputs) provide the means for organisms to meet these challenges. Thus, demographic, environmental, and genetic factors should all be considered when evaluating the persistence of populations and species.

Many of the higher ideals in conservation reflect a proactive approach. Proactive conservation attempts to maintain populations at a size that not only guarantees buffering against short-term environmental perturbations, but also maintains the integrity of ecological and evolutionary processes. In proactive conservation, the solution to all three forms of stochasticity is the same: Maintain conservation landscapes that preserve ecological systems and viable population sizes for all species (Reed et al., 2003c). The most "important" factor becomes the factor or combination of factors that sets the minimum viable population size for attaining our most lofty conservation goals.

ENVIRONMENTAL STOCHASTICITY AND EVOLUTION

Environmental Stochasticity II

A more complete understanding of the role of environmental variation in determining population dynamics is crucial for our ability to conserve and manage wildlife and predict extinction risk. It has been shown experimentally that increasing amounts of environmental variation decrease persistence times (Drake & Lodge, 2004); however, modeling of environmental variation is not straightforward. *Environmental stochasticity* is an umbrella term that disguises considerable heterogeneity in the patterns of perturbations that influence temporal variation in demographic rates, and numerous types of endogenous and exogenous forces shape population dynamics (Bjørnstad & Grenfell, 2001; Lande et al., 2003; Turchin, 2003).

One type of environmental stochasticity is completely random fluctuations in the environment. Fluctuations of this sort are generally viewed as high-frequency, small-scale, and low-impact disturbances. The effects of these types of perturbations on birth and death rates are temporally uncorrelated and are often modeled as white noise. Most real-world environmental stochasticity is dominated by positive temporal autocorrelations—that is, environmental conditions present at time *t* are dependent on the environmental conditions one or more time steps prior to time *t*. Temporal autocorrelation in environmental variables is due in large part to cycles in the abiotic environment. Examples of such an autocorrelation include the El Niño/Southern Oscillation (ENSO) and short-term (for example, 22-year) cycles in the sun's energy output.

However, fluctuations in abiotic factors can also cause temporal correlations in the biotic environment. One such example is the dependence of malarial infection rates on the density and biting activity of mosquito vectors, which in turn depends on temperature and rainfall levels. Positive correlations among biotic and abiotic factors will generally decrease population persistence times by increasing the likelihood of a series of years that are sufficiently severe to drive a population to extinction. Just such a consequence of autocorrelations has been demonstrated theoretically and also experimentally (Pike et al., 2004).

Temporal correlations in environmental quality also limit our ability to predict the fate of populations. The reason is because sampling of population growth rates must be long enough in duration to estimate temporal variation accurately (Reed et al., 2003c). Thus, understanding the details of autocorrelation among environmental variables is of great importance in any attempt to manage or conserve populations.

On the other end of the spectrum from white noise are cycles that are very low-frequency, large-scale (often global), high-impact disturbances (Etterson, this volume). These cycles occur so infrequently compared with the generation length of the organism that they may be perceived by the organism as deterministic environmental changes. Shifts in the axial tilt and orbit of the earth are examples of perturbations that operate on long time spans, perhaps over tens of thousands of years. Other environmental changes, so-called *discrete changes* (Boulding, this volume; Boulding & Hay, 2001), may occur relatively quickly and have nearly permanent effects. Discrete changes might include global warming (which occurs on a time scale of decades), introduction of exotic species to a habitat (Carroll & Fox, this volume), habitat fragmentation (Sambatti et al., this volume), and industrial activities or accidents. These discrete changes can often be viewed as a permanent lowering of the quality of the habitat.

The final category of environmental perturbation is perhaps the most important and also the most poorly understood. Catastrophes are brief periods of environmental stress that have very large negative impacts on population size. The probability of extinction is disproportionately affected by rare perturbations of large magnitude, to the point where small-scale random perturbations might be inconsequential to most populations.

The definition of catastrophe given here leaves a lot to be desired because it does not specify what a "brief" period of time is and what a "very large" effect is. Many workers have described catastrophes as stemming from a different source than from "normal" environmental variation, and have suggested that if one looks at birth and death rates, the presence of catastrophes will be evident from the bimodal distribution in these vital rates through time. Some authors have found such bimodal distributions. However, I have reviewed distributions of population growth rates and population sizes for literally thousands of populations (Heering & Reed, 2005; Reed & Hobbs, 2004; Reed et al., 2003b, c) and have found very little evidence for such bimodal distributions. In my opinion, it is arbitrary where the line is drawn between what is and what is not a catastrophe. The truth of that statement depends to some degree on how one defines a catastrophe. For example, the complete reproductive failure of a population of organisms in a given year would generally be considered a catastrophe. And yet the impacts of this failure on population size over the next few generations might be inconsequential if survival of juveniles is density dependent or the organism is long-lived (for example, trees and vertebrates with mass > 20 kg).

I believe the general impression of what a catastrophe is has been influenced in part by anthropocentrism. The average biologist might classify earthquakes, floods, hurricanes, tornadoes, and volcanic eruptions as examples of catastrophic environmental changes. All these phenomena are very dramatic and cause much human suffering. However, with the exception of hurricanes, it is unclear how important such catastrophes are in plant and animal populations. The most common forms of catastrophe in plant and animal populations seem to be drought, starvation, and disease. Severe winters are another fairly common catastrophe in temperate regions. And on a local scale, severe bouts of predation are a surprisingly common source of extreme variation in demographic rates (see, for example, Brooks et al., 1991; Festa-Bianchet et al., 2006).

Catastrophes have several things in common. First, catastrophes are often sources of environmental variation that industrialized nations have ameliorated or nearly nullified. For instance, widespread access to clean drinking water and vaccination programs has halted the spread of many infectious diseases. Thus, biologists may often underestimate such forms of stress on nonhuman populations. Second, catastrophes are an arbitrary point in the tail end of the distribution of environmental perturbations (Reed et al., 2003b). Third, different sources of catastrophic perturbations are likely to act synergistically. Notably, the probability of famine and plague are not independent of each other. Presumably, malnourished individuals during the famine have weakened immune systems and are more likely to be infected by and/or suffer mortality from a given infectious disease. And lastly, the frequency of catastrophes, along with other demographic parameters, scales with the generation length of the organism.

Thus, modeling catastrophes is important to predicting extinction risk. Understanding the mechanisms underlying the commonalities listed here may provide insight into the nature of catastrophes and enlighten management decisions designed to ameliorate the frequency and severity of cata strophes.

Complex Interactions Involving Population Size

When thinking of the interactions among environment, evolution, and population size, it is important to account for their complexity. Figure 2.3 is an attempt to illustrate some of this complexity. This directed graph shows the environment as having direct effects on fitness, genetic diversity, and population size. That the environment determines fitness and population size is fairly obvious; however, the long-run environment also helps (in conjunction with mutation) shape the amount and type of population genetic variation.

Fitness has obvious effects on population size. However, it is often overlooked how fitness contributes to the evolutionary potential of a population by molding the amount and types of genetic variation present both directly and indirectly through effects on population size. Boulding and Hay (2001) noted this relationship explicitly in their models in which populations with high fecundity were more likely to produce individuals with extreme phenotypes and, therefore, adaptive potential increased with per-capita fecundity.

Population Size and Fitness

The relationship between long-term effective population size and fitness is an important one. It is often assumed that genetic stochasticity only affects populations of very small size (50 or fewer individuals). Yet, as I have emphasized, population size influences fitness through several mechanisms, and many of these effects are not limited to small populations. The notion that populations of a few hundred individuals, a common recovery goal, are safe from inbreeding depression and genetic stochasticity is refuted by empirical studies. For example, a study by Briskie and Mackintosh (2004) found that egg-hatching rates were significantly reduced in New Zealand birds that had passed through population bottlenecks of 150 or fewer individuals.

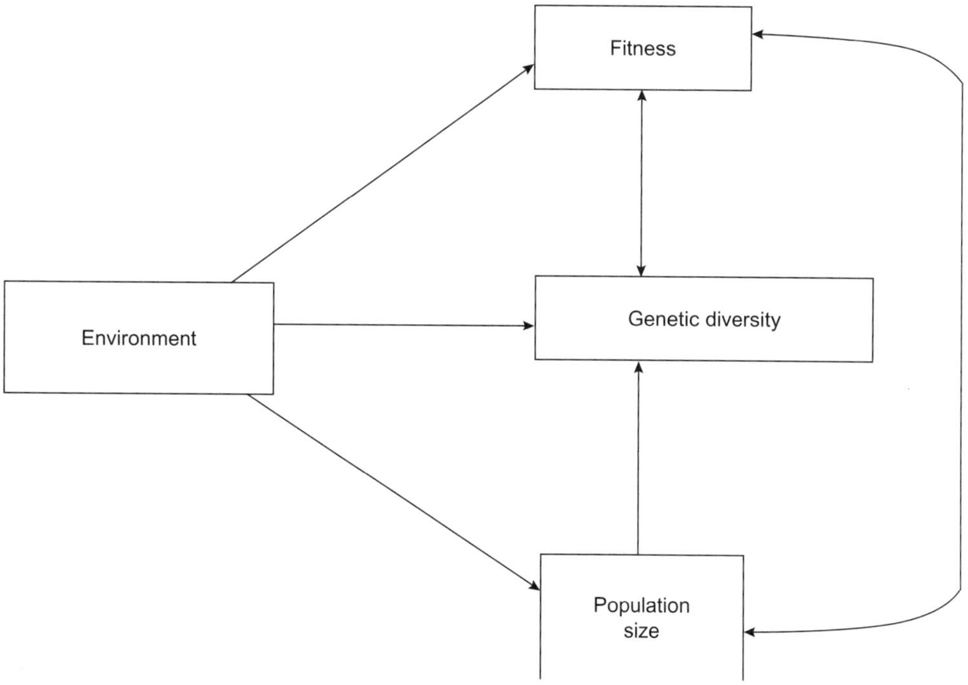

FIGURE 2.3 Directed graph showing direct and indirect impacts on population size and population viability, and how they are influenced by the environment.

Figure 2.4 demonstrates the log-linear increase in hatching failure rates with decreasing size of the bottleneck in 15 species of birds, with comparisons being made between the source population for the introduction and the introduced population. Other studies have consistently found a log-linear relationship between fitness and population size in natural populations (Reed, 2005, Reed et al., 2007) (Fig. 2.5). Thus, the small-population paradigm and too much attention on inbreeding depression may not be beneficial to conservation efforts, as other aspects of genetic stochasticity have received less attention and little emphasis has been placed on defining meaningful population size criteria in the context of overall conservation concerns (for example, the interaction between genetic and environmental stochasticity).

How much fitness is enough to guarantee persistence? The answer to this question will be complex. We have already seen that persistence depends, in part, on where the population begins on the x-axis of Figure 2.1. Density-dependent reproduction or mortality could render moot questions concerning decreases in fitness components. If an organism produces hundreds of offspring and only 1% survive density-dependent forms of mortality to reach fecund species, but many of the effects will be indirect and difficult to detect. As pointed out by Puurtinen and colleagues (2004), for the case of populations partially regulated by ecological and density-dependent factors, the gradual erosion of fitness via genetic stochasticity is particularly insidious because it may not be noticed until it is too late.

Evolution under Different Types of Environmental Stochasticity

The different forms of environmental stochasticity that I described are not just technical obstacles to overcome in modeling population viability. Stochasticity also has ramifications for the evolutionary response of populations to environmental variation. I will briefly describe how environmental stochasticity might involve underappreciated evolutionary dynamics.

Random, high-frequency changes in the environment might seem impossible to adapt to. However, the primary evolutionary response to this type of environmental stochasticity occurs via an increase in phenotypic plasticity for ecologically relevant

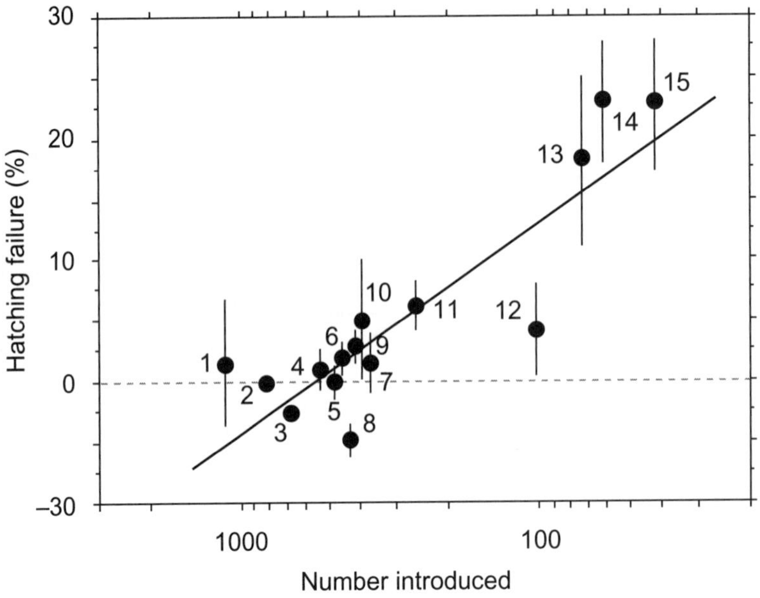

FIGURE 2.4 Increase in differences in the rate of hatching failure between each introduced population in New Zealand (after the bottleneck) and their source population (before the bottleneck) for 15 species of introduced birds with data in both localities. (Adapted from Figure 3 in Briskie and Mackintosh [2004]. Copyright 2004 The National Academy of Sciences of the USA.)

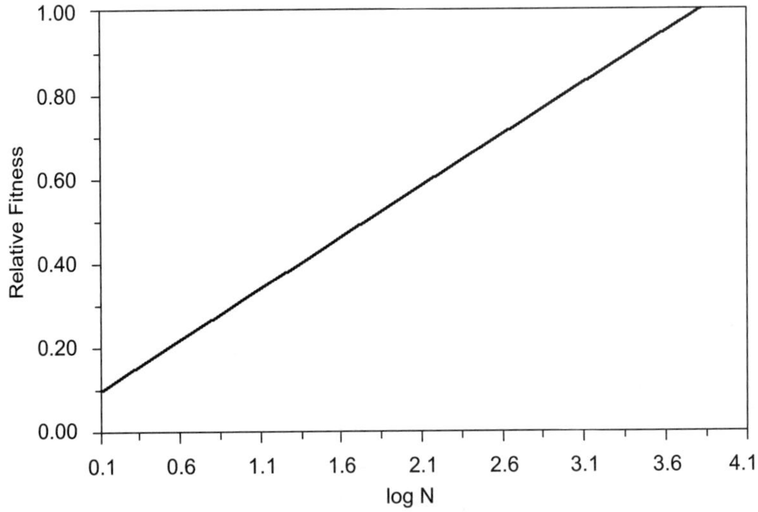

FIGURE 2.5 The log-linear relationship between population size and relative fitness. The intercept and slope were estimated from 16 studies (18 data sets) and is updated from the data presented in Reed (2005).

traits. Developmental plasticity enables populations to maintain fitness homeostasis despite variable environmental conditions. The ability to deal with this type of environmental stochasticity might have less to do with genetic variation and more to do with the absence of developmental and phylogenetic constraints on evolving a plastic phenotype. However, it is likely that genetic variation is positively associated with the ability to evolve phenotypic plasticity (see, for example, Paschke et al., 2003), and limits and costs to phenotypic plasticity make the maintenance of pools of potentially adaptive genetic variation very important. The importance of genetic variation should become even greater as the frequency and magnitude of the perturbations becomes longer compared with the generation length of the organism.

Long-term deterministic changes will depend partly on the standing genetic variation but also on the amount and types of mutational inputs into the population. The rate at which the changes in the environment occur will have a huge impact on the likelihood of extinction and on the relative importance of standing genetic variation versus novel genetic variation produced by mutation during the rapid adaptation phase. The faster the rate of change in the environment, the more important the initial levels of genetic variation. Discrete changes in the environment of large magnitude will be particularly challenging to species and may be most influenced by the amount of genetic variation present in mutation–selection–drift balance and by the genetic correlations between fitness and the trait(s) under selection.

Part of the title of this chapter is "From Mutation to Environmental Catastrophes." This emphasis reflects my thoughts that population size mediates extinction risk all the way from the molecular level to the ecosystem level, and it reflects my ambition to integrate and understand how the environment and genes interact to affect extinction risk. Although such a goal may be never fully realized, some light may have already been shed on the topic. Parasitic infection and starvation are two frequent and powerful stresses that can affect population persistence times. Quantitative trait loci (QTL) mapping has been conducted for the first of these traits in *Tribolium castaneum* (Zhong et al., 2005) and for the second in *Drosophila melanogaster* (Harbison et al., 2004). Both studies found or suggest that genes conferring resistance to parasitism or starvation have negative effects on other aspects of fitness. Such trade-offs are commonly reported in the literature (see, for example, Luong & Polak, 2007). Knowledge of whether evolution is unable to overcome such trade-offs and just how ubiquitous such trade-offs are among species, will help in understanding the persistence of populations during times of extreme stress.

CASE STUDY—ENVIRONMENTAL AND GENETIC CONTRIBUTIONS TO EXTINCTION

The seminal work examining multiple factors affecting extinction risk is still that of Saccheri and colleagues (1998). Their study examined the effects of inbreeding level and other factors on extinction risk. The study was conducted using several hundred, sometimes occupied, habitat patches that form a metapopulation of the Glanville fritillary butterfly. Inbreeding levels were assessed by estimating heterozygosity at seven allozyme loci and one microsatellite locus. Among subpopulations, extinction risk increased with decreasing heterozygosity (Fig. 2.6). The upper panels of Figure 2.6 show that the proportion of heterozygous loci in extinct populations was, on average, lower than in surviving populations and that the probability of extinction is predicted with more accuracy (in other words, more variance is explained) when heterozygosity levels are included in the model. Lower panels show the relationship between the local risk of extinction and heterozygosity predicted by two different models (see Saccheri and colleagues [1998] for details). This elegant study is widely cited as establishing a strong relationship between inbreeding level and the probability of extinction.

The success of the study by Saccheri and colleagues (1998) is the result of a number of factors, including the large number of populations observed; the manageable size of populations, which in turn permitted extinction to be observed within short study periods; complementary results from laboratory-based studies of inbreeding depression; and the integration of demographic, environmental, and genetic data. Although often unnoticed in the literature, a number of other explanatory variables besides heterozygosity were significant in the model of Saccheri and colleagues (1998). In particular, population size and heterozygosity were positively correlated (as predicted), but population size explained additional independent variation

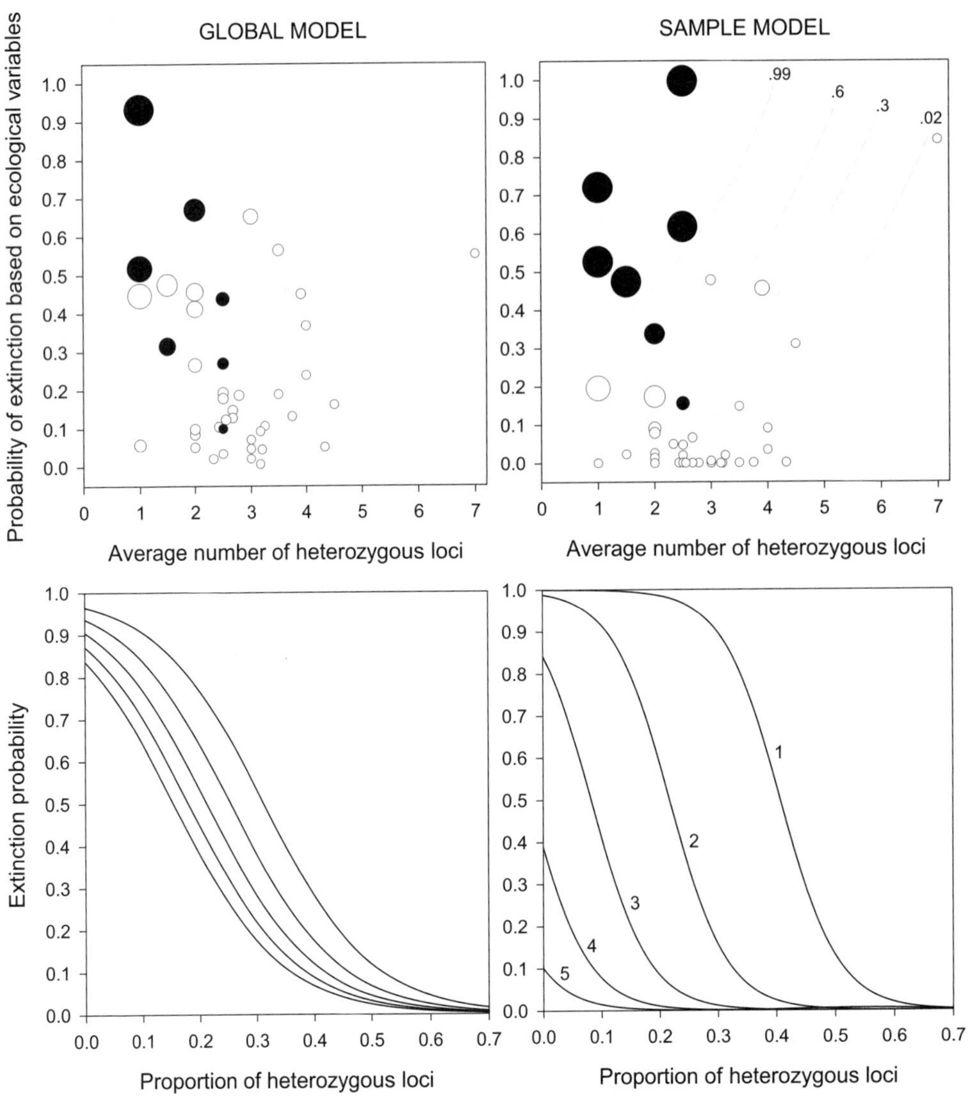

FIGURE 2.6 For both global and sample models, the upper panels show (1) the observed average number of heterozygous loci in extinct (filled circles) and surviving (open circles) populations, (2) the probability of extinction predicted by the models without heterozygosity compared with the observed heterozygosity, and (3) the probability of extinction predicted by the full model, including heterozygosity (proportional to circle size). For the sample model, isoclines have been drawn for extinction risk predicted by the model, including ecological factors and heterozygosity. These models illustrate that both ecological factors and heterozygosity influence extinction risk. Lower panels show the relationship between the local risk of extinction and heterozygosity predicted by the global and sample models. (Adapted from Figure 2 of Saccheri and colleagues [1998].)

that was likely the result of demographic stochasticity. Population size of the nearest neighbor was also a highly significant explanatory variable. This pattern reflects the superior contribution of immigrants from larger populations compared with smaller populations. Flower abundance, reflecting resource availability, was also a significant factor in the model. Thus, both genetic and environmental factors were important in determining extinction risk in this butterfly metapopulation.

Can these results be generalized to other taxa and populations? The weakness of the study is also one of its strengths. The subpopulations investigated by Saccheri and coworkers (1998) were very small (many being only one sib group); thus, the results could not address how environmental and genetic stochasticity interact in populations of hundreds or thousands of individuals. There has been surprisingly little follow-up work of this sort being published. Vergeer and colleagues (2003) studied the performance of 17 populations of the annual plant *Succisa pratensis* and concluded that both the direct effects of genetics and habitat deterioration (namely, eutrophication) are important for population persistence over even short time frames. However, Vergeer and coworkers (2003) also concluded that "habitat quality and environmental stochasticity are of more immediate importance than genetic and especially demographic processes in determining population persistence." Cassel and colleagues (2001) found persistent effects of food quality on fitness components of a rare butterfly (the scarce heath) and showed that population size, a proxy for inbreeding, was important under stressful environmental conditions. And lastly, results from a 3-year study on 14 field populations of wolf spider (Reed et al., 2007a, b) suggest that genetic variation and changes in prey availability contribute almost equally to the temporal variation in fitness among spider populations that ranged in size from 50 to 20,000 individuals. There were strong inbreeding–environment interactions for these populations, with smaller populations faring disproportionately worse during times of stress.

FUTURE DIRECTIONS

A central question in biodiversity conservation is: How large do populations need to be to avoid the combined effects of genetic and environmental stochasticity? Without knowledge of how population size and the risk of extinction scale with each other over relevant time frames, we will not be able to make intelligent choices in planning protected areas or in assessing risk from increasing human population size and consumption rates. Answering questions about population size and population viability will require long-term genetic (molecular and quantitative) and ecological monitoring of populations and their fitness in the wild. Evolutionary processes determine the fate of populations and species, and these processes are tied to the interactions between ecological and genetic processes. An explanation is needed for the log-linear relationship between population size and fitness components. Understanding the mechanisms underlying this pattern will inform long-term conservation decision making. Along the same lines, more experiments are needed to elucidate the rates at which beneficial mutations arise (Houle & Kondrashov, 2006), and the correlation in the direction and magnitude of selection coefficients for such mutations across ecologically relevant environments. Mutation provides the raw material for evolution, and models predicting long-term evolutionary patterns require better estimates of the parameters involved.

Efforts at conserving biodiversity will also be enhanced by increased dialogue between theoreticians, experimentalists, and field workers. To some extent, scientists in each of these areas work in a vacuum. Conservation theory has provided us with a strong basis for conservation actions; however, the ultimate test of theory needs to come from the field. Conservation efforts will also benefit from less territoriality and increased communication among ecologists and geneticists.

SUGGESTIONS FOR FURTHER READING

Shaffer (1981) provides the original introduction to stochastic processes in small populations and their importance to conservation biology. Willi and colleagues (2006) provide an excellent review of how small population size limits evolutionary potential. Kristensen and colleagues (2006) as well as Swindell (2006) provide outstanding examples of how to apply modern molecular methods to estimate the number of genes promoting stress resistance and their interactions.

Kristensen, T. N., P. Sørensen, K. S. Pedersen, M. Kruhøffer, & V. Loeschcke. 2006. Inbreeding by environmental interactions affect gene expression in *Drosophila melanogaster*. Genetics 173: 1329–1336.

Shaffer, M. L. 1981. Minimum population sizes for species conservation. BioScience 31: 131–134.
Swindell, W. R. 2006. The association among gene expression responses to nine abiotic stress treatments in *Arabidopsis thaliana*. Genetics 174: 1811–1824.
Willi, Y., J. Van Buskirk, & A. A. Hoffmann. 2006. Limits to the adaptive potential of small populations. Annu Rev Ecol Evol Syst. 37: 433–458.

3

Demographics versus Genetics in Conservation Biology

BARRY W. BROOK

Conservation biology is a mature and multidisciplinary science, underpinned by more than 25 years of theoretical and empirical development (Avise, this volume; Dobson et al., 1992; Frankham et al., 2002; Morris & Doak, 2002). The science arose during the late 1970s as an amalgam of concepts and tools developed across a wide span of theoretical and applied fields, including population ecology, demography, and life history theory; quantitative and population genetics; community and ecosystem ecology; and wildlife management and captive breeding, among other fields (Frankel & Soulé, 1981). Conservation biologists are ultimately concerned with how to manage threats to avoid the biological extinction of populations, species, and clades. The systematic pressures humans exert on the natural world through their direct and indirect actions, such as habitat loss and fragmentation, overexploitation, invasive species, and environmental pollution, are generally agreed to be the overarching causes of most modern (and many historical and recent prehistoric) declines of once-pristine populations (Lande, 1998; Wilson, 1992). Yet identifying the "nail in the coffin" of populations—that is, the factor(s) driving the final phase of population decline—remains a hotly debated problem. Indeed, a controversy almost as old as the discipline of conservation biology itself concerns the relative role of demographic versus genetic factors in determining the fate of small populations teetering on the brink of extinction.

Many prominent papers have argued that genetic factors are relatively unimportant in species extinctions. The selection of quotations given in Box 3.1 should dispel any doubts that the question of the role of genetics in conservation warrants a chapter in this book. These quotations illustrate the frequency, chronology, and context within which the importance of genetics in determining extinction risk have been questioned in the scientific literature. The common basis of these arguments are threefold: (1) healthy populations become threatened populations as a result of deterministic human impacts, often operating over multiple scales; (2) during the final stages of decline of natural populations, the ravages of demographic and environmental stochasticity will overwhelm the effects of inbreeding, loss of genetic diversity, and other genetic hazards such that genetics, for most practical purposes, can be considered irrelevant to conservation management; (3) the evidence for important genetic impacts on population viability are weak, piecemeal, and require more convincing integration; and (4) there are examples of apparently healthy populations that are genetically depauperate. Much of the contemporary opinion that genetic factors play little or no role in the extinction of populations stems from a selective interpretation of Lande (1988). Yet Lande's fundamental point was not that genetics has no influence on extinction risk, but that population size required to buffer against nongenetic sources of variation (for example, environmental fluctuations)

BOX 3.1 Arguments against a Role for Genetics in Extinction

Lande (1988): "...demography may usually be of more immediate importance than population genetics in determining the minimum viable size of wild populations."

Pimm (1991): "For most species the threat from losing genetic variability should be a minor, if additional, problem of being rare."

Young (1991): "Smaller populations are at greater risk of extinction than larger populations, but this risk is more likely to come from demographic constraints than from genetic constraints. There are no examples of wild populations in which either inbreeding depression or a lack of genetic variability is known to have been primarily responsible for significant reductions in population size, much less extinction. On the contrary there are numerous examples of wild populations that are both monomorphic and apparently healthy."

Wilson (1992): "For species passing through the narrows of small population size, the Scylla of demographic accident is more dangerous than the Charybdis of inbreeding depression."

Caughley (1994): "... no instance of extinction by genetic malfunction has been reported whereas the examples of driven extinction are plentiful."

Caro and Laurenson (1994): "The effects of inbreeding and loss of genetic diversity on the persistence of populations in the real world are, however, increasingly questionable. Although inbreeding results in demonstrable costs in captive and wild situations, it has yet to be shown that inbreeding depression has caused any wild population to decline. Similarly, although loss of heterozygosity has detrimental impacts on individual fitness, no population has gone extinct as a result."

Brookes (1997): "There are plenty of successful species with little detectable genetic variation ... by the time inbreeding becomes a problem in wild populations, a freak storm or a motorway extension will just as likely seal the fate of those few remaining individuals. Preaching about the genetic risks of species extinction detracts from more familiar and prosaic truths: habitat destruction, pesticides, pollution and so on remain the biggest threats to biodiversity.... Conservation and genetics don't mix. A swift divorce should leave both science, and what's left of life on Earth, in better shape."

Elgar and Clode (2001): "... there is compelling evidence that inbreeding depresses individual reproductive success in typically out-breeding domestic plants and animals and captive populations of vertebrate wildlife. But drawing general conclusions from individual species requires quantitative comparative analyses, rather than appeals to particular studies."

are expected, on theoretical grounds, to be considerably larger than those required for resilience to genetic threats.

Notwithstanding the considerable skepticism illustrated by the comments in Box 3.1, there is in fact substantial evidence from case studies and multitaxa meta-analyses of the importance of inbreeding depression and loss of genetic diversity to the short- and long-term viability of wild populations (reviewed most recently by Frankham [2005b]). Inbreeding depression has been documented in all well-studied outbreeding species in captivity (Lacy, 1997), and there is clear evidence for its impact on natural populations of mammals (for example, lions, golden lion tamarins, deer mice), birds (for example, Mexican jay, red-cockaded woodpecker, greater prairie chicken, great tit, American kestrel, song sparrows), a snake, several populations of fish, a snail, a butterfly, and plant species (Crnokrak & Roff, 1999; Frankham, 1997; Frankham et al., 2002; Keller & Waller, 2002). Inbreeding depression has also been observed in many domesticated species, a fact known well before the advent of modern genetics (Darwin, 1876). Furthermore, genetic

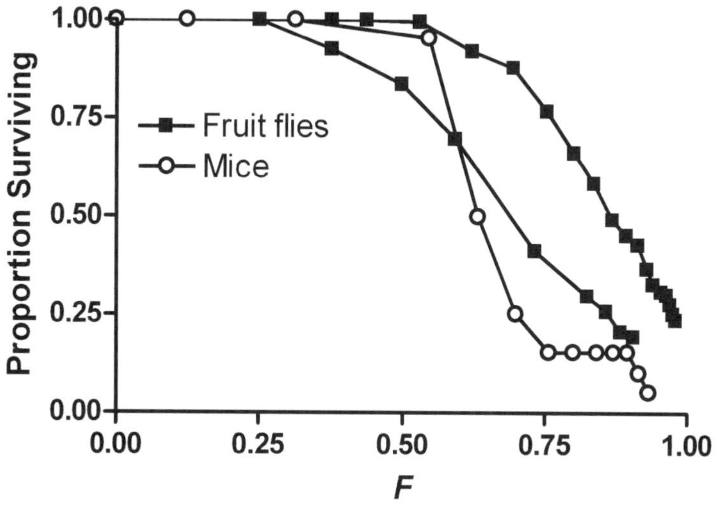

FIGURE 3.1 Decreasing probability of population persistence (expressed as increment per generational survival) with increasing inbreeding *F* for *Mus musculus* and each of two *Drosophila* species. (Figure modified from Frankham [1995b], with permission.)

factors have been significant in the decline and extinction of both laboratory (Fig. 3.1) (Frankham, 1995b; Reed et al., 2002) and natural populations, even after accounting for relevant ecological factors (Keller et al., 1994; Madsen et al., 1999; Saccheri et al., 1998; Westemeier et al., 1998). Genetic parameters are also known to hasten markedly the time to extinction in population viability analyses (Brook et al., 2002; O'Grady et al., 2006).

Interestingly, arguments in support of the preeminence of demographic factors continue to emerge in parallel with studies showing the importance and relevance of genetics. In this chapter I review the basis for these disparate viewpoints, describe each of three case studies that illustrate key evidence for genetic effects on extinction risk, and provide a perspective on how the demographic and the genetics schools of thought in conservation biology can be more usefully integrated.

CONCEPTS

Theory and evidence regarding stochastic demographic and genetic effects on small populations are well developed and relatively uncontroversial (for some excellent and comprehensive reviews, see Caughley, 1994; Frankham et al., 2002; Morris & Doak, 2002; Young & Clarke, 2000). I do not attempt to recapitulate these concepts in detail (a more thorough explanation is given in Reed, this volume), but will instead provide a sufficient overview of the principles necessary to frame properly the debate about the importance of demography and genetics in determining extinction risk.

Demographic and Environmental Stochasticity

When reduced to relatively few individuals by deterministic threats, natural populations are then vulnerable to *stochastic hazards*. These are probabilistic effects in which the underlying processes and variability may be known, but uncertainty exists in predicting when and how such events unfold. Stochastic factors can be grouped into four broad categories: demographic (intrinsic), environmental (extrinsic), natural catastrophes (extrinsic), and genetic fluctuations (intrinsic, see next section).

Demographic stochasticity involves random fluctuations in vital rates at the individual level, and arises from the fact that individuals are discrete entities (Morris & Doak, 2002). For example, although the mean fecundity of population may be two offspring per female, any given individual may produce [0, 1, 2, 3...] offspring. Similarly, the mean survival rate in a large population may be 75%, but each entity can only liveor die, and so with four individuals there is a high probability that the realized survival rate is, by chance, 0%, 25%, 50%, or 100%. Skews can also arise in the sex ratio at birth. Theory predicts that demographic stochasticity may have significant impacts on populations with fewer than 100 individuals, but that this process is of negligible importance to large populations (Lande, 1998).

Environmental stochasticity includes effects of random fluctuations in extrinsic processes on vital rates, population growth, and habitat suitability, and is usually driven by external forces such as weather or interactions with other species (Caughley, 1994). Because most sources of environmental stochasticity operate independently of population size, it can affect populations of any size, and is usually most important in variable or seasonal environments and in small-size or short-lived species.

Catastrophes are often considered a special case of environmental stochasticity (Reed, this volume). This is because catastrophes represent extreme and rare events such as severe storms and wildfires, or even the emergence of a disease pandemic or the arrival of a novel predator in a pristine environment. Catastrophes thus have the potential to devastate even large populations.

Genetic Threats

Genetic stochasticity can be defined as the random allocation of alleles, under the effects of inbreeding, from individuals in one generation to individuals in the next generation. Inbreeding depression, loss of genetic diversity, mutational accumulation, self-incompatibility, and outbreeding depression are the genetic factors that could contribute to extinction risk. For details and numerous examples, refer to Young and Clarke (2000) and Frankham and colleagues (2002).

Inbreeding—the mating of genetically related individual—increases the frequency of homozygotes (in other words, individuals with two copies of the same allele), thereby increasing the probability that rare, recessive, and often deleterious mutations will be expressed. Although the impact of inbreeding varies across different life history stages, taxa, populations, environments, and geographic settings (Frankham, 1997; Keller & Waller, 2002; Mills & Smouse, 1994; Saccheri et al., 1998), inbred individuals generally tend to show suppressed reproductive fitness, or inbreeding depression. Thus it is not surprising that most organisms have evolved inbreeding avoidance mechanisms (Frankham et al., 2002). Indeed, animal and plant breeders have long understood that mating close relatives leads to a loss of productivity, and more than a century ago the father of modern evolutionary biology documented inbreeding-induced reduction in viability in a large number of plant species (Darwin, 1876).

Genetic theory predicts that extinction risk is related to the intrinsic population growth rate r, a demographic measure of population resilience, and the level of inbreeding F, which represents the proportion of alleles that share the same ancestral sequence without any intervening mutation (in other words, the proportion identical by descent). Inbreeding is inversely related to the genetically effective population size N_e (Box 3.2), and increases with the number of generations of inbreeding. The magnitude of depression in reproductive output and survival is related proportionally to F (Falconer & Mackay, 1996). As such, inbreeding will be greatest when the effective population size is small and the number of generations of inbreeding is large. The magnitude of inbreeding depression may be partially ameliorated by selective purging of recessive deleterious or lethal alleles that are more readily exposed to selection because of increased homozygosity resulting from inbreeding. However, experimental evidence suggests that selective purging is often weak (summarized in Frankham et al., 2002). Effective population size is a function of the number of individuals within a population, sex ratio, variance in family size, and degree of overlap among generations. Usually, N_e is a small fraction of census N (Frankham, 1995a). Fluctuations in population size are caused by population dynamic processes (for example, density dependence and unstable age structure), demographic and environmental stochasticity, and catastrophes. Thus, a range of interacting factors have the capacity to decrease r (even if temporarily), increase F, and therefore amplify both the immediate and long-term risk of extinction (see next section).

In large (in other words, most nonthreatened) populations, adaptation via natural selection prevails as the dominant evolutionary force (Boulding, this volume). However, in small (often-threatened) populations, stochastic processes, including inbreeding and genetic drift, predominate (Falconer & Mackay, 1996). Genetic drift may

BOX 3.2 Census and Effective Population Sizes

A population can be broadly defined as a discrete group of individuals of a given species that inhabit a defined area. Population size can be determined directly, by counting all individuals, or by sampling a fraction of individuals and using statistical inference to estimate the size of the missing portion. Either method results in an estimate of the actual number of individuals present, which, in the conservation genetics literature, is termed the *census population size* (N). However, this metric ignores details about population structure and composition. For instance, most species have two sexes, not all individuals may be reproductive, and some individuals may contribute fewer or greater numbers of offspring to succeeding generations than others. These complications need to be accounted for if one is to estimate the genetically effective population size (N_e). In simple terms, N_e is the number of individuals in a population who contribute offspring to the next generation. More precisely, N_e can be defined as the number of breeding individuals in an idealized population that would show the same dispersion of allele frequencies under random genetic drift, or the same level of inbreeding, as the population under consideration (Frankham et al. 2002) An *idealized population* is characterized by random mating, discrete generations, and constant size, with all individuals reproducing and the variation in offspring number following a Poisson distribution. The population is closed (no migration or gene flow), and mutation and selection do not occur. Effective population size is usually considerably lower than census population size because unequal sex ratios, overdispersed variation in family size, fluctuations in population size, and other factors cause variation in the genetic contribution of individuals to the next generation (for example, age structure). A meta-analysis by Frankham (1995a) found the median N_e/N ratio for 102 species was just 0.11.

become important at small population sizes, when alleles are lost from a population by chance rather than selection (as a result of random segregation of alleles or nonreproductive individuals), which can eventually result in some alleles being "fixed" (homozygous in all individuals) and others becoming "extinct." Fixation of deleterious alleles is a concern in the short-term because individuals within the population at large are thus condemned to mate with others of the same (fixed) genotype, resulting in population-level inbreeding depression.

Drift also causes an erosion of genetic variation that is particularly pronounced in small populations (Vucetich & Waite, 1999), a matter of long-term concern. Because inbreeding and drift lead to a loss of quantitative genetic variation (heterozygosity) affecting reproductive fitness, these processes also undermine the capacity of populations to evolve adaptations to changing environments (Boulding, this volume; Gilchrist & Folk, this volume; Lacy, 1997) and disease (Altizer & Pederson, this volume; Spielman et al., 2004b). For example, many species of forest birds on Hawaii have been decimated by introduced diseases, a loss due in part to a lack of variability in host immune response (O'Brien & Evermann, 1988). Moreover, meta-analyses have shown that endangered species, island populations, and those populations that have undergone past demographic disturbance have less genetic variation than their nonendangered or mainland counterparts (Frankham, 1997; Garner et al., 2005; Spielman et al., 2004a). Because low genetic variation is an indicator of past inbreeding or population size bottlenecks, many endangered but persisting species are likely to be suffering from detrimental genetic impacts.

Synergies and Ratchets: The Extinction Vortex

Regardless of the cause of population decline from a large to small sizes, unusual (and often detrimental) events assume prominence at low abundances. For instance, intraspecific competition is reduced at low densities and can induce a density-dependent population recovery (Morris & Doak, 2002). However, a countervailing phenomenon known as *inverse density-dependence*—the *Allee effect* in animal ecology or *depensation* in fisheries science (see Courchamp et al., 2008)—draws populations toward extinction by disrupting behavioral patterns (for example, herd defense against predators or reproductive "lekking"), reducing the capacity of organisms to control their environment (for example, causing soil erosion when vegetation cover is reduced), and making it difficult to find available or reproductively fit mates. Genetical threats such as inbreeding depression, the random loss of genetic diversity, and the accumulated fixation of deleterious mutations of small effect via drift (Lynch et al., 1995a) can also be considered Allee effects, as well as stochastic phenomena that interact with and reinforce other nongenetic Allee effects like demographic stochasticity. Populations dominated by stochastic factors are often considered to have breached their *minimum viable population size*.

The notion of a synergy between stochastic hazards was developed in detail by Gilpin and Soulé (1986), who coined the term *extinction vortex* to describe the positive feedback that was conjectured to be an inevitable result of synergistic effects. The path to extinction under the *vortex model* is theorized to proceed along roughly the following lines. An event or systematic pressure (habitat loss, new predator, and so forth) causes population size (N) to decrease and variability in the population rate of change (r) to increase via Allee effects such as demographic stochasticity. Alternatively, the spatial distribution of the population may be fragmented such that even if total N remains constant, the population size in each habitat patch will be reduced and gene flow is thus decreased or halted (Sambatti et al., this volume). The rate and magnitude of inbreeding increases in the reduced population, and genetic diversity is concomitantly lost at a greater pace via drift, causing reproductive output to diminish, mortality rates to increase, and hence r, the demographic "sum" of these vital rates representing the mean fitness of the population (Lacy, 1997), to decline. A positive feedback loop is set in train (Fig. 3.2), and when r is driven to less than zero, the population nosedives to extinction. Moreover, because the inbreeding coefficient is the genetic equivalent of a ratchet, the only way the population can recover lost fitness and adaptive potential in the short term is through the introduction of unrelated

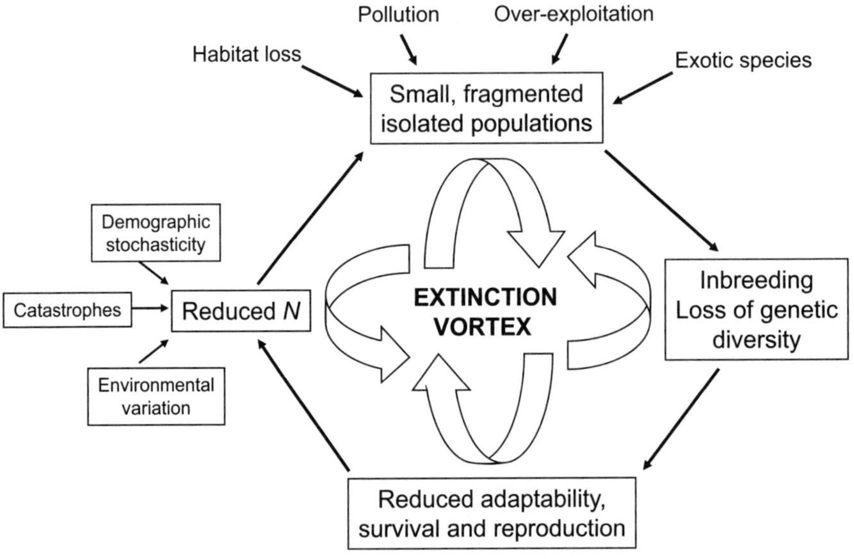

FIGURE 3.2 A conceptual representation of the *extinction vortex*. Illustrated is the interaction between demographic, environmental, and genetic hazards, and the resulting positive feedback loop that is established after a reduction in population size. (Figure modified from Frankham and colleagues [2002], with permission.)

individuals to the population (for example, via dispersal or translocation [Vergeer et al., this volume]). As such, past transgressions to low population size, even if transitory, will leave a genetic legacy that incrementally reduces the intrinsic recovery potential of the population (Morris & Doak, 2002; Tanaka, 1997).

Recent study of natural vertebrate populations declining toward extinction provide important empirical support for the extinction vortex concept. For instance, Fagan and Holmes (2006) examined the demise of 10 wild populations to show convincingly that equivalent-size populations become less "valuable" over time in terms of staving off decline, and that both variability in *r* and the rate of decline accelerates as populations approach extinction.

If extinction vortices operate as Gilpin and Soulé (1986) postulate, some important implications emerge. First, the cause of the final extinction of a population may be unrelated to the cause of its decline, provoking Caughley (1994, p. 239) to describe the extinction vortex as "the physiology of a population's death rattle." Reinforcing this disconnection, Brook and colleagues (2006) showed that a number of ecological and life history attributes (for example, geographic range, body size) were able to explain more than 50% of the variation in World Conservation Union (IUCN) species list status as "threatened" or "in decline" (list available at www.redlist.org/). Yet these same correlates explained only 2.1% of the deviance in cross-species estimates of minimum viable populations—the population size required to avoid an imminent stochastic extinction. Second, acknowledgment of the reality of the complicated interactions and feedbacks inherent in the extinction vortex make the search for any single cause of extinction both quixotic and rather obtuse.

CASE STUDIES OF GENETICS, DEMOGRAPHY, AND EXTINCTION

As mentioned earlier, a wealth of examples now exist on the pervasive impact of inbreeding and other genetic hazards on reproductive fitness, including some very elegant field tests on natural populations (Keller & Waller, 2002). In this section I describe three studies that demonstrate alternative methodological approaches for deciphering the impact of genetics on population viability and extinction risk—the issue that lies at the core of the demography-versus-genetics debate in conservation biology.

Correlative Studies

Genetic Restoration of Declining Populations

Isolated to a small strip of grassy meadow by encroaching agriculture, the adder *Vipera berus* of Smygehuk on the Baltic coast of Sweden suffered a substantial decline in abundance about four decades ago, and thereafter suffered from severe inbreeding depression and low genetic diversity (Madsen et al., 1999). In 1992, when the population teetered on the edge of extinction (Fig. 3.3), 20 adult males were collected from a large and genetically healthy northern population and then used to supplement the genetic diversity of Smygehuk adders. Northern males readily bred with resident females, and restriction fragment length polymorphism analysis later confirmed that outbred offspring showed much higher genetic variability than the average for Smygehuk. About 3 years later, northern males were returned to their native population, thus removing any direct demographic influence of northern males on the Smygehuk. After a 4-year lull in population growth, outbred juveniles reached sexual maturity and the abundance and recruitment rate of the formerly declining Smygehuk population increased dramatically (Fig. 3.3). Interestingly, growth of the population occurred despite a decrease in the total number of females attempting to breed during the recovery period.

In a longitudinal study of the greater prairie chicken, *Tympanuchus cupido pinnatus*, the size of the Illinois population declined from more than 25,000 chickens in 1933 to approximately 2,000 in 1972. By 1992 there were fewer than 50 individuals. The cause of the initial decline in density was contraction and fragmentation of the prairie chicken habitat, but later a decline in genetic diversity and reproductive fitness (for example, egg fertility and hatching success) also occurred (Westemeier et al., 1998). Despite intensive management efforts, which included restoration of grasslands targeted at arresting the original threat, the population continued to tumble toward extinction. It was only after the translocation of unrelated birds from the large populations of Minnesota, Kansas, and Nebraska that reproductive fitness, genetic diversity, and

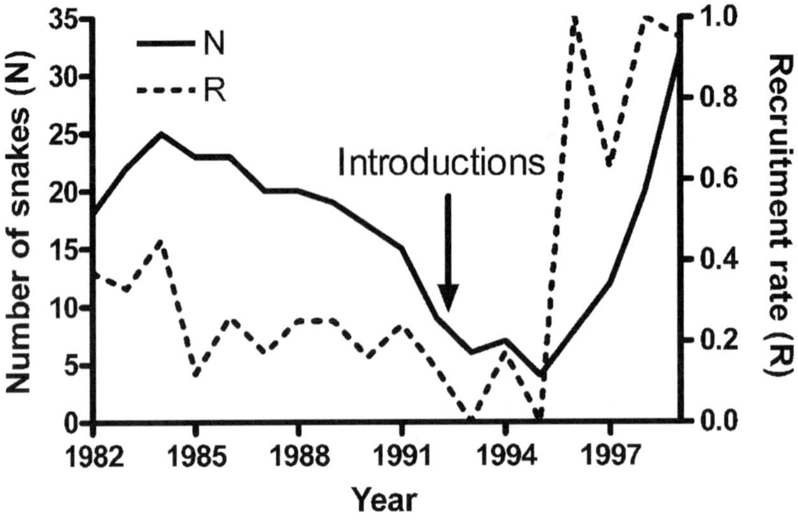

FIGURE 3.3 Recovery history of a formerly declining population of the adder *Vipera berus* in Sweden after the introduction of 20 males from a large, nearby population (introduced animals are not plotted). Both population size and recruitment rate show a clear, positive response to the introduction of new genetic diversity. (Figure modified from Madsen and colleagues [1999], with permission.)

subsequently abundance of the Illinois population began to increase. Environmental and climatic conditions were not unusual during these years and were thus excluded as a cause of the recovery.

These examples and others (for example, Vilà et al., 2003) of natural populations monitored throughout a protracted decline in abundance and subsequent recovery provide important messages for conservation biology. In each case, reproductive fitness of small and declining populations had been demonstrably damaged by inbreeding depression and lack of genetic diversity, with the injection of new blood (in other words, new genotypes) corresponding to an increase in population viability. These results support the real-world applicability of theoretical and experimental work linking increased inbreeding and reduced genetic variation with lower fecundity, higher mortality, and compromised demographic and population-level resilience. Moreover, this work also shows how a timely application of the principles of conservation genetics provides practical management solutions for threatened species that suffer from serious genetic problems. Vergeer and colleagues (this volume) give a detailed account on the conservation application of reintroductions to counter the effects of inbreeding and loss of genetic diversity.

Meta-Analyses

In the majority of cases, the full (and usually complex) circumstances surrounding the decline of formerly abundant species or populations are not known. However, a strong signal of genetic problems preceding eventual extinction has been found. First, there is evidence of reduced genetic variability in threatened species. Many authors have asserted that species are usually driven to extinction before genetic factors have had time to affect populations (Box 3.1). If this assertion is true, there will be little difference in genetic diversity between threatened and taxonomically related, nonthreatened species. Yet contrary to this expectation, a meta-analysis of 170 pairwise comparisons of genetic diversity in a threatened and closely related nonthreatened species showed that, in 77% of cases, the putatively more extinction-prone IUCN-listed species had lower genetic diversity (Fig. 3.4) (Spielman et al., 2004a). An independent meta-analysis of 108 mammals also found consistently lower genetic

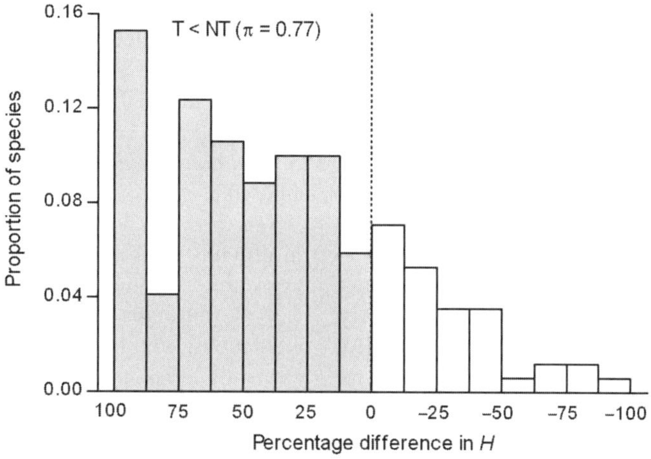

FIGURE 3.4 Histogram of percentage differences in genetic diversity (measured as microsatellite or allozyme heterozygosity [*H*]) between paired couplets of threatened (T) and taxonomically related nonthreatened (NT) taxa. The shaded region indicates those taxa for which T < NT, and the dotted line marks the point of equality. (Figure reprinted from Spielman and colleagues [2004b], with permission.)

variation in populations that had suffered demographic shocks for both rare and common species (Garner et al., 2005). As discussed previously, low genetic diversity indicates both the occurrence of past fluctuations in population size, inbreeding, drift, and thus reduced population fitness and evolutionary potential. But is the expected reduction of fitness important enough to affect extinction risk significantly?

In many cases the differences in genetic diversity observed by Spielman and colleagues (2004a), for which the median heterozygosity was 40% lower in threatened taxa than in related nonthreatened ones, are indeed of sufficient magnitude to cause increased probability of extinction in experimental populations of fruit flies, mice (Fig. 3.1), and many other taxa (Frankham et al., 2002). Furthermore, Crnokrak and Roff (1999) showed that 90% of the 34 taxa in 157 published data sets showed deleterious effects of inbreeding in the wild. Moreover, the magnitude of the effect of inbreeding depression on demographics in more stressful natural environments is seven times larger than in benign, captive environments. Reed and colleagues (2002) provide a neat experimental demonstration of a similar demographic effect of inbreeding depression.

If the results described here are coupled with the meta-analytical findings that show that (1) the genetically effective size of wildlife populations is about 11% that of the observed (census) size (Frankham, 1995a) (Box 3.2) and that (2) population fitness is correlated with the level of genetic diversity (Reed & Frankham, 2003), then the basis for arguing that genetics are important for species threatened with extinction—but not yet extinct—seems remarkably robust.

Population Modeling of Real Cases, with and without Inbreeding Depression

Notwithstanding the synergies implied by the extinction vortex, separating the genetic and nongenetic components of extinction in natural populations (rather than just the processes leading up to extinction) is notoriously difficult. The difficulty arises because teasing apart the causes of extinction requires long-term monitoring of replicate extant and extinct populations, as well as estimates of the relative impact of all sources of risk. In addition, constraints on time and resources means that labor-intensive field studies have concentrated on only a few high-profile species.

Population viability analysis (PVA) (see Allendorf & Ryman, 2002) offers a powerful alternative for investigating the role of inbreeding depression in extinction. The advantages of PVA are that it can be performed quickly for many species and that it provides a framework to analyze genetic factors in concert with demographic and environmental stochasticity and catastrophes (otherwise impossible except in tightly controlled laboratory microcosm experiments). Although most management-oriented PVAs do not consider genetic effects (Allendorf & Ryman, 2002), a number of published studies have used such computer simulations to demonstrate that inbreeding negatively affects population growth and so accelerates extinction, even with many ecological, demographic, or community factors considered.

For instance, Mills and Smouse (1994) used PVA to predict that inbreeding would have an impact on the viability of a range of generalized mammalian life histories (rodent, ungulate, carnivore), especially those with slow intrinsic growth rates. They also showed that skew in sex ratio exacerbates low-population viabilities. Dobson and coworkers (1992) showed that the effect of inbreeding on an endangered rhinoceros was strong but depended on population size, whereas Thévenon and Couvet (2002) highlighted the importance of interactions among density, environmental carrying capacity, and repeated bouts of inbreeding. Tanaka (1997) emphasized the importance of synergistic interactions between inbreeding depression and demographic disturbances, whereas the model of Vucetich and Waite (1999) predicted that even in the absence of inbreeding depression, many populations will lose almost all of their neutral genetic diversity (evolutionary potential) before extinction.

The two most comprehensive simulation studies used PVA to examine the potential impact of inbreeding on the extinction risk of a broad range of taxa. Brook and colleagues (2002) considered genetic impacts on 20 threatened species (including mammals, birds, reptiles, amphibians, fish, invertebrates, and plants). Applying only a conservative level of inbreeding depression to juvenile survival (3.14 lethal equivalents; the median value derived from a meta-analysis of 40 captive-bred mammalian species analyzed by Ralls and colleagues [1988]), they showed that inbreeding hastened times to extinction by almost one third in populations with 50 to 250 individuals, and by a one quarter in those as large as 1,000 individuals. The only circumstance in which inbreeding proved unimportant was when populations were already declining deterministically toward extinction. Starting with an initial population size of 1,000 individuals and a carrying capacity (K) of twice that size (reflecting a limit to available habitat), the average extinction probability across the 20 species was only 11% without inbreeding but 89% with inbreeding imposed.

Yet there are strong (albeit indirect) empirical grounds for believing that even the results of Brook and colleagues (2002) are likely to underestimate the full impact of genetic factors on natural populations. For instance, inbreeding depression affects all components of an organism's life cycle (Frankel & Soulé, 1981; Frankham et al., 2002), not just juvenile survival, and is substantially greater in natural environments that are more stressful than captive ones (Crnokrak & Roff, 1999; Reed et al., 2002). On this basis, O'Grady and coworkers (2006) undertook a meta-analysis of the literature to determine the full impact of inbreeding depression on the fitness of noncaptive species over their entire reproductive life. They used the information derived therein (12.3 lethal equivalents estimated from more than 14,000 individuals distributed among 10 species) to reevaluate the impact of inbreeding depression on the simulated extinction risk of 18 mammalian and 12 avian example species. Under these conditions, the average reduction in median time to extinction was 41% for an initial population size of 1,000 individuals ($K = 2000$).

There are at least two compelling implications suggested by the results of Brook and colleagues (2002) and O'Grady and colleagues (2006). The first is that conservation managers will have between one quarter to one half *less* time to avoid extirpation of imperiled populations than they might otherwise have believed had they ignored the effects of inbreeding depression. The second implication is that populations that appear demographically stable may actually be drawn into an extinction vortex that includes genetic feedbacks.

Expected Effect of Genetics in Population Viability Analysis Depending on the Demographic Context

One question left unanswered by simulation studies is the *relative* importance of factors that might enhance or mitigate inbreeding depression via synergistic feedbacks that occur through density dependence, catastrophes, environmental stochasticity, purging, number of lethal equivalents, and N_e. To consider this issue, I constructed a hypothetical life history in the PVA package VORTEX 9.0 (Brookfield, Illinois, USA; www.vortex9.org/vortex.html). This example had discrete generations, a substantially positive growth rate ($r = 0.12$), moderate environmental variation (CV in vital rates = 15%), and a carrying capacity (K) of twice the initial population size (N). Inbreeding was, as in Brook and colleagues (2002), conservatively modeled to affect only juvenile mortality (in other words, age zero individuals), with 3.14 lethal equivalents, one half of which were recessive lethal alleles subject to purging, and the other half of which were slightly deleterious, mildly recessive alleles. Surviving individuals bred at age one and then died. All scenarios (Table 3.1) were started with an initial N of 50 individuals and then run for 25 generations.

The baseline (standard) scenario of the hypothetical life history showed a substantially higher risk of extinction when inbreeding depression was included versus excluded (Table 3.1), as expected based on the previous results (Brook et al., 2002; O'Grady et al., 2006) (Fig. 3.5). Increased fluctuations in population size resulting from catastrophes and year-to-year environmental fluctuations greatly magnified the impact of inbreeding. For instance, extinction risk was five times higher than the standard (in other words, noninbreeding) scenario when only catastrophes were modeled, but was more than 20 times greater if inbreeding was also included. Conversely, inbreeding had a lesser (but still considerable) impact when the simulated population was forced to cycle deterministically via strong overcompensatory density dependence. This interesting outcome was presumably the result of the strong "rescue effect" that occurs at low population densities when intrinsic r is substantially positive. A high number of lethal equivalents per individual (5.0) almost doubled the extinction risk compared with the relatively conservative estimate used as the default—a result similar to that found by Allendorf and Ryman (2002) for grizzly bears (but see O'Grady and colleagues [2006] for a model with a more realistic number of lethal equivalents).

TABLE 3.1 Effects of Inbreeding and Demographics on Extinction Risk

Scenario	% Extinct	Difference from Standard*
No inbreeding depression		
1. Standard: discrete generations, $r = 0.12$, $N = 50$, $K = 2N$, environmental stochasticity (CV) $= 0.15$	3	N/A
2. Cycling density dependence†		−1
3. Catastrophes (1/25 probability, 75% extra mortality)		14
Inbreeding depression on mortality (3.14 lethal equivalents, 50% recessive lethals)		
1a. Standard (with inbreeding)	47	44
2a. Cycling density dependence†		12
3a. Catastrophes		62
4. No environmental stochasticity		23
5. No K		34
6. No purging (0% recessive lethal alleles)		75
7. Complete purging (100% recessive lethal alleles)		16
8. 5.0 lethal equivalents		82
9. $N_e/N = 0.11$		97

*Percent extinct of each scenario less the percent extinct of the standard model (no inbreeding).

†Mortality $= 1 - (d/[1 + (aN)^b])$, where $d = 1.0$, $a = 0.00762$, and $b = 5$.

Top of table: Sensitivity analysis of the interaction between inbreeding, form of density dependence, and presence of catastrophes; bottom of table: relative importance of different levels of inbreeding depression (number of lethal equivalents per diploid genome), level of environmental stochasticity, effectiveness of purging, and ratio of effective to census population size (N_e/N). The standard simulation was projected using the PVA package VORTEX9.0 for 25 generations with an initial population size (N) of 50 and a ceiling carrying capacity (K) of 100. A simple hypothetical life history with nonoverlapping breeding generations was used.

CV, coefficient of variation; N/A, not applicable.

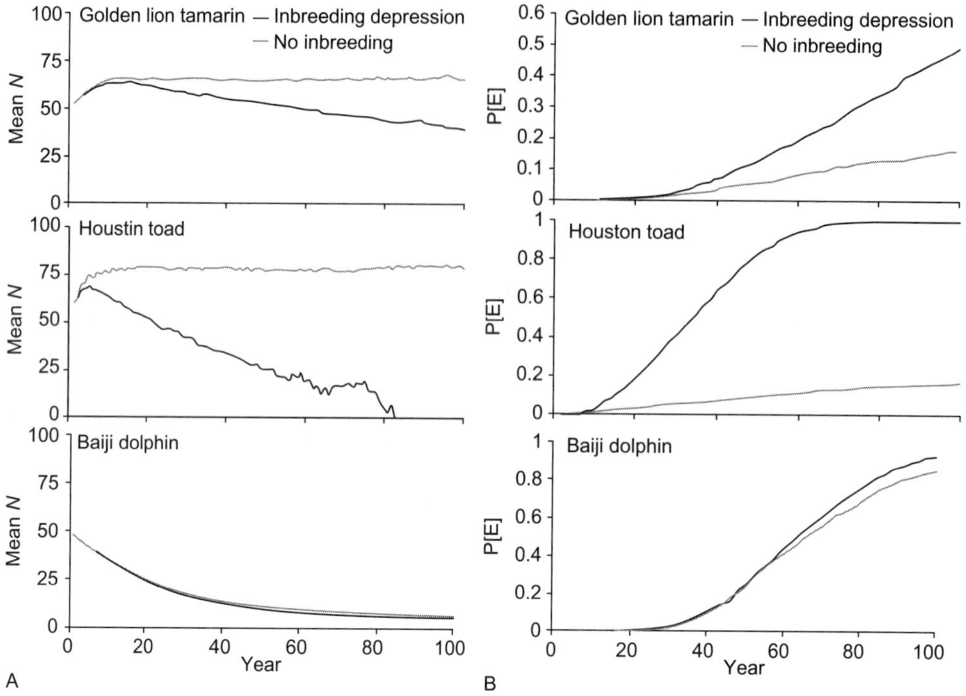

FIGURE 3.5 (A, B) Population viability analysis simulations showing the mean size of persisting populations (A) and the cumulative probability of extinction with and without inbreeding depression for three representative threatened species (B). In addition to the genetic effects, all demographic, environmental, and catastrophic effects were also operating in these simulations. (Figure reprinted from Brook and colleagues [2002], with permission.)

Reduction in the N_e/N ratio from the standard scenario of 83% (resulting from environmental and demographic fluctuations in N) to 11% (achieved by limiting the number of breeding males) led to guaranteed extinction within only 13 generations. Because many endangered species have a very low N_e/N ratio (Frankham, 1995a), even populations with relatively large census sizes are likely to be vulnerable to the deleterious effects of inbreeding.

In the absence of homozygous lethal alleles, inbreeding effects on population viability were extremely severe compared with when the entire genetic load was subject to purging. This result suggests that already inbred or naturally inbreeding populations are likely to be less susceptible to genetic threats if most genetic load is subject to purging. Yet either extreme is probably unrealistic for most species. Recent experimental studies have detected only a limited impact of purging (summarized in Frankham, 2005b), suggesting that most inbreeding depression is the result of slightly deleterious alleles rather than a high proportion of recessive lethals.

Because most species lack relevant data, Morris and Doak (2002) advocate excluding genetics in PVAs. Their alternative is to set a quasi-extinction threshold large enough to minimize the chances that genetic problems dramatically change PVA conclusions (they suggest $N_e > 50$). To quote Morris and Doak (2002, p. 43): "This solution seems better to us than including wild guesses about genetic effects that are open to endless argument and can weaken the credibility of the whole analysis." Yet based on the N_e/N ratio of 1/10 (Frankham, 1995a), this equates to a census size of about 500 individuals. Moreover, when initial N was increased from 50 to 500, a very low extinction risk was observed in all scenarios, with or without inbreeding (results not shown). However, if the intrinsic population growth was cut in half under this scenario, from 0.12 to 0.06, or alternatively, if the number of generations is doubled from 25 to 50, the same pattern as in the $N = 50$ scenarios emerged. Brook and colleagues (2002) and O'Grady and colleagues (2006) also observed a strong impact of inbreeding on extinction risk at census sizes of 1,000 individuals. As such, I suggest that rather than trying to set a quasi-extinction threshold sufficient to avoid worry about genetic issues, conservation modelers are better advised to consider how sensitive their results are to scenarios with low and high inbreeding depression.

FUTURE DIRECTIONS

Evidence for a genetic role in extinctions, from individual case studies of threatened species, meta-analysis of large databases, experimental replicate populations, and computer simulations of concurrent demographic and genetic processes is, I argue, now unequivocal. However, a number of arguments need to be brought to the wider attention of ecologically oriented conservation biologists before the book on demography versus genetics in conservation biology can be closed. For instance, contrary to Young (1991), Caughley (1994), Caro and Laurenson (1994), and others, there is no necessity for or merit in adopting the Popperian standpoint that to establish the credibility of genetic threats to extinction one must prove that genetics was solely or primarily responsible. Similarly, without knowledge of the frequency of failure (which by definition will be unobserved) or history of cumulatively more muted recoveries, observations that some populations have survived past bottlenecks cannot be used to argue against the impact of inbreeding depression (Allendorf & Ryman, 2002). I should note that quite a number of authors agree with these sentiments (Box 3.3), and offered useful suggestions for mapping a future research path.

The consensus that emerges from these viewpoints and the examples discussed throughout this chapter can be summarized as follows:

1. It is pointless arguing for a single cause of the stochastic extinction of any given population because of the inherent and inextricable synergies that exist among demographic, environmental, and genetic effects.
2. Genetic hazards can substantially reduce population fitness well before demographic extinction occurs, and indeed can dramatically hasten the final demise. And conversely, because threatened populations can accumulate a large genetic load, mating among unrelated individuals is expected to result in an immediate and large increase of vigor and fecundity (see, for example, Madsen et al., 1999).
3. Populations must maintain *census sizes* of at least a few hundred individuals to avoid suffering inbreeding depression, but thousands to tens of thousands to retain evolutionary potential and to avoid mutational meltdown (Lynch et al., 1995a). Similarly, although less than 100 individuals are usually sufficient to minimize the impact of demographic stochasticity, many thousands may be required to ensure resilience against environmental fluctuations (Frankham et al., 2002).
4. Genetic problems are observed in the majority of cases of endangered populations, with the constraint being that few studies have actually monitored or experimented on natural populations in a way that teases apart genetic influences from demographic influences. All of the previous points are important, but tackling this point more adequately and regularly would be the most obvious future direction that research in this area will take (Frankham et al., 2002; Oostermeijer et al., 2003).

In what circumstances are genetic factors likely to affect extinction risk, and when will these factors be less important? Spielman and colleagues (2004a) and Frankham (2005b) outline at least three conditions under which genetic hazards are less important: (1) in rapidly growing populations (which may continue to increase even if factors such as inbreeding depression act to suppress the rate of recovery) or, conversely, in populations declining precipitously as a result of ongoing deterministic threats (which will be driven to extinction regardless of whether growth rates are reduced by genetic stochasticity); (2) in large populations that have not suffered from repeated or extended historical bottlenecks or large fluctuations in population size, and thus retain substantial genetic variation and adaptive potential; and (3) in large-bodied species with long generation times. Given that inbreeding operates over generations, such species are less likely to manifest genetic problems over the duration of species management programs. In the big picture of conservation biology, however, the first two circumstances are unusual for

BOX 3.3 Arguments Supporting a Role of Genetics in Extinction

Keller and colleagues (1994): "Inbreeding depression was expressed in the face of an environmental challenge . . . . We suggest that environmental and genetic effects on survival may interact and, as a consequence, that their effects on individuals and populations should not be considered independently."

Mills and Smouse (1994): "Counter to the current fashion, which downplays the importance of inbreeding in stochastic environments, we conclude that, while inbreeding depression is not necessarily the primary cause of extinction, it can be critical."

Lacy (1997): "Genetic threats to population viability will be expressed through their effects on and interactions with demographic and ecological processes. Theoretical analyses, experimental tests, field studies, and conservation actions should recognize the fundamental interdependency of genetic and non-genetic processes affecting viability of populations."

Lande (1998): "All factors affecting extinction risk [including genetics and demography] are expressed, and can be evaluated, through their operation on population dynamics."

Allendorf and Ryman (2002): "The disagreement over whether genetics should be considered in demographic predictions of population persistence has been unfortunate and misleading. Extinction is a demographic process that will be influenced by genetic effects under some circumstances."

Oostermeijer et al. (2003): ". . . demographic studies can detect changes in vital rates in small populations, but cannot reveal underlying genetic causes. Fitness and demographic studies are also well represented in the literature, but remarkably few studies have attempted to integrate empirical demographic and genetic studies . . . . We conclude that demography and demographic–genetic experiments should play a central role in plant conservation genetics."

Frankham (2005b): "Thus, there is now sufficient evidence to regard the controversies regarding the contribution of genetic factors to extinction risk as resolved."

threatened species, and the third is evolutionarily short-sighted.

Lastly (and somewhat ironically, given the debate discussed herein), it is worth noting that genetic methods have proved to be powerful tools for inferring a suite of largely unobservable demographic and biogeographic information (Frankham et al., 2002; Young & Clarke, 2000). This includes population-level insights into sex ratios, abundance and survival (for example, genetic mark–recapture studies using biopsies or naturally discarded tissue or hair samples), long-term historical changes in genetic diversity and phylogeographic substructuring of populations, estimates of dispersal rates over a range of temporal and spatial scales, and improved taxonomic resolution for more clearly defined conservation management. The potential for future developments in these areas, especially if integrated with population and evolutionary models, is simply enormous. Genetics should be seen not only as a tool to describe past demographic events, but also as an important process that synergizes with demographics to determine the fate of a threatened species.

Summary of Major Points

Formerly large and viable natural populations become threatened primarily because of extrinsic threats that cause substantial declines in range and abundance. The threats most relevant to conservation biology are widely agreed to be driven by direct and indirect human activities and include habitat loss, overexploitation, invasive species, and pollution. What remains surprisingly controversial is the relative role played by demographic versus genetic factors in the extinction of threatened populations. Although several well-studied taxa show that inbreeding and loss of genetic diversity demonstrably reduce reproductive fitness, many conservation biologists maintain that demographic and environmental hazards operate with sufficient strength and rapidity to render irrelevant any genetic impacts on extinction risk. I have argued in this chapter against the view that any one factor plays a dominant role in the population dynamics of threatened species. This is primarily because of the many inevitable synergies and feedbacks inherent in the extinction vortex, coupled with the often overlooked (yet persistent) genetic legacy associated with past population fluctuations and declines.

SUGGESTIONS FOR FURTHER READING

The chapter by Gilpin and Soulé (1986) was the first to formalize the concept of the extinction vortex. Indeed, the book in which it is found (*Conservation Biology: The Science of Scarcity and Diversity*) is a seminal work of conservation science. Lande (1988) wrote a highly influential review that provided a credible argument that genetic factors may play little or no role in the extinction of populations (although later work refuted this conclusion). Caughley (1994) wrote a deliberately controversial (and highly cited) opinion paper, written shortly before the author's untimely death, that attempted to provoke conservation biologists into thinking more deeply about ways to confront theory with empirical data and experiments (especially for declining populations). The volume edited by Young and Clarke (2000) provides a wealth of interesting case studies that illustrate the relationship between genetics and demography in threatened populations. Frankham (2005b) penned a concise, up-to-date summary of the evidence for genetic impacts on population viability.

Caughley, G. 1994. Directions in conservation biology. J Anim Ecol. 63: 215–244.
Frankham, R. 2005b. Genetics and extinction. Biol Conserv. 126: 131–140.
Gilpin, M. E., & M. E. Soulé. 1986. Minimum viable populations: Processes of species extinction (pp. 19–34). In M. E. Soule (ed.). Conservation biology: The science of scarcity and diversity. Sinauer Associates, Sunderland, Mass.
Lande, R. 1988. Genetics and demography in biological conservation. Science 241: 1455–1460.
Young, A. G., & G. M. Clarke (eds.). 2000. Genetics, demography and viability of fragmented populations. Cambridge University Press, London, UK.

4

Metapopulation Structure and the Conservation Consequences of Population Fragmentation

JULIANNO B. M. SAMBATTI
ELI STAHL
SUSAN HARRISON

Rapid fragmentation of wild populations is a ubiquitous consequence of human population growth. Although some contiguous areas of wild habitat remain—for example, large tracts of the rainforests of Amazonia are still untouched—rates of fragmentation around the planet have not been slowed by ongoing political focus on the problem, and thus the prognosis is not encouraging. Therefore, much of conservation biology will continue to be devoted to understanding the biology of fragmented species. We argue here that certain pervasive characteristics of population fragmentation make it particularly amenable to metapopulation theory (Levins, 1970), and that this theoretical framework can be used to derive general principles of species conservation.

A metapopulation refers to a set of local populations that exist within a network of empty and occupied patches of habitat where migrants are exchanged among populations and where each population may face local extinction. Thus, in a metapopulation, species persistence is the dynamic result of extinction of local populations and recolonization of empty patches. Moreover, in metapopulations, demography, genetics, and selection interact in ways that are often not considered in demographically stable species. Since the seminal work of Levins (1970), the metapopulation concept has been incorporated into fields, including population genetics (Slatkin, 1977), population and community ecology (Hanski, 1998), and evolutionary biology (Olivieri et al., 1995).

Human-caused fragmentation is characterized by a drastic and rapid reduction in the number of individuals within a species, along with a radical modification of its spatial distribution. A fragmented species can be seen as a collection of local populations that may remain connected to each other by migration. After fragmentation, smaller local populations become more prone to extinction as a result of chance demographic or environmental events, and populations occupying different patches may become more isolated. Extinction of a fragmented species occurs by the accumulated extinctions of its component populations.

The impacts of fragmentation go beyond demography. The genetic consequences that arise from demographic reduction are aggravated by the increasing isolation of small populations (Higgins & Lynch, 2001). Theory predicts that small populations may be particularly susceptible to loss of genetic diversity. Both theoretical and empirical work demonstrate that lack of genetic variation limits response to the selection pressures caused by fragmentation itself, global environmental change, or the normal vagaries of a harsh environment.

Because the genetics of a species reflects its demography, the study of neutral genetic variation in a fragmented species may help us to monitor it. Many recent molecular and analytical techniques are available for this important area of molecular ecology. Here we describe some general principles and point to some limitations that conservationists must be aware of when using genetic data. We illustrate these ideas with an application of molecular metapopulation genetics to the fragmented sunflower species *Helianthus exilis*.

METAPOPULATION CONCEPT AND CONSERVATION

The classical metapopulation concept is based on the theoretical work of Levins (1970). His theoretical model represents an infinite population of identical semi-isolated patches, a proportion p of which is occupied by local populations, each subject to a constant probability of extinction e. The likelihood of colonization of an empty patch is given by cp, a constant colonization rate c multiplied by the patch occupancy (which is proportional to the number of available colonizers), so that empty patches are colonized at a higher rate as overall patch occupancy increases. The rate of change in patch occupancy is given by

$$\frac{dp}{dt} = cp(1-p) - ep \qquad (4.1)$$

which has the nontrivial equilibrium $1 - \hat{p} = \delta = (e/c)$ if $e < c$. The proportion of empty patches is determined by the ratio of extinction and colonization rates. Lande (1987) used Levins' model to derive the number of patches necessary for a metapopulation to survive after all but a proportion h of the habitat is destroyed. The decrease in habitat in turn decreases colonization probability, yielding the new equilibrium patch occupancy $\hat{p} = 1 - (\delta/h)$, and the metapopulation will survive if $h > \delta$. Although this classic model has had an important impact on how ecologists and evolutionary biologists view the demography of many species, and has provided initial insights toward the conservation of fragmented species, it has several unrealistic assumptions. First, most species show significant heterogeneity in patch sizes, population sizes, and distances between patches, leading to variation in the probabilities of extinction and colonization (Harrison & Hastings, 1996). It is well known, for example, that extinction probability is a negative function of population size and that colonization capacity of a local population may depend on its density (Hanski, 1998). Second, the assumption of an infinite number of patches is always violated, especially in systems relevant to conservation. The infinite-patch assumption makes the original Levins model deterministic even when describing e and c as random variables. This makes the model unable to address relevant questions such as the expected time to metapopulation extinction or the probability of extinction within a certain time.

When the infinite-patch assumption is relaxed, stochasticity in extinctions and colonizations generates a nonzero probability of metapopulation extinction. For a metapopulation to persist, the expected number of new populations generated by a single population, during its lifetime, in an otherwise empty patch network (in other words, the metapopulation replacement rate) must be greater than one, giving a nonnegative metapopulation growth rate (Hanski, 1998). As a result of stochasticity, a finite metapopulation with an expected positive growth rate will also show variation in the growth rate around its expected value. This variation may lead to a very small metapopulation with a positive expected growth rate to extinction because, by chance, the metapopulation growth rate can be negative in a particular year. Furthermore, because the metapopulation growth rate depends on intrinsic attributes of the species such as reproductive potential and dispersal capability, and on the spatial configuration of the habitat, the stochastic process by which colonization and extinction probabilities are "sampled" from a heterogeneous landscape could give rise to a "bad" parameter combination that drives the metapopulation extinct (Brook, this volume; Reed, this volume).

From a conservation perspective, incorporating heterogeneity and stochasticity into metapopulation theory opens up the possibility for analysis of real systems. Hanski and colleagues (Hanski, 1998; Hanski & Ovaskainen, 2000; Hanski et al., 1996a, b; Ovaskainen & Hanski, 2003) have developed *spatially realistic* Levins models, in which the patch network is finite and transition probability matrices incorporate patch-specific size and connectedness. Thus, large patches support large populations that are less extinction prone and produce more potential colonizers, and colonization is more likely to occur in empty patches that are closer to large, occupied patches. These model parameters can be estimated using a snapshot of metapopulation distribution in space.

In the spatially realistic models described here, a fragmented landscape can be characterized by a single parameter λ_M, the metapopulation capacity, which is the leading eigenvalue of a matrix of patch occupancy transition probabilities. These probabilities depend on the area and connectivity of patches in the landscape. A population can persist in a given landscape if $\lambda_M > \delta$, where δ is the ratio of extinction and colonization rate parameters (Hanski & Ovaskainen, 2003). The expected fraction of patches occupied at equilibrium becomes $p_\lambda = 1 - (\delta/\lambda_M)$. This result is analogous to the original Levins model, but takes heterogeneity into account (Hanski & Ovaskainen, 2000, 2003). Metapopulation capacity (λ_M) provides a measure by which the ability of different landscapes to support a given species can be compared (Ovaskainen & Hanski, 2003). Because λ_M is a weighted sum of individual patch contributions, it is also possible to quantify the impact on λ_M when a particular patch is added to or subtracted from the metapopulation (Hanski & Ovaskainen, 2000, 2003). A related approach is to describe a metapopulation in terms of its effective size, or the size of a homogeneous metapopulation with the same properties as the given heterogeneous metapopulation, a method akin to the effective size N_e in population genetics (Ovaskainen, 2002) (see "Metapopulation Genetics" later in this chapter).

Although spatially realistic metapopulation models can, in principle, be parameterized with a snapshot of the pattern of patch occupancy, such an approach assumes that the metapopulation is at equilibrium. Fragmentation violates this assumption by causing changes in extinction and colonization rates. Patch occupancy is likely to be larger immediately after fragmentation than at the new equilibrium. Thus, using equilibrium model parameter estimates to make conservation decisions would be misleading; an existing metapopulation may already be moving toward extinction even though its nonequilibrium patch occupancy and capacity may indicate a viable population (Hanski, 1998). The rate of equilibration is an important issue for both demographic and population genetic analyses (discussed later), particularly for species with high longevity, such as trees. How long this time to equilibration is and the characteristics of these transient metapopulation dynamics are subjects of active research (Ovaskainen & Hanski, 2002; Tilman et al., 1994; Vellend et al., 2006).

Equilibration time depends critically on the habitat patch network in a metapopulation. This time increases with the strength of the perturbation and with a ratio of metapopulation capacity and a threshold value that is characteristic of each species, (δ/λ_M) (Ovaskainen & Hanski, 2002). The more abundant the species (small (δ/λ_M)), the faster the system moves toward equilibrium (Ovaskainen & Hanski, 2002). This transient time also increases with the characteristic turnover time of a species, scaled by $(1/e)$, in a given metapopulation, which is particularly long when the metapopulation is composed of few large patches (Ovaskainen & Hanski, 2002).

Habitats subjected to fragmentation are expected to undergo a gradual loss of species after habitat destruction and are said to have an *extinction debt* (that is, species that are still present but are doomed to extinction) that will be paid in time (Tilman et al., 1994). Thus, within the metapopulation framework, an important question from a conservation perspective is whether extinction is the fate of a fragmented species despite an apparently healthy present-day abundance. Although theories such as that of Ovaskainen and Hanski (2002) have been developed to address this problem, answering this question with an empirical snapshot of the metapopulation is quite risky (Thomas et al., 2002), and it has been particularly challenging even when historical data are available (Vellend et al., 2006). The question is whether there is an extinction debt and the rate associated with the debt. Vellend and colleagues (2006) parameterized a patch occupancy model for several plant species with historical data gathered in one region to make predictions of patch occupancy of the same species in a different region of fragmented forest in England. Although quantitative predictions were poor, they obtained some reasonable qualitative predictions. For example, habitat effects that influenced a species' patch occupancy in one landscape also influenced patch occupancy in the second landscape. Their study showed an extinction debt two centuries after habitat fragmentation. Because theory predicts that the rate of extinction debt should not be constant across species, several species may have gone extinct immediately after fragmentation. It is possible that the species used in their study were already a sample biased toward species with a rate of extinction that is low. A timely and productive empirical agenda would be to conduct comparative studies to determine which phenotypic and ecological attributes of species correlate with mean extinction times.

In Box 4.1 we present a simple nonequilibrium metapopulation model illustrating some

BOX 4.1 Modeling Metapopulation Genetics

We formulated a metapopulation model with internal colonization (empty patches are colonized at a higher rate as the fraction of occupied patches increases) and a partial rescue effect (populations go extinct at a somewhat slower rate as the occupancy fraction increases), so that the change in occupancy fraction f per unit time $(df/dt) = p_c f(1-f) - p_e f(1-rf)$, where p_c and p_e are the colonization and extinction rate parameters, respectively, and $r < 1$ determines the rescue effect strength. This model yields the equilibrium occupancy fraction $\hat{f} = (p_c - p_e)/(p_c - rp_e)$ when $p_c > p_e$.

We take this model to represent a fragmented species, with initial conditions that represent the state of the species before fragmentation. We will provide the simulation code in Perl on request. Box Figure 4.1 shows random simulated trajectories for large, small, and very small metapopulations (1024, 100, and 16 patches, respectively) under colonization/extinction parameters (p_c, p_e) equal to (0.12, 0.06), (0.1, 0.1), and (0.06, 0.12). These (p_c, p_e) parameter sets yield an equilibrium proportion of 0.53 of patches occupied, eventual extinction, and more rapid extinction, respectively. All our simulations started with 53% of patches occupied, and colonizations and extinctions randomly generated were at discrete time steps. The colonization/extinction parameters strongly affect the metapopulation extinction rates. In addition, because $f = 0$ is an absorbing boundary with internal colonization and a finite number of patches, smaller metapopulations also have faster extinction rates (Box Fig. 4.1). Indeed, finite metapopulations have nonzero extinction probabilities even with parameters expected to lead to metapopulation persistence.

To study metapopulation genetic variables, we simulated the coalescent (Hudson, 1983) conditional on random metapopulation histories. The metapopulation history is taken as the fragmentation phase of the sample history, with an earlier prefragmentation phase modeled by a large panmictic population. Coalescent simulations begin at the present (time of sampling), and a series of times since fragmentation was considered (for those metapopulation histories that persisted to the sampling time). Briefly, the coalescent (Hudson, 1983, 1990) is a statistical model that describes the tree or network by which haploid individuals sampled from a population are related, where common ancestry (going backward in time) occurs at a rate inversely proportional to population size for ancestral lineages that happen to be present in the same populations. Migration as well as population colonization disperses lineages among populations. Colonization founding events also force ancestry among all lineages (if present) within populations, and colonization and extinction events change the numbers of source populations for dispersal. Infinite alleles mutation (Kimura, 1969) was modeled along the coalescent tree branches (for consideration of microsatellite or allozyme data), and two loci were simulated to consider the linkage disequilibrium generated by metapopulation dynamics.

For purposes of illustration, we used plausible values for all model parameters. For genetic parameters, we assumed a small within-population effective size $N_e = 50$, a mutation rate of 10^{-4} (giving $\theta = 2N_e u = 0.01$), recombination fraction $\rho = 0.5$ between unlinked loci (giving $\rho = 2N_e \rho = 50$). We used a migration parameter $2\ N_e m = 1$. Colonization and extinction events for each metapopulation time step were implemented at coalescent time increments of 0.1 (in units of $2N_e$ generations); this assumption was made considering the computer algorithm and memory limitations, and essentially makes for weaker metapopulation dynamics than might be realistic (implying that p_c and p_e would be

(continued)

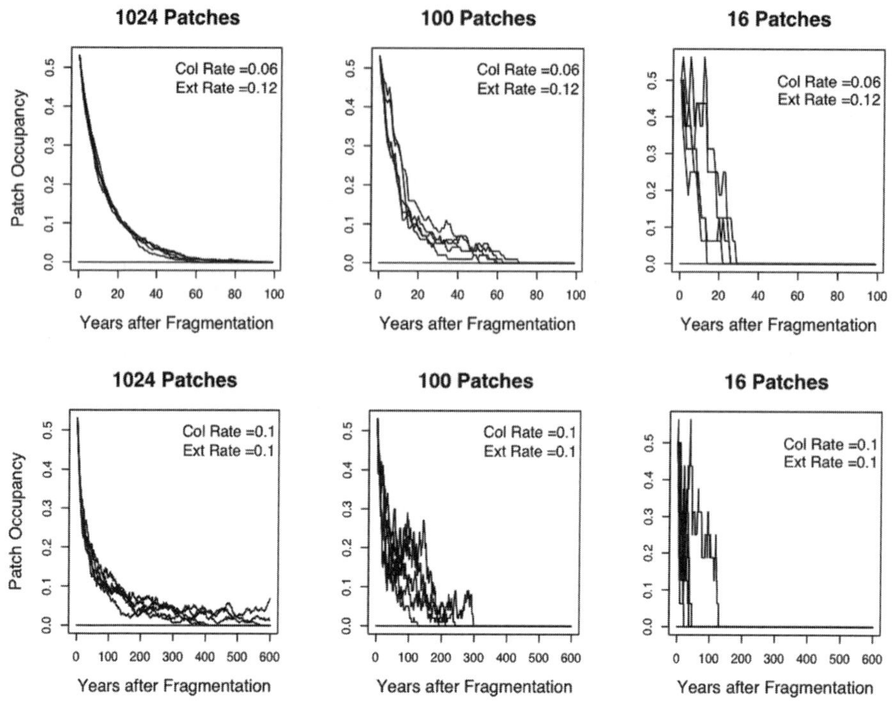

Box Fig 4.1 Simulation results showing five patch-occupancy trajectories per parameter set after a large continuous species was fragmented and started to behave as a finite metapopulation. Each chart shows the fragmented metapopulation trajectory in a demographic stochastic scenario with different metapopulation parameters—in other words, different extinction and colonization rates ($c = 0.06$ and $e = 0.12$ or $c = 0.10$ and $e = 0.10$). From the right to the left, simulations represent a fragmented population with 16, 100, and 1024 available patches, respectively. Variance in patch occupancy is a function of the number of patches. Sixteen patch species tend to have a larger variance in patch occupancy, leading to a higher likelihood of extinction than greater patch number species. These simulations were used as a template to perform coalescent simulations and to calculate population genetic parameters.

colonization/extinction rates per 10 years rather than per year, for an annual plant or animal). The size of the prefragmentation panmictic population was assigned N_e times the large-metapopulation equilibrium number of occupied patches, or 13,568. Modifications of Hudson's (1990) simulation code were used, and a more extensive treatment of the model and simulation results will be presented elsewhere.

For samples of 10 individuals from each of two populations, we calculated the total number of alleles, the population fixation index $F_{ST} = 1 - H_W/H_T$ (where H is heterozygosity, within populations and in total), and the within-population linkage disequilibrium R (the allele–frequency-weighted average squared correlation coefficient between alleles at different loci) (Sambatti & Sickler, 2006).

important principles for the conservation of fragmented species. This model considers colonization (Levins, 1970) and extinction with a partial rescue effect—in other words, prevention of local extinction resulting from the arrival of immigrants from neighboring populations (Brown & Kodric-Brown, 1977). Because the risk of extinction is higher in smaller metapopulations, we are particularly interested in cases in which fragmentation leads to a small number of patches, with colonization and extinction rates leading to inevitable metapopulation extinction. Simulations allow for straightforward analysis of the probability distributions of metapopulation persistence and extinction time. The probability distribution of time to extinction depends strongly on both metapopulation size and rates of colonization and extinction. For example, metapopulations with a small but realistic number of patches have high probabilities of extinction within the span of a human generation. Very small metapopulations have substantial extinction probabilities, even with $c > e$ and therefore nonzero equilibrium expected patch occupancies (Box 4.1). At the end of the following section, we incorporate genetic data into these fragmented metapopulation scenarios.

Metapopulation-Level Stochasticity

Species can be affected by environmental factors with demographic consequences (see, for example, Harper & Peckarsky, 2006; Reed, this volume). Environmental factors such as water availability and temperature vary randomly in time and are correlated with each other in space. A dry year, for example, will affect all local populations simultaneously within a region. Such metapopulation-level stochasticity can create temporally correlated risks of extinction among populations, which is expected to reduce metapopulation persistence (Hanski, 1998; Higgins & Lynch, 2001). Thus, from a conservation perspective, the decision concerning the optimal distribution of preserved fragments represents a tradeoff between maximizing connectivity to allow between-population dispersal, and maximizing environmental independence among populations (Hanski, 1998). Metapopulation-level stochasticity may also affect the sequence of extinction and colonization events and may considerably alter the patch occupancy model predictions described earlier (Frank, 2005). For example, according to Frank (2005), increasing heterogeneity in the colonization abilities always reduces the effective number of patches and increases the risk of metapopulation extinction. Moreover, it can be shown that a greater number of patches is required for metapopulation persistence to increase in a general stochastic environment (Bascompte et al., 2002). Thus, increasing the number of patches with high colonization potential is, at least in theory, an important conservation measure in a metapopulation-level stochasticity scenario.

Metapopulation and the Population Local Dynamics

One limitation of patch occupancy models is the absence of local population dynamics. However, including local population dynamics into modeling approaches usually requires the development of simulation models with a large number of parameters that may not be easily estimated in the field. Moreover, the cost of greater realism in models is the loss of generality (compare with Lopez & Pfister, 2001). Lopez and Pfister (2001) evaluated the effect of violating the homogeneous Levins model assumptions when local population dynamics are taken into account. They found that the Levins model usually overestimates patch occupancy when local population dynamics are not considered. Modern patch occupancy models have, nonetheless, the ability to describe metapopulation heterogeneities roughly and have been shown to predict reasonably well the dynamics of certain fragmented species with a limited number of parameters (but see Schtickzelle & Baguette, 2004). The extent to which local population dynamics need to be considered to make reasonable predictions in models of fragmented populations remains an unresolved issue (Hanski, 2004).

METAPOPULATION GENETICS

Population genetics and demography are tightly interwoven disciplines. Regardless of historical demography, a completely new demography is imposed on a species after fragmentation, along with predictable genetic consequences. We present here a summary of the theory that describes these genetic consequences and the conservation lessons one can draw from it. We use population genetics theory to assess the extent to which neutral molecular markers can be used to monitor the demography of a fragmented species.

Another important aspect of the interplay between genetics and demography is to ascertain whether genetic consequences of fragmentation feed back on demography and contribute to extinction. For some years, demography alone was the immediate focus of conservation practices (Brook, this volume; Lande, 1988). However, it has been shown both theoretically (Higgins & Lynch, 2001) and empirically (Saccheri & Hanski, 2006) that natural selection has demographic consequences and must be taken into account in conservation programs (see also Boulding, this volume; Brook, this volume; Reed, this volume). A maladapted species may have a suboptimal demographic growth rate (Saccheri & Hanski, 2006). The efficacy of natural selection for favorable (and against deleterious) mutations, the effectiveness of recombination, and the number of mutations that arise in a species all depend on (effective) population size and the connectedness of local populations in a fragmented species. For example, Higgins and Lynch (2001) proposed that a highly connected metapopulation is less extinction prone because natural selection can act more efficiently to eliminate deleterious mutations. Therefore, it is very important to understand how fragmentation and fragment connectedness influence effective population size, because this parameter links evolution and ecology.

This section also provides the basis to understanding the last section of this chapter, where we discuss the nature of selection pressures that emerge as a result of metapopulation demography in a fragmented species.

Genetics of Population Structure and Metapopulation Dynamics

Considerable effort has been devoted to developing tools to understand demographic dynamics based on genetic variation within and among populations. In a fragmented metapopulation, one would like to know the patterns of gene flow that link populations, the effective size of the metapopulation and of local populations, and the implications of these quantities for demographic phenomena such as population size reductions, changes in rates of extinction or colonization, and so forth. Available methods usually use theoretical models that incorporate the main evolutionary forces affecting genetic variation, and assume certain properties to make the theoretical problem tractable.

The classic island model of Sewall Wright (1931), for example, depicts a system of semi-isolated populations in demographic equilibrium where local effective population sizes do not fluctuate. This model is the basis of a popular method to estimate gene flow between semi-isolated populations indirectly. Wright derived the formula $F_{ST} \sim (1 + 4N_e m)^{-1}$, where F_{ST} (Box 4.2) measures the deficit of heterozygotes observed in this system of semi-isolated populations when compared with the frequency of heterozygotes expected in a single population of the same size in Hardy-Weinberg equilibrium (Hartl & Clark, 1997). Heterozygote deficit can also be estimated as the proportion of the genetic variation distributed among populations compared with the total genetic variation, such that a larger F_{ST} indicates a higher degree of genetic differentiation among populations (Box 4.2). In this system of semi-isolated populations, m is the mean proportion of immigrant alleles arriving in a local population each generation. If among-population differentiation results from the balance between the loss of variation resulting from genetic drift (associated with local population size N_e) and the rate of transfer of genetic variation among populations m, then the mean number of immigrants per generation, $N_e m$, can be also estimated.

Although estimation of F_{ST} is used extensively in field studies, its demographic equilibrium assumptions are, by definition, violated in a metapopulation with extinction and recolonization. Moreover, using population genetics to infer nonequilibrial demographic events is problematic because usually a compound parameter, such as $N_e m$, is estimated. For example, using F_{ST} alone, it is impossible to distinguish equilibrium migration from complete isolation that started at some time in the past (Wakeley, 1996). In this case, the migration fraction m is not constant, and more ancient isolation (longer time during which $m = 0$) increases F_{ST} just as decreasing m does under an equilibrium island model.

Extinction and recolonization redistribute genetic variation in a metapopulation. This redistribution depends critically on where founding colonists come from and how many colonists found a new colony. Two extreme possibilities are that each founder originates from anywhere in the metapopulation independently (the migrant pool model) or

BOX 4.2 F_{ST} as a Measure of among-Population Differentiation

Wright's F_{ST} estimates the deficit of heterozygotes in a (meta)population resulting from population subdivision (Hartl & Clark, 1997; Wright, 1931). The distribution of genetic diversity of a species can also be estimated with F_{ST}, now interpreted as the standardized genetic variance among populations. If two alleles, a_1 and a_2, of a single locus are segregating in a species, $F_{ST} = (\text{var}(p))/(\bar{p}(1-\bar{p}))$, where var($p$) is the variance of the frequency p of the allele a_1 across populations, and $\bar{p}(1-\bar{p})$ is the mean within-population allelic frequency variance (Hartl & Clark, 1997; Pannell & Charlesworth, 2000). Thus, the larger the F_{ST}, the more genetically differentiated are the local populations in a metapopulation.

For demographic studies, such as the ones described in this chapter, one is usually interested in estimating F_{ST} of neutral loci—in other words, loci that are not under the effects of natural selection. Natural selection can change patterns of genetic variation and makes the analysis of allelic variation of loci under selection less useful to demographic studies. Because F_{ST} is a function of the between-population connectedness, it is also seen as a measure of the degree of population isolation in a metapopulation (see main text).

Box Figure 4.2 illustrates various cases of how the genetic variation of a diallelic locus (alleles a_1 and a_2) can be distributed among three populations. The frequency of allele a_1

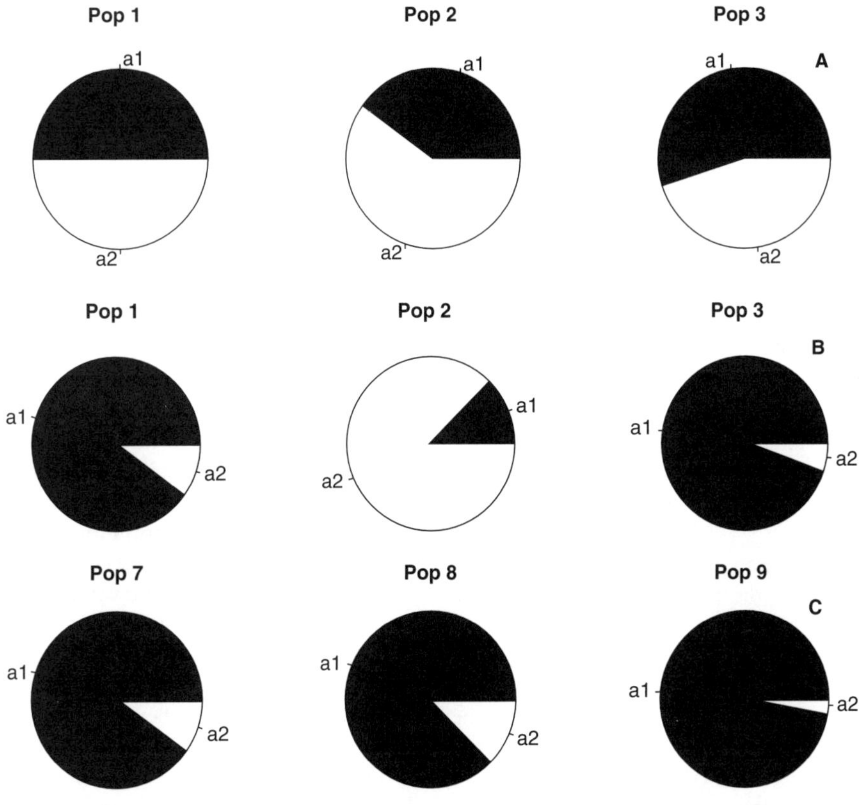

Box Fig 4.2 (A–C) Distribution of genetic variation among populations. *(continued)*

BOX 4.2 F_{ST} as a Measure of among-Population Differentiation *(cont.)*

in a population is depicted as the dark area in the circle, and the frequency of allele a_2 is the complementary white area. In Box Figure 4.2A, total and within-population variation are high—the maximum variation of a diallelic locus occurs when the allelic frequency is 0.5, as in Pop 1. However, population differentiation is low in this case (in other words, low F_{ST}). All three populations (1, 2, and 3) have similar allelic frequencies. In Box Figure 4.2B, the within-population allelic variation is low, but the among-population variation is high (high F_{ST}). Although allele a_1 predominates in populations 4 and 6, allele a_2 predominates in population 5. Box Figure 4.2C shows a case with low within- and among-population variation (low F_{ST}). In this case, allele a_1 predominates in all populations (7, 8, and 9).

that all founders come from a single, perhaps a neighboring, population (the propagule pool model [Slatkin, 1977]). In general, F_{ST} increases with increasing colonization rates under a propagule pool model if founder numbers are small relative to the migration parameter $2N_em$ (Whitlock & McCauley, 1990). At the same time, F_{ST} generally decreases with increasing colonization rates under a migrant pool model because colonization is a form of gene flow. This effect is enhanced when the number of founders is large and when the extinction rate e is comparable in magnitude with the migration fraction m. When $e << m$, the metapopulation approaches the equilibrium island model and $F_{ST} \sim (1 + 4N_em)^{-1}$, (Pannell & Charlesworth, 1999). Attempts to distinguish migration from extinction in a metapopulation based solely on F_{ST} can be difficult as well; multiple combinations of extinction and recolonization rates can explain the same amount of within-population diversity. However, the allelic frequency distribution in a metapopulation differs markedly when extinction *versus* migration approaches zero (Wakeley & Aliacar, 2001).

Constant immigration from stable populations may contribute to the persistence of populations in unfavorable habitats that would otherwise go extinct. This has been termed *source–sink dynamics* (Pulliam, 1988). In these cases, the magnitude of genetic variation within sink populations and the degree of genetic differentiation between source and sink populations are functions of migration from the source population, and genetic drift and new mutation within the sink population (Gaggiotti & Smouse, 1996). The effects of migration, drift, and mutation on genetic variation in sink populations are regulated primarily by the rate at which sink populations decay toward extinction. Short-lived populations will resemble the source, but long-lived populations will acquire some independence after accumulating new mutations and undergoing genetic drift (Gaggiotti & Smouse, 1996). Likewise, the ages of local populations affect differentiation in a metapopulation. During both colonization and population senescence, F_{ST} will tend to be higher as a result of higher genetic drift associated with lower effective populations sizes at these stages (Giles & Goudet, 1997).

Effective Population Size

Effective population size, N_e, represents an idealized, random mating population with a finite number of individuals with a level of genetic diversity that corresponds to the equilibrium between the loss of allelic variation resulting from genetic drift and the increase of allelic variation through new mutations (Whitlock & Barton, 1997). The rate at which a population loses its genetic variation is inversely proportional to N_e. Ecologically, N_e reflects the number of individuals that actually contribute to the next generation, and is equal in a single population to the census number of individuals divided by the variance in offspring number (Donnelley & Tavaré, 1995); N_e is virtually always smaller than the census population size. Effective population size is the parameter that links ecology and evolution, and it is proposed that management practices maximize its value as a way of maximizing the capacity of a population to respond adaptively to environmental challenges (Nunney, 2000).

Effective population size of a metapopulation is expected to be smaller than that of a single demographically stable population of the same size, because of the reduction in the population census size, as well as the variance in reproductive success among individuals resulting from local extinctions; $N_e \approx (k/4(m+e)F_{ST})$, where k is the number of local populations (Nunney, 2000; Whitlock & Barton, 1997). The magnitude of N_e is independent of the total population size, but it increases with k (Nunney, 2000). Consequently, total genetic diversity will be reduced in a metapopulation in the majority of circumstances. Because of the reduction in total genetic variation, variation within local populations will also tend to be reduced. In a propagule pool model with a small number of colonists, reduction in within-population diversity will be reinforced because migrants are sampled from already genetically poor populations (Pannell & Charlesworth, 1999, 2000).

Fragmentation from a Metapopulation Perspective

One very important issue when studying fragmentation from a metapopulation perspective is the rate at which measured parameters approach equilibrium after fragmentation. Some results suggest that the amount and distribution of genetic variation slowly achieves a new equilibrium after a fragmentation event. Pannell and Charlesworth (1999) conducted simulations in which the number of local populations is suddenly reduced from 1,000 to 10—a situation analogous to a fragmentation event. The number of generations required for F_{ST} (an estimate of the distribution of genetic variation) to reach halfway to a new equilibrium can vary between 600 to 35,000 generations, depending on whether the source of migrants is from the propagule or the migrant pool, respectively. Levels of within- and among-population diversity take between 8,000 to 600,000 generations to reach equilibrium, again depending on the migration model and number of colonists. Generally, a recently fragmented population will have an amount of genetic variation greater than expected for its census size at equilibrium.

As a result of the increasing among-population isolation and the likely reduction in local population size, genetic drift will be the dominant force in defining the new local and total metapopulation equilibrium genetic diversities. The new equilibrium will be approached at a rate inversely proportional to the reduced effective population size N_e^{-1} (Pannell & Charlesworth, 2000). Thus, F_{ST} may at first increase steeply, but will then approach its new equilibrium asymptotically (Pannell & Charlesworth, 1999).

The practical implications of these results are striking. In recently fragmented systems, F_{ST} and overall genetic diversity (interpreted under an equilibrium model) may represent a demography that no longer exists, or a demography that never existed at all. Estimates of current gene flow or effective population sizes based on expected F_{ST} or heterozygosity under equilibrium models will be misleading. This problem can, in principle, be solved by using estimators of effective population size and gene flow that are insensitive to nonequilibrium dynamics or that are derived under nonequilibrium models of fragmentation. Alternatively, these models can estimate effective population sizes and gene flow directly.

Linkage disequilibrium (LD), the nonrandom association between alleles of different loci, can be highly sensitive to population demography. Linkage disequilibrium can be expressed as the average correlation between pairs of alleles of different loci, with an expectation that is a function of effective population size, $E[r^2] = (1+4N_e\rho)^{-1}, \sim (4N_e\rho)^{-1}$ for large populations, where ρ is the recombination rate between pairs of loci (Vitalis & Couvet, 2001). A bottleneck resulting from a founding event, for example, will increase LD by virtue of its effect on N_e. Note that this is true even among nonlinked loci because LD decay resulting from recombination is only effective when the two loci are heterozygous. Thus the recombination rate between nonlinked loci ρ has an upper limit of 0.5 in a random mating population (Hartl & Clark, 1997). Interestingly, LD decays, at a rate $(1 - \rho)$ per generation (Hartl & Clark, 1997); therefore, LD is an indicator of relatively recent past demographic events. Furthermore, because the rate of LD decay decreases as the recombination rate ρ between loci declines, LD between genetically and closely linked loci can be used as an indicator of less recent demographic events. In metapopulations, LD within local populations is expected to increase when migration decreases and, to a lesser extent, the number of demes increases (Wakeley & Lessard, 2003).

Recently, Zartman and colleagues (2006) showed empirical evidence that F_{ST} and genetic diversity in the liverwort *Radula flaccida* was not altered after a recent experimental fragmentation of the species in the Amazon. However, within-population LD was significantly higher in fragmented populations. This suggests that LD may be a better statistic to describe recent demographic events in a population.

To explore levels and patterns of genetic variation under fragmentation from a metapopulation perspective, we simulated the coalescent histories of a population sample (Hudson, 1983) with microsatellite mutation, under nonequilibrium metapopulation histories (see previous section and Box 4.1). The coalescent is a statistical model for the relatedness of a population sample of individuals (gametes, chromosomes, haplotypes) based on its genealogy, which can include mutation and recombination to study realistic molecular population genetic data. This model can be used as a tool to make inferences about a population's demographic history from molecular population genetic data. For samples of 10 individuals from each of two populations, Figure 4.1 shows the total numbers of alleles, F_{ST}, and LD for a large population that was suddenly fragmented into 16 local fragment patches undergoing metapopulation dynamics. All summary statistics change rapidly during the first few tens of generations and then gradually approach new equilibria. The same pattern can be observed for fragmentation events that lead to 100 (instead of 16) fragments (data not shown), suggesting that the local population effective size (set to 100 in our simulations) determines the rate at which genetic variation reaches a new equilibrium. As expected, the effect of metapopulation dynamics on population genetic statistics is to decrease genetic variation and increase LD. Interestingly, F_{ST} levels reach a maximum and then gradually decline. This is correlated with a general reduction in genetic diversity and total heterozygosity (Pannell & Charlesworth, 1999, 2000), which suggests that loss of genetic variation resulting from local population extinction also reduces F_{ST} under our fragmentation scenario.

Simulations such as these allow researchers to explore how much genetic data can reveal about the demography of fragmented species within a timescale relevant to management practices. Using our simple model and small sample size, genetic data do not perform well in distinguishing fragmented metapopulations with different extinction probabilities (results not shown). However, further exploration of nonequilibrium models is an important goal in conservation genetics. The use of genetic measures to infer demography may be most effective when combined with additional knowledge about the history and biology of the species in question (discussed later).

Case Study: Limits and Possibilities of Population Genetic Analyses in a Metapopulation of the Serpentine Sunflower *Helianthus exilis*

Inferring demographic events from genetic data alone is problematic. First, there are always competing demographic hypotheses and, second, the rates at which population genetic variables reach equilibrium may lead to misinterpretations. Our inferences can become more accurate when the ecology and history of a taxon are well studied. In this case, it may be possible to formulate hypotheses regarding the demographic scenario that can be tested with population genetic data.

Helianthus exilis is an obligate outcrossing sunflower species that thrives in serpentine and nonserpentine seeps in northern California (Sambatti & Rice, 2006). Two surveys 20 years apart revealed considerable turnover in the distribution of populations among seeps—a pattern suggesting metapopulation dynamics. Alternatively, seed dormancy and recruitment from seed banks could explain apparent extinction and colonization. To distinguish between these hypotheses, we used an approach that combined ecological and population genetic data with coalescent simulation modeling.

We sampled 10 independently segregating microsatellite loci in individuals from each of seven populations that had either been persistent in time or had an unknown history. We also sampled these same loci from each of eight putatively recolonized populations. We predicted that if populations differed in their founding histories, then recolonized populations should show lower microsatellite variation and greater LD. For each population we calculated the number of alleles per locus, the frequency of the most common allele, and, as an estimate of LD for each population, the weighted correlation between locus pairs, *R*. We also adapted an estimator of LD that does not require knowing the gametic phases of alleles (Sambatti & Sickler, 2006). A population that suffered

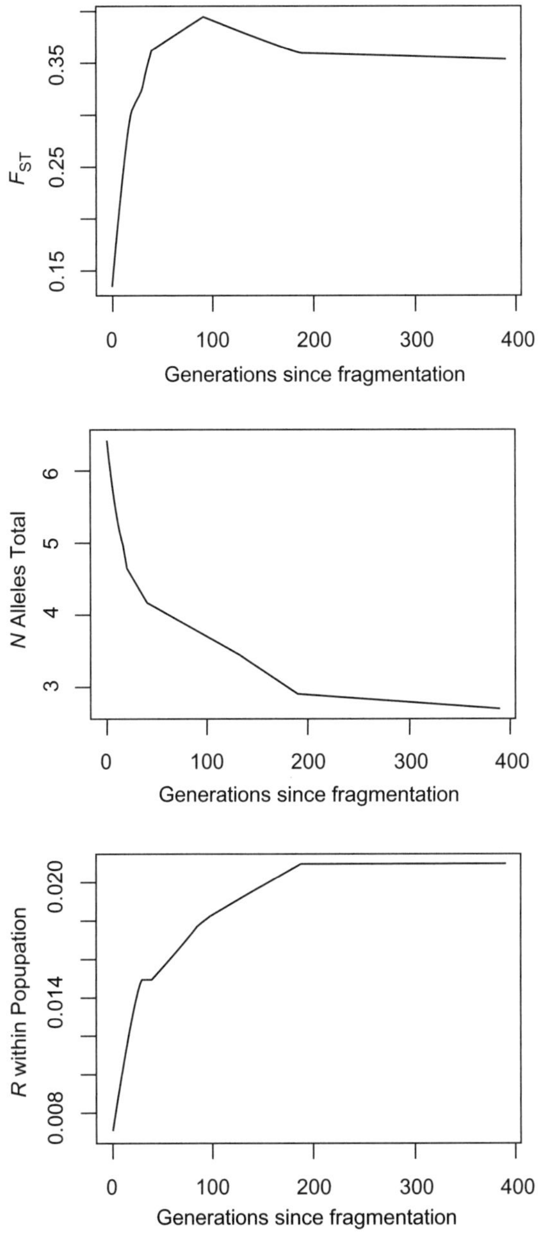

FIGURE 4.1 These graphs depict simulation results for a large nonsubdivided species suddenly fragmented into a very small metapopulation. Here the metapopulation has 16 patches linked by migration ($M = 2Nm = 1$) with probabilities of recolonization and extinction of 0.12 and 0.06, respectively. See Box 4.1 for a complete description of the model. Mean values for F_{ST}, the total number of alleles (N Alleles Total), and mean LD (R) are shown over a range of generations since fragmentation. Means were calculated over 1,000 simulations; sample sizes were reduced to as low as 104, where lack of polymorphism prevented calculation of F_{ST} and R. As generations since fragmentation exceeds 400, F_{ST} values continue to decrease and R continues to increase (data not shown).

a recent bottleneck will likely have fewer alleles with highly uneven frequencies, whereas a stable population of the same size will likely have more alleles with small but more evenly distributed frequencies.

We modeled a scenario focusing on a recently colonized population that receives immigrants from throughout the metapopulation (a migrant pool model). Using a migrant pool model is a conservative approach because the effect of a recent colonization on microsatellite variation is erased faster than when using a propagule pool model. Coalescent simulations were performed that generated many independent data sets using the model. Parameters needed in these simulations include time since the founding event, number of founders, $\theta = 4N_e\mu$ (where μ is the per-locus mutation rate), and $4N_em$, the migration rate. With aboveground census sizes of the largest populations close to 10,000 and a mean mutation rate of 10^{-4} estimated for plant microsatellites (Thuillet et al., 2002; Vigouroux et al., 2002), we chose $\theta = 5$, which is likely an overestimate because N_e is usually much smaller than the census population. This also makes the expectations conservative because high mutation erases the signal of a recent colonization faster. For each simulation we calculated the number of alleles, the frequency of the most common allele, and the correlation coefficient between locus pairs (R). We chose a value of $4N_em = 100$, from the upper portion of the distribution of migration rates estimated among populations that had either a history of persistence or no recorded history assuming demographic equilibrium. We then performed simulations varying the number of founders.

We found that to achieve the observed number of alleles (Fig. 4.2) and levels of LD, the populations must have been founded by more than 15 individuals. This could have happened through a rain of seeds or by recruitment from the seed bank. A reciprocal transplant experiment (Sambatti & Rice, 2006) showed that the number of seeds required to fall in a site is, on average, about 10 times larger than the effective number of founders (in other words, more than 150 seeds would be required). Because sunflowers have no obvious mechanism to promote long-distance dispersal, the seed bank scenario seems to be the more parsimonious explanation.

Other genetic data are consistent with these results. The mean and 95% confidence interval of F_{ST} estimates, 0.10 (0.08, 0.13), indicates weak differentiation among populations, consistent with a large number of founders or an absence of local extinction. Also, deviations from Hardy-Weinberg equilibrium resulting from an excess of homozygotes are consistent with seed banks (Vitalis et al., 2004). The population inbreeding coefficient, F_{IS}, expresses this excess of homozygotes (Hartl & Clark, 1997). Estimates of inbreeding did not differ significantly between populations (overall mean and 95% confidence intervals were 0.22 [0.12, 0.33]). This level of inbreeding is consistent with the presence of a seed bank. (F statistics were calculated with Microsatellite Analyser [Dieringer & Schlötterer, 2002].)

Ideally, estimation of migration rates should be achieved through independent methods such as paternity analysis or mark–release–recapture methods. Thus, the results must be taken with some caution. However, we now know that if there is extinction and recolonization in *H. exilis*, large amounts of seed movement are likely to be necessary. Our analysis demonstrates that complementing genetic analyses with multiple sources of information on ecology and population history is essential to narrow down possible alternative demographic scenarios.

Coalescent analyses may be used to describe the fragmentation process, create expectations regarding genetic parameters given a fragmentation event, and make sense of the observed genetic variation in a nonequilibrium demography. Because many fragmentation events have occurred recently, we often have an idea of the population census size before and after fragmentation, as well as the time since fragmentation. This information will be useful for adding to the power of coalescent simulations to elucidate demographic scenarios.

ADAPTIVE EVOLUTION IN A METAPOPULATION: PERSPECTIVES FOR FRAGMENTED SPECIES

Metapopulation dynamics may generate a variety of selective pressures and evolutionary outcomes that differ from those observed in stable populations. In a conservation context, perhaps one of the most important issues is inbreeding depression and its potential to increase the risk of local and metapopulation extinction. Metapopulation dynamics leads to lower effective population sizes,

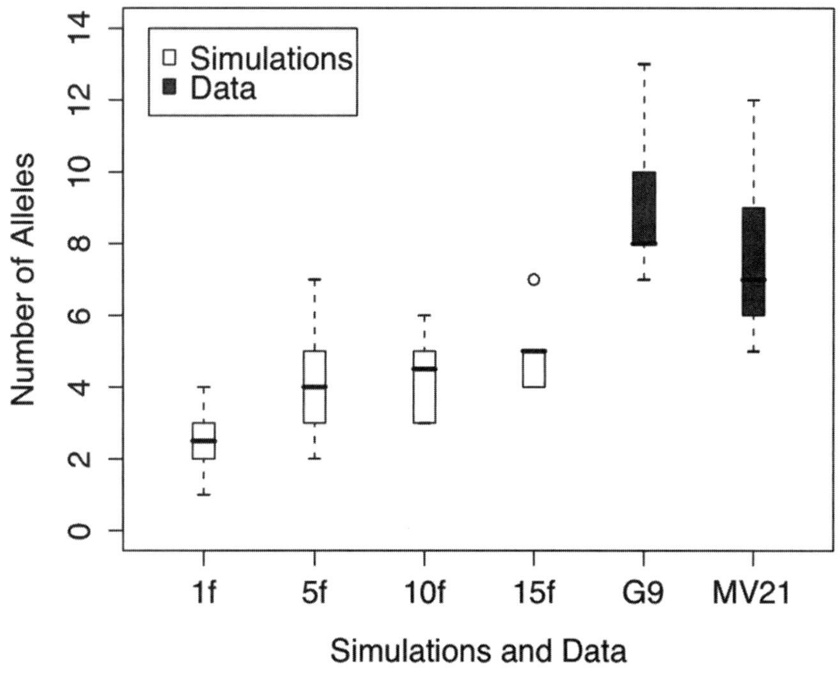

FIGURE 4.2 Mean and 95% confidence intervals (n=10) of the number of microsatellite alleles observed in a sample of 20 individuals in simulated (open boxes) and field populations (filled boxes) of *Helianthus exilis*. Simulated populations were founded by 1, 5, 10, and 15 founders (1f, 5f, 10f, and 15f); grew exponentially to the size of a field population (around 10^4 individuals); and received migrants from all other populations in the species. Populations G9 and MV21 are from the field and have approximately 10^4 individuals each. More field populations existed, but their means and 95% confidence intervals are encompassed by those shown in the chart. For this parameter set, a very large number of founders, a higher migration rate, or larger values of θ are necessary to create a signal (in other words, simulations = field populations) of a recent colonization event.

which in turn increases homozygosity, may fix deleterious mutations in different populations, and may reduce the potential for local adaptation (Ives & Whitlock, 2002). Selection becomes weaker in small populations because its efficacy is proportional to the effective population size (Boulding, this volume; Hartl & Clark, 1997; Reed, this volume). In other words, to achieve the same amount of phenotypic change, selection must be stronger in smaller populations. Ebert and colleagues (2002) empirically demonstrated the occurrence of inbreeding depression in natural *Daphnia* metapopulations. *Daphnia* spp. inhabit temporary, semi-isolated rock pools that experience frequent extinction and recolonization. By comparing the fitness of hybrids between foreign and resident individuals with that of resident individuals, Ebert and coworkers (2002) found that hybrids showed more than 30 times the fitness of local population residents. These results suggest high levels of inbreeding depression, a condition that may in turn affect the rate of local population extinction. Inbreeding depression is also important in the evolution of certain traits in a metapopulation.

Different mating systems and alternative sexual forms within a species have the potential to change the magnitude of inbreeding depression (see Dudash & Murren, this volume). Selfing in plants, for example, may increase inbreeding depression because it increases both population homozygosity and the probability that deleterious recessive mutations are expressed (Hartl & Clark, 1997). Contrary to the belief that deleterious mutations are purged by selection, inbreeding depression is often found in selfing plants (Barrett, 2003). However, in a metapopulation, a crucial constraint for sexual species is the mating limitation in a recent colonized population resulting from small numbers of individuals. In plants, it is not uncommon for an asexual reproductive system and/or the possibility of selfing to evolve for reproductive assurance. Selfing may be favored when patch occupancy is low as a result of high extinction rates and/or low numbers of colonists, whereas self-incompatible or otherwise outcrossing genotypes are often found in species with many successful colonizers and when patch occupancy is high (Pannell & Barrett, 1998).

When reproductive assurance is required, metapopulations of outcrossers will persist if they disperse a greater number of seeds than "selfers" (Pannell & Barrett, 1998). However, when reproductive assurance is not obligate (such as in plants with heterostyly), or if the available pollen is used for self-fertilization prior to outcrossing, and if selfers are more effective in using their own pollen, bottlenecks during colonization may further increase the probability of fixation of selfing alleles (Pannell & Barrett, 2001). However, in the presence of inbreeding depression, frequent population turnover will reduce effective population sizes, and outcrossers may produce and disperse more viable seeds. In this scenario, higher extinction rates may favor outcrossing genotypes (Pannell & Barrett, 2001). Moreover, if extinction rates are high, there may be selection for facultative outcrossers (Pannell & Barrett, 2001).

Similarly, alternative sexual forms can be selected in a metapopulation depending on the degree of population turnover and inbreeding depression. The evolution of androdioecy—the coexistence of males and hermaphrodites within a population—seems to be one of these cases (Pannell, 2002). Because males need a female to reproduce, males need to fertilize more than twice the number of female flowers compared with hermaphrodites to spread in a population. This effect is aggravated if hermaphrodites partially self-fertilize. Therefore, evolution of males in a hermaphrodite population can only occur if inbreeding depression (resulting from selfing) is very high. However in a metapopulation in which reproductive assurance is necessary because few individuals found local populations, hermaphrodites will be selected. There may be opportunities for male invasion as the founded population grows and if allocation to male fertility in hermaphrodites—for instance, changing allocation to pollen production—is substantially inferior to fertility of the pure male form. Thus, variation in extinction rates, the number of founders establishing new colonies, and migration rates will interact and potentially influence sex ratios within and among populations, and will affect the stability of androdioecy itself (Pannell, 2002). High migration will tend to homogenize sex ratios across the metapopulation, whereas low migration and rapid population turnover may completely eliminate males from the metapopulation.

The ability to disperse is perhaps the most important condition for a species to exist in a metapopulation context, and traits associated with dispersal are expected to be under strong selection. Olivieri and colleagues (1995) showed that in a stable, locally adapted population, dispersal may be selected against because individuals may be maladapted to new sites—assuming that dispersal carries a cost in terms of local recruitment. However, metapopulation dynamics should tip the balance toward dispersal, if even locally adapted populations go extinct. Conversely, competition within a local population increases as the local population ages and grows toward its carrying capacity, and dispersal alleles should become relatively more favored. The extinction rate and the local population life span determine the proportion of local populations at carrying capacity in a metapopulation. If this proportion is high and competition is also high, then dispersal is selected against. However, if there are many empty patches, recently colonized local populations will be overrepresented, and a complex interaction among factors will determine selection on dispersal and its outcome.

The correlation between population age and the frequency of dispersal alleles was studied empirically in the Glanville fritillary butterfly *Melitaea cinxia* (Haag et al., 2005). More mobile individuals carry a specific allele of the metabolic enzyme phosphoglucose isomerase (*pgi* enzyme). In a metapopulation established after fragmentation, these alleles were found in higher frequencies in newly founded populations than in older populations, supporting the hypothesis that selection for dispersal ability is dependent on the population age and contributes to maintaining a variety of dispersal phenotypes in a metapopulation.

Similarly, evolution of reproductive effort, or the proportion of biomass allocated to reproduction, depends on metapopulation dynamics (Ronce & Olivieri, 1997). Competition usually favors allocation to survivorship when local populations approach carrying capacity, whereas reproduction is favored in a growing population. The balance of selective pressures depends on the proportion of occupied sites at carrying capacity, and on rates of extinction, dispersal, and population growth. High dispersal rates may modulate selection on reproductive strategies by increasing the proportion of populations at carrying capacity. Moreover, in highly competitive environments, hierarchies of plant size and reproductive effort may be established—such that reproduction is concentrated in fewer larger individuals—further decreasing the effective population size (Sambatti & Rice, 2006; Weiner, 1990).

These recent theoretical and empirical results suggest several issues of concern from a conservation perspective. Besides the genetic and demographic risks associated with reductions in population size, evolutionary dynamics within a metapopulation can reinforce adverse circumstances conducive to species collapse. For example, there may be selection for selfing under a reproductive assurance scenario. If outcrossing genotypes are completely eliminated from new colonies, inbreeding depression may lead to the rapid extinction of these new colonies. Alternatively, dispersal genotypes may be eliminated in small populations when extinction rates are high. The good news, however, is that we now have some clues about which selection pressures may be operating in a fragmented metapopulation. With some knowledge of the species' biology, it should be possible to identify and monitor traits that may allow the species to survive in a fragmented landscape.

CONCLUSIONS

Metapopulation theory provides a useful framework to describe the demographic and genetic dynamics of fragmented populations. However, equilibrium approaches derived from analytical models have limited predictive power when dealing with fragmented populations. Only by acquiring an idea of the nonequilibrium dynamics from both demographic and genetic perspectives can one develop the basis for sound management practices. From a genetic perspective, molecular markers can be useful, but we need to use caution in interpreting their variation and distribution. New analytical techniques, such as coalescent analyses, have the potential to help our understanding of observed molecular variation patterns in a nonequilibrium context, but additional ecological and historical information is fundamental to the successful application of these techniques.

Demography and evolutionary processes are inextricably linked in ways vitally important to conservation. In particular, demography affects the evolutionary response of a species to environmental challenge. Lack of evolutionary response may contribute to species extinction. A synthesis of ecological and evolutionary methods is required to understand this interplay, and more studies need to be conducted in systems of conservation interest.

A better understanding of the unique selection pressures that arise from fragmentation and metapopulation dynamics is starting to emerge from theory and empirical work. There is a great need to continue exploring the evolutionary dynamics of fragmented populations and their implications for conservation.

FUTURE DIRECTIONS IN METAPOPULATION/ FRAGMENTATION RESEARCH

Transient Dynamics and the Extinction Debt

Different species possess different associated extinction risks as a result of their intrinsic characteristics. Thus, after fragmentation, some species should move faster toward extinction than others. Comparative studies should be performed to evaluate which class of species tends to have shorter transient equilibration rates. This information is paramount to prioritize groups of species at higher immediate risks of extinction so that urgent conservation measures are taken.

Seed or Propagule Banks and Metapopulations

Local extinction and subsequent recolonizations are assumptions of the classic metapopulation theory. Plants and animals often store part of the population in "propagule banks," a strategy that may prevent local extinctions when environmental conditions are unfavorable. Seed banks in plants are the most common example. In a bad year, for instance, individuals in a local population may die before reproduction, but the propagule bank will provide new individuals for subsequent years. In a metapopulation context, the question is: If propagule banks are part of the species' biology, can the absence of aboveground individuals in a local population be interpreted as a local extinction? In our view the answer is no. A local extinction implies the elimination of a population with all evolutionary characteristics acquired during its history. Local adaptation, for example, will be lost with an extinction event. When an aboveground population is not observable, the evolutionary information is stored in the propagule bank population. An extinction followed by a recolonization is usually associated with a significant genetic bottleneck. Propagule banks can be considered a source of local recolonization and may thus prevent drastic reductions in census population sizes, as well as contribute to increasing the effective population size (Nunney, 2002). Incorporating propagule banks into the metapopulation theory means, among other things, accounting for a reduction in local extinction rates. We argue that propagule banks are an important component of metapopulations and need to be incorporated more effectively in empirical and theoretical work.

The Cost of Natural Selection and the Risk of Extinction

Small and fragmented populations face heightened difficulties in adapting to environmental change, including the changes caused by the fragmentation process itself. As we have seen, metapopulation dynamics have the potential to affect evolutionary trajectories. Moreover, the cost of natural selection may lead small populations to extinction (Nunney, 2003). To what extent can small, fragmented species respond evolutionarily to environmental challenges without increasing the risk of extinction? What are the conditions to minimize the costs of selection? Metapopulation theory can potentially contribute to answering these questions.

Estimating Demographic Parameters with Population Genetics Data

Parameter estimation is a major source of concern in applying the methods we have described. In many cases many equally likely solutions are possible, which makes it impossible to maximize likelihood functions effectively. It may be that when information from other sources (ecology, history) is incorporated into the estimation process, so that the number of possible scenarios is restricted, the estimation procedure may be more effective. Bayesian approaches should be developed to improve the estimation of demographic parameters from genetic data. These Bayesian methods should allow researchers to combine ecological and genetic data in a unified framework.

Estimating Effective Population Sizes

The problem of estimating effective population sizes in a nonequilibrium situation is also a problem of timescale. Using population genetic models to estimate effective population size may be a problem when equilibrium rates are too slow. It might be possible for certain species to use ecological methods of estimating effective sizes. Comparing ecological and population genetic estimates may be an alternative approach to estimating effective population size, and another method to determine how far population genetic parameters are from equilibrium.

SUGGESTIONS FOR FURTHER READING

Hanski (1998), and Hanski and Ovaskainen (2000) are excellent references for a deeper understanding of metapopulation theory and its specific application to fragmented landscapes. Frank (2005) provides a recent and relevant discussion about the role of environmental stochasticity on metapopulation persistence. Schtickzelle and Baguette (2004) and Hanski (2004) discuss the importance of local population dynamics in models of metapopulation persistence. Whitlock and McCauley (1990), Whitlock and Barton (1997), and Pannell and Charlesworth (1999, 2000) explain the theory of population genetics in a metapopulation. Nunney (2000) offers a conservation perspective on various aspects of this theory. Lastly, after an introduction to coalescent theory in Hudson (1990), Wakeley and Aliacar (2001) and Pannell (2003) are the key references on the emerging theory of gene genealogies in a metapopulation.

Frank, K. 2005. Metapopulation persistence in heterogeneous landscapes: Lessons about the effect of stochasticity. Am Nat. 165: 374–388.

Hanski, I. 1998. Metapopulation dynamics. Nature 396: 41–49.

Hanski, I. 2004. Metapopulation theory, its use and misuse. Basic Appl Ecol. 5: 225–229.

Hanski, I., & O. Ovaskainen. 2000. The metapopulation capacity of a fragmented landscape. Nature 404: 755–758.

Hudson, R. R. 1990. Gene genealogies and the coalescent process. Oxford Surveys in Evolutionary Biology 7: 1–44.

Nunney, L. 2000. The limits to knowledge in conservation genetics: The value of effective population size. Evolutionary Biology 32: 179–194.

Pannell, J. R. 2003. Coalescence in a metapopulation with recurrent local extinction and recolonization. Evolution 57: 949–961.

Pannell, J. R., & B. Charlesworth. 1999. Neutral genetic diversity in a metapopulation with recurrent local extinction and recolonization. Evolution 53: 664–676.

Pannell, J. R., & B. Charlesworth. 2000. Effects of metapopulation processes on measures of genetic diversity. Phil Trans R Soc Lond. 355: 1851–1864.

Schtickzelle, N., & M. Baguette. 2004. Metapopulation viability analysis of the bog fritillary butterfly using RAMAS/GIS. Oikos 104: 277–290.

Wakeley, J., & N. Aliacar. 2001. Gene genealogies in a metapopulation. Genetics 159: 893–905.

Whitlock, M. C., & N. H. Barton. 1997. The effective size of a subdivided population. Genetics 146: 427–441.

Whitlock, M. C., & D. E. McCauley. 1990. Some population genetic consequences of colony formation and extinction: Genetic correlations within founding groups. Evolution 44: 1717–1724.

5

The Influence of Breeding Systems and Mating Systems on Conservation Genetics and Conservation Decisions

MICHELE R. DUDASH
COURTNEY J. MURREN

After habitat preservation, an understanding of the reproductive biology of an endangered, threatened, or invasive plant species is one of the first research steps that should be taken to identify conservation priorities. The goals of this chapter are to explore from an ecological–genetic perspective the essential roles that the breeding system and mating system play in conservation initiatives, and to encourage further research in this area. Many excellent reviews exist on the biology of plant mating and breeding systems (see, for example, Barrett, 2003); thus, we instead describe the relevance of such systems to the conservation of threatened and endangered populations and the management of invasive taxa. We have chosen to focus on plants because they exhibit a great diversity in breeding systems and mating systems. Many of these principles apply to animals as well, and for further details we direct our readers to a recent review on animal systems by Jarne and Auld (2006).

We begin by examining methods for assessing breeding systems, mating systems, and inbreeding depression. We take a close look at plasticity of breeding and mating system traits in response to environmental heterogeneity, and discuss how these traits contribute to long-term population persistence of taxa of conservation concern. We use three case studies from our own work with plants to highlight the complexity of interactions among breeding and mating systems and the environment, and show how these interactions affect population persistence. Lastly, we summarize information that we believe is needed to examine the possible ecological–genetic consequences of the environment-by-breeding system interaction. We then conclude with several recommendations for conservation-specific issues and discuss suggested avenues for future research.

BREEDING AND MATING SYSTEMS

The first of the two major attributes of the reproductive biology of any plant species is the breeding system. The breeding system represents characteristics of the flowers within an individual plant that may influence gamete transfer among conspecifics (Barrett, 2002). Breeding system traits are responsible for preventing interference between pollen removal from anthers and pollen deposition onto stigmas. In addition, these traits also influence the frequency of outcrossing (Barrett, 2003; Lloyd & Webb, 1986; Webb & Lloyd, 1986). Thus, the first categorization that needs to be determined is whether a plant species is self-compatible or self-incompatible. However, self-incompatibility is complex, and determining the type of incompatibility is time-consuming. Thus we often use results in the literature on closely related taxa to infer whether a taxon of interest is self-incompatible. Terminology commonly used to

TABLE 5.1 Terms Commonly Used in Plant Reproductive Biology.

Mating system	The within-population proportion of matings between related individuals or with self compared with matings between unrelated individuals
Self-compatibility	Able to mate with self
Self-incompatibility (SI)	Genetic incompatibility that prohibits self matings and mating between individuals that share the same self-incompatibility (S) alleles
Gametophytic SI	Genotype of haploid pollen grain dictates SI reaction
Sporophytic SI	Genotype of diploid parent plant dictates SI reaction
Geitonogamy	Transfer of pollen between two flowers on the same plant
Chasmogamy or xenogamy	Flowers open to external, nonself pollen sources; cross-pollination
Cleistogamy	Flowers closed to external, nonself pollen sources; self-pollination occurs within closed flower
Autogamy	Within-flower selfing of a chasmogamous flower
Apomixis	Within-flower reproduction without fertilization such that the progeny are genetically identical to their mother
Breeding system	Attributes of the flowers within an individual plant that may influence gamete transfer among conspecifics
Protandry	Pollen dehisces prior to stigma receptivity
Protogyny	Stigmas are receptive prior to pollen dehiscence
Herkogamy	Physical separation between male and female function within a flower
Dichogamy	Temporal separation between male and female function within a flower
Hermaphrodite	Male and female function within a single flower or between flowers of an individual plant
Monoecy	Male and female flowers on the same individual
Gynomonoecy	A population composed of females and monoecious individuals
Andromonoecy	A population composed of males and monoecious individuals
Dioecy	Separate male and female individuals within a population
Gynodioecy	Coexistence of hermaphrodites and female plants within a population
Androdioecy	Hermaphrodites and male plants within a population
Monostyly	A single length of style present within a population
Distyly	Two style morphs present within a population
Tristyly	Three style morphs present within a population
Monomorphic	One style morph within a population
Polymorphic	More than one style morph present within a population
Life histories	
Semelparity	Individuals reproduce no more than once and then rapidly senesce
Iteroparity	Individuals potentially reproduce multiple times before senescence
Annual	A plant population that completes its life cycle within one season
Biennial	A plant population that completes its life cycle within two seasons
Perennial	Living longer than 2 years

describe plant breeding system traits is summarized in Table 5.1.

Breeding system assessment begins with careful observation of sexual expression among flowers within an individual. The breeding system researcher typically wishes to resolve several important issues early on in the study; for instance, ascertain whether a species is hermaphroditic or dioecious, estimate the timing of stigma receptivity relative to pollen shedding (dehiscence), and establish whether stamens mature and release pollen before the stigma is receptive (protandry) or vice versa (protogyny). Additionally, data from inquiries into these questions allow us to garner insights into the potential for within-plant versus between-plant mating opportunities, which ultimately influence the mating system (discussed later). Many plant taxa exhibit multiple breeding system traits, suggesting that independent selection pressures may favor different traits simultaneously. Long-term phenotypic selection studies show that selection acts on plant populations in multiple ways to achieve efficient gamete transfer between conspecifics (see, for example, Fenster et al., 2004). Plant sexual expression may vary within and among species, and new variation can be introduced through hybridization between congeners (see Weller et al., 2001). Thus, variation in the breeding system can influence population persistence (Barrett, 2003).

A classic illustration of the interaction between population size and breeding system is the breakdown of tristyly to distyly and homostyly in heterostylous taxa (see, for example, Husband & Barrett, 1993; Weller 1979). In the tristylous annual plant *Eicchornia paniculata* (Pontederiaceae), Husband and Barrett (1993) demonstrated that populations in Brazil often consist of three female plant morphs represented by three different stigma lengths, each of which is accompanied by two whorls of stamens of dissimilar lengths that preclude self-pollination (Fig. 5.1). This arrangement of reproductive parts inhibits gamete transfer within and between flowers of the same floral plant morph. Through careful monitoring of numerous populations in Brazil and Jamaica (where the species has migrated), Husband and Barrett (1993) showed that small populations in Jamaica often lose the short morph (Fig. 5.1). One dominant allele is responsible for the presence of the short morph within the population (see, for example, Barrett, 1988). The Jamaica population thus exemplifies the ease with which an allele can be lost in a small population when the breeding system is affected by invasion into a new habitat. Studies of other tristylous taxa have also shown morph loss via drift in small populations, and demonstrated that gene flow among populations can prevent morph loss in small populations (see, for example, Andersson, 1994b; Eckert & Barrett, 1992; Eckert et al., 1996). Thus, in a fragmented native habitat with limited gene flow, drift may remain a concern.

The number of census individuals (N) often differs from the effective population size (N_e) within a population. Therefore, different breeding and mating strategies may influence the number of individuals contributing progeny to the next generation (Table 5.2). A general trend is that whenever a breeding/mating system restricts matings between individuals (for example, restriction through self-incompatibility), then a population is likely to experience a decrease in N_e. In contrast, when matings occur more easily among individuals within a population because, for instance, plants are self-compatible, then the effective population size is increased because the population of potential mates is greater. However, inbreeding may also occur in these latter situations (discussed later).

The second major attribute of the reproductive biology of a plant species is its mating system. The mating system describes the proportion of matings between related individuals or the proportion of self-matings compared with matings between unrelated individuals within a population (Barrett, 2002). Selfing occurs when both the pollen (sperm) and the ovule (egg) are produced by the same individual. This type of hermaphroditism is more common in plants than animals, thus the selfing rate in plants is much greater (see, for example, Dudash & Fenster, 2000). However, in a recent analysis of published counts, Jarne and Auld (2006) report that the incidence of mixed mating (outcrossing estimates, t, between 0.2–0.8) in animals was 47% compared with 42% in plants.

In comparison with outbred lineages, selfing lineages exhibit a 50% decline in heterozygosity per generation. In outbred dioecious taxa, for example, the decline in heterozygosity after matings between related individuals is significantly less than for selfing organisms (Falconer, 1981), with the rate of loss of heterozygosity depending on the extent of coancestry among alleles. In natural populations, inbreeding has at least two significant consequences: a decline in number or phenotypic quality of offspring, and a decline in heterozygosity that in turn reduces a population's ability to adapt to a changing environment. In both plants and animals, there are documented declines in vigor after matings between related individuals and the expression of inbreeding depression (reviewed in Crnokrak & Roff, 1999; Dudash & Fenster, 2000; Fenster & Dudash, 1994). The effects of inbreeding depression are especially aggravated in small and fragmented populations where matings among related individuals are more likely to occur and thus precipitate the onset of population decline (Charlesworth & Charlesworth, 1987).

The mating system of a plant can be estimated (1) ecologically through observations of within-plant pollinator movements versus between-plant pollinator movements (see, for example, Dudash, 1991; Dudash & Fenster, 2001), aided by the use of fluorescent dye powders, visual inspection, and videorecorded observations (see, for example, Dudash, 1991); (2) genetically via allozymes (Soltis et al., 1983) or other molecular genetic markers (Carr & Dudash, 2005); and

(3) through the use of common-garden experiments that compare performance of progeny from selfed, outcrossed, and open-pollinated parents (Charlesworth & Charlesworth, 1987; Dudash, 1987, 1990).

Study of mating system variation within and among populations has been an area of active research since the advent of isozyme electrophoresis, from which we can readily and repeatedly obtain estimates of genetic variation for many individuals and populations (Hamrick & Schnabel, 1985). Using genetic data, the outcrossing rate parameter t can be estimated. Several alternative methods for estimation of t have been implemented (Ritland, 2002). The outcrossing rate parameter t ranges from zero to one, where a value of zero indicates that all matings result from selfing events and a value of one indicates that all matings are between totally unrelated individuals. One method to estimate t utilizes seed progeny arrays for which the maternal genotype is known and all individuals (mother and progeny) are scored at multiple isozyme loci or other codominant markers. A robust sampling protocol is 10 offspring from each of 30 maternal plants per population of interest. However, in many biological cases, the maternal genotype may not be known, and in those cases sampling of adult tissue randomly throughout the population can be utilized (Crow & Kimura, 1970). From a sample of adult tissue, one can obtain an estimate of the population-level outcrossing rate by calculating the inbreeding coefficient F for each locus where, under random mating, $F = (1 -$ observed heterozygosity/expected heterozygosity). Outcrossing rate can be estimated as $t = (1 - F)/(1 + F)$. To estimate the outcrossing rate with reasonable confidence, at least 30 individuals within a population and as many variable loci as possible should be used for an assessment (Dudash & Fenster, 2001). The mean and standard errors of F statistics can be calculated by jackknifing estimated F values across each locus used to calculate t (Sokal & Rohlf, 1985). The population-level approach, therefore, has the financial benefit of requiring fewer individuals to be genotyped in comparison with the progeny array approach. The population-level outcrossing rate via F statistics also has a biological benefit in that it

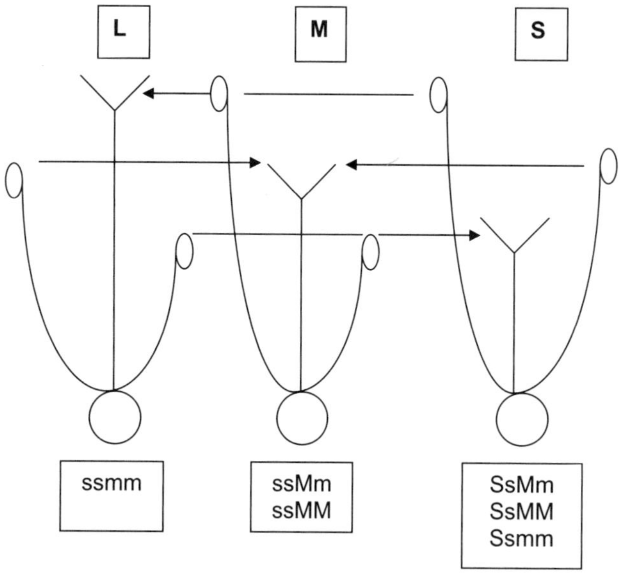

FIGURE 5.1 Schematic representation of relative positions of styles and stamens in the three floral morphs (L, long; M, mid length; S, short) of *Eicchornia paniculata*. Genotypes of the floral morphs under a two-locus model (S, M) for the inheritance of tristyly are shown. Arrows represent compatible pollen transfers. (Redrawn from Barrett [1988].)

TABLE 5.2 How Breeding System and Mating System May Affect N_e (the Effective Number of Reproducing Individuals within a Population).

Type of System	Effect on Effective Population Size		
	Decrease	No change	Increase
Mating	Self-incompatibility (SI)	Self-compatibility	Chasmogamy
	Gametophytic	Selfing	Cleistogamy
	Sporophytic	Apomixis	Autogamy
	Geitonogamy		
Breeding	Tristyly (if SI)	Hermaphrodite	Protandry
	Distyly (if SI)	Monostyly	Protogyny
	Polymorphic (if SI)	Monomorphic	Herkogamy
			Dichogamy
			Monoecy
			Gynomonoecy
			Andromonoecy
			Dioecy
			Gynodioecy
			Androdioecy

captures variation across the lifetime of individuals within a population and is thus less sensitive to variation among seasons in reproductive success.

Population outcrossing estimates can also be based on performance comparisons among hand-self, hand-outcross, and naturally open-pollinated progeny (described in Charlesworth & Charlesworth, 1987). Because the performance of selfed and outcrossed progeny are already being quantified with this combination of mating procedures, inbreeding depression can be evaluated as well. In our opinion, this approach has been underutilized and can have great utility in assessing the conservation status because we are able to estimate both population-level outcrossing rate and inbreeding depression through a crossing program that can be readily conducted in the field. Greenhouse or common-garden space is needed for growing progeny, and in many circumstances the method may be more cost-effective and more accessible to stewards and land managers than are techniques that use genetic markers. Obtaining estimates of mating system parameters and the magnitude of inbreeding depression can help stewards assess the vulnerability of populations of concern. In other words, one can assess the probability of a population experiencing extreme effects of inbreeding along with a decrease in population size because of habitat fragmentation or edge effects, and whether it is more vulnerable to extinction in the presence of environmental stochasticity.

Mixed Mating

Schemske and Lande (1985) observed that in most plant populations, mating systems were either highly selfing (range in t: 0.0–0.2) or highly outcrossing (t: 0.8–1.0), with few plant species in the middle range. More recently, other authors (see, for example, Barrett & Eckert, 1990; Goodwillie et al., 2005; Vogler & Kalisz, 2001) have also noted bimodality in plant mating system estimates and consider this repeatable pattern suggestive of a biotic or abiotic (wind, water) vector for gamete transmission. However, mixed mating systems (defined broadly here as populations with t in the range of 0.2–0.8) still persist in many animal-pollinated plant taxa (see, for example, Goodwillie et al., 2005; Vogler & Kalisz, 2001) and animal fauna (Jarne & Auld, 2006). The presence of mixed mating systems in animals is just beginning to be explored and requires a more balanced phylogenetic sampling within the animal kingdom.

The frequency of occurrence of mixed mating systems in nature is not fully understood. However, we are gaining insight into the factors that influence persistence of mixed mating systems in plants, including factors such as variable expression of inbreeding depression, variation in life history strategy, and breeding system traits (Goodwillie et al., 2005). These factors can vary as a result of both genetic and environmental variation within and among populations (see, for example, Elle & Hare, 2002). To complicate the mix further, the use of maternally based seed or seedling progeny arrays versus population-level adult tissue approaches to estimate *t* can influence observed values (described earlier). Using a maternally based estimate of outcrossing rate, variation in estimates of *t* (in other words, among mothers) may in part reflect the fact that all or most seeds within a fruit can readily germinate in the greenhouse and then be used in the molecular genetic sample. However, in the field some seed may not germinate and therefore may not be present in the sample for the population-level adult tissue approach. Discrepancies in *t* observed in greenhouse and natural field populations may be the result of differential expression of inbreeding depression later in the life history (Husband & Schemske, 1996). Thus, adults that flower in any given population likely represent a fraction of those individuals that were produced initially via seed within a population. Moreover, pollinator fauna varies within and among seasons in both abundance and efficiency (see, for example, Kalisz et al., 2004), which may in turn influence seed quality and quantity, and ultimately the mating system. Data from self-compatible, hermaphroditic snails exemplify the within- and among-population variation in mating systems and expression of inbreeding depression (see, for example, Jokela et al., 2006; Trouve et al., 2003). Together these data suggest that mixed mating systems are indeed present in the field, and likely vary temporally, but additional attention is required to determine their exact nature and to find the selective influences that maintain their presence.

Population Genetics and the Mating System

The mating system also contributes to how genetic variation is partitioned within a population. In general, selfing species tend to have lower within-population genetic variation than populations of outcrossing species. Additionally, there is greater genetic differentiation between populations of selfing species than populations of outcrossing species (see, for example, Hamrick & Godt, 1989). These patterns are explained in part by higher overall homozygosity in selfing species than outcrossing species, thus causing recombination and segregation to play a more important role in enhancing genetic variation within an outcrossing population (Falconer, 1981). Subsequently, the inbred mating system more readily allows the purging of genetic load of some traits more than others within a population (Dudash & Carr, 1998). In contrast, for populations with an outcrossing mating strategy, the genetic variation of the population is represented both within each individual family and among families within the population. Consequently, an outcrossing mating strategy reduces the opportunity to purge deleterious recessive alleles from the population, and we expect a greater genetic load (held in the heterozygous state) in the outcrossing than the selfing species. For instance, if a population of a selfing species and a population of an outcrossing congeneric species were each reduced from 500 individuals to 50 individuals, one would expect that the 50 individuals remaining in the outcrossing population would represent a greater proportion of the original population genetic variation than that remaining in the selfing congener. This prediction stems from the genetic variation both within and among maternal families in the outcrossing populations and reflects significant maternal line variation commonly detected within populations (Dudash et al., 2005). Additionally, it may turn out that phenotypic plasticity among maternal families may also be quite variable (see, for example, Peperkorn et al., 2005) and is yet another reason why it is important to track variation in the maternal line whenever feasible (see, for example, Murren et al., 2006).

Ecological Influences on the Mating System

Although the scale and aspect of environmental heterogeneity may vary widely, environmental heterogeneity is present in most plant populations (Gurevitch et al., 2002). Environmental variation can occur at both spatial and temporal scales and have differential effects on breeding and mating system traits. If environmental variation occurs at spatial scales between populations, selection may favor different breeding or mating systems in different populations. In contrast, if environmental variation occurs within a population, selection may favor multiple characteristics of the breeding or mating system simultaneously. Fluctuations of the environment on temporal scales may contribute to mixed mating systems within populations (see, for example, Holsinger, 1993). Abiotic factors such as precipitation, temperature, and soil may directly and indirectly influence plant reproduction. For example, reproduction of the garlic mustard *Alliaria petiolata* was greatest at the edges of well-lighted, mesic forests that were more susceptible to invasion than drier, upland forest interiors (Meekins & McCarthy, 2001). From a biotic point of view, many plant species are dependent on mutualists to reproduce successfully, and the activity and efficiency of these mutualists can be heterogeneous (see, for example, Kalisz & Vogler, 2003). Antagonistic interactions with herbivores may also be heterogeneous in their mode and impact. Herbivores may reduce resources available for reproduction, thus creating reproductive variation from another source (Carr & Eubanks, 2002; Ivey et al., 2003; Steets et al., 2006). Given the biotic and abiotic variation noted among populations, it is not surprising that we have detected variation in mating system estimates among plant populations and across years (Ashman et al., 2004; Goodwillie et al., 2005).

From a biotic perspective, pollination environments can vary spatially and temporally, and this variation can influence the mating system. One particularly dramatic example includes populations in habitat fragments after landscape-level fragmentation. The isolation of patches and reduction of habitat size associated with fragmentation can lead to the reduction in size of some native plant populations, and can increase habitat for other edge or weedy species. Several challenges face small remnant populations of native flora that remain in forest fragments. In order for successful pollination and reproduction to occur, animal visitors first need to be attracted to a site, and animal attraction may also be dependent upon whether other conspecifics are flowering in the same habitat space. Small populations of native plants that remain in forest fragments may be influenced by both pollinator assemblage and visitation rate. Temporal variation of pollinator activity can further affect the reproductive success and mating system of these small populations (see, for example, Johns & Handel, 2002). On the other hand, nonnative invasive plant species are a threat to native biodiversity in forest fragments in a number of ways, and tend to have their greatest presence in disturbed areas such as the edges of forest fragments. Both intentionally and accidentally introduced invasive plants outcompete native plant species and disrupt native plant pollinator mutualisms (see, for example, Ghazoul, 2004; Traverset & Richardson, 2006), and facilitate the spread of invasive pollinators (Hanley & Goulson, 2003). Facilitation among native and invasive taxa in attracting pollinators may also occur in some natural populations (Moeller, 2004); moreover, if invasive taxa make a particular area more attractive to both native and nonnative taxa, both may benefit in terms of enhanced reproductive success.

Plant responses to environmental heterogeneity may potentially influence both the mating system and the breeding system. However, few studies in plants have examined how phenotypic plasticity and mating system are related (Dudash et al., 2005). However, in the water flea, *Daphnia*, sexual and asexual lineages exhibited differences in morphological characters, but did not differ in amounts of phenotypic plasticity (Scheiner & Yampolsky, 1998). In a study that compared populations of an inbred plant, *Sporobolus cryptandrus* (sand dropseed), with the closely related but outcrossing *Panicum virgatum* (switchgrass), researchers concluded that both species exhibited significant variation in pattern and amount of phenotypic plasticity among populations, and that *Panicum* exhibited greater plasticity than *Sporobolus* for a variety of traits (Quinn & Wetherington, 2002). This study further showed that observed patterns of phenotypic variation were consistent with the extent of environmental heterogeneity within the native ranges of the two species.

Although few studies have been conducted on the relationship between plant phenotypic plasticity and the nature of mating and breeding systems, we suggest here several ideas that we find important to consider for species of conservation concern. Because seeds of endangered or threatened species are seldom available, experimental evidence of plasticity is rarely at hand for such taxa. However, species that are able to maintain populations (even if in low census population numbers) after habitat fragmentation or other environmental disturbance may in fact be those populations that harbor the ability for many of their phenotypic traits to express plasticity when exposed to new environmental conditions and may thus express adaptive plasticity. Taxa that exhibit opportunistic plasticity are those that can maintain maximal fitness while moving from native to novel environments or across novel environments, and at the same time have not had sufficient time to adapt across generations to either a particular habitat or a new source of heterogeneity (Dudash et al., 2005). In discussion of contrasts between expression of phenotypic plasticity in native and invasive species, researchers have hypothesized that invasive species may have greater plasticity and plasticity of a greater number of traits. In *Lythrum salicaria*, Mal and Lovett-Doust (2005) demonstrated significant phenotypic plasticity for an array of vegetative and morphological traits in response to experimental changes in soil moisture. Quite relevant to the discussion of mating and breeding systems here, *L. salicaria* exhibited a genotype-by-environment interaction in reproductive traits, suggesting that soil moisture may modify positions of anthers relative to stigmas. For this highly invasive species in North America, the continued investigation of the interaction between breeding and mating systems and response to environmental heterogeneity may lead to important insights into the invasion success of this species.

Environmental heterogeneity can influence the expression of inbreeding depression, an important component of mating system evolution (Hedrick & Kalinowski, 2000). From an abiotic perspective, Dudash (1990) detected significant variation in inbreeding depression across three environments: the field, a garden plot, and the greenhouse in the native species *Sabatia angularis*. Specifically, Dudash found a greater expression of inbreeding depression in the field environment and the least expression of inbreeding depression in the greenhouse. Furthermore, when CVs for reproduction of maternal families were assessed across environments, the inbred progeny exhibited greater values than the outcross progeny. These data support the idea of buffering via genomic heterozygosity, such that increased heterozygosity in the outcross progeny improved the ability of these plants to maintain more consistent performance across the three environments than the self progeny (Falconer, 1981; Lerner, 1954). More recently, Daehler (1999) demonstrated that competitive environments exerted greater inbreeding depression by assessing a high-nutrient environment and a low-nutrient environment for populations of *Spartina alternaflora* invading San Francisco Bay. For this species, variation in inbreeding depression among environments and within populations may significantly influence the invasion success of this species.

CASE STUDIES

Variation in Mating System, Expression of Inbreeding Depression, and Purging of Genetic Load in *Mimulus* Congeners

Understanding the role of inbreeding depression in the evolution of mating systems remains an overarching question in our quest to explain the immense diversity of breeding and mating systems documented among the angiosperms. The feasibility of purging the genetic load from a population is an important management concern because populations purged of deleterious alleles tend to increase in fitness, thus enhancing the probability of population persistence in a relatively constant environment. Thus, to address these issues we took a multifaceted approach. First we assessed the differential performance of self and outcross progeny in each of two hermaphroditic *Mimulus* species with contrasting mating systems. Second we determined whether we could purge the genetic load in two complementary ways by first performing self-pollinations by hand ("hand-self") across five generations of selfing and outcrossing (Carr & Dudash, 1997; Dudash & Carr, 1998; Dudash et al., 1997). Next, we used the same progeny and parents in a North Carolina 3 (NC3) breeding design (Comstock & Robinson, 1952) to calculate the level of dominance responsible for the observed inbreeding depression in these two taxa. This multifaceted approach has implications for species of conservation concern when stewards are trying to assess congeners that exhibit variable breeding and mating system traits.

Mimulus micranthus is primarily a selfing taxon resulting from the close proximity of stigmas to dehiscing anthers, whereas *M. guttatus* exhibits spatial separation between male and female function (in other words, herkogamy) and is generally dependent on animal visitors for outcross pollination. However, in older *M. guttatus* flowers, the anthers may brush over the sensitive stigma and induce self-pollination, a process called *corolla dragging* (Dole, 1992).

In comparison with outcrossed replicates, five generations of enforced selfing caused a loss of 50% of the *M. guttatus* maternal lines and differential expression among traits in inbreeding depression (Dudash et al., 1997). Overall, the mixed mating species *M. guttatus* exhibited a greater magnitude of expression of inbreeding depression across a larger number of traits than did the selfing species *M. micranthus*. Interestingly, the traits most closely associated with reproductive fitness—ovule and pollen production—showed no inbreeding depression when self progeny were compared with outcross progeny in *M. micranthus*. In stark contrast, the larger flowered *M. guttatus* exhibited strong inbreeding depression in gamete production (Carr & Dudash, 1997). Furthermore, results of the quantitative genetic breeding experiment found that gamete production in *M. micranthus* was additive, with little evidence of dominance effects, whereas in *M. guttatus* there was evidence of dominance-based inbreeding depression (Dudash & Carr, 1998).

This study demonstrates that the investment of time and the loss of maternal lines makes purging of inbreeding depression a less-than-ideal management tool. However, a promising application of this body of research is to compare the breeding system (and infer potential mating systems when possible) between congeners without conducting extensive breeding programs. Using knowledge of the breeding and mating systems, we can infer an expected level of inbreeding depression that will help estimate population vulnerability in the face of declining habitat and unpredictable environmental heterogeneity.

This body of research also revealed the potentially important role of maternal variation in the expression of inbreeding depression, thus helping to explain the fluctuations in mating patterns among individuals observed within a population. Variation among individuals within a population in expression of inbreeding depression also contributes to the persistence of mixed mating in nature (Dudash et al., 1997). Other researchers have showed that significant maternal variation within populations is common in nature (see, for example, Goodwillie et al., 2005; Pico et al., 2004). Tremendous variation among maternal lines has conservation significance and again highlights the need to collect seed for germplasm and empirical experiments. We also caution researchers against collecting seeds haphazardly with respect to maternal identity. Thus maternal variation may be the key to maintaining the evolutionary potential of many species in peril and deserves further investigation (Dudash et al., 2005; Fenster & Dudash, 1994).

Catasetum viridiflavum Reproduction and Gene Flow across a Fragmented Forest Landscape

Habitat fragmentation and population isolation are considered to affect pollinator visitation rate negatively and thus increase the potential for inbreeding and reduced reproductive success in habitat fragments (Laurance & Bierregaard, 1997), particularly for obligately outcrossing species. The question that one of us (Murren) set out to answer was whether a plant species with a species-specific pollination system and an epiphytic life history would show increased inbreeding and lowered reproduction in response to habitat fragmentation (Murren, 2002).

Catasetum viridiflavum is an epiphytic orchid endemic to central lowland Panama. It has an unusual breeding system that entails functional dioecy and sex switching. Female *C. viridiflavum* are present in high-resource environments. The euglossine bee *Eulaema cingulata* is its sole pollinator. Male bees visit flowers and collect volatile substances produced in both male and female flowers. When the bee hits a whisker located inside male flowers, the spring-loaded pollinia are slapped on its back, and some of them are carried by the bee to female flowers.

Islands created during construction of the Panama Canal served as replicate forest "fragments." These forest islands are advantageous for many reasons, including the fact that experimental replicates are each surrounded by a nearly identical matrix: water. For comparison, large tracts of mainland forest were also chosen. For 3 years, plants on each of 10 islands and each of five mainland sites were sampled to estimate correlates of male and female reproductive success (pollinia removal and fruit set, respectively). During 1 year of the study, male and female estimates of reproductive success were not significantly different between island and mainland sites. However, fruit set was reduced on island sites in comparison with mainland sites in the other 2 years. Two pieces of ecological evidence suggest that plants in forest fragments and mainland sites continue to be interconnected: (1) pollinia were viable for long periods of time and (2) pollinators moved frequently among sites. Although fruit set data indicated some impact of fragmentation, among-year variation in reproductive success suggested interconnection among sites, with pollinators possibly playing a key role in maintaining connection among fragments (Murren, 2002).

In a companion genetic study, Murren (2003) detected temporal genetic structure but minimal among-site variation. However, inbreeding coefficients (in other words, the proportion of the variance in the subpopulation contained in an individual) were large in magnitude and thus consistent with a mating system that has frequent among-relative matings and high rates of seed movement. Although *C. viridiflavum* has a very specific breeding system, the observed patterns of mating seen in a heterogeneous forest fragment nevertheless exhibited patterns more similar to temperate or tropical trees with generalist pollination systems. Highly selfing species with a shorter dispersal distance would have caused a very different pattern of response to fragmentation (for example, a high level of differentiation among populations in different fragments). In summary, study of the mating and breeding systems of plant taxa that differ in pollinator specificities and life history strategies may provide novel insights into the potential responses of plants in habitats of conservation concern. These data are valuable guides when outlining the complex set of possible outcomes during the creation of management plans when a multiyear study is not feasible.

Role of Breeding System and Inbreeding Depression in the Maintenance of an Outcrossing Mating Strategy

Understanding the relationship among the breeding system, mating system, and expression of inbreeding depression of a long-lived perennial allows one to infer the reproductive dynamics of a species, and to examine both the causes and the consequences of selfing. Dudash and Fenster (2001) utilized this combined approach and we suggest here that conservation initiatives could readily apply the method discussed next to assess the reproductive status and potential for population growth of many at-risk plant taxa.

Silene virginica (Caryophyllaceae) is a protandrous long-lived iteroparous species native to eastern North America. The average numbers of flowers open on any given day on an individual is two, such that geitonogamy is possible between flowers within the same plant. Each flower is first male, with the first whorl of anthers presented on day 1 of flower opening, and another set of five anthers presented on the second day. Each flower then normally spends 1 day in a neuter phase, where pollen has been removed and styles are emerging from the functional corolla tube. About 4 days after opening, the flower enters the female phase. Flowers are visited diurnally by pollinators, with the ruby-throated hummingbird as the most important and most effective pollinator of *S. virginica* (Dudash & Fenster, 1997; Fenster & Dudash, 2001; and unpublished data).

In the field, careful breeding experiments were first conducted to determine the timing of pollen presentation and stigma receptivity. Whole-plant relative sex ratios (female-to-neuter-to-male-stage flowers) were also sampled daily. Observations of pollinators were performed to assess within-plant visits (potential selfing via geitonogamy) versus between-plant visits (likely resulting in matings between less related individuals). Next, hand-self and outcross pollinations on field plants were performed to generate progeny to assess inbreeding depression in the greenhouse. Lastly, researchers estimated a population-level outcrossing rate with adult tissue (described earlier).

Strong congruence among mating system parameters was observed among the cumulative inbreeding depression estimate derived from sampling germination rate and flower production, the potential for geitonogamy, and the estimated population-level outcrossing rate of $t = 0.89$. Together these results readily explained the observed relationship among breeding system, mating system, and the expression of inbreeding depression (Dudash & Fenster, 2001). We advocate the use of this multiprong approach to learn basic breeding and mating system information about species of concern so that knowledgeable decisions can be made when planning conservation and restoration initiatives.

MANAGEMENT AND CONSERVATION CONSIDERATIONS OF BREEDING AND MATING SYSTEMS

The conservation implications of mating and breeding systems are broad and diverse, thus we highlight only a few key issues. Attributes of breeding and mating systems have direct effects on the probability of a population persisting over a given period of time. These attributes can be of striking importance when a small population is at risk of extinction. This is because the census population size N (total number of individuals within a population independent of reproductive status) does not necessarily take into account which individuals actually survive to mate. Thus, additional information may be required when assessing populations that are in decline. The effective number of reproducing individuals, N_e, is a more appropriate measure than N as a guide to assessing the potential for a change of population size (Brook, this volume; Falconer, 1981; Reed, this volume). Additional reductions in N_e may occur when only a subset of flowering plants produce mature fruit, further compromising vulnerable small populations (Dudash and Murren, unpublished data). Study of the breeding system, the number of reproductive individuals within a focal population, and estimates of the opportunities to mate with self and others, can provide insights into the long-term likelihood of persistence of a population in question. In Table 5.2 we summarized the potential effects that variation in the breeding system and mating system may have on N_e within a plant population.

Inbreeding can lead to inbreeding depression, and reduced quantity and quality of offspring are not desirable traits for endangered species. Thus, in restoration efforts where small populations consist of genetically related individual plants, we should consider lessons from captive animal breeding programs. Maximizing genetic diversity is one of the primary goals of zoo breeding programs (see, for example, Fiumera et al., 2004). By analogy, restoration attempts in natural field conditions of plants with purged genetic load may be those with the greatest chance for success.

Management and Conservation Recommendations to Assess Mating and Breeding Systems

Our goal for this section is to highlight what we consider "best practices" with respect to breeding and mating systems that in our view need to be compiled to assess factors influencing the persistence of endangered species, to assist in restoration efforts, and to combat spread of invasive taxa.

Breeding and Mating System Considerations

As we described earlier, the breeding system can be used as a predictor of mating system, can contribute to estimates of N_e, and can illuminate many other critical population parameters. First one needs to assess breeding system traits to understand features such as type of breeding system (for example, herkogamy, dichogamy, heterostyly, dioecy), schedule of pollen dehiscence and stigma receptivity, likelihood of pollen movement within a flower versus between flowers on the same plant, and the overall phenology for individual plants and the local population (Dudash & Fenster, 2001). The descriptive biology that emerges from the completion of these tasks will assist in determining whether and how mating system studies should be pursued. Second, perform hand-self pollinations to determine whether the species is self-compatible. If a species is self-incompatible, it may be more susceptible to population decline as a result of habitat fragmentation. If a species is self-compatible, then one can perform pollinator observations to quantify within-plant (potential for selfing) versus between-plant pollinator movements. We also suggest that researchers perform emasculation experiments to determine whether the species is apomictic (see, for example, Dupont, 2002). Apomixis results in the production of seed identical to the maternal parent, and thus new genetic variation arises in apomictic populations through mutation. The relative role of inbreeding and outbreeding depression should be considered and tested with controlled hand pollinations if possible (Dudash & Fenster, 2001), such that appropriate seed stocks can be maintained and restoration efforts can later target plant lineages that maximize the chance of success.

Within-Population Considerations

Whenever possible, seed collection initiatives should maintain maternal-line independence (Dudash et al., 2005). This consideration is important because maternal lines are often found to contain significant amounts of genetic variation within a population. Thus, maintenance of independent maternal lines allows the researcher to examine among-family variation in traits to assess local adaptation and variation among populations in a quantitative way. The information can later be ignored if it is uninformative, but after bulk collection occurs, one cannot recover this information unless new seed is collected.

Efforts to determine the number of maternal lines that initiated populations of invasive species are useful to guide management and eradication efforts. All else being equal, populations of nonnative species initiated from multiple maternal lines should be eradicated first because they likely house greater quantitative genetic variation (including genetic variation for plasticity) that can respond more rapidly to selection, and thus harbor greater potential for invasion success. When seed are available to conduct restoration efforts, we could simultaneously gather valuable information regarding phenotypic plasticity among maternal lines for a variety of ecologically important traits. Estimates of variation in phenotypic expression across microenvironments may guide choice of sites for restoration efforts, some of which may be beyond the region where the population previously resided. Such advance planning for restoration projects on particular rare or endangered species may shed light on the relatively little studied interaction among mating system, breeding system, and phenotypic plasticity.

Population-Level Considerations

Whenever possible, conservation managers should use a number of different source populations for separate reintroduction projects to maximize genetic variation in the system for rare species (Murren et al., 2006). This bet-hedging approach also minimizes the risk of mixing distant gene pools that may result in outbreeding depression of native taxa (see, for example, Dudash & Fenster, 2000). By contrast, the goal for management of invasive taxa should be to minimize mixing of seed sources, a process that may otherwise maximize genetic variation and contribute to a heterozygote advantage for invasives.

Repeated introductions of new genetic material in a restoration project may be needed to maintain populations for the long term at the restoration site. Repeated introduction of new material would maximize both genetic diversity and the potential for establishing families that exhibit complementary phenotypic plasticity over the environmental heterogeneity of the restoration site. These same reasons that could contribute to predicting the success of a restoration effort are attributes that are widely described as important in nonnative species establishment. Furthermore, assessment of differential performance in native and nonnative areas may also provide novel insights into potential interactions between phenotypic plasticity, and the mating and breeding systems (Dudash et al., 2005; Murren et al., 2006) that contribute to the success of both invasive and native species in novel field restoration sites.

The quantification of abiotic habitat heterogeneity (for example, water level and light) of source and target populations (Murren et al., 2006) of species of concern may provide additional insights into the interaction between abiotic pressures and the expression of the breeding and mating systems. These abiotic attributes may directly contribute to variation in breeding and mating system estimates in populations of endangered species. We recommend, therefore, that, if possible, more than one population of a threatened species or an invasive be monitored. Using detailed and replicated monitoring programs we can assay variation in population responses and highlight whether gene-by-environment interactions contribute to population success (Murren et al., 2006).

FUTURE DIRECTIONS

Although the conceptual links between mating and breeding systems and conservation have been discussed for some time, we find that there are numerous future directions that would clarify our understanding and strengthen our predictions regarding population persistence. Basic research on breeding and mating systems and phenotypic plasticity are needed to allow informed assessments of how their interactions influence plant population persistence. Surveys that document pollinator abundance on native and nonnative sympatric species can alert land managers to the need for replacing the invasive species with sufficient numbers of native taxa that flower at similar times to ensure adequate pollinator visitation to the natives (and perhaps outcompete the invasive). For highly invasive species, the continued investigation of the interaction between breeding/mating systems and response to environmental heterogeneity may lead to important insights into invasion success and thus assist efforts to thwart continued population expansion. Further work on the effects of fragmentation on species persistence is also desirable, and researchers should focus on quantifying the species' breeding system and mating system (if feasible) along with other abiotic and biotic attributes of these vulnerable populations.

Too often policy recommendations are made with little or no data on the species in question. We could benefit greatly from a universal database that allows researchers around the globe to share information through a data network. Bioinformatics modeling approaches will only become more robust with the compilation of more empirical data. Further communication efforts are needed between academicians and land managers, such that exchange of perspective, expertise, and insights may lead to successful conservation strategy. We need to advocate for funding nationally and internationally to protect biodiversity and natural resources.

SUGGESTIONS FOR FURTHER READING

Barrett (2003) provides an overview of the variation in mating strategies in flowering plants. Goodwillie and colleagues (2005) explore the evolution of selfing and the persistence of mixed mating in plants, and Jarne and Auld (2006) review the presence of selfing among hermaphroditic animals; both papers discuss how biotic and abiotic ecological factors likely contribute to the persistence of mixed mating systems in nature. Dudash and Fenster (2000) and Dudash and colleagues (2005) provide the link between basic research on mating system evolution and how these studies can provide insights to current obstacles facing conservation and restoration initiatives.

Barrett, S. C. H. 2003. Mating strategies in flowering plants: The outcrossing-selfing paradigm and beyond. Phil Trans R Soc Lond B. 358: 991–1004.

Dudash. M. R., & C. B. Fenster. 2000. Inbreeding and outbreeding depression in fragmented populations (pp. 55–74). In A. Young & G. Clarke (eds.). Genetics, demography, and viability of fragmented populations. Cambridge University Press, Cambridge, UK.

Dudash, M. R., C. J. Murren, & D. E. Carr. 2005. Using *Mimulus* as a model system to understand the role of inbreeding in conservation and ecological approaches. Ann Miss Bot Gard. 92: 36–51.

Goodwillie, C., S. Kalisz, & C. Eckert. 2005. The evolutionary enigma of mixed mating systems in plants: Occurrence, theoretical explanations, and empirical evidence. Ann Rev Ecol Evol Syst. 36: 47–79.

Jarne, P., & J. R. Auld. 2006. Animals mix it up too: The distribution of self-fertilization among hermaphroditic animals. Evolution 60: 1816–1824.

Acknowledgments We thank Chuck Fox and Scott Carroll for their joint efforts behind this book and for their thoughtful editorial comments. We also thank a great copy editor, Mike Loeb, and an anonymous reviewer for their helpful comments. This chapter was funded by SC Sea Grant to C. Murren and was partially funded by National Science Foundation grant DEB-0108285 to M. Dudash.

II

CONSERVING BIODIVERSITY WITHIN AND AMONG SPECIES

The four chapters comprising this part of the book address the fundamental issue of conserving biodiversity within and among species. Two primary themes run through these chapters. The first is the need to conserve the process of evolution, as well as the pattern, to allow future adaptive change. The second theme is the importance of recognizing the distinction between genetic variation within and among conspecific populations.

Smith and Grether (chapter 6) consider the importance of integrating evolutionary processes into management decisions regarding the rank and priority given to different geographic regions that may be slated for conservation. They argue for preserving diverse, locally adapted populations within a species to maximize the potential of the species to respond to future environmental changes. An interesting twist in this chapter is the consideration of sexual selection, as well as natural selection, in promoting adaptive diversity. The authors provide a case study in which understanding natural and sexual selection across environmental gradients in sub-Saharan Africa has provided important insight into conservation planning and reserve design. The take-home message of Smith and Grether is that increased emphasis in conservation planning should be placed on preserving environmental gradients.

In a view complementary to that of Smith and Grether, Faith (chapter 7) recommends integrating evolutionary processes into conservation decision making by considering phylogenetic diversity. Faith provides an approach that goes beyond earlier recommendations that species that are taxonomically distinct deserve greater conservation priority. He argues that the phylogenetic diversity approach provides two ways to consider maximizing biodiversity. First, considering phylogeny as a product of evolutionary process enables the interpretation of diversity patterns to maximize biodiversity for future evolutionary change. Second, phylogenetic diversity also provides a way to infer biodiversity patterns better for poorly described taxa when used in conjunction with information about geographic distribution.

Vergeer and coworkers (chapter 8) consider the application of genetics to the use of introductions as a restoration tool in conservation. They review costs and benefits of alternative genetic strategies for maximizing the probability of successful introduction. They focus on the balance between increasing within-population genetic variation (by intentional hybridization of distinct populations) versus maximizing the possibility of local adaptation. Vergeer and coworkers recommend that the benefits of multiple-source introductions (which will maximize within-population genetic variation) will usually outweigh the advantages of introduction from a single source that is more likely to contain locally adapted genotypes.

Rhymer (chapter 9) argues that the human-caused increase in rates of hybridization pose an important risk to future conservation of biodiversity both within and among species. She advocates application of recently developed modeling approaches to predict the effects of hybridization. Rhymer presents a practical review that considers the delicate balance between recognizing hybridization and introgression as natural evolutionary processes while taking a conservative approach when managers consider translocating species or individuals into the range of conspecific populations or related species. This balance is analogous to that Vergeer and coworkers describe in the previous paragraph.

The issues addressed in Part II go back to the very foundations of conservation genetics (Avise, chapter 1 of this volume). In 1974, Otto Frankel published a landmark paper titled "Genetic Conservation: Our Evolutionary Responsibility" that set out conservation priorities from a genetic perspective:

> First... we should get to know much more about the structure and dynamics of natural populations and communities.... Second, even now the geneticist can play a part in injecting genetic considerations into the planning of reserves of any kind.... Finally, reinforcing the grounds for nature conservation with an evolutionary perspective may help to give conservation a permanence which a utilitarian, and even an ecological grounding, fail to provide in men's minds.

Frankel's first recommendation was to consider both the pattern (structure) and process (dynamics) of evolution in guiding conservation actions—the primary theme of this section. Frankel's second point anticipated the recommendation by Vergeer and co-authors that genetic considerations should play a crucial role in the design of restoration programs and species introduction.

PROMOTING ADAPTATION

The take-home message of Part II is the importance of conserving evolutionary process to allow future adaptive evolutionary changes. However, some recent papers have gone beyond this recommendation and suggest that conservation biologists should sometimes actively direct adaptive change (Norris, 2006). This proposal is somewhat analogous to efforts in medicine to use the evolutionary process to promote evolution of reduced resistance to antibiotics, not just reduce the rate of the evolution of resistance (Ewald, 1994). For example, Kilpatrick (2006) has proposed establishing breeding programs to select for resistance to malaria in Hawaiian bird species and then releasing the genetically resistant birds back into the wild.

The hope is that promoting adaptation would provide a more sustainable long-term strategy than alternative short-term efforts. For example, controlling—or attempting to remove completely—invasive species is extremely expensive and generally an unsustainable long-term strategy. A more long-term solution is to promote the evolution of adaptations that allow native species to persist along with introduced species (Schlaepfer et al., 2005; Carroll and Watters, chapter 12 of this volume).

PHYLOGENIC DIVERSITY AND POPULATION GENETICS

The second theme of this section is the important distinction between within and among population genetic variation. According to the UN Convention on Biological Diversity (1992), biodiversity includes diversity within species, among species, and among ecosystems. However, a fourth component of biodiversity—genetically distinct local populations—is arguably the most important level for focusing conservation efforts. The conservation of multiple genetically distinct populations is necessary to protect the functioning of ecosystems and to ensure long-term species survival (Hughes et al., 1997; Luck et al., 2003).

As discussed in this section, identifying populations and describing population relationships is crucial for conservation and management (for example, planning translocation strategies, identifying local adaptations, and predicting the effects of hybridization). Population relationships are generally assessed using allelic frequency data at multiple loci to cluster genetically similar populations as represented using a dendrogram or tree.

Population and species trees look similar to each other, but they are designed to communicate fundamentally different types of information (see Wilson and colleagues [1985] for an early and exceptionally lucid discussion of this dichotomy). Species trees represent phylogenies that show the time since the most recent common ancestor between species. Such phylogenies show relationships among taxa that have been reproductively isolated for many generations. Population trees, in contrast, generally identify groups of individuals that have similar allelic frequencies because of ongoing genetic exchange (in other words, gene flow). The concept of time since the most recent common ancestor is not meaningful for populations with ongoing gene flow. Populations with greater gene flow will have similar allelic frequencies and will cluster together in population trees. Avise (chapter 1, this volume) points out the distinction between population and phylogenetic trees in the context of the dichotomy between a traditional population genetic approach and a phylogeographic approach in his consideration of intraspecific geographic variation.

Genetically distinct populations often are given highest conservation priority. This is analogous to the phylogenetic diversity approach of giving species that are more taxonomically distinctive greater conservation priority. However, there is potential danger in applying the criterion of distinctiveness to conservation rank of intraspecific populations. It is important to recognize the three primary *processes* responsible for the genetic divergence of populations: natural selection, gene flow, and genetic drift. Genetic drift in small or bottlenecked populations can rapidly result in substantial genetic divergence. Therefore, assigning higher conservation rank to genetically distinct populations may result in placing high conservation priority on small, inbred, isolated populations that have relatively little future adaptive potential (see, for example, Johansson et al., 2007).

Fred W. Allendorf
Missoula, Montana

6

The Importance of Conserving Evolutionary Processes

THOMAS B. SMITH
GREGORY F. GRETHER

Preservation of biodiversity is the primary goal of conservation programs. However, biodiversity is a complex, multifaceted concept that is difficult to define. As a result, developing effective criteria for the measurement of biodiversity has been a challenge and has often been reduced to relying on simple indices. These indices typically include identifying areas for preservation based on levels of endemism, species richness, and degree of threat. Although such a "hotspot" approach is often a valuable first step in ranking regions for preservation, it may fail to capture essential evolutionary processes that promote and sustain diversity. In particular, biodiversity indices may not identify regions where natural and sexual selection have been important in shaping adaptive diversity (Cowling & Pressey, 2001; Crandall et al., 2000; Nicholls, 1998; Smith et al., 1993). In this chapter we illustrate and explore the importance of integrating evolutionary process into conservation decision making. We examine the importance of assessing the roles of natural and sexual selection in promoting adaptive variation and consider how a greater understanding of these evolutionary processes can inform efforts to preserve regional variation in biodiversity.

HOTSPOT APPROACHES TO PRESERVING BIODIVERSITY

Rank ordering regions for conservation efforts based on species richness, levels of endemism, and degree of threat have figured prominently in efforts to conserve biodiversity (Myers, 2002; Myers et al., 2000a). Using such indices to define hotspots is attractive because obtaining the necessary information is relatively easy, especially for vertebrates with distributions that are relatively well described. Scientists can visit a region, conduct field surveys, tabulate numbers of species, or simply compile existing data sets, and then evaluate levels of threat. For example, Meyer and colleagues (2000a) defined biodiversity hotspots based on levels of endemism and threat with the objective of maximizing the number of species protected per dollar invested, thereby allowing conservation organizations to concentrate efforts in regions of greatest need. In using this approach, they found that 44% of all species of vascular plants and 35% of the vertebrate taxa, including birds, mammals, reptiles, and amphibians, were confined to 25 hotspot areas comprising only 1.4% of the earth's land surface. The hotspot approach, based on levels of endemism and threat, still determines conservation priorities for some of the major conservation organizations (Mittermeier et al., 1999). Defining hotspots in this way does have advantages; for example, an analysis that rank orders regions can be conducted relatively quickly, with the result that conservation efforts focus on defined geographic regions.

How effective are such indices in capturing biodiversity pattern and process? Do regions defined in terms of endemism also capture regions that are important for species richness? Recent work suggests that the answer to the second question may be no. Orme and colleagues (2005), focusing on birds, examined the congruence of three indices of biodiversity on a worldwide scale. They found that three indices–species richness, richness of threatened species, and endemic species richness–predicted each other poorly. In fact, only 2.5% of hotspot areas were predicted by all three indices, and any single index explained less than 24% of the total variation in the other indices. If taxa other than birds had been included in the analysis (for example, butterflies, trees, and fungi), a scheme that rank orders regions using the aforementioned indices might perform even more poorly (Orme et al., 2005).

Methods to describe and characterize patterns of biodiversity continue to be revised and improved (Ferrier et al., 2000). Priority-setting approaches may now incorporate the concepts of complementarity, which is a measure of the contribution that each area makes to a particular conservation goal (Pressey et al., 1993), and irreplaceability (Pressey et al., 1994), to describe the overall importance of an area to achieving a conservation target. For example, in the succulent Karoo biome of South Africa, irreplaceability has been used to refer to the number of species found only in one particular area and nowhere else (Ferrier et al., 2000; Lombard et al., 1999). Regions with low irreplaceability are less likely to be required for achieving a conservation target, whereas regions of high irreplaceability are likely to be core areas of conservation activity. Although the lexicon of new terms and the analytical approaches for defining patterns of biodiversity expands, there is a growing realization that regions must also be capable of capturing evolutionary processes (Smith et al., 1993).

THE VALUE OF INTEGRATING EVOLUTIONARY PROCESSES INTO CONSERVATION PRIORITIES

Emphasis on biodiversity pattern does not explicitly consider the processes that preserve adaptive variation and allow populations to evolve with changing environments. Hotspot approaches to species preservation that do not take into account ecological and evolutionary processes are problematic, particularly when applied at small geographic scales. Although it is true that most hotspot approaches are applied across larger geographic scales (Araujo, 2002), preserving populations across uniform habitats such as coniferous forests is analogous to building an investment portfolio made up of a single stock. Conserving populations from diverse habitats may ensure that adaptive variation is maximized. However, a more integrated conservation approach would be to include regions important to the generation and maintenance of biodiversity, regardless of whether these areas harbor endemic species or are particularly rich in species.

With climate change threatening large-scale shifts in species distributions and the habitats on which they depend (Balmford & Bond, 2005), today's hotspots will most certainly shift. Clearly, no approach will be successful unless the effects of multiple stresses from anthropogenic and climatic causes are taken into account. However, by identifying regions in which adaptive variation is maximized, it may be possible to preserve the evolutionary response to changing climate and environmental conditions. Populations are being lost at a much higher rate than species. When populations across diverse habitats go extinct, novel adaptations crucial for meeting future environmental challenges may be lost with them.

New approaches are needed for preserving the adaptive diversity represented by the range of populations within a species, thus ensuring the maximum potential of the species to respond to future environmental conditions. One strategy for preserving the maximum amount of adaptive variation is to identify regions that are particularly important for the generation of new diversity and speciation. But where might such *evolutionary hotspots* occur and what might be their role in promoting natural and sexual selection? These and other questions are discussed in this chapter.

THE ROLE OF NATURAL SELECTION IN PROMOTING ADAPTIVE DIVERSITY

By-product Speciation

Given the central role that natural selection plays in producing adaptive variation and biodiversity, it is surprising that conservation efforts have not focused more on conserving features of the environment that promote selection. Recent reviews of the processes that promote speciation suggest a dominant role for natural selection in leading to divergence and the evolution of reproductive isolation (Coyne & Orr, 2004; Schneider, 2000). Natural selection may promote speciation in various ways. One dominant way can be summarized by the term *by-product speciation*, where ecological differences between populations result in divergent selection on morphological, behavioral, or physiological traits. Reproductive isolation is achieved as the by-product of natural selection acting on traits that are associated with pre- or postzygotic isolation (Rice & Hostert, 1993). For example, populations occurring in distinctly different habitats would be expected to experience divergent selection regimes. As populations adapt to their respective habitats, traits important in reproductive divergence are expected to accumulate genetic changes (Coyne & Orr, 2004; Rice & Hostert, 1993). This may happen either via pleiotropy or hitchhiking, where traits important in reproductive isolation are correlated with those on which divergent selection is acting (Rice & Hostert, 1993). Thus, populations from different environments are expected to be more reproductively divergent from one another than populations from similar habitats, regardless of their recent evolutionary histories. This may happen singly, when a population invades a new habitat, or repeatedly, as in the case of parallel speciation, when an ancestral population invades two distinct habitats (Fig. 6.1). Speciation will occur most rapidly with a complete cessation of gene flow, although divergence and speciation is predicted to occur even under moderate levels of gene flow (Gavrilets et al., 2000).

An excellent example of ecologically driven divergence and incipient speciation can be found in the walking-stick insect *Timema* (Nosil et al., 2002). These insects occur in chaparral habitats of the western United States and Mexico where

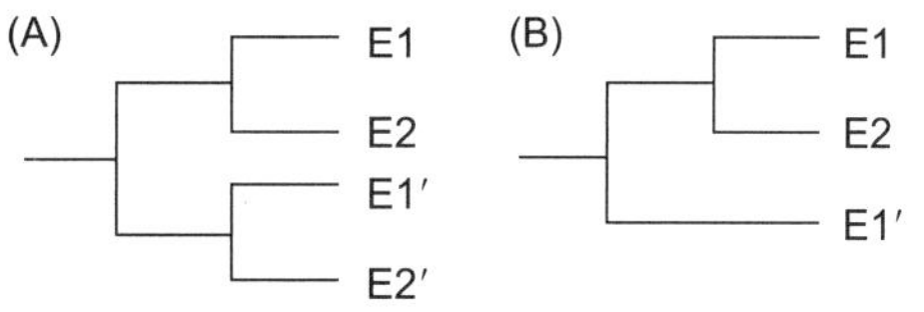

FIGURE 6.1 (A) Parallel speciation (Schluter, 2000). E1 and E2 represent populations adapted to different ecological environments. The phylogeny shows that different populations adapted to the same ecological environment (for example, E1 and E1′) are more distantly related than populations adapted to different environments (for example, E1 and E2). A role of ecological divergence in the evolution of reproductive isolation is supported by the inequalities RI (E1 × E2) > RI (E1 × E1′) and RI (E1 × E2) > RI (E2 × E2′), where RI is a measure of the reproductive incompatibility (for example, assortative mating and hybrid inviability) between populations. The reverse inequalities would support the role of some mechanism unrelated to the environmental difference under study. Tests can be repeated by comparing RI (E1′ × E2′) with RI (E1 × E1′) and RI (E2 × E2′). (B) In this case, E1′ is ancestral and only a single population (E2) has invaded a new habitat. RI (E1 × E2) > RI (E1 × E1′) would support a role of ecologically driven reproductive isolation. The advantage of the model in (B) is that it does not require repeated colonization events within a taxon and might therefore apply to a greater number of study organisms. The disadvantage of this model is that only one independent comparison of reproductive isolation can be made. Additional independent comparisons could be achieved by repeating the test in (B) across numerous taxa (for example, fish, insects, birds). Combined results would address the general role of ecology versus drift in speciation across taxa. (Modified from Orr and Smith [1998].)

they show a genetically determined color–pattern polymorphism in which different morphs are associated with different host plants. Predation by birds and lizards is intense and has resulted in divergent selection for crypsis on respective host plants. In *T. cristinae*, striped morphs are more common on the chamise, *Adenostoma fasciculatum*; whereas unstriped morphs are more common on the greenbark ceanothus, *Ceanothus spinosus*. Local adaptation to host plants has also resulted in divergence in other morphological traits such as body size and shape, host preference, and resting behavior. Furthermore, phylogenetic analyses indicate

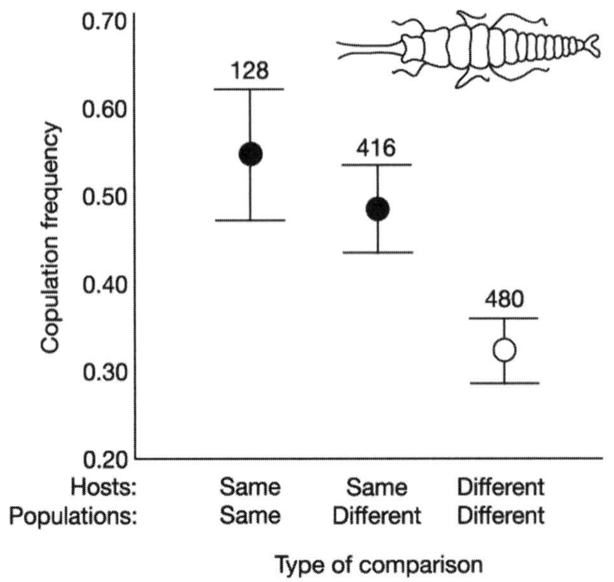

FIGURE 6.2 Copulation frequencies for ecologically similar and different populations of the walking insect *Timema cristinae* on similar and different host plants. Numbers of mating trials for each pairing is shown above each bar. (Modified from Nosil and colleagues [2002].)

that populations using the same host plants do not form monophyletic groups, thus suggesting differentiation and reproductive divergence has occurred repeatedly across the range, revealing a pattern of parallel evolution. Moreover, copulation frequencies are higher for individuals using the same host plants than for individuals using different hosts (Fig. 6.2). Thus, reproductive divergence in *Timema* has apparently evolved as a by-product of adaptation to different hosts (Nosil et al., 2002).

Divergence with Gene Flow

Another way to assess the role of natural selection in promoting divergence and speciation is to examine divergence in adaptive traits as a function of gene flow. If natural selection is a potent force leading to adaptive divergence and, potentially, to speciation, one would predict that between-habitat differences in adaptive traits would show greater divergence than within-habitat comparisons per unit level of gene flow or genetic distance (Fig. 6.3). A pattern of greater divergence in between-habitat versus within-habitat comparisons is a central prediction of the divergence-with-gene-flow model of speciation (Rice & Hostert, 1993), a form of by-product speciation in which the likelihood of speciation depends on the magnitude of selection and the level of gene flow. The more intense the selection and the weaker the gene flow, the greater the likelihood of speciation. This model of speciation, which centers on the balance between selection and gene flow, contrasts with the simple dichotomy presented by sympatric and allopatric speciation.

What features of the environment favor speciation? A growing number of studies suggest that ecological gradients play a particularly important role. Ecological gradients, resulting in divergence and incipient speciation, have been implicated in a diverse array of wild populations, including birds (Smith et al., 2005a), fish (Hendry et al., 2002; Lu & Bernatchez, 1999; Maan et al., 2006), and lizards (Calsbeek & Smith, 2003; Jordan et al., 2005; Ogden & Thorpe, 2002; Schneider et al., 1999). In addition, recent theoretical studies (Doebeli & Dieckmann, 2003; Gavrilets, 2000b) indicate that natural selection may be particularly important in leading to divergence along gradients. In the case

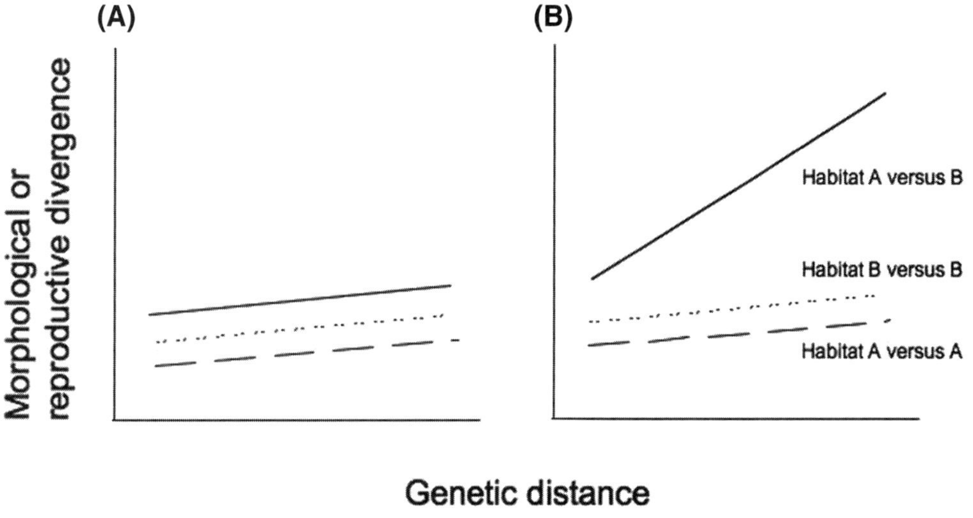

FIGURE 6.3 The divergence-with-gene-flow model (Rice & Hostert, 1993) predicts that as the intensity of divergent selection increases and gene flow decreases, the likelihood of speciation increases. One way to document the influence of habitat using this model is to contrast traits between populations that differ ecologically. The figure compares values of trait divergence (y-axis) and genetic distance, a measure of gene flow (x-axis) between populations from different habitats (habitat A vs. habitat B, unbroken line), and between populations from the same habitat (habitat A vs. habitat A, dashed line; habitat B vs. habitat B, dotted line). (A) This view shows the pattern when ecological differences between habitats are unimportant. Comparisons within and between habitats show a similar slope, which is predicted if ecological differences between habitats do not result in divergence. (B) When ecological differences between habitats lead to differential selection and divergence, the slope of between-habitat comparisons (habitat A vs. habitat B) should be positive and larger than within-habitat comparisons (habitat A vs. habitat A or habitat B vs. habitat B). In both (A) and (B), within-habitat comparisons (and between-habitat ones in the case of [A]) show a slightly positive slope because of the interaction of drift and genetic distance. Because measuring morphological divergence between populations is generally easier than measuring divergent selection, the approach has wide application. This same approach could be used to assess the strength of reproductive divergence. Indices of reproductive divergence could be obtained from mate choice experiments or vocal differences (if they are important in mate choice). In all instances, it is important to ascertain a genetic basis for the trait(s) under study. (Modified from Orr and Smith [1998].)

studies presented later in this chapter we examine how gradients may be studied and integrated into conservation planning.

THE ROLE OF SEXUAL SELECTION IN PROMOTING ADAPTIVE DIVERSITY

Sexual selection arises from competition for mates and operates in three basic modes: mate choice, intrasexual competition, and intersexual conflict (for a comprehensive review, see Andersson, 1994a). As a biodiversity-generating process, sexual selection is potentially more important than most other forms of natural selection. The traits affected by sexual selection can contribute directly to prezygotic isolation, which is thought usually to be the first step toward speciation. Sexual selection may also contribute to postzygotic isolation by reducing the mating success of hybrids. Thus, human activities that interfere with sexual selection can stop or even reverse the speciation process.

The traditional view is that the influence of sexual selection on biological diversity in general, and speciation in particular, is largely decoupled from ecology. The modern roots of this view can be traced to a mathematical model showing that a process of sexual selection originally described by Fisher (1930) could lead to rapid speciation (Lande, 1981). In the "Fisherian" process, linkage disequilibrium (a genetic association) develops between a female mate preference and a male secondary sexual character, leading to a positive feedback loop in which the female preference and male character coevolve unpredictably. Reproductive isolation could arise between populations merely as a by-product of this process. Evolutionary conflicts of interest between the sexes (in other words, intersexual conflict) can also cause sexual traits to evolve in ecologically arbitrary directions and promote speciation (Arnqvist et al., 2000; Gavrilets, 2000a).

Endler (1992) was among the first to emphasize the multitude of ways in which the strength and direction of sexual selection could be influenced by the environment. He coined the term *sensory drive* to refer to the idea that sensory systems and sensory conditions in the environment "drive" evolution in particular directions. The evolution of male courtship displays, for example, could be influenced by biases in the visual system of females, ambient light conditions at the times and locations where courtship occurs, and the visual systems and activity patterns of predators. Sensory drive was offered not as an alternative to the ecologically arbitrary processes of sexual selection, but instead as a context within which these and other processes are likely to occur in natural systems.

From a conservation standpoint, the relative importance of ecologically driven versus ecologically arbitrary processes is critical. If speciation is largely a product of ecologically arbitrary processes, then conservation efforts should be directed toward preserving the most phylogenetically divergent populations of a species, regardless of whether such populations are the most divergent ecologically. In contrast, if speciation is largely a product of ecologically driven processes, then conservation efforts should be directed toward preserving the most ecologically divergent populations. Given that both types of processes appear to be operating in nature, phylogenetic and ecological divergence should both be taken into account when setting conservation priorities. Determining how much weight should be placed on these two factors is an important topic for future research (the answer will probably vary by taxonomic group). It should also be noted that the traditional hotspot approach to conservation, which focuses on standing levels of biodiversity, takes neither ecological nor phylogenetic divergence into account.

Resolving the extent to which ecologically arbitrary versus ecologically driven processes are responsible for variation in sexual traits and mate preferences is an open area of study. Our primary focus here is to show how environmental gradients can promote prezygotic isolation by accelerating divergence between populations in sexual traits and mate preferences. Another way that environmental gradients could cause prezygotic isolation is by causing divergence between populations in the timing or location of mating activity.

Processes that can cause divergence in sexual traits or mate preferences along environmental gradients fall into three broad categories: (1) selection arising from changes in the local optima of secondary sexual traits or mate preferences, (2) indirect selection on secondary sexual traits or mate preferences caused by changes in the local optima of genetically correlated traits, and (3) plastic changes in the development or expression of environmentally sensitive secondary sexual traits or mate preferences. These categories are not mutually exclusive because changes in the environment can have multiple effects. Here we describe each group of processes in general terms and provide supporting examples.

Selection Arising from Changes in the Local Optima of Secondary Sexual Traits or Mate Preferences

Much of the diversity in secondary sexual characters can be explained as a product of direct selection on these traits in response to changes in the physical or biotic environment (reviewed in Andersson, 1994). Although much less well studied, environmental factors can also affect the selective optima of mate preferences (Boughman, 2002). Mate choice frequently involves time and energy costs, and may also increase vulnerability to predators (see, for example, Gibson & Bachman, 1992). Environmental gradients that influence the costs of mate preferences could cause populations to diverge in mate choice criteria. Environmental variation may also affect the benefits of mate preferences by altering the relationship between sexual traits and mate quality (Grether, 2000). In theory, this could cause losses, gains, or shifts in the magnitude of mate preferences and, in turn, alter the evolutionary trajectory of sexual traits.

Prospects for ecological speciation are enhanced when mate preferences and sexual traits evolve in parallel with niche divergence. One illuminating example is in the lakes of British Columbia, Canada, where the threespine stickleback occurs as two ecologically and morphologically distinct species pairs, or *ecotypes*. One ecotype is benthic and forages in the littoral zone whereas the other ecotype is limnetic and forages primarily on zooplankton in open water. After the last retreat of the glaciers some 10,000 to 12,000 years ago, species pairs in each of several lakes evolved independently from their marine ancestor, *Gasterosteus aculeatus*. Limnetic and benthic ecotypes within a lake are more closely related to each other genetically than they are to fish of the same ecotype in different lakes (Rundle et al., 2000). Benthics and limnetics within a lake are reproductively isolated in the wild. In laboratory mating trials, benthics and limnetics from different lakes show a degree of prezygotic isolation similar to that of benthics and limnetics from within the same lake. Within an ecotype, however, fish from different lakes mate and hybridize readily in the laboratory, a result suggesting that the same prezygotic isolating barriers evolved independently in different lakes.

In a study of three different lakes in British Columbia, Boughman and colleagues (2005) confirmed that male coloration and body size have diverged between limnetics and benthics in the same direction in all three species pairs. Furthermore, in all three species pairs, prezygotic isolation was caused by the same two factors: size-assortative mating and asymmetrical female choice based on male color. Color divergence between benthics and limnetics can be explained, at least in part, by differences in the color of the water in their respective nesting habitats. Compared with limnetics, benthics tend to nest in deeper parts of the lake with more vegetative cover and where the light transmission spectrum of the water is "red-shifted" to longer wavelengths. Boughman (2001) found a strong negative relationship between the total area of red coloration on males and the degree to which the water was red-shifted. Boughman (2001) also reports that across each of six stickleback populations there have been parallel changes in sensitivity of females to red light, and in the strength of female preference for red coloration. Moreover, population-level differences in male coloration and strength of female preference correlated positively with degree of prezygotic isolation between populations. These results are consistent with the sensory drive hypothesis and, in any case, argue strongly against the hypothesis that prezygotic isolation arose through ecologically arbitrary processes. What makes this example more compelling than most is that sexual traits and mate preferences have evolved in parallel with morphological changes associated with divergence in foraging niches.

Indirect Selection on Secondary Sexual Traits or Mate Preferences Caused by Changes in the Local Optima of Genetically Correlated Traits

Genetic correlations between phenotypic traits can arise from linkage disequilibrium or pleiotropy. Pleiotropy refers to the fact that genes often have multiple phenotypic effects. Examples in this section illustrate how populations could diverge in sexual traits or mate preferences as a by-product of divergent selection on genetically correlated traits.

Radiation of Darwin's finches into different ecological niches of the Galapagos Islands involved divergence in beak morphology associated with naturally occurring variation in food sources (for example, large seeds, small seeds, insects, and so forth). Beak morphology places biomechanical and acoustic constraints on song production, and thus song has diverged in parallel with morphology (Podos, 2001). Reproductive isolation between sympatric species is largely the result of female choice based on male song (Grant & Grant, 1996), but whether the specific song features affected by beak morphology contribute to reproductive isolation is not yet known. Cultural divergence in song may also contribute to reproductive isolation in this group (Grant & Grant, 1996).

Seddon (2005) contrasted predictions based on pleiotropy and local adaptation and found evidence for both mechanisms in the songs of Neotropical antbirds. As predicted from the biomechanics of song production, pitch and temporal patterning of songs correlated with body mass and bill size, respectively. After controlling for the effects of body mass, however, song pitch correlated with acoustic transmission properties of the forest strata in which antbirds typically sing—specifically, higher pitch songs in the midstory compared with the understory and canopy. Thus, both biomechanical constraints (pleiotropy) and sensory drive (a direct selection hypothesis) appear to have shaped the evolution of Neotropical antbird song.

When two traits compete for limited resources during development, a change in the environment that favors increased investment in one trait may cause a reduction in the other as a correlated response. This may explain some of the spectacular diversity in the horns that male *Onthophagus* dung beetles use to compete for females in underground tunnels. *Onthophagus* spp. vary in horn size and shape as well as in the position of the horns on the exoskeleton. Much of horn diversity appears to be unrelated to ecology and may be a product of ecologically arbitrary processes such as random genetic drift or selection favoring novelty per se (novel horns may confer a tactical advantage), but some evolutionary changes in horns are associated with changes in ecology (Emlen et al., 2005). Developing dung beetle larvae are constrained by the finite amount of food provided by their parents in the form of a dung ball. Horns are expensive structures that negatively affect development of nearby structures. Thus, species with large thorax horns tend to have small wings, and species with large head horns tend to have small eyes or antennae. Using a molecular phylogeny, Emlen and colleagues (2005) tested for statistical association between changes in ecology and evolutionary gains and losses of horns at specific morphological positions. They found that gains of horns on the thorax usually occurred in lineages characterized by very high population densities, whereas loss of horns from the head was associated with shifts from diurnal to nocturnal flight. These trends can be explained in terms of the relative strength of selection on thorax horns versus wings and head horns versus eyes, respectively. Because beetle horns are used in fights between males rather than in courtship, horn divergence is unlikely to contribute directly to prezygotic isolation. Nevertheless, horn divergence could result in unidirectional gene flow (for example, if males with one horn type outcompeted males with a different horn type) and could favor reinforcement of any existing prezygotic barriers by reducing the fitness of male hybrids.

Some mate preferences appear to be derived from sensory biases that evolved in an ecological context. In some cases, these preferences still appear to serve their original sensory functions. For example, in the water mite *Neumania papillator*, females assume a particular posture (*net stance*) to detect vibrations produced by copepod prey. When a male *N. papillator* detects a female via chemical cues, he vibrates his legs at a frequency that mimics copepod vibrations. Females orient to and clutch trembling males as though they were prey, which puts the males in a good position for presenting spermatophore packets. The response of females to the male leg-trembling display appears to be nothing more than an unmodified adaptation for ambushing prey (Proctor, 1991). Phylogenetic analysis indicates that the leg trembling evolved concomitantly with (or after) the evolution of the female net-stance posture (Proctor, 1992). Presumably, if a change in the environment led to further changes in female predatory behavior, this would select for further changes in male courtship.

Plastic Changes in the Development, or Expression of Environmentally Sensitive Secondary Sexual Traits or Mate Preferences

Secondary sexual characters typically are not expressed fully until sexual maturity and tend to be unusually sensitive to environmental perturbations of development (Andersson, 1994; but see Cotton et al., 2004). We know less about the environmental sensitivity of mate preferences, but behavioral traits in general tend to be phenotypically plastic. Depending on the nature of phenotypic effects, changes in the environment that alter expression of sexual traits or mate preferences may reduce or increase gene flow between populations.

When changes in the environment weaken expression of mate preferences, this can lead to hybridization between closely related species and, in extreme cases, a complete breakdown of species boundaries and a loss of biodiversity. This appears to have happened in Lake Victoria, the largest of the African Great Lakes. Hundreds of species of cichlids are endemic to the lake, and some species are genetically isolated from each other only by female preferences based on male coloration. In recent times, human activities have caused eutrophication in parts of the lake. Several lines of evidence indicate that species (and color) diversity has decreased through hybridization in the turbid parts of the lake because the transmission spectrum of the water is too narrow for females to express color preferences (Seehausen et al., 1997). This represents partial reversal of the processes responsible for the extraordinarily high rate of speciation in this taxon.

When a parasitic species colonizes a new species of host, the abrupt change in host environment can trigger immediate changes in sexual characters and mate preferences. Under the right conditions, such host shifts may even cause sympatric speciation. The classic, albeit controversial, example is that of phytophagous insects colonizing new species of host plants (reviewed in Berlocher & Feder, 2002). Brood-parasitic indigobirds (*Vidua* spp.) provide another interesting case. Female indigobirds lay their eggs in the nests of particular host species where indigobird nestlings later imprint on the songs of their host. As adults, male indigobirds mimic host songs and females use these songs to choose mates and to pick which nests to parasitize (Payne et al., 2000). Thus, when novel hosts are parasitized, new host-specific species of indigobirds may arise suddenly. Conversely, when a female parasitizes a host normally used by a different indigobird species, hybridization is the expected result. Molecular genetics and behavioral observations provide support for this model of sympatric speciation with occasional introgression (Sorenson et al., 2003).

When changes in the environment perturb the development of sexual characters away from local optima, selection may favor genetic changes that restore the ancestral phenotype in the new environment. This process, known as *genetic compensation,* could reduce gene flow between populations because hybrids are likely to develop suboptimal phenotypes in both environments (Grether, 2005).

A clear example of the consequences of genetic compensation is provided by the Pacific salmon, which occurs as either the anadromous sockeye or the nonanadromous kokanee. Sockeye "residuals" (in other words, individuals that remain in freshwater lakes or streams throughout their lives instead of migrating to the ocean), develop green coloration at sexual maturity whereas sockeye that mature in the ocean develop red coloration (Craig et al., 2005, and references therein). Red color is produced by carotenoid pigments, compounds that animals in general cannot synthesize. Thus, residuals are green because carotenoid availability is lower in lakes and rivers than in the ocean. Kokanee, which may have evolved from sockeye residuals multiple times in different drainages, are red despite developing in freshwater lakes because they have evolved higher carotenoid assimilation rates than sockeye. Hybrids between kokanee and sockeye have been found in lakes where the two ecotypes spawn sympatrically, but mate choice tests show that green color is a disadvantage and thus hybrids, which are green at maturity, are expected to have low mating success (Craig et al., 2005). If kokanee had not reevolved red coloration, then presumably they would not discriminate against residuals or hybrids on the spawning grounds. The counterintuitive conclusion is that reevolution of the ancestral (sockeye) phenotype in kokanee has reduced gene flow between sockeye and kokanee.

This is just a sample of the ways in which environmentally induced changes in sexual characters and mate preferences could foster (or hinder) adaptive divergence and speciation.

CASE STUDY

Important information on the pattern and process of natural and sexual selection across gradients and in different regions can be incorporated into conservation planning successfully. Here we describe a case study in which evolutionary studies have helped to inform conservation planning and reserve design.

Sub-Saharan Africa has a rich fauna and flora, and harbors many endemic species, especially in mountainous regions. Efforts to conserve this region's biota for the past several decades have largely been driven by hotspot approaches based on theories of refugial isolation and speciation. For example, the Pleistocene Forest Refugia Hypothesis (Mayr & O'Hara, 1986) has been an important conceptual tool for establishing protected spaces, and has led specifically to the identification of three large refugial areas thought to be isolated during the Pleistocene glacial periods. Essentially a vicariant or allopatric model of speciation, the Pleistocene Forest Refugia Hypothesis stresses the role that large forest refugia isolated during glacial periods have played in isolating populations and ultimately in generating new taxa and high species diversity. Numerous reserves and national parks have been established to capture this diversity, and current strategies for conserving biodiversity have targeted these areas as hotspots and priorities for conservation (Myers et al., 2000a).

In contrast, little attention has focused on the conservation importance of ecological gradients formed between regions of rainforest and savanna. Evaluating the role of gradients in generating biodiversity is especially important in sub-Saharan Africa, where deforestation rates are higher than for any other tropical region (Achard et al., 2002). This transition zone, or ecotone, formed by the border of forest and savanna, can be more than 1,000 km wide and in total comprises more than 8,000,000 sq. km of sub-Saharan Africa (Millington et al., 1992). The ecotone is a mosaic of habitats and is characterized by forest fragments embedded in savanna, with fragment size decreasing as one moves away from the central rainforest. Forest fragments found in the ecotone differ ecologically from contiguous rainforest in many ways. Annual rainfall is typically two to three times more variable in fragments than in the rainforest, and the vegetation structure is different, with forest fragments in ecotones having lower forest canopies. Moreover, species assemblages and available foods differ (Chapin, 1954), as does the prevalence and diversity of some pathogens (Sehgal et al., 2001).

Chapin (1954) was one of the first to recognize that ecotones are ecologically dynamic. He identified many species and subspecies of birds that appeared to have their contact zones in these regions and noted many morphological differences in species across this savanna–forest gradient. Furthermore, in an evolutionary context, Endler (1982) showed that 52% of the avian contact zones between species occurred in the ecotone, with 39% within, and only 9% existing between, purported refugial areas. If refugial isolation was the engine driving speciation, one would have expected the majority of contact zones to be concentrated between refugia. Instead, the majority are concentrated in the gradient between savanna and forest. In support of Ender's work, Arctander and Fjeldså (1994), using phylogenic data, found recently diverged taxa were concentrated in transitional zones, such as ecotones, and in mountainous regions. Collectively, these studies suggest that the ecotone formed by the transition between forest and savanna may form a selection gradient that fosters divergence and speciation.

To examine the importance of ecotones in divergence and speciation, we have been examining patterns of morphological, genetic, and behavioral variation in the little greenbul *Andropadus virens*, a small passerine bird common to both the rainforest and the ecotone (Slabbekoorn & Smith, 2002; Smith et al., 1997, 2005a). Examining morphological divergence with genetic differentiation, we tested a central component of the divergence-with-gene-flow model (Fig. 6.3). We contrasted divergence in morphological traits known to be important in fitness (including wing, tarsus, and tail length, and bill length and depth) with genetic distance estimated with 10 microsatellite markers (Smith et al., 2005a). We examined relative divergence across four different habitats in Lower Guinea, including forest, ecotone, mountain, and island. Bivariate plots of genetic divergence (estimated from either $F_{ST}/(1 - F_{ST})$ and Nei's genetic distance) against normalized Euclidean distance of morphological characters revealed that ecotone–forest and forest–mountain comparisons were more divergent than comparisons within habitat, including forest–forest, ecotone–ecotone, and mountain–mountain (Fig. 6.4). Morphological divergence per unit genetic distance was greatest between forest and ecotone populations. The only between-habitat comparison to show little morphological divergence was between mountain and ecotone. However, subsequent habitat analyses using remote sensing data have revealed that the vegetative structure of mountain and ecotone habitat, as measured by canopy cover, does not differ, which appears to explain the lack of morphological divergence between mountain and ecotone (T. Smith, unpublished data). Genetic divergence between mountains tended to be higher, whereas morphological divergence was lower. In fact, the two mountains separated by only 91 km were more genetically divergent from each other than forest populations more than 800 km apart. Nevertheless, morphological divergence between mountains was very low (0.94). As with sticklebacks, genetic isolation of little greenbul populations within similar habitats contributed little to morphological divergence.

To what extent are differences in habitat that yield differences in morphology also producing differences in secondary sexual traits important in reproductive isolation? To answer this question, we analyzed song variation of male little greenbuls from ecotone and forest habitats (Slabbekoorn & Smith, 2002), where song may play a powerful role in reproductive isolation. We recorded greenbul songs from six ecotone habitats and six rainforest habitats. Little greenbul song is complex, with four alternative song types. Moreover, we found statistically significant differences in frequency measures and song-note delivery rates from the two habitats (Slabbekoorn & Smith, 2002). Further investigations of song transmission rates showed that these

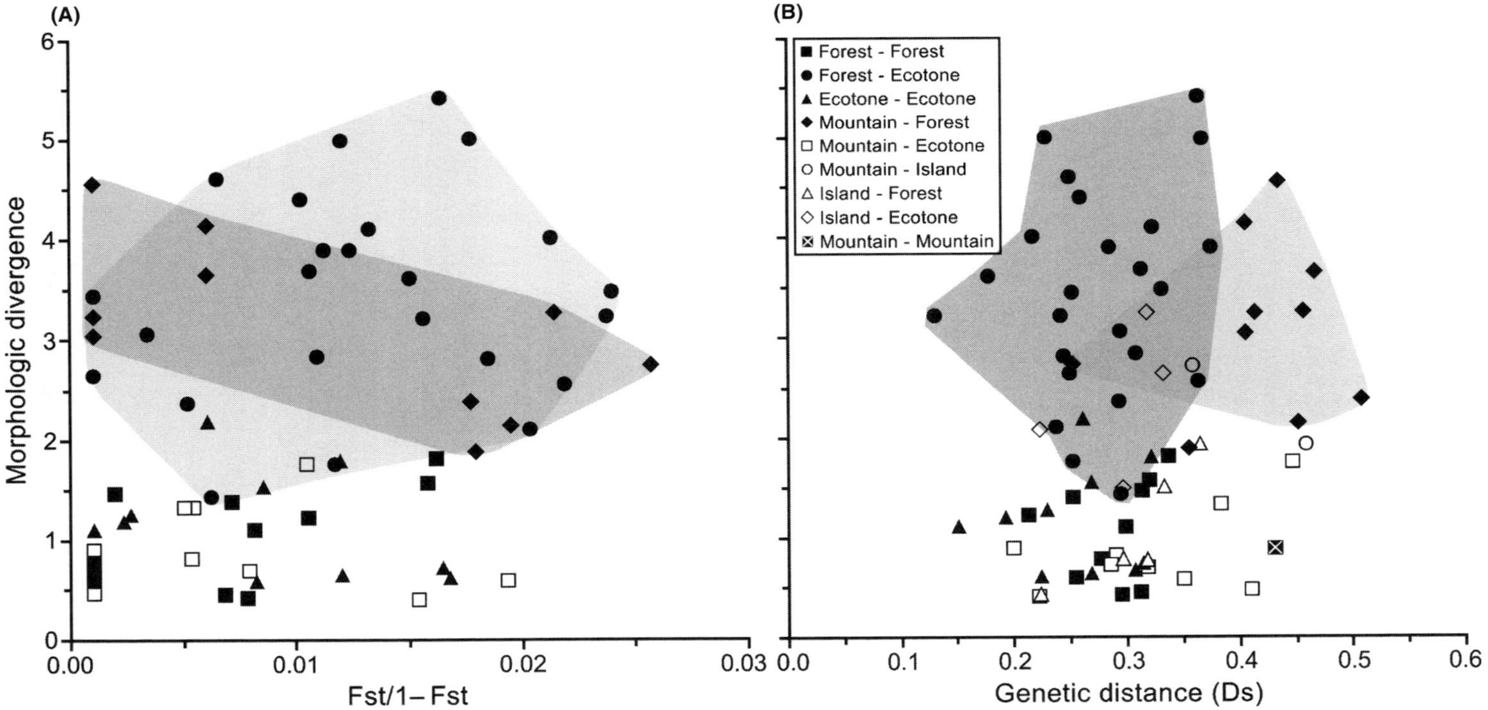

FIGURE 6.4 (A, B) Plot of normalized euclidean distance of morphological character divergence against $F_{ST}/1 - F_{ST}$ (A) and Nei's genetic distance (B) for habitats in Lower Guinea. Shaded areas highlight the two habitats that exhibit the highest divergence (for example, ecotone–forest and mountain–forest). Island habitats are similar to forest habitats on the mainland and show little divergence between them. (Modified from Smith and colleagues [2005].)

parameters were not differentially influenced by the physical structure of the habitat. However, ambient noise levels were found to differ significantly between forest and ecotone—a difference that may explain variation in song.

Do males and females from different habitats respond differently with respect to song? Preliminary results from song playbacks directed at male little greenbuls in Cameroon rainforests show that these birds responded more aggressively to both Cameroon forest and Uganda forest song than to Cameroon ecotone song (Alexander Kirschel, personal communication, October 2005). These preliminary results are particularly salient because they showed males were responding to songs from the same habitat more strongly than they were to songs from different habitats, even if those similar habitats were thousands of kilometers away, as in the case of Uganda.

In addition to these studies, we have also found strikingly similar patterns of divergence in other bird and lizard species, suggesting that patterns of divergence across the forest–ecotone boundary may be ubiquitous across taxa, as Chapin (1954) and others have suggested. If we are interested in preserving not only patterns of diversity, but also the processes that generate and maintain them, then clearly preserving the ecotonal region of Central Africa should be a priority. Although current hotspot approaches still focus largely on regions of high endemism and species richness in regions under the greatest threat, the direction and scope of conservation efforts are nevertheless beginning to change.

In 1997, after publishing the results of one our first papers on the role of ecotones in generating rainforest biodiversity (Smith et al., 1997), we were contacted by environmental planners for the World Bank. Our publication generated media attention and had been circulated among members of the conservation community. The World Bank had recently joined a consortium of oil companies and governments in Central Africa to develop an oil pipeline connecting oil fields in southern Chad with the coast of Cameroon where oil could be loaded onto tankers and shipped to world markets. A prerequisite for World Bank participation in the project included development of mitigation projects focused on biodiversity and impacts on indigenous communities. As part of mitigation efforts related to biodiversity, the consortium was interested in establishing new national parks in Cameroon. Previously, much of the emphasis on park development focused on establishment of either rainforest or savanna parks. Little or no conservation efforts had been directed toward establishing ecotone parks. Through many meetings and discussions, we helped convince planners from the World Bank of the importance of preserving a portion of Cameroon's ecotone and, in 2000, the government of Cameroon, in collaboration with the World Bank, established one of the first ecotone parks in Central Africa (Nadis, 2005). The recently gazetted 42,000-ha Mbam-Djerem National Park spans both rainforest regions to the south and savannah to the north, and is the largest national park in Cameroon. Interest in establishing other ecotone parks is increasing. Conservation International is developing new reserves in Brazil that incorporate portions of Brazilian ecotone or Cerrado (Nadis, 2005), and efforts are also focused on the protection of elevation gradients in the Andes and elsewhere.

FUTURE DIRECTIONS

How can evolutionary processes be taken into account when choosing regions for conservation? To preserve both the pattern of biodiversity and the processes that produce and maintain it, conservation decision makers must take a more integrated approach. It will not be sufficient to identify biodiversity hotspots based solely on species richness and levels of threat. Dynamic regions where evolutionary processes are occurring at high rates will also need to be given high priority. Moreover, given climate change and the likelihood of a 3- to 5-degree increase in global mean temperature, the hotspots of today will likely not be the hotspots of tomorrow, as habitats and populations shift to adjust.

Although biodiversity hotspots are fairly easy to identify (based on survey data), intensive research is needed to identify *evolutionary hotspots*. When decisions are made about which populations of a species to protect, genetic divergence and ecological divergence should both be taken into account. Genetic distance is usually measured at "neutral" loci, which means that it may not provide an accurate representation of the degree of adaptive divergence between populations. Even with moderate rates of gene flow, populations in different environments can diverge in ecologically significant ways. As discussed earlier, prezygotic isolation, and thus speciation, can arise merely as a by-product of ecological divergence. As a general rule of thumb, the most phenotypically divergent populations (for example, with respect to coloration, morphology, behavior, or physiology) are likely to be the furthest along in the speciation process. For any given taxonomic group, however, some phenotypic traits are likely to be more important than others as barriers to interpopulation gene flow. As the Pacific salmon example illustrates, adaptive divergence between populations may be masked by genetic compensation. Individuals from populations that appear phenotypically identical may not be able to develop normally in the other population's environment. Common-garden or cross-fostering experiments may be required to detect cases of genetic compensation and to determine whether phenotypic differences between populations are genetic or environmentally induced.

Further research is needed to evaluate the relative importance of geographic isolation versus environmental gradients as agents of speciation. As a first step, regions might be ranked in terms of genetic and ecological uniqueness. Populations that are high on both scales should be given the highest conservation priority and, conversely, populations that are lowest on both scales should be given the lowest priority. Research efforts could then be directed at the subset of populations that score high in either genetic or ecological uniqueness. New approaches that allow genetic and adaptive phenotypic data to be mapped onto a landscape (Manel et al., 2003) permit different regions to be compared with regard to their genetic and adaptive features. Integrating these data with levels of species richness and endemism, coupled with environmental layers gained from remote sensing, would be one way to integrate pattern and process into conservation planning. After adaptive and genetic features are mapped, several modeling techniques are now available (see, for example, Phillips et al., 2006), making it possible to correlate them with various types of environmental data. This in turn allows one to make predictions regarding how the distribution of adaptive and genetic traits may change with climate warming. This would allow for the creation of parks and reserves that maximize preservation of biodiversity pattern and process under both current and future climates. Specific steps involved in establishing new protected areas might include: (1) examining and quantifying regional biodiversity, ideally at all levels in the biological hierarchy (from genes to ecosystems) in a reserve network; (2) integrating across all levels of biological organization to quantify as many ecological and evolutionary processes as possible, including phenotypic and genetic divergence among populations as well as the geographic context of diversification; (3) quantifying the correspondence among regions identified as centers of species diversity with regions important to adaptive and genetic diversity; and (4) quantifying current and historical socioeconomic factors that might affect the priority and feasibility of establishing parks or reserves (Smith et al., 2005b).

As illustrated in this chapter, greater emphasis on preserving environmental gradients is paramount for two reasons. First, natural and sexual selection along ecological gradients are powerful drivers of adaptive variation and, under the right conditions, speciation. Second, given the reality of climate change, preserving gradients (and their associated adaptive variation) may offer a bet-hedging approach—the hope that at least some portion of the population will be adapted to new climate conditions.

SUGGESTIONS FOR FURTHER READING

Smith and colleagues (2005) provide an overview of how one might integrate pattern and process into conservation planning, and Crandall and colleagues (2000) provide an excellent review of the steps important in rank ordering regions according to adaptive variation. For an excellent summary primer on ecological speciation, see Albert and Schluter (2005), and for greater in-depth treatment see Schluter (2000). Endler (1992) provides an excellent starting point for delving deeper into the literature on sensory drive and related processes. For more examples of genetic compensation, and its potential importance for conservation, see Grether (2005).

Albert, A. Y. K., & D. Schluter. 2005. Selection and the origin of species. Curr Biol. 15: R283–R288.

Crandall, K. A., O. R. P. Bininda-Emonds, G. M. Mace, & R. K. Wayne. 2000. Considering evolutionary processes in conservation biology. Trends Ecol & Evol. 15: 290–295.

Endler, J. 1992. Signals, signal conditions, and the direction of evolution. Am Nat. 139: 125–153.

Grether, G. F. 2005. Environmental change, phenotypic plasticity and genetic compensation. Am Nat. 166: E115–E123.
Schluter, D. 2000. The ecology of adaptive radiation. Oxford University Press, Oxford, UK.
Smith. T. B., S. Saatchi, C. H. Graham, et al. 2005b. Putting process on the map: Why ecotones are important for preserving biodiversity (pp. 166–197). In A. Purvis, J. Gittleman, & T. Brooks (eds.). Phylogeny and conservation. Cambridge University Press, Cambridge, UK.

Acknowledgments We thank S. P. Carroll, B. Larison, D. M. Shier, and two anonymous reviewers for suggestions that improved the quality of this chapter. Portions of the research presented were supported by grants from the National Geographic Society and the National Science Foundation, grants DEB-9726425 and IRCEB9977072.

7

Phylogenetic Diversity and Conservation

DANIEL P. FAITH

A goal of this book is to explore and promote evolutionary thinking in conservation biology. This chapter focuses on how evolutionary thinking can help address one of the biggest challenges faced by conservation biology—the conservation of *overall biodiversity*. I adopt the common usage of biodiversity to describe the variety of living forms on the planet, extending from genes to species to ecosystems. Overall biodiversity (or total diversity) (for example, see Faith, 2005; Millennium Ecosystem Assessment, 2005a, b) is intended to reflect the large amount of variation, over all these levels, that remains unknown to science. This knowledge gap extends further; we also do not know whether future societies will value these same biological components. Therefore, strategies for conservation of overall biodiversity must first estimate general patterns of variation and then conserve as much of that estimated variation as possible, thus preserving components that may be valued in the future (in other words, retention of *option values*) (for a discussion, see Faith, 1992a; IUCN, 1980).

A calculus for overall biodiversity would allow useful estimation of overall gains and losses in biodiversity in different locations and from different actions or threatening processes. The Millennium Ecosystem Assessment (2005a, b) called for fresh efforts toward development of a global biodiversity calculus to serve conservation planning and setting of priorities. An important challenge lies in making the best possible use of available information to estimate overall biodiversity patterns (in other words, the biodiversity surrogates problem). One simple strategy is to count up what we can observe and then assume these counts reflect more general quantities. To understand how observed variation may extend to more general cases, a better strategy is to incorporate information about *processes* generating variation. Phylogenetic pattern—the Tree of Life—reflects evolutionary processes of speciation, and can play an important role in estimating overall biodiversity patterns.

Smith and Grether (this volume) illustrate and explore the importance of integrating evolutionary processes when rank ordering regions for conservation (see also Moritz, 2002). They argue that procedures that focus on representation of variation (*representativeness*) for rank ordering regions for preservation may "fail to capture essential evolutionary processes that promote and sustain diversity" (p. 85). Thus, it is important to preserve not only the pattern of biodiversity but also the evolutionary processes that produce and maintain it.

My perspective in this chapter complements Smith and Grether in arguing that integration of process is critical even for the case when the focus of planning and priority setting is representativeness. The reason is that inference of patterns for overall biodiversity is boosted by the incorporation of such information.

This chapter shows how phylogenetic pattern and process help accomplish this process-based inference at two biodiversity levels. First, it provides a way to talk about the overall feature diversity of sets of taxa. Phylogeny as a product of evolutionary processes enables the inference of feature diversity patterns (reflecting, in turn, evolutionary potential of the set; discussed later). Second, the phylogeny for one set of taxa provides a way to infer biodiversity patterns better for other, unobserved taxa. Here, shared history, inferred through phylogenetic pattern for one set of taxa, predicts patterns over geographic areas for other taxonomic groups.

In another chapter in this book, Avise argues that "A basic notion is that unique (in other words, long-separated) evolutionary lineages contribute disproportionately to the planet's overall genetic diversity, such that their extinction would constitute a far greater loss of biodiversity than would the extinction of species that have extant close relatives. Although phylogenetic considerations can be important in particular instances, my own guess is that they will seldom override more traditional criteria used by societies to decide which species or biotas merit greatest protection" (p. 11).

This chapter presents a more optimistic perspective about the prospects for conservation priorities based on phylogeny. Phylogeny can contribute in two ways: The phylogeny of a set of taxa helps us talk about biodiversity not only *within* that set (feature diversity), but also *beyond* that set (diversity with respect to other taxa). I address these themes in turn, and show that an approach based on phylogenetic diversity serves both tasks.

PHYLOGENETIC DIVERSITY AND FEATURE DIVERSITY

Avise (2005) summarized the challenge posed by the Tree of Life as follows: "The fundamental challenge for conservation biology is to promote the continuance of the outermost tips in the tree of life: to promote the vigorous as well as the most tender of the extant shoots so that, in this latest instant of geological time, humanity does not terminate what nature has propagated across the aeons." Given that conservation biology faces high current extinction rates, combined with limited conservation resources (Millennium Ecosystem Assessment, 2005a, b; Weitzman 1992), we have to ask the question: How do we proceed when we cannot protect *all* the tips?

Some time ago, the IUCN (1980) considered this issue and recommended that a species that is taxonomically "distinctive" is deserving of greater conservation priority. Subsequent linking of this idea to genetic or feature diversity (Faith, 1992a, b, 1994) helped to justify the idea of such a differential weighting of species. One philosophical objection was that we know so little about differential values of species over time that we could not afford to weight species differentially. However, when biodiversity encompasses features, or characters, of taxa, we arrive at the opposite conclusion (Faith, 1994): If features, in principle, are equally weighted, then preserving feature diversity *obligates* us to weight species differentially.

Setting priorities to maintain feature diversity addresses option-value arguments for conservation; that is, we maximize retention of possible future values. However, surrogate or proxy information is needed for such assessments because all the features of possible interest cannot be directly observed and counted. Phylogenetic pattern, however, can play this important role. Phylogenetic theory explains how shared ancestry may account for shared features, and this model has been used to link phylogenetic pattern to expected feature diversity of sets of taxa (Faith, 1992b). Any subset of taxa that has greater phylogenetic diversity should also represent greater feature diversity.

I quantify this idea by considering branch lengths on a phylogenetic tree as estimates of relative magnitude of character change in different lineages. The *spanning path* on a phylogeny indicates relative feature diversity of any given set of taxa. Stated differently, if we connect up taxa on the tree, the spanning path shows how much of the tree has been traveled over (Fig. 7.1). The total phylogenetic diversity of any subset of lineages or taxa (for example, species, haplotypes, or other terminal taxa) from a phylogenetic tree is given by the total branch length spanned by the members of the subset. This measure called "PD" indicates the relative feature diversity of phylogenetic subsets (Faith, 1992a, 1994). As noted in the original examples (Faith, 1992a, b), the total feature diversity of a set of taxa will normally count all the branches spanned by the set, including those to the root of the tree (Box 7.1) (for a discussion, see Faith and Baker, 2006). PD reflects the total evolutionary history of the set of taxa (Faith, 1994), and thus PD can be equated with *evolutionary potential* (see, for example, Forest et al., 2007). Furthermore,

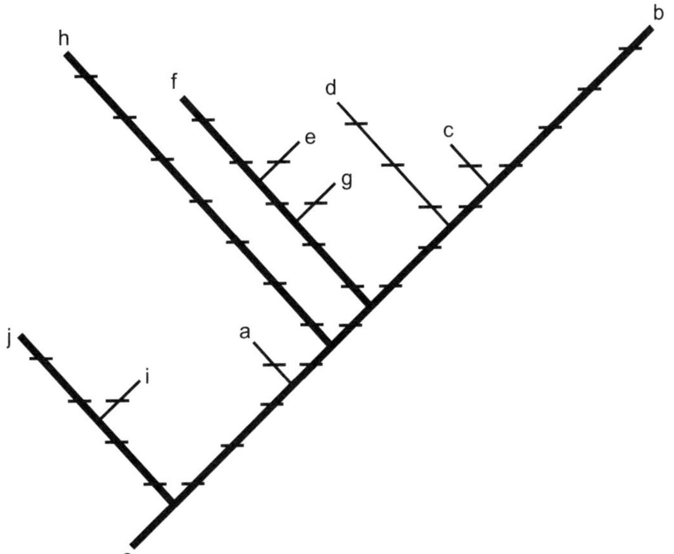

FIGURE 7.1 A hypothetical tree or cladogram for taxa a through j and outgroup O. The inferred derivations of new features from the data matrix of Faith (1992a) are recorded by tick marks along branches. The bold lines show the set of four taxa—b, f, h, and j—that have greatest PD. (Redrawn from Faith [1992a, b].)

PD allows for a range of assumptions about meaningful branch lengths, including the possibility of using existing taxonomy as a rough indicator of divergence amounts (Faith, 1994).

Examples of PD Applications

Two examples illustrate the PD links to feature diversity and option values. Figure 7.1 shows a hypothetical tree or cladogram from Faith (1992a, b; see also Faith, 2006a) for taxa a through j and outgroup O. The inferred derivations of new features from a original data matrix (Faith, 1992a, b) are recorded by tick marks along branches. Given these branch lengths, PD calculations reflect numbers of features for different sets of taxa. For example, species j on its own would have a PD score of 4, reflecting the four new features in addition to all those represented by the outgroup. The path length traced with the bold line shows the PD for the best set of four species, with a total PD of 28. If only four species could be retained, this set is predicted to be the one having greatest feature diversity.

A second example is from Forest and colleagues (2007), who carried out a phylogenetic analysis of the flora of a global biodiversity hotspot: the Cape of South Africa. Their study used a large, biomewide phylogeny estimated using an exemplar species for each of 735 of the 943 genera of angiosperms that occur in the Cape. The genetic marker used was the plastid rbcL exon. Forest and coworkers (2007) asked whether maximizing PD can be expected to maximize retained option values. To explore this question, they identified all those genera in the Cape with species of known medicinal or economic importance (designating three types of use: food, medicine, and other). They determined that the different categories of use are clustered in different parts of their phylogeny, suggesting that simple preference of taxa for one use will not always capture other uses. To address the issue of option values, the researchers asked whether PD-based priorities would have been a good way to capture all the identified uses, assuming that these uses had not yet been known. They found that choosing a set of taxa based on PD would have maximized the probability of having representatives of each of the three classes of use. The genera chosen to maximize PD contained a higher number of useful genera (over all use classes) than the same number of genera selected at random. This study also found that selection of sets of localities in the Cape based on maximizing PD was

BOX 7.1 Phylogenetic Diversity Definitions

Although there are now many applications of the PD measure of Phylogenetic Diversity (sensu Faith, 1992), terminology and definitions are not standardized in the biodiversity literature (Faith, 2006a). For example, in *Phylogeny and Conservation,* where these topics are reviewed (Purvis et al., 2005), *PD* is sometimes used to indicate PD sensu Faith (1992), as a general term indicating a kind of *PD,* and as an abbreviation of a different quantity: *phylogenetic distinctiveness.* At the same time, some measures discussed that are equivalent to PD are given other names, with correspondence to PD not made clear.

Lewis and Lewis (2005) introduced a new term *exclusive molecular phylodiversity* to describe what is equivalent to PD for a set of taxa sharing some characteristic of interest. But Faith and colleagues (2004b), for example, discuss the application of PD to other collections of taxa, as in the PD value of those taxa found in a designated ecotype.

The terminological problem is made worse in considering the need to discuss gains and losses, and not just total diversity amounts.

- *PD of a set of taxa* is the total phylogenetic branch length spanned (in other words, represented) by its member species (including the root of the tree).
- *PD complementarity* of a species is measured by the additional branch length it represents that is not spanned by a reference set of species (Faith, 1992a).
- *PD endemism* refers to the PD complementarity value of a species when the reference set includes all other species. This unique PD contribution of the species can be thought of as *endemism* at the level of features-within-species (rather than species-within-areas).

PD Applied to Localities or Other Collections of Taxa

The *PD endemism* of a locality is the amount of branch length (phylogenetic diversity) or *evolutionary history* (Faith, 1994) uniquely represented by that locality. An example is seen in Figure 7.4, where locality p1 uniquely has taxa f, g, and h, and has PD endemism including branch z and all descendant branches.

The *PD complementarity* of one or more localities is the additional branch length collectively contributed by those species that are in the localities but not in a reference set. Such calculations involve a twofold application of the principle of complementarity: Not only do we disregard species that are already represented, but we also exclude branches represented by a new species if they are already spanned by other species in the reference set. An example is shown in Figure 7.4, where the phylogenetic diversity complementarity of locality p1, given locality p2, includes branch z but none of the deeper branches shared with locality p2.

The *PD endemism of a locality* is its PD complementarity value for the case when the reference set is defined by the species found in all other localities.

Probabilistic PD

These definitions extend to the case when species have probabilities of extinction or probabilities of absence, for example, derived from a predictive model (see also Box 7.2). The *expected PD* is defined as the sum, over all possible species combinations (C), of P(C) × PD(C), where P(C) is the probability that combination C will not go extinct, and PD(C) is the PD value for combination C. The *PD complementarity* of a species, resulting from a change in its probability of extinction, is then defined as the change in the expected PD value when recalculated with the new extinction probability. This extends to multiple species and localities.

(continued)

Probabilistic PD can be defined generally as the framework for a range of PD calculations incorporating extinction probabilities. For example, it may be used to calculate the probability of obtaining particular PD values, such as values less than some nominated threshold. Furthermore, PD(C) may be replaced by an alternative PD-based calculation for each species combination. For example, the *expected PD endemism of a locality* is the sum, over all possible species combinations (C), of P(C) × PDE(C), where P(C) is the probability that combination C will not go extinct, and PDE(C) is the PD endemism of the locality based only on those species in combination C.

Adapted from Faith [1992] and Faith and colleagues [2004]. See also http://www.edgeofexistence.org/forum/forum_posts.asp? TID=13.

more effective in sampling all useful genera compared with results when localities were selected to maximize number of taxa. The conclusion was that PD provides an effective way to maximize option values at the level of features.

This fundamental link from PD to feature diversity and option values counters the early rejection (Takacs, 1996) of the supposedly unnecessarily "intricate" calculations used in phylogenetically based valuations of species. The objection was that such methods adopt one of many arbitrary ways of differentially valuing species. However, arbitrariness disappears when the focus is on biodiversity units at a lower level of features, and differential values for species are a product of the *equal* values for those lower-level units (for a discussion, see Faith, 1994). Thus, PD is more a calculus of feature diversity than species-level diversity. This is apparent on occasions when PD may be applied without reference to species designations (Faith, 1994) (discussed later).

PD Complementarity and Endemism

Although PD provides a measure of total diversity (total evolutionary history or total feature diversity) of individual or sets of taxa that may be delimited in various ways, it also defines a family of calculations relating to gains, losses, and endemicity. For conservation planning such marginal gains and losses are typically more useful than are total diversity values (Faith, 1992a, 1994). Because PD operates as if it is counting up features, PD complementarity and PD endemism are feature analogues of the normal species-level versions of these terms (Box 7.1).

Imagine a case in which a single taxon (for example, species) is "pruned" by extinction from the phylogenetic tree. The degree of loss in feature diversity is reflected in the branch length lost from the spanning set of branches. This reflects the PD complementarity of the taxon, defined in general as the additional branch length the taxon provides relative to any reference set of taxa. PD endemism is the PD complementarity value when the reference set is all other species (Box 7.1) (Faith, 1992a; Faith et al., 2004b). In the example of Figure 7.1, the PD endemism of b is 4 units, whereas that of h is highest, at 7 units. However, PD complementarity of b is 12 units, if only the outgroup is given as the reference set, but is only 7 units if we also have taxon f. Taxon f captures deeper, shared branches. Complementarity may also be assessed for a whole collection of taxa—for example, those found in a given locality (for other examples, see Faith & Williams, 2006; Faith et al., 2004b).

I noted that PD complementarity for a taxon or for a locality clearly is not a fixed value, and depends on the other taxa in a reference set (for example, those that are not extinct, or those found in a protected area system). As illustrated in the comparison of complementarity values for taxon b in Figure 7.1, PD complementarity will be smaller as more taxa are included in the reference set. If we begin selecting species from a phylogeny to maximize PD, initially phylogenetic diversity complementarity values—gains in total PD—will be high, reflecting capture of some deeper branches. But as more species are selected, PD complementarity values are *smaller*, because deeper branches are already represented.

Because of this drop-off in PD gains, PD thus defines a species–phylogenetic diversity curve that is analogous to the well-known species–area curve (see Faith & Williams, 2006) when number of species sampled is plotted against the PD value of the set.

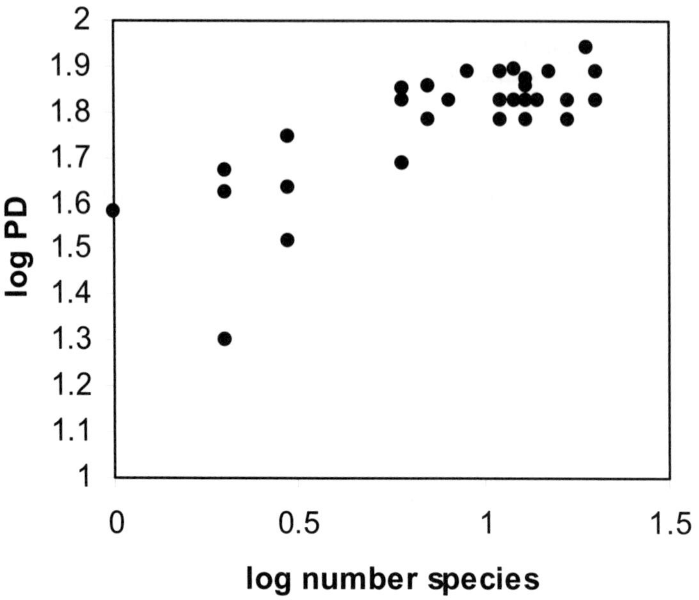

FIGURE 7.2 PD defines a species–PD curve analogous to the well-known species–area curve. A simple example of such a curve is derived here from a study by Pillon and colleagues (2006). Random taxon samples of different sizes from a given phylogenetic tree produce a roughly linear relationship in log–log space. In this example, the sets of taxa correspond to those defined by different localities. This nonrandom sampling may account for the apparent cases when PD has lower values for the given number of taxa. This may reflect phylogenetic clumping within localities.

Figure 7.2 provides a simple example of such a curve, derived from a study by Pillon and colleagues (2006). This characteristic curve means that there is inevitably some correspondence to be expected between numbers of species and the PD value. The relative PD of different sets of localities tends to reflect differences in the number of member taxa (see, for example, Mace et al., 2003). Although such a correspondence, when observed empirically, has sometimes been taken as evidence that conservation planners need only maximize species numbers (and not bother with phylogeny and PD), conservation planning considerations point to differences. For example, Forest and colleagues (2007) focused on real-world scenarios in which a new locality is to be added to a protected area system. In this case, complementarity values matter for decision making, and these researchers showed that actual gains in PD are decoupled from observed species-level complementarity values (Fig. 7.3). Thus, species-level assessment would not help preserve PD.

Phylogenetic Clumping and Dispersion

The magnitude of PD gains and losses will be decoupled from species counts particularly when losses are "clumped" on the phylogenetic tree. In Figure 7.4, the PD loss implied by the loss of taxon h will be greater if taxa f and g already have been lost. The loss of the three taxa does not imply a PD loss simply equal to the sum of the individual losses. Instead, losing all three sister taxa also implies loss of the branch or lineage linking them. Such phylogenetic clumping of losses, and consequent inflation of PD loss, has been documented in various studies. For example, an assessment of Indonesian fauna (Mooers & Atkins, 2003) showed that species vulnerability to extinction is phylogenetically clumped and, consequently, the amount of PD at risk is significantly greater than would be found for the same number of species selected at random from the phylogeny.

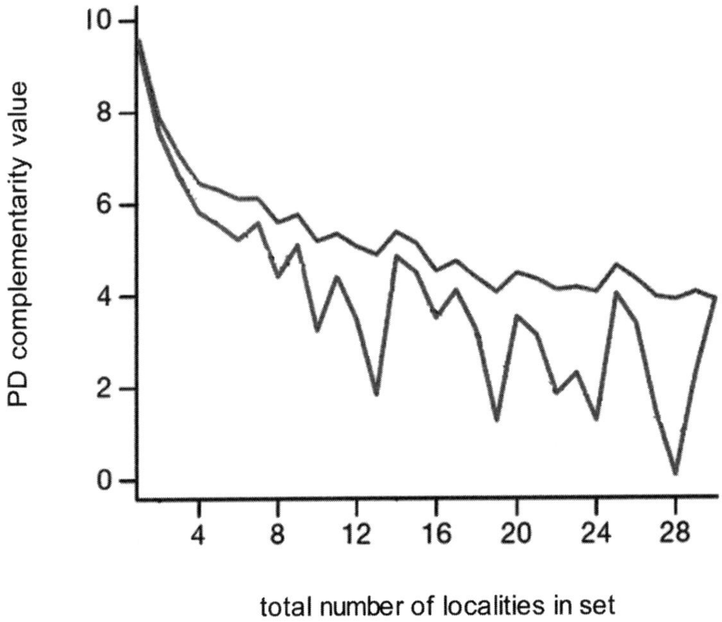

FIGURE 7.3 Redrawn results from Forest and colleagues (2007) showing the decoupling of taxon-based loss and PD complementarity (plotted along the vertical axis). The horizontal axis shows the accumulation of localities in the Cape selected according to an algorithm that builds up a set of localities by choosing the locality adding the greatest number of additional taxa at each step (values along the axis indicate the total number of localities in the set at each step). The lower curve shows the corresponding PD complementarity of the chosen locality at each step. The upper line shows the PD complementarity for that alternative locality at each step that would have provided maximum PD complementarity. The conclusion is that use of taxon complementarity is a poor predictor of localities that would provide high PD gains. (Redrawn from Forest and colleagues [2007].)

Forest and coworkers (2007) demonstrated that the flora of the western region of the Cape is phylogenetically clumped, and that this results in a lower PD score for localities for a given level of taxon diversity. In contrast, the eastern region showed phylogenetic dispersion of member taxa, producing higher PD values than taxon numbers might have indicated. The authors used this to argue that the eastern Cape region might deserve increased conservation attention.

Localities, therefore, can have a disproportionately large contribution relative to species counts when there is phylogenetic clumping within localities (for a review, see Faith and Williams, 2006). This also affects PD endemism estimates: The clumped effect can mean a disproportionate PD loss for a given number of species. For example, Sechrest and colleagues (2002) found that about one third of the total PD of primates and carnivores was contained only in the recognized global biodiversity hotspots. They found that the amount of PD lost, assuming that biodiversity hotspots were eliminated, would be significantly greater than for a random selection of the same number of species. In a similar spirit, follow-up work to that of Forest and colleagues (2007) in the Cape hotspot could explore whether the high total PD values of eastern localities also implies high PD endemism.

I conclude that, although the species–PD curve defines a general correspondence, there are good reasons why PD sometimes gives different answers, relative to species counting, in assessing the biodiversity contributions of localities. Faith and Williams (2006) discuss other examples of

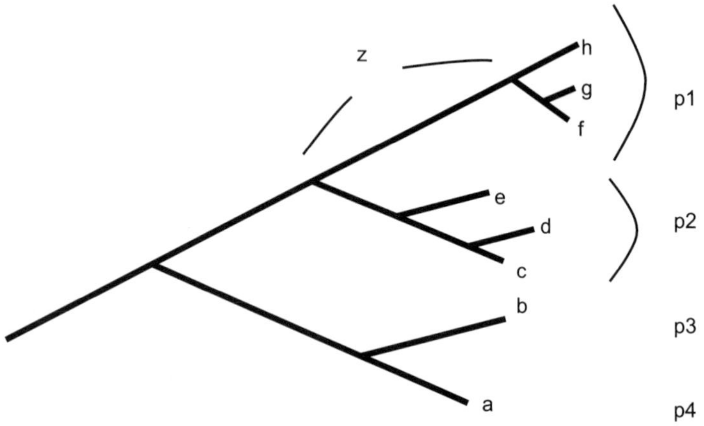

FIGURE 7.4 A phylogenetic tree example, redrawn from Faith and Williams (2006), for taxa a through h, found in localities p1 through p4. Taxa f, g, and h occur uniquely in locality p1, so that any loss of p1 would mean loss not only of the proximal connecting branches, but also the loss of the deeper branch z. Such phylogenetic clumping results in larger PD losses.

phylogenetic clumping, and a revealing case study is presented next.

Example: Phylogenetic Diversity and Global Warming Impacts

Phylogenetic clumping versus dispersion is also important when considering the effects of threats on species. These factors, and the contrast between PD and species-counting approaches, is highlighted by studies that have begun to move beyond species and on to examining impacts of global warming on biodiversity. In an interesting PD study, Yesson and Culham (2006) found phylogenetic dispersion of those *Cyclamen* taxa with lowest probability of extinction resulting from potential climate change impacts. This pattern implied that the potential loss of PD and evolutionary potential resulting from climate change was *smaller* than might be expected based on species counting. To quote Yesson and Culham (2006), "while many individual species are at high risk, each major lineage is seen to contain at least one species with a reasonable chance of survival. This pattern lowers the overall risk to phylogenetic diversity" (Fig. 7.5).

This PD study of climate change impacts raises important questions for future studies. First, is the most meaningful measure of biodiversity impacts not conventional species-loss estimates, but rather estimates of loss of phylogenetic diversity and evolutionary potential? Second, will other studies similarly find relatively low PD loss, or will they find the opposite—high PD loss, resulting from affected (or persisting) species that are clumped on the phylogenetic trees (Fig. 7.5)?

Probability of Extinction

In the cyclamen case study of Yesson and Culham (2006), degree of potential loss of phylogenetic diversity depended on the phylogenetic pattern of survival or extinction arising from the threat of climate change. More generally, different species may have estimated probabilities of extinction arising from a number of threats rather than a single all-or-nothing score. Such probabilities of extinction may be estimated, for example, from a survey of habitat loss (Faith et al., in press). We can assess potential PD losses taking these probabilities into account, based on simple modifications of PD calculations that incorporate probabilities of extinction (Witting & Loeschke, 1995; but see also Faith, 1996, 2002; Faith & Walker, 1996a).

Probabilistic PD can be interpreted as an extension of Weitzman's *expected diversity* calculations (Weitzman, 1992). Weitzman defined a general measure of expected diversity for sets of species, taking into account the estimated probability that

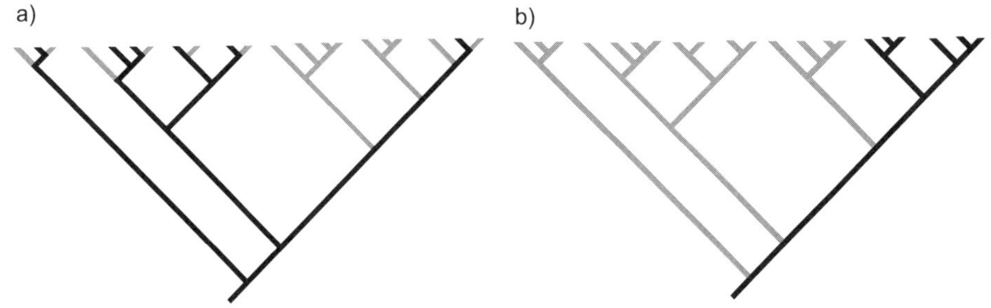

FIGURE 7.5 A schematic drawing of a phylogenetic tree for cyclamen species from the study of Yesson and Culham (2006). (A) Dark branches indicate species relatively unaffected by climate change and lighter branches indicate those affected. This pattern corresponds to that reported by Yesson and Culham (2006), where the dispersion of six persisting species implies that a large amount of PD persists. (B) This tree presents a hypothetical contrasting result, where the same number of persisting species is phylogenetically clumped, with the result that the loss of PD, for the same level of species loss, is much greater. A major challenge for research on the affect of climate change will be to determine whether potential loss of PD, and thus evolutionary potential, is large or small.

each species will go extinct. The expected diversity retained is given by the sum, over all possible combinations of species C, of P(C) × D(C), where P(C) is the probability that combination C will not go extinct, and D(C) is some measure of diversity of combination C. Witting and Loeschcke (1995) applied Weitzman's expected diversity formula with PD as the nominated diversity measure, D, providing a framework for estimating *expected PD* given extinction probabilities for the different species (Box 7.1).

Box 7.2 illustrates basic calculations for expected PD. Expected PD makes sense when interpreted as a statement about expected feature diversity. The expected PD corresponding to any assignment of extinction probabilities to species could be calculated as the sum of the probabilities of presence of all features. Because all features are not directly observed, this summation is made over branch lengths, under an assumption that an appropriately chosen branch length is proportional to the number of derived features along that lineage, with features then shared by all descendants (see Faith, 1992b). For a given branch, the probability of presence, PP, of any corresponding feature, j, is one less its probability of loss or extinction. This probability of loss is equal to the product of the extinction probabilities of all descendant species (assuming independence). The sum of the probabilities of presence over all the features represented by that branch is the branch length multiplied by PP. The total expected PD is the sum of these values over all branch lengths.

Box 7.2 also illustrates how changes in the probabilities of individual species can change the overall expected PD (thus providing the probabilistic version of PD complementarity). Therefore, a change from some current probability of extinction to a probability near zero, as a result of conservation action targeted to that species, may yield a large or small change in expected PD depending not only on the initial probability, but also on the degree to which other species are closely related, and the degree to which they are vulnerable to extinction. Priority setting may therefore look for opportunities to increase expected PD, either through a single species action or through a set of species that collectively create a gain in overall expected PD.

This rationale for using expected PD suggests that, if a species of concern is part of a phylogenetic clump of related endangered taxa, then its priority should reflect the potential to lose deeper branches when compared with a species whose closest-related taxa are mostly secure. It is apparent that the status of other species matters, but when setting priorities for individual species, the nomination of estimated extinction values for the other species is worth considering very carefully. One approach is simply to use current extinction probabilities for all species. However, this maximization of expected

BOX 7.2 Calculating Expected PD

Calculations are illustrated for a hypothetical tree for species A, B, and C; and branch lengths x, y, and z (Box Figure 7.1). The probability of extinction of species A, for example, is designated as pA.

Expected PD (see also Box 7.1) can be calculated as the sum, over all branches, of the branch length multiplied by its probability of persistence. The probability of persistence for a given branch is one less the product of the probabilities of extinction of all species that are descendants of that branch.

For this simple example,

$$\text{Expected PD} = (1 - pA)x + (1 - pB)x + (1 - pA \times pB)y + (1 - pC)(x + y) + (1 - pA \times pB \times pC)z$$

PD complementarity ("gain") can be defined, for example, for the case in which an initial extinction probability changes to 0.0 as a result of conservation action.

$$\text{A gain} = pA \times [x + pB \times y + pB \times pC \times z]$$

$$\text{B gain} = pB \times [x + pA \times y + pA \times pC \times z]$$

$$\text{C gain} = pC \times [x + y + pB \times pA \times z]$$

Examples

Suppose that pA is 0.3, pB is 0.666, and pC is 0.0. Further suppose that branch lengths for x, y, and z are all 1.0. Then the PD gain for A is 0.3 times [1.0 + 0.666(1.0)] = 0.5. This example also can be reevaluated under a risk aversion approach. Using a "min/max" risk analysis, consider the case in which the high probability of extinction of B in the worst case implies its loss. Setting pB = 1.0, the gain in expected PD for protection of A would now be 0.3 times [1.0 + 1.0(1.0)] = 0.6. The min/max approach means that a scenario as described here now implies a greater priority for species A. For further discussion, see main text and Faith (in press).

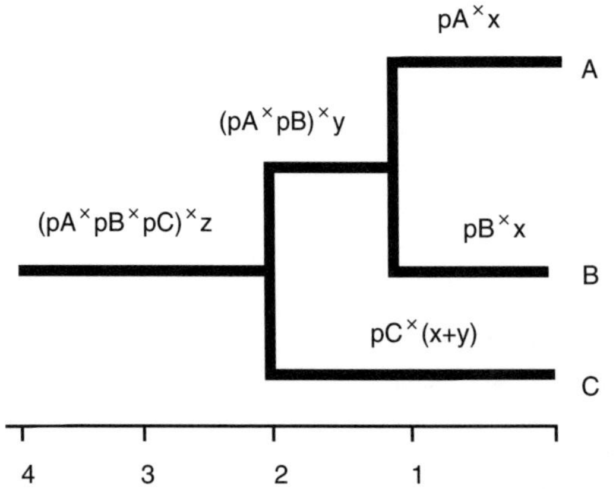

Box Fig 7.1 Hypothetical cladogram for illustrating how to calculate PD.

PD may not be the best way to minimize risks. The problem is analogous to that recognized for simple maximization of the expected number of species in conservation planning (for example, see O'Hanley et al., 2007). Any expected PD value, given the associated probabilities, in fact allows for a range of variation around that expected value. However, we may want to avoid the possibility of especially low PD values, given the irreversibility of PD losses.

Standard risk analysis can provide a more conservative approach to conservation strategies that will place a premium on preventing especially high PD losses. Box 7.2 presents an example that adopts a "min/max" approach (for example, see O'Hanley et al., 2007) for application to the PD context. Specifically, we seek to minimize the maximum loss of PD under worst-case patterns of species loss. In this example we assume that highly vulnerable species B is lost in the worst case. Loss of B would mean that taxon A is now the only representative of a deeper branch. Under the risk exposure-based analysis, we would now give a higher priority to protection of species A (Box 7.2).

It is important, during these assessments, that PD complementarity be taken into account, both in selecting individual species for action and in identifying sets of priority species. There are several recent high-profile proposals for phylogeny-based priority setting, intended to take into account both probability of extinction and potential loss of evolutionary history (Isaac et al., 2007; Redding & Mooers, 2006). However, these approaches largely ignore the longstanding probabilistic PD methods, and fail to take complementarity into account. For example, the EDGE program (short for *evolutionarily distinct and globally endangered*; [see Isaac et al., 2007]) uses a static apportioning of the phylogeny to different species to create a simple scoring system.

In Box 7.3, an example contrasting EDGE-type scores and PD is described. Such PD complementarity scores recognize that securing species C conserves its long ancestral branch, which has no protection from other species. At the same time, PD scores show that protecting more vulnerable species A would in fact imply a smaller gain in expected PD, because its long ancestral branch is already well conserved through secure species B. In contrast, the EDGE-type score gives high priority to A at the expense of C. This preference arises because the method ignores the security of A's close relative B. The failure to take extinction probabilities of sister taxa into account is a serious shortcoming of EDGE-type indices (a similar problem arises in the method of Redding and Mooers [2006]).

The EDGE program is attracting worthy support for conservation action for threatened and evolutionarily distinctive species. However, in largely ignoring the longstanding framework for integrating PD and extinction probabilities, the program may misdirect scarce conservation funds and, worse yet, unnecessarily condemn portions of evolutionary history to extinction.

THE SURROGATES PROBLEM: INFERENCES BEYOND THE OBSERVED SET OF TAXA

PD definitions and calculations, as illustrated in Box 7.1 and Figure 7.1, make sense in the context of a given phylogenetic tree. But our interest in *overall* biodiversity patterns requires a broader focus on (1) PD values over several trees and (2) inferences about more general PD gains and losses based on PD calculations for observed taxa—the surrogates issue. Such inferences are the basis for PD contributions to conservation planning and priority setting for overall biodiversity. This extends the role of PD for setting priorities for a given taxonomic group, to the broader role of providing surrogates information for overall biodiversity.

As background, note that a family of computer-based methods known as *systematic conservation planning* (see, for example, Margules & Pressey, 2000) relies on surrogates information for strategies ranging from selection of new protected areas to the targeting of conservation payments to private landowners. Surrogates approaches must make the best possible use of our current knowledge base so that the known can somehow speak for the unknown.

Effective surrogates in conservation planning require congruence among taxonomic groups in complementarity and endemism values (*predicting complementarity* [see Williams et al., 2005b]). These are the values that are assessed for localities in conservation planning exercises. For effective surrogates strategies, species complementarity values of different localities for one set of species predict general complementarity patterns.

PD already has well established links to location-based systematic conservation planning methods (Faith, 1994; Faith & Walker, 1996a; Moritz &

BOX 7.3 EDGE-Type Species Scores Compared with Phylogenetic Diversity*

For comparison with Box 7.2, the corresponding formulas for gains for an EDGE-type scoring are as follows:

A score = pe(A) × [x + 0.5y + 0.33z]
B score = pe(B) × [x + 0.5y + 0.33z]
C score = pe(C) × [w + 0.33z]

Example

Using the tree shown in Box 7.2, we assume y about 10, x = 1, z = 100, and

pe(A) = 0.8
pe(B) = 0.01
pe(C) = 0.7

1. For phylogenetic diversity, we obtain the following:
    - A gain = 1.44
    - B gain = 0.65
    - C gain = 8.26
    - C is of much higher priority.
2. For EDGE-type scoring, we obtain the following:
    - A gain = 31.44
    - B gain = 0.393
    - C gain = 27.51
    - A is of highest priority.

Phylogenetic diversity gain scores recognize that securing species C conserves its long branch. Phylogenetic diversity gain scores also recognize that protecting species A, while addressing a more vulnerable species, yields a smaller gain because the long branch of length 10 is already well conserved through secure species B. In contrast, the EDGE-type score gives arguably undeserved high priority to A at the expense of C. It does so because it ignores the security of A's close relative B, and also because its fixed multipliers of branch lengths do not reflect actual probabilities.

*Definitions follow Box 7.2.

Faith, 1998). New localities contribute additional amounts of PD to that already represented by a given set of localities. Such PD complementarity values can be used in real-world planning methods that incorporate economic trade-offs and other considerations. However, although it is clear that PD complementarity values can be used in systematic conservation planning methods, and that this can provide different information relative to species-level complementarity (Fig. 7.3) (Forest et al., 2007), it is less clear whether PD complementarity improves surrogacy.

Surrogacy has power relative to simple species counting, because PD gains for localities may reflect area relationships mirrored in phylogenies of other groups. Thus, historical relationships among

BOX 7.4 DNA Bar Coding

DNA bar coding is defined as the use of a small, standardized portion of DNA sequence for the purpose of species identification and discovery. A DNA bar code based on COI has been proposed as a way to boost the discovery of species dramatically, and the documentation of species' geographic distributions information (see The Consortium for the Barcode of Life at www.barcodinglife.com). Such an approach assumes that this standard region of a gene can distinguish among different species over a wide range of taxonomic groups. The Barcode of Life Data System (Ratnasingham & Hebert, 2007) illustrate the potential for an extensive bar coding database for biodiversity assessment, with links to species names, museum voucher specimens, phylogenetic relationships, and geographic distribution information.

areas captured by PD calculations suggest improved prediction of general biodiversity patterns (Faith, 1992a; Faith et al., 2004b; Moritz & Faith, 1998).

A case study based on freshwater macroinvertebrates in rivers of New South Wales (N.S.W.), Australia (Baker et al., 2004; Faith & Baker, 2006), illustrates this potential. PD applications in N.S.W. have extended work establishing patterns of distribution of freshwater macroinvertebrates in the Sydney water supply catchment region of southeast N.S.W. (Baker et al., 2004). Conservation strategies in this region must respond to a number of potential threats to biodiversity, including construction of new dams and mining operations.

One of the taxa studied by Baker and colleagues (2004) are the spiny crayfish, *Euastacus* spp. Most *Euastacus* have highly restricted distributions in localities that are particularly sensitive to habitat disturbance. This research group examined phylogenetic patterns for closely related species from the group using sequence data from the mitochondrial COI gene. They determined that a number of potential species (including newly discovered cryptic species) occurred within the group, each of which has a quite restricted geographic distribution. Thus, different lineages on the *Euastacus* COI tree were represented by only a small number of locations within the region.

In the context of impact scenarios based on recent events in N.S.W., Faith and Baker (2006) examined PD assessments of conservation priorities. Mining activities had affected sites in the upper Nepean River where one cryptic *Euastacus* lineage was found. PD analysis based on *Euastacus* phylogenetic and distribution information suggested a consequent higher conservation priority for another location (the upper Georges River) that is threatened (but not yet affected) by mining activities. This location uniquely had a lineage that was a phylogenetic sister to the lineage from the Nepean. This is shown schematically in the top of Figure 7.6, with lineages labeled according to their unique locality. Faith and Baker (2006) argued that a PD analysis would now assign the upper Georges River locality higher priority because the overall PD loss if *both* lineages were eliminated would now be higher because of the loss of a shared, deeper branch (the extended diagonally striped branch in the bottom of Fig. 7.6).

The patterns for two other invertebrate groups examined by Baker and colleagues (2004) strengthen the potential arguments for higher priority for the Upper Georges locality and thus illustrate congruence in PD complementarity findings. Each of the three trees (based on COI) from their study agreed in the implied area relationships among Upper Nepean, Upper Georges, and Upper Shoalhaven (Fig. 7.6). The PD complementarity observed for *Euastacus* for the Upper Georges populations is predictive of that for other groups. In each case, loss of the Upper Nepean populations implies that the contribution to PD from the Upper Georges taxa is now greater. Furthermore, likely species designations for these taxa would not appear to provide the same congruent complementarity values (Faith & Baker, unpublished).

A full PD analysis may reveal other congruent patterns, including possible congruence in PD endemism scores for another locality, the Upper Shoalhaven (Fig. 7.6) (Baker et al., 2004).

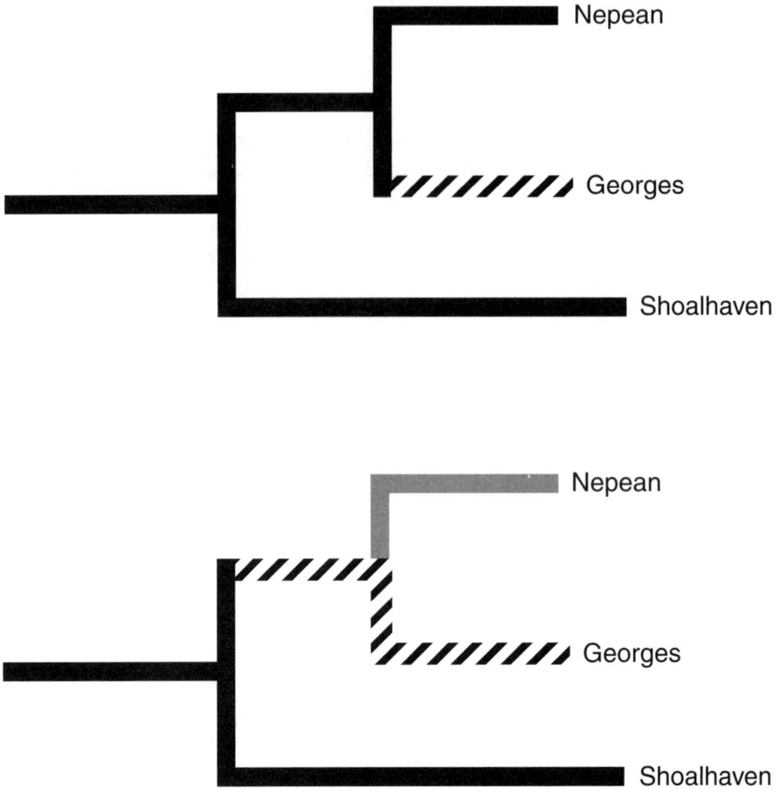

FIGURE 7.6 A schematic drawing of the phylogenetic trees derived by Baker and colleagues (2004) based on gene sequence data for the mitochondrial cytochrome *c* oxidase I gene. The taxa samples are labeled with geographic distribution identifiers and indicate the common pattern over three groups: the spiny crayfish (*Euastacus* spp.) plus leptophlebiid mayflies (*Atalophlebia* spp.) and atyid shrimp (*Paratya* spp.). The human impacts on the Nepean River locality suggests that the PD contribution of the Georges River locality may change from the scenario shown in the top tree drawing to that of the bottom tree (with the PD contribution of the Georges River locality shown in diagonal striping, grey shading indicates loss of the Nepean locality). The PD calculations indicate potential greater conservation priority for the Georges River locality, even in the absence of species designations. The congruent shift in PD complementarity value for the locality over these observed taxa gives some confidence that this may be valid for other taxa as well. For further information, see the main text.

The freshwater macroinvertebrate case study, in using phylogenetic patterns derived from COI, mimics the data that might have been generated through a DNA bar coding program (Box 7.4). The following section discusses the prospects for using PD to take advantage of this potential wealth of new data for assessment of overall biodiversity patterns.

FUTURE DIRECTIONS

DNA Bar Coding and PD

PD and phylogenetic pattern can provide complementarity and related values for conservation planning, without the need for identifying and counting species (Faith, 1992a). Moreover, the use of phylogenetic pattern potentially may boost the prediction of patterns of overall biodiversity. However, any practical use of PD in conservation assessment requires not only good phylogenies but also geographic distribution information for corresponding lineages. Any surrogates approach will perform poorly when there is limited coverage of different taxonomic groups that are confounded by limited geographic distribution information.

Recent work on large-scale DNA bar coding (Box 7.4) suggests fresh prospects to overcome this problem, with more taxa and more geographic information (Hebert et al., 2003b). At the same time, this effort has generated controversy about species delimitation and discovery. For bar coding, a problem of definition arises through the use of a divergence threshold based on COI (or other) sequence differences. Arguably, one divergence threshold is not suitable for all species (Moritz & Cicero, 2004; but see Hebert et al., 2004). At the extreme, it has been argued that bar coding sequence data may fail to distinguish, at any threshold, among "true" species (for a discussion, see Will and Rubinoff, 2004).

PD offers a way for the conservation biologist to take advantage of this wealth of information while at the same time side-stepping species identification problems. A large-scale bar coding program may provide data for more taxonomic groups, sampled at more places, and thus could improve our predictions about differences in places in overall biodiversity. DNA bar coding offers a potential explosion of new taxonomic and distribution data, and PD offers a new way to look at this information.

These considerations suggest a role for PD applied to phylogenetic estimates (or hierarchical patterns) that accompany bar coding programs (Faith & Baker, 2006). A simple example will illustrate how geographic data and associated phylogenetic information from the public Barcode of Life Database (BoLD; based on a 648-base pair section of COI) can be used for rapid PD-based conservation planning exercises. As a simple example using a small data set, a study of arctic Collembola (Hogg & Hebert, 2004) incorporated into BoLD has samples from 19 species, with geographic distribution information summarized as seven localities (Table 7.1). I will assume that the neighbor-joining tree from BoLD (using the Kimura two-parameter distance model and all 54 sequences in BoLD for the 19 species) provides an estimate of phylogenetic relationships suitable for PD analyses.

The PD analysis used a module of an existing "rapid biodiversity assessment" software package (DIVERSITY, v. 2.1) (Faith & Walker, 1996 a&b; Walker & Faith, 1994). This software allows PD-based selection of localities to be combined with costs so that, for example, a set of localities with some target level of total PD and minimum cost can be identified. The software reads in phylogenetic tree and geographic distribution information and then selects localities for conservation priority under different user-defined constraints (Faith & Walker, 1996a).

Suppose first that all seven localities are selected by the software (Table 7.1, set e). The complementarity values of the localities then reflect their PD endemism. Locality 3 has highest PD endemism (184 units). Locality 5, in contrast, has the lowest PD endemism (12 units). The PD analyses in Table 7.1 also describe hypothetical loss of localities (for example, to nonconservation uses). Moving from right to left in Table 7.1, localities are lost from the initial complete set. The new PD complementarity values of members of a set are now generally greater relative to the initial PD endemism value. These results highlight how complementarity values are dynamic in nature.

It is also possible to assess PD gains if nonmembers are added to a given set. For example, given set a, the addition of locality 6 (creating set b) is valuable, given its high complementarity of 107 to the initial set units. Such complementarity values are important in practice because these gains and losses at the "margins" are the values that are compared with costs of conservation (Margules & Pressey, 2000). DIVERSITY software for PD may be used to select a set of localities that maximize overall PD (say for a fixed set size). The algorithm uses dynamic complementarity values in adding members to a set. These values can be compared with costs. Practical analyses using DIVERSITY or other PD software might produce sets of alternative solutions. Conservation planning often considers the implications of different land-use scenarios, and contrasts are made among the corresponding solutions.

The potential advantage of a large-scale DNA bar coding program for biodiversity conservation planning would be that the PD analyses for data over many taxonomic groups would give us more confidence that estimated complementarity values

TABLE 7.1 Arctic Collembola PD Complementarity Values for Sites within Nominated Sets.

Site No.*	Sets of Sites†				
	a	b	c	d	e
1	154	154	154	152	152
3	267	226	226	226	184
7	201	201	201	201	161
6		107	107	41	41
2			47	47	47
4				41	41
5					12

*Sites are listed in the order of selection by algorithm within DIVERSITY v.2.1software (Faith & Walker, 1996b). 1, Somerset Island R11; 2, Somerset Island R9; 3, Cornwallis Island R2, R21; 4, Cornwallis Island R6; 5, Cornwallis Island R7; 6, Igloolik Island I14; 7, Igloolik Island I9, I10, I16. For further information, including the neighbor-joining tree, see BoLD at www.barcodinglife.com/views/login.php and Table 1 in Hogg and Hebert (2004).

†Set b, for example, contains sites 1, 3, 7, and 6; removal of site 3 from that set would reduce represented PD by 226 units. As more member sites are added moving to the right in the table, the complementarity value of an existing member may decrease, which is what happens to site 3 when site 5 is added to the set. PD complementarity values are based on branch lengths expressed in an arbitrary rescaling of the neighbor-joining tree branch length units.

reflect those for overall biodiversity. The freshwater case study described earlier illustrated the degree to which PD locality values from a single phylogeny might indicate values for the phylogenies of other groups (see also Moritz & Faith, 1998). The case for PD bar coding-based planning will be stronger if we can document cases in which calculated PD values predict the PD values for other taxonomic groups.

A limitation in applications of PD to DNA bar coding trees will be the potential poor recovery of true phylogenetic patterns. There is some evidence for congruence in the branch lengths (and phylogenetic pattern) from different character sets relative to COI (see, for example, Baker et al., 2004). There is also the rationale that conservation planning applications using large-scale DNA bar coding will not be worried about the status of any particular tree; only the overlay of many trees will determine complementarity values of localities. This suggests that the hierarchical trees from bar coding are linked to geographic patterns of biodiversity in the same way that characters are linked to cladograms, where any one character may well be misleading, but the overall signal from all characters reveals the underlying pattern. PD may therefore overcome inherent limitations of DNA bar coding and thus help bar coding data to fill the huge gap in information available for biodiversity conservation.

The integration of such biodiversity information into systematic conservation planning may help address the globally recognized 2010 target for a significant reduction in the rate of loss of biodiversity (for discussion, see Faith, 2005). Table 7.1 illustrated how good planning may identify, among all current intact places, those localities with lower complementary values, such that ongoing take-up of non-conservation land use at these localities implies a minimal rate of biodiversity (PD) loss (see also Faith, 2006b).

SUGGESTIONS FOR FUTURE READING

DIVERSITAS, the international organization for biodiversity science, and its core project, Biogenesis (www.diversitas-international.org/), provide current material about projects linking phylogenetics and conservation, including work on PD. *Phylogeny and Conservation* (Purvis et al., 2005) reviews strengths and weaknesses of PD and discusses many related topics (see also a review of that book by Faith [2007]). Mace and colleagues (2003) also review the utility of PD as a strategy to "preserve the tree of life." Weitzman (1998) explores PD in a practical framework called the "Noah's Ark Problem," which not only considers probabilities of extinction but also costs and trade-offs. There also has been important related work on PD computational issues; see Steel (2005) and Minh and colleagues (2006). These papers address the problem of finding optimal sets of species or localities and have begun to include many real-world planning constraints.

Shaw and colleagues (2003) compare PD with other measures, and Andreasen (2005) discusses advantages and disadvantages of PD for setting conservation priorities for threatened taxa, with a particular focus on the representation of genetic diversity. Also, at the genetic level, work by Rauch and Bar-Yam (2004) explores PD-type measures to describe within-species variation and provide a theoretical justification for such patterns.

Andreasen, K. 2005. Implications of molecular systematic analyses on the conservation of rare and threatened taxa: Contrasting

examples from Malvaceae. Conserv Gen. 6: 399–412.
Faith, D. P. 2007. Phylogeny and conservation. Syst Biol. 56: 690–694.
Mace, G. M., J. L. Gittleman, & A. Purvis. 2003. Preserving the tree of life. Science 300: 1707–1709.
Minh, B. Q., S. Klaere, & A. Von Haeseler. 2006. PD within seconds. Syst Biol. 55: 769–773.
Purvis, A., T. Brooks, & J. Gittleman (eds.). 2005. Phylogeny and conservation. Cambridge University Press, Cambridge, UK.
Rauch, E. M., & Y. Bar-Yam. 2004. Theory predicts the uneven distribution of genetic diversity within species. Nature 431: 449–452.
Shaw, A. J., C. J. Cox, & S. B. Boles. 2003. Global patterns in peat moss biodiversity. Mol Ecol. 12: 2553–2570.
Steel, M. 2005. PD and the greedy algorithm. Syst Biol. 54: 527–529.
Weitzman, M. L. 1998. The Noah's Ark Problem. Econometrica 66: 1279–1298.

8

Genetic Considerations of Introduction Efforts

PHILIPPINE VERGEER
N. JOOP OUBORG
ANDREW P. HENDRY

Use of species introduction as a management tool is not yet common practice, but its popularity is growing. However, despite expanding interest, scientific arguments underlying introduction strategies have rarely been investigated in detail. Genetic considerations in particular may play a dominant role. In this chapter we discuss the costs and benefits of several genetically based strategies for introduction. We start by considering different types of introductions, including supplementing a small population, reestablishment of an extirpated population, and founding of an ecological equivalent of an extinct species. We next discuss two major points of concern when considering source material for introduction programs: genetic variation and genetic integrity. We then consider specific introduction decisions that bear on these concerns, particularly with regard to inbreeding depression, outbreeding depression, and local adaptation. Lastly we examine several case studies of introductions and close with a synthesis.

CONCEPTS

Introduction as Restoration Tool

A distinction must be drawn between three different contexts for introduction, for which we use the terms *translocation, reintroduction,* and *rewilding.* The following paragraphs explain each context, and we will use the terms throughout this chapter as defined therein. The term *introduction* is used when a statement can apply to multiple contexts.

Translocation refers to situations in which indigenous individuals are still present at a focal site, but the population size is small or declining. Translocation of individuals from elsewhere is aimed at strengthening the demographic and genetic viability of the focal population (in other words, *genetic reinforcement*). Additional restoration efforts in this context often focus on habitat quality, the number of potential habitat patches, connectivity among remaining populations, and within-population genetic variation. These efforts are broadly aimed at population enlargement, which should counteract the negative effects of small population sizes (discussed later).

Reintroduction refers to situations in which a species has completely disappeared from a focal site but can still be found elsewhere. The goal of reintroduction is to colonize a site that is supposedly suitable but is currently unoccupied, perhaps because the species is unable to recolonize the site naturally. Efforts are aimed at restoring the diversity of local species diversity. After the initial reintroduction has taken place, additional reintroductions fall under the context of translocations, as discussed earlier. Introductions of species in nature development areas (in other words, areas that are made suitable for the species but where the species did not previously inhabit) are here considered a part of the reintroduction context.

We use the term *rewilding* to refer to situations in which a species has gone extinct everywhere. Here the only option for introduction is to use a different species, often one that is either closely related to the extinct species or an apparent ecological equivalent. The goal in this context is often to restore community stability, ecosystem function, or utility to human populations. This context for introduction has been taken to extremes with the idea that contemporary equivalents of extinct Pleistocene megafauna might be reintroduced to North America (Foreman, 2004). Despite the engaging intellectual exercise offered by this last context, our chapter focuses on the more common and realistic scenarios considered in the first two contexts.

The three contexts for introduction provide a profitable backdrop for introducing the concept of genetic integrity, a major theme of our chapter. *Genetic integrity* refers to the maintenance of a natural gene pool, commonly at the species level, but sometimes at the population level as well. When genetic integrity is a priority, introduction strategies must sometimes consider both accuracy and functionality (Clewell, 2000; Falk et al., 2001). High accuracy means that the gene pool of the introduced population well represents the gene pool of the original population. High accuracy is often not possible if the original population went extinct because at least some genotypes would have been unique. In these situations, functionality may be more important. Functionality can be expressed in terms of survival, viability, or persistence; the higher these qualities, the higher the functionality. If functionality is given a higher priority than accuracy, different genotypes from various regions might be the best source material. Final success would then be judged by the number of established and successfully reproducing individuals, rather than their specific genotypes or phenotypes.

Genetic Problems in Small Populations

The conservation value of reintroductions is obvious, but what of translocations into existing populations? One concern here is that small populations may be susceptible to extinction as a result of demographic stochasticity. Another reason is that population size can influence the genetic "health" of a population by determining the strength of random fluctuation in allelic frequencies over time (in other words, genetic drift) and inbreeding caused by consanguineous mating. Both these processes will have larger impacts on population growth rate and persistence (in other words, population fitness) as population size decreases (Leimu et al., 2006; Ouborg et al., 2006; Reed, this volume). Genetic drift will lead to the random loss of alleles, which in turn causes a decrease in genetic diversity and an increase in homozygosity. Although inbreeding does not in itself lead to a loss of alleles, it does cause an increase in homozygosity. Even if there is random mating, homozygosity levels may increase in a finite population as a result of the increase in average relatedness among individuals in finite populations (in other words, *inbreeding effect* [Crow & Kimura, 1970]). Elevated homozygosity may cause increased expression of recessive alleles. Because these recessive alleles often have deleterious effects, average fitness of the population may decrease, a phenomenon known as *inbreeding depression* (Barrett & Kohn, 1991; Young et al., 1996). As predicted by these arguments, several studies have revealed negative relationships between population size and genetic variation (overviews by Booy et al., 2000; Vergeer et al., 2003; Young et al., 1996), and between population size and average fitness (see, for example, Fischer & Matthies, 1998; Ouborg & Van Treuren, 1995; Vergeer et al., 2004).

Processes such as habitat fragmentation or a reduced propensity for dispersal may cause declines in population size and increased genetic isolation (Fahrig, 2003). Increased isolation of small populations can have two important consequences. First, gene flow among populations will decrease, which in turn reduces the chance of *genetic rescue*—a process that might otherwise alleviate the negative effects of drift and inbreeding (Ingvarsson, 2001). Second, because isolated populations will be demographically and evolutionarily independent, genetic drift will cause an increase in among-population differentiation. Increasing diversification can, in theory, be beneficial, detrimental, or neutral, and thus translocations intended to increase gene flow deserve careful consideration (Hedrick, 1995).

Especially in small populations of randomly mating individuals, the level of inbreeding is expected to increase through time. Thus, incrementally decreasing fitness might have a serious effect on population persistence. Managers charged with protection and recovery of small and fragmented populations therefore need to consider strategies that might increase population size, reduce inbreeding, and possibly restore gene flow. The obvious strategy is to undertake some sort of translocation, for which the selection of source materials (in other words, animals, plants, or seeds) becomes a critical consideration. Many strategies are possible and each might have different effects on long-term population viability. Selection of source materials therefore requires consideration of both genetic variation and genetic integrity, as we will now discuss.

Genetic Variation

Genetic variation can be greatly influenced by the genetic composition of source materials and any remnant native individuals, as well as by gene flow with adjacent populations. Gene flow is generally expected to be less important in this context, simply because suitable unoccupied sites are often not located within the dispersal range of other populations. And even if they are within dispersal range, Allee effects or other stochastic factors may prevent natural colonization. Genetic variation is therefore expected to reflect the original sources, coupled with any subsequent drift and inbreeding. Thus, the genetic source of individuals used for introductions therefore demands substantial deliberation.

One major consideration is whether to increase genetic variation in the area under study by using a mixture of individuals from each of several nonlocal source populations. One benefit of this approach is that the probability of introducing related individuals is lower and thus using a mixture of nonlocal genetic sources decreases probability of inbreeding. Another benefit is that a genetically diverse population may include individuals adapted to different conditions and environments, enabling the new population to survive under a wider range of conditions. Genetic variation may also enhance adaptation to changing conditions that may typify highly fragmented or deteriorated landscapes. High levels of genetic variation can therefore be important for the establishment of a self-sustaining population.

A caveat to this assertion is that nonadaptive genetic variation might negatively affect the population by increasing genetic load. This effect will be considered in more detail later. Another caveat is that population persistence in a new or changing environment may be more closely tied to phenotypic plasticity than to genetic variation (Price et al., 2003). And yet, a loss of genetic variation may coincidentally reduce the potential for phenotypic plasticity. Current experimental data, however, are insufficient to distinguish between two possible scenarios: either plasticity buffers against the effects of loss of genetic variation on fitness, or the capacity for plasticity itself is affected by this loss.

Another major consideration is the possibility of founder effects that can occur when small numbers of individuals are used to establish a new population. For practical reasons, it is uncommon that large numbers of individuals are introduced (at least for low-fecundity animals or highly threatened plants), and so founder effects are plausible. In such cases, a reintroduced population may start with low levels of genetic variation regardless of its source. Even when many genetically diverse individuals are reintroduced, the population may experience a genetic bottleneck as a result of an initial reduction in population size. Indeed, reintroductions are likely attended by high mortality, particularly in the early generations, and so only a small fraction of the reintroduced individuals may actually survive. Hence, consideration of genetic variation does not end after choice of source materials has been made.

Genetic Integrity

Genetic integrity can be disrupted when genetically dissimilar individuals are translocated into an existing population. Many conservationists now favor high accuracy in translocations by aiming to keep the restored population as similar as possible to the original genetic stock (Knapp & Rice, 1994). However, it is not always possible to achieve high accuracy because of a lack of local material, degraded site conditions, or financial constraints. In addition, high accuracy may increase the risk of inbreeding and reduce genetic variation, especially in small and isolated populations. In this case, it may be better to forgo high accuracy and instead use one or more nonlocal source populations that can genetically rescue a small and inbred population from inbreeding.

Genetic rescue has been demonstrated in several instances of translocation (Hedrick, 1995; Madsen et al., 1999; Westemeier et al., 1998; Willi & Fischer, 2005). In these situations, the increase in fitness is thought to be mainly the result of interbreeding between genetically different individuals, which has the effect of increasing average heterozygosity. Increased heterozygosity may increase fitness because deleterious recessive alleles may become hidden in heterozygotes, or because heterozygotes just generally have higher fitness than homozygotes (Tallmon et al., 2004). Translocation strategies emphasizing high levels of within-population variation, rather than genetic integrity, might therefore be preferred in small, inbred populations. It also may be possible to avoid inbreeding and maintain some genetic integrity simultaneously by translocating individuals from different populations that are adapted to similar environments. The likely benefits of this strategy will be a recurring point of discussion throughout this chapter.

Despite the previous arguments, it is important to recognize that low levels of genetic variation are not always a problem, and that inbreeding does not always have negative effects (Hamrick & Godt, 1996; Prodöhl et al., 1997). Furthermore, low levels of genetic variation do not necessarily indicate inbreeding. For many species, rarity is a normal condition, and these species may have long occurred in small populations with low levels of genetic variation (Huenneke, 1991). Such populations may have purged recessive deleterious mutations as a result of a long history of inbreeding. Moreover, naturally rare species may be highly adapted to their specific local environments. Populations on isolated mountaintops are a classic example, being both small and genetically divergent from other populations. To generalize, a population in a stable environment may have a fully adapted, but relatively narrow, range of genotypes. In these cases, high levels of genetic variation may not be needed and may even reduce population fitness (Lande & Shannon, 1996). Such populations might thus benefit by maintaining a narrow range of genotypes that are adapted to the local conditions.

A consideration of genetic integrity thus indicates that increasing genetic variation should not be the only focus of restoration projects. Indeed, it will not always be the best strategy to restore small and isolated populations. Increased within-population genetic variation may thus result in either increased or decreased fitness, depending on the degree of local adaptation, environmental heterogeneity, and current and past inbreeding.

Outbreeding Depression

Outbreeding depression can occur when genetically dissimilar individuals interbreed. One form of outbreeding depression occurs when individuals are genetically incompatible irrespective of the specific ecological environment. This effect can arise when unconditionally incompatible alleles become fixed in different populations. Genetic incompatibilities of this sort are usually not considered in restoration programs because it is thought to require long periods of isolation and an absence of gene flow (as in allopatric speciation by the Dobzhansky-Muller process [Coyne & Orr, 2004]). However, it is at least theoretically possible that genetic incompatibilities can arise in only hundreds of generations (Gavrilets, 2003). An interesting property of this form of outbreeding depression is that it can occur between populations adapted to similar environments (Coyne & Orr, 2004; Gavrilets, 2003).

Another form of outbreeding depression more commonly considered in the context of restoration occurs when individuals from different populations perform poorly in each other's environment because of genotype-by-environment interactions (in other words, local adaptation). In this case, individuals transferred between environments and any resulting hybrids will have an adaptive disadvantage and therefore low success. This sort of incompatibility can evolve on very short time frames and in the presence of gene flow (Hendry, 2004). Despite reduced success of introduced individuals in this context, some of their maladapted genes may still introgress into the native population, causing a breakdown of local adaptation and a decrease in population fitness (Fenster et al., 1997). If only half the alleles in the first generation of offspring (F_1) come from the local environment, breakdown in adaptation may become apparent immediately (Hufford & Mazer, 2003). Alternatively, if recombination disrupts epistatic interaction between locally adapted alleles, breakdown may not occur until later generations (F_2 onward). In either case, the introgression of maladaptive genes can dramatically reduce local population viability—perhaps even to the point of extinction (Boulding & Hay, 2001; Tufto, 2001).

Outbreeding depression can seriously affect introduction success, as well as the viability of any remnant native population. Therefore, any introduction of nonlocal individuals should be considered carefully. In some cases, the risks of outbreeding depression appear to overwhelm the potential benefits of nonlocal source material (Hufford & Mazer, 2003; Keller et al., 2000). Unfortunately, no unassailable generalizations are possible. One possibility is to reduce the risk of outbreeding depression and preserve genetic integrity by using local material only, such as from a remnant population, a seed bank, or a captive population. In the latter case, however, one would hope that the captive population had not adapted to the captive environment, thereby reducing its adaptation to natural conditions. Moreover, the use of local material may increase the risk of inbreeding and, potentially, inbreeding depression (as discussed earlier).

A promising compromise between the benefits and costs of local versus nonlocal material is to reintroduce individuals from different populations that are adapted to similar environments (in other words, similar ecotypes). Although this strategy might simultaneously reduce the possibility of both inbreeding and outbreeding depression, it is still attended by some caveats. One is that seemingly similar ecotypes from different locations may differ cryptically in fitness-related traits, or they may manifest genetic incompatibilities. As an example, odd- and even-year pink salmon lineages (the species has a strict 2-year life cycle) that are adapted to the same stream show substantial fitness reductions when F_2 hybrids are artificially generated and released back into the stream (Gharrett & Smoker, 1991).

Local Adaptation

We have alluded several times to the importance of local adaptation, and it now seems profitable to consider this topic in greater depth. Local adaptation is generated by strong divergent selection between environments, coupled with low gene flow (Kawecki & Ebert, 2004). For this reason, populations isolated in stable environments may maintain adaptively low levels of genetic variation. Introduction of individuals from populations adapted to different environments might then reduce adaptation and thereby decrease fitness. Conservationists are here confronted with the dilemma of restoring within-population genetic variation at the expense of maintaining local adaptation. In this scenario, the origin and nature of source material becomes critical. Even when local material is preferred as a result of strong local adaptation, it is usually unknown which alternative populations can be considered sufficiently local in origin. Unfortunately, there is no unambiguous determination of local and nonlocal, either in general or for specific species.

Locality may be related to geographic scale, with populations closer to each other having more similar genetic characteristics. From an ecological–genetic point of view, however, locality may be more closely associated with adaptation to ecological factors that may vary consistently with distance. On the one hand, ecology and distance might coincide in the case of broad and reasonably continuous spatial gradients such as latitude, altitude, or climate. On the other hand, adaptation may be correlated with patchily distributed habitat types. In this case, more distant populations in similar habitats may respond as if they were more local than nearby populations in different habitats. In this situation, locality may be better defined based on specific habitat features. Because local adaptation can seriously affect introduction success, insight into the process of local adaptation within a given species is crucial.

Considering this theme in more detail, Hufford and Mazer (2003) have argued that proper decisions about the source material for introductions depend on the genetic processes responsible for population differentiation. In a heterogeneous environment, isolated populations may be genetically differentiated for at least two reasons. The first is genetic drift proceeding independently in each population. The second is natural selection leading to local adaptation. In the case of an introduction, the two underlying causes of population differentiation may lead to very different effects, particularly depending on whether native individuals are still present at the introduction site.

First consider the translocation of individuals into a remnant, native population with reduced levels of within-population genetic variation. When population differentiation is caused by genetic drift, hybridization between translocated and native individuals may often lead to an increase in offspring fitness (in other words, heterosis). An exception, of course, is when genetic incompatibilities have accrued, although this is reasonably unlikely on small spatial and temporal scales. On the other hand, when population differentiation is caused by local adaptation, hybridization might lead to outbreeding depression and thus a decrease in offspring fitness. In these situations, it is important to determine the cause of low genetic variation in the recipient, native population. If the within-population genetic variation has always been adaptively low (which is more likely in the latter case), an increase in genetic variation may have detrimental effects by reducing local adaptation.

Next consider reintroduction into a site where the native population has been extirpated. If a variety of source populations are used, then the previous considerations about heterosis and outbreeding depression may still hold. When a single source population is used, hybridization issues are less relevant. In this case, differentiation resulting from drift may not have a large impact on reintroduction success, provided introduced individuals have sufficient genetic variation. However, differentiation resulting from local adaptation will not cause hybrid breakdown, but the introduced individuals might be less fit in the new site than in their native site (unless fitness is strongly density dependent). As already discussed, this problem might be mitigated by using individuals adapted to similar environments, or simply by a genetically diverse population of individuals that could rapidly adapt to the new environment. A possible trade-off in this last strategy is that substantial time and money may be required to maintain the population during the period of intense selection.

Although the theoretical predictions are well worked out, it is often difficult in practice to distinguish between different causes of population differentiation. Such efforts usually require insight into both genetic drift and local adaptation. A good approach is to combine surveys of (neutral) molecular markers that can be used to infer genetic drift, neutral genetic variation, and gene flow, with transplant studies that can be used to infer local adaptation, genotype–environment interactions, and quantitative genetic variation. Such experiments, however, are often not possible, and so generalizations would be helpful. Such generalizations are difficult, however, because of variable results in existing studies. For example, between-population crosses in plants sometimes reveal heterosis (see, for example, Fenster & Galloway, 2000; Ouborg and Van Treuren, 1994; Van Treuren et al., 1993) and other times reveal outbreeding depression (see, for example, Keller et al., 2000; Sheridan & Karowe, 2000; Waser et al., 2000).

It has been suggested, for plants at least, that there might be an optimal outcrossing distance at which the benefits of heterosis and the cost of outbreeding depression offset each other and generate an optimal net effect on fitness (Waser & Price, 1994). It is difficult to draw a general conclusion, however, because the various studies were undertaken with very different perspectives and using different definitions of locality. Moreover, some studies only investigated the immediate effect in F_1 progeny, whereas others follow fitness over additional generations. Unfortunately, it is still not clear to what extent heterosis or outbreeding depression may affect the success of introductions and, if both effects occur together, which might be stronger. Studies analyzing the risks of outbreeding and inbreeding depression are therefore strongly encouraged.

CASE STUDIES

The following case studies are provided to illustrate how some introduction efforts have dealt with genetic considerations.

Our first case study represents a situation in which a local population has been severely depleted and is then supplemented by individuals from a nonlocal population. This is the case of the Florida panther, *Puma concolor coryi*. Reduced to a population size of less than 50 individuals, the Florida panther showed very low levels of genetic variation (Culver et al., 2000), as well as a high frequency of individuals with deleterious traits, including undescended testicles, kinked tails, and poor sperm quality (Roelke et al., 1993). A likely origin of these problems was inbreeding depression caused by recessive deleterious mutations. It was suggested that these mutations might be masked after interbreeding with introduced nonlocal panthers, for which the most suitable source appeared to be a related subspecies from Texas. To evaluate the potential genetic costs and benefits of this plan, Hedrick (1995) used population genetic models to estimate the amount of gene flow from Texas cougars that would reduce inbreeding depression without causing substantial outbreeding depression. The best trade-off was predicted to occur at 20% gene flow in the first generation followed by 2.5% each generation thereafter. Based in part on Hedrick's analysis, eight female Texas panthers were translocated to Florida in 1995. Five of these females are known to have mated with Florida males and to have produced viable F_1 and F_2 offspring (Land & Lacy, 2000). After these translocations, the population now has a lower frequency of undescended testicles and kinked tails, as well as improved sperm quality. This example shows how severely inbred populations can be rescued by genetic supplementation from related populations.

Our second case study represents a situation in which a local population was extirpated and nonlocal populations of the same species were used to found a new population at the same site. Although we could have reviewed a well-known restoration

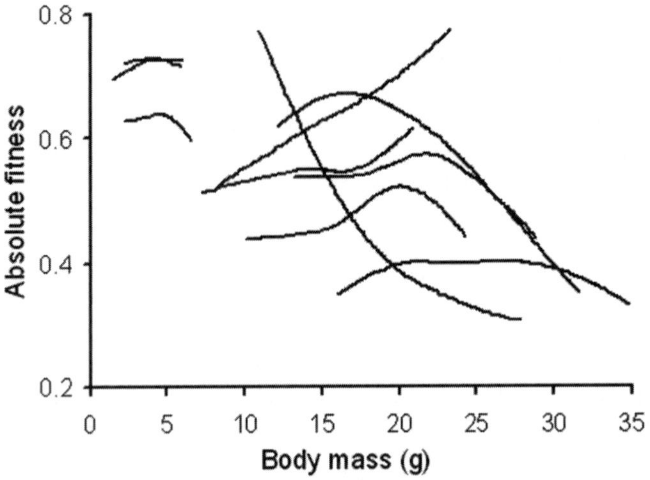

FIGURE 8.1 Viability selection acting on body mass in a reintroduced population of Connecticut River Atlantic salmon studied using mark–recapture procedures in a particular tributary stream. Different lines represent cubic splines relating absolute fitness (survived = 1; died = 0) to body mass for various combinations of cohort, season, and age class. If general trends are difficult to discern, this is the point: Variation among lines shows that consistent directional selection is lacking. Selection coefficients for these and other traits and samples were generally low, even in comparison with natural populations. (After Hendry and colleagues [2003].)

program, such as Yellowstone wolves, black-footed ferrets, or California condors, we will instead introduce the reader to a different story. This is the case of Connecticut River Atlantic salmon (*Salmo salar*) in the northeastern United States. Most salmon have an anadromous life cycle during which breeding takes place in freshwater, most juveniles migrate to the ocean, and maturing adults then return to breed in freshwater. The Connecticut River was historically home to a robust population of salmon that was nevertheless extirpated by 1798 (Gephard & McMenemy, 2004). Extirpation was largely the result of harvesting, pollution, and dams. Restoration efforts began in 1967, and by 1993 fish had been reintroduced from at least 18 different source rivers (although most came from the Penobscot River in Maine [Rideout & Stolte, 1989]). Introductions ceased by 1994 and the population has since been maintained through stocking of hatchery-reared juveniles that are either offspring or grandoffspring of adults that returned naturally to the river (Gephard & McMenemy, 2004). Less than 100 fish return on average each year despite massive stocking efforts—a poor showing that has motivated a continuing investigation of factors that might compromise success of restoration efforts. Dams have been made passable, harvesting is rare, and pollution is greatly reduced, which leaves the possibility that genetics may be the source of the problem.

A series of studies have now considered how genetic factors might influence restoration of Connecticut River Atlantic salmon. No definitive solution has yet emerged, but the various approaches are instructive. As the Florida panther case study illustrates, inbreeding depression is frequently a concern in restoration efforts that include only a few breeding adults. For Connecticut River Atlantic salmon, genetic variation likely started high, because a heterogeneous set of source material was introduced. However, the returning population is very small, and thus inbreeding is strongly implicated.

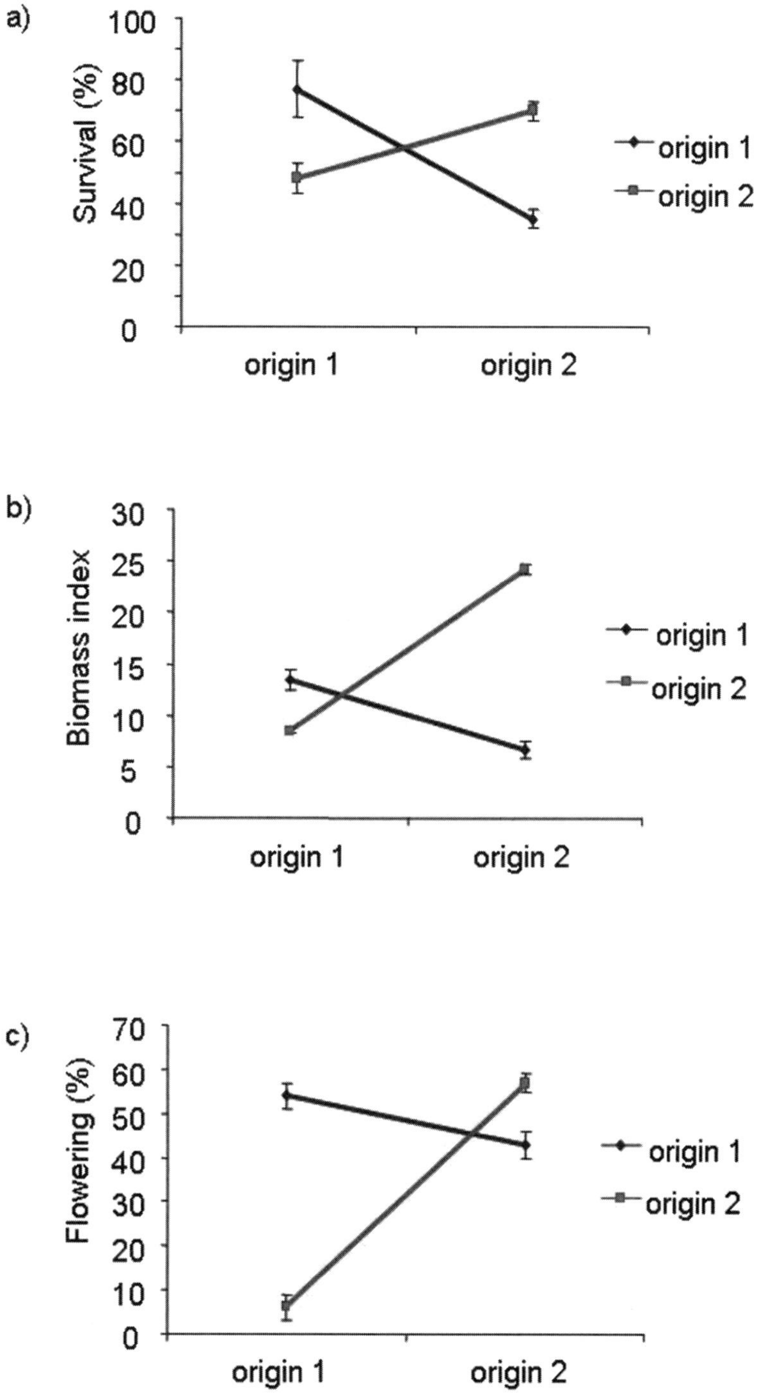

FIGURE 8.2 (A–C) Mean survival percentage (A), mean biomass index (B), and mean flowering percentage (C) of *Succisa pratensis* planted reciprocally at two sites. The data shown are from 1.5 years after transplantation. Biomass index was calculated by multiplying the number of leaves by length of longest leaf by width of widest leaf.

A breeding program was therefore designed to avoid mating between relatives. All returning adults are genotyped at microsatellite loci and genetically similar individuals are not mated together (Letcher & King, 2001). This program maintains genetic variation at healthy levels and reduces the chance of inbreeding. However, the population has not yet turned the corner to sustainability.

Reintroduced fish came from nonlocal populations located further north; thus, another concern is maladaptation of recovering stocks. Concern over maladaptation is reinforced by the observation that salmon typically demonstrate strong adaptation to local spawning environments (Quinn, 2004; Taylor, 1991). If maladaptation was indeed a problem, one might expect strong selection on Connecticut River fish in the wild, as well as the beginnings of phenotypic divergence from the source populations. Hendry and colleagues (2003) tested the first prediction by using mark–recapture experiments to examine selection on size, condition, and growth of stocked juvenile salmon in a Connecticut River tributary. Although some slight trends were evident, strong and consistent directional selection was absent (Fig. 8.1). This finding suggests several possibilities: (1) the specific study tributary was anomalously benign, (2) Connecticut River fish are now well adapted, or (3) selection is acting on other traits. Regarding the second possibility, Obedzinski and Letcher (2004) used a common-garden experiment to test whether Connecticut River and Penobscot River juveniles show genetic differences in developmental timing, growth rate, and development rate. They found that some traits had diverged between populations, suggesting that at least some adaptation may have taken place since introduction.

Future work on Connecticut River Atlantic salmon will examine selection and adaptation for other traits, as well as other explanations for the slow recovery. In particular, acidic streams or reduced quality of ocean-rearing conditions may be strongly hampering recovery. If so, the important lesson may be that although genetic concerns are important, the best laid plans and efforts can fail in the face of unforeseen ecological changes.

Our third case study concerns reintroduction of the devilsbit *Succisa pratensis* Moench in the Netherlands. *Succisa pratensis* is a perennial, self-compatible species that typically inhabits nutrient-poor grasslands and heathlands. Although this species is still rather common in the Netherlands, its range has decreased by 74% since 1935 (Van der Meijden et al., 2000), and it is disappearing from many of its traditional habitats. In hopes of preventing further decrease, reintroduction was suggested as a restoration tool. An experimental study therefore investigated the effects of various possible source materials. At first, possible effects of local adaptation in two potential source populations were tested by means of a reciprocal transplant experiment between two field sites. Figure 8.2 shows that plants from site 1 performed better in terms of survival, biomass, and flowering percentage when transplanted to their original site (site 1) than when transplanted to the other site (site 2). The expected reciprocal results were observed for plants originating from site 2. Statistically significant origin-by-site interaction reveal a home-site advantage, and thus suggested local adaptation.

Experimental studies were then conducted to test the relative importance of inbreeding, heterosis, and local adaptation. Several artificial populations were created: one from a local population, one from a nonlocal population that was small in size, one from a nonlocal population that was large in size, and one from a mixture of nonlocal populations (both small and large). These treatments were then planted in a field at similar census population sizes. However, effective population sizes (N_e) were varied by adjusting relatedness of individuals and therefore the level of inbreeding: Each artificial population received a self-pollination and a within-population cross-pollination. Eighteen months after planting, the performance of all treatments and fertilization types was compared (Vergeer et al., 2004). Germination was significantly affected by type of pollination, being much lower in self-pollinated plants that were expected to be inbred (Fig. 8.3). Effects of source material origin (the four categories just mentioned) were evident for biomass, which was much higher for plants of local origin (Fig. 8.4). Significant effects of both origin and original population size were observed for the proportion of plants that flowered (Fig. 8.5). Here an effect of population size was evident in the reduced flowering of nonlocal, small populations relative to nonlocal, large populations. Furthermore, plants originating from the local population and from a mix of nonlocal populations showed higher flowering percentages than plants from single, nonlocal populations, large or small.

These results show that (1) an outbred, local source is the best to use for reintroduction; (2) performance is enhanced by increased genetic variation, which is strongly influenced by origin and relatedness among source materials;

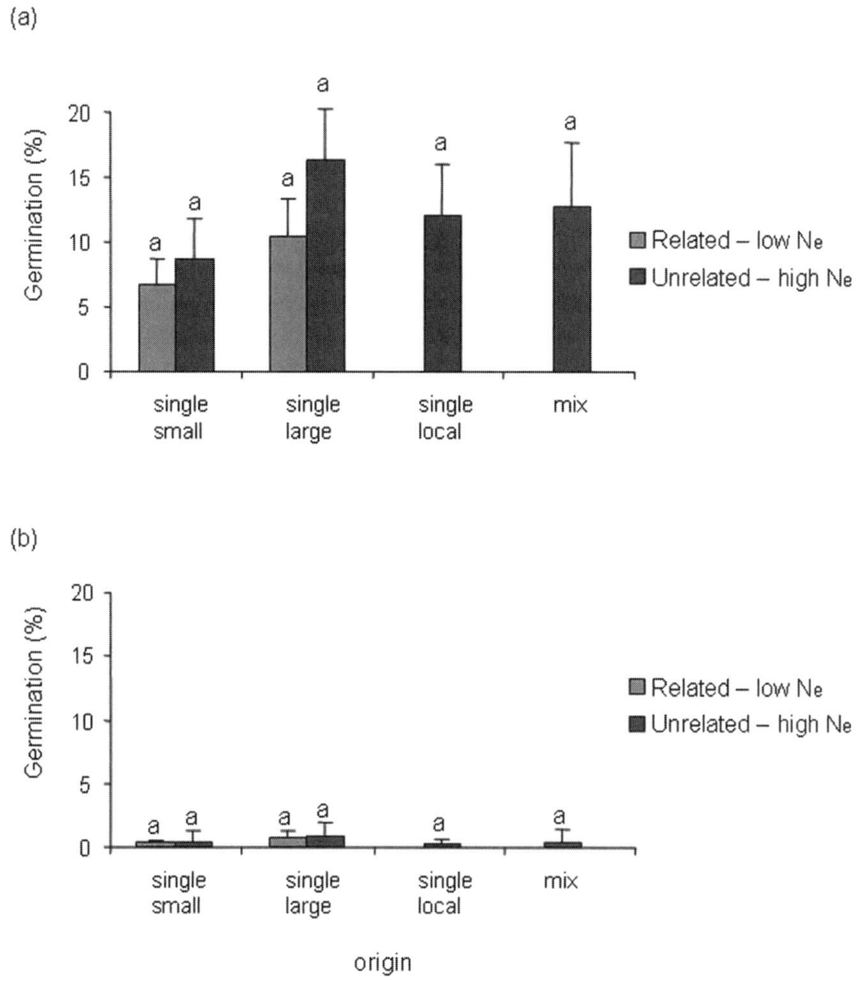

FIGURE 8.3 (A, B) Mean germination of *Succisa pratensis* in relation to origin (nonlocal small, nonlocal large, local, and a mix of nonlocal populations), original population size (nonlocal large and nonlocal small), and relatedness (related and unrelated individuals) of the source material after within-population crossing (A) and selfing (B). Significant differences within the level of origin are indicated with different letters (Tukey's multiple-comparisons test, $\alpha = 0.05$). (After Vergeer and colleagues [2004].)

(3) inbreeding can dramatically reduce seed production and germination percentage; and (4) the use of several large populations may yield high fitness. The benefit of multisource introductions is evident in the increased seed production after interpopulation crosses (in other words, an unrelated mix). This heterosis suggests that at least some of the source populations were already subjected to genetic erosion and that interpopulation crossing increased genetic variation and thereby enhanced progeny fitness. Lastly, (5) the origin-by-site interaction effects found in this species suggest that material taken from populations of different habitats may perform worse than those taken from populations of similar habitats. Further problems may arise if nonlocal individuals reproduce with local individuals, thus disrupting locally adapted genotypes and potentially causing outbreeding depression.

No detrimental effects of outbreeding depression were observed in the experiment. Outbreeding depression, however, is often manifest only in the

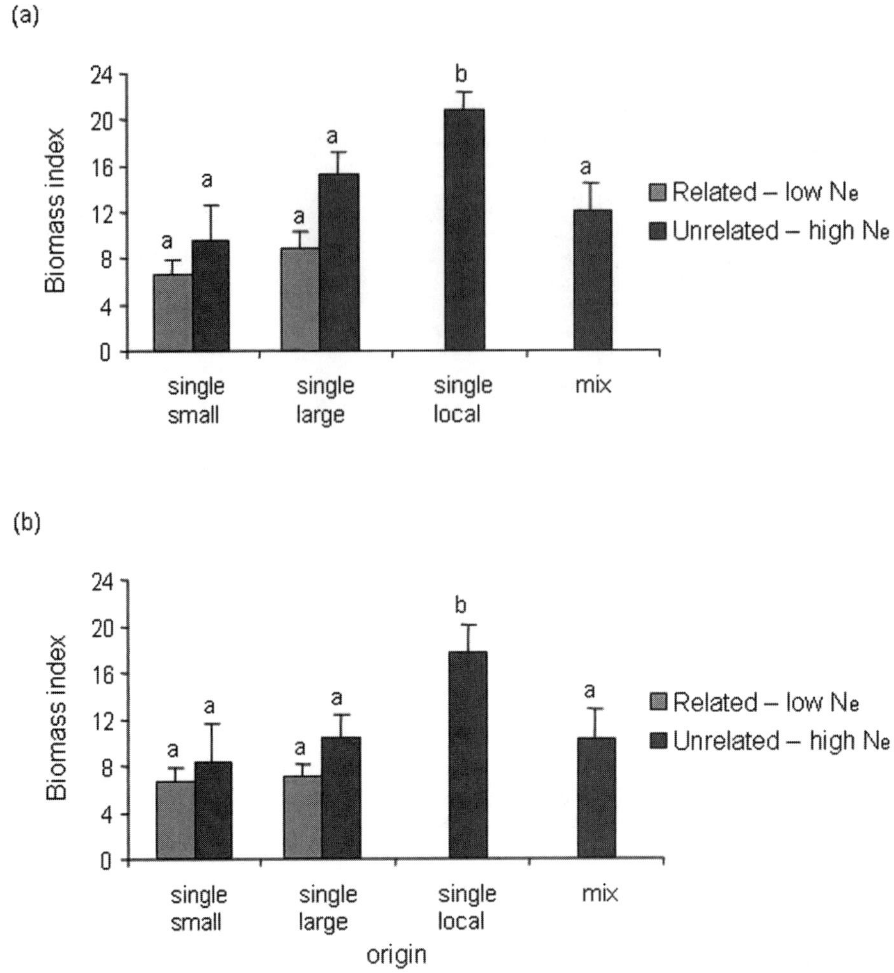

FIGURE 8.4 (A, B) Mean biomass index of *Succisa pratensis* in relation to origin (nonlocal small, nonlocal large, local, and a mix of nonlocal populations), original population size (nonlocal large and nonlocal small), and relatedness (related and unrelated individuals) of the source material after within-population crossing (A) and selfing (B). Significant differences within the level of origin are indicated with different letters (Tukey's multiple comparisons test, $\alpha = 0.05$). Biomass index was calculated by multiplying the number of leaves by length of longest leaf by width of widest leaf. (After Vergeer and colleagues [2004].)

F_2 generation or later (see, for example, Gharrett & Smoker, 1991). Because the *S. pratensis* study only observed effects in the F_1 generation, the relative effects of outbreeding versus inbreeding depression are not yet certain. The authors, however, expect that outbreeding depression will be of little concern because the benefits of growing in the home site were not as large as the detrimental effects of severe inbreeding. In this case at least, the extinction risk in genetically eroded populations resulting from outbreeding depression may be less severe than that resulting from inbreeding. Of course, the conclusion could change in comparisons of different source populations. If they hold, however, introduction of individuals originating from nonlocal populations would perhaps be a good option in

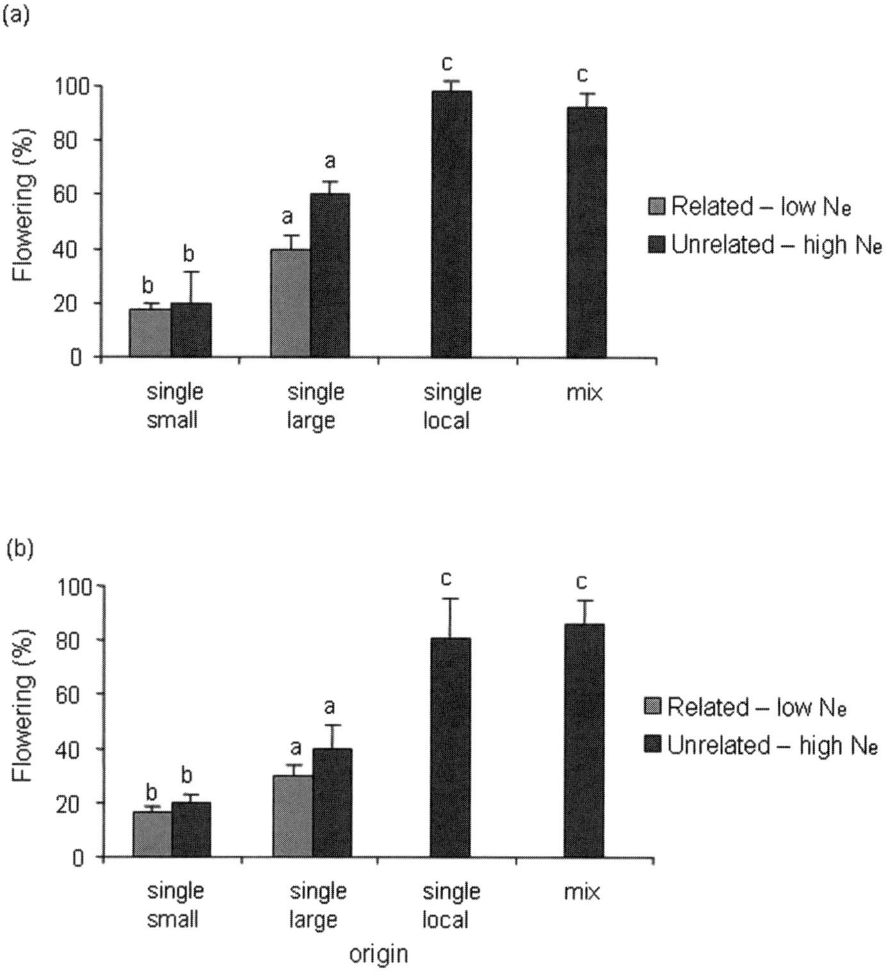

FIGURE 8.5 (A, B) Mean proportion of *Succisa pratensis* flowering in relation to origin (nonlocal small, nonlocal large, local, and a mix of nonlocal populations), original population size (nonlocal large and nonlocal small), and relatedness (related and unrelated individuals) of the source material after within-population crossing (A) and selfing (B). Significant differences within the level of origin are indicated with different letters (Tukey's multiple comparisons test, $\alpha = 0.05$). (After Vergeer and colleagues [2004].)

situations in which a viable local population is lacking.

FUTURE DIRECTIONS

We still have a long way to go before we can confidently specify the best genetic strategy in introduction efforts, and, indeed, the optimal strategy will often be system specific. Fortunately, some general recommendations may be tentatively advanced. On the one hand, the risks of inbreeding and reduced evolutionary potential support strategies that increase genetic variation, such as the translocation of nonlocal individuals into a remnant local population. On the other hand, the risks of outbreeding depression support strategies that use only local material, which is likely best adapted to local conditions. Multisource introductions can increase genetic variation and reduce inbreeding, but they can also lower genetic integrity and provoke outbreeding depression. In general, we suggest that a good general strategy may be the use of multiple, nonlocal populations that are found in similar environments to the focal site, provided these populations do not exhibit outbreeding depression.

In general, then, specifying the best strategy for introduction in a given situation will require a detailed understanding of the relative importance of factors influencing population fitness. Our case studies are instructive in this regard. In the Florida panther, inbreeding was the major concern and so the use of nonlocal source material was clearly optimal. In Connecticut River Atlantic salmon, however, inbreeding is unlikely and so local adaptation may be more important. Indeed, introduction of diverse nonlocal material has failed to establish a self-sustaining population; however, in this case the ultimate problems may also be nongenetic. In the perennial plant *Succisa pratensis*, both local adaptation and avoidance of inbreeding appear important, but outbreeding depression does not. Although detrimental effects of outbreeding depression could not be entirely excluded, benefits of heterosis are likely to be higher than detrimental effects of both inbreeding and outbreeding depression.

It has become clear that one of the major challenges in designing introduction strategies is to find a balance between the effects on within-population genetic variation versus local adaptation, thus also avoiding both inbreeding and outbreeding depression (Hedrick, 1995). A balance of competing evolutionary–genetic considerations will vary among species and specific circumstances as determined by population history, recent inbreeding, local adaptation, and genetic divergence between introduced and native individuals. It will also depend on the diversity of the introduced genotypes. In general, we expect the benefits of multisource introductions to exceed those of single-source introductions, because the former are less likely to manifest inbreeding depression. We are aware of the risks of outbreeding depression, when maladapted genotypes dilute adaptation in the target population. However, if only a few maladapted genotypes are transferred, these genotypes and deleterious genes are likely to be eliminated in the following generations by natural selection. When a wide variety of genotypes is used, it might be expected that, eventually, the fittest ones will persist and proliferate. The reestablished population might thus exhibit an initial reduced fitness resulting from outbreeding depression followed by a gradual recovery in subsequent generations.

We would also like to emphasize the important point that species can be rare in different ways (Rabinowitz, 1981) that each bear on optimal introduction strategies. Generally speaking, species that are localized on a small geographic scale with narrow ranges of habitat conditions are likely to be strongly adapted to their specific conditions. Many of these species are becoming increasingly imperiled as a result of habitat loss or other effects of human activities. Interbreeding of these species with nonlocal individuals can be of high risk because of the disruption of genetic integrity and outbreeding depression. Therefore, in these situations, much attention has to be paid to maintenance of genetic structure and identity.

The alternative extreme is when formerly widely distributed and abundant species are forced into small and isolated populations. These species may be less likely to be highly adapted to their local environment, and may instead suffer more from the recent effects of reduced population size. In this case, restoration strategies should focus more on the enlargement of population size and the increase of genetic variation. However, one has to bear in mind that the longer populations have been isolated, the more genetically differentiated they generally are. One consequence of this population genetic reality is that more care has to be taken with translocation or introduction of genetically dissimilar individuals.

Development of general strategies that incorporate restoration of suitable habitat conditions, knowledge of species abundance, history, and genetics, and the way in which a species can be rare, is a challenge that should have high priority in future conservation genetic research.

SUGGESTIONS FOR FURTHER READING

In 2003, Hufford and Mazer reviewed numerous field and greenhouse studies that have direct implications for the effects of translocation of plants and plant community restoration. They also clarified differences among alternative genetic phenomena that might result in outbreeding depression, caused either by dilution in the F_1 offspring or by the breakdown of epistatic interactions between locally adapted alleles in later generations.

For further reading on outbreeding depression we suggest Fenster and Galloway (2000). Thier paper provides an excellent overview of the concept of outbreeding depression and heterosis, illustrated by experimental results. By using interpopulation crosses along the North American range of the partridge pea, *Chamaecrista fasciculata*, Fenster and Galloway (2000) were able to analyze the effects of heterosis and outbreeding depression on performance. This study differentiates itself from most other similar studies by conducting experimental crosses through the third generation and measuring plant performance under field conditions.

The review of Kawecki and Ebert (2004) covers the most important conceptual issues in local adaptation. It provides a comprehensive overview of the theoretical issues relevant for local adaptation. They advocate multifaceted approaches to the study of local adaptation and stress the need for experiments explicitly addressing hypotheses about the role of particular ecological and genetic factors that affect local adaptation.

Fenster, C. B., & L. F. Galloway. 2000. Inbreeding and outbreeding depression in natural populations of *Chamaecrista fasciculate* (Fabaceae). Conserv Biol. 14: 1406–1412.

Hufford, K. M., & S. J. Mazer. 2003. Plant ecotypes: Genetic differentiation in the age of ecological restoration. Trends Ecol Evol. 18: 147–155.

Kawecki, J., & D. Ebert. 2004. Conceptual issues in local adaptation. Ecol Lett. 7: 1225–1241.

Acknowledgments P. Vergeer was supported by the Dutch Technology Foundation (STW). A. Hendry was supported by the Natural Sciences and Engineering Research Council of Canada. We thank Roy Peters and Leon van den Berg for their assistance with pollination and field experiments with *S. pratensis*. We also thank Scott Carroll and two anonymous reviewers for their valuable comments on an earlier version of this chapter.

9

Hybridization, Introgression, and the Evolutionary Management of Threatened Species

JUDITH M. RHYMER

Hybridization and introgression among taxa are natural evolutionary processes, but there is growing concern about the role of these processes in biotic homogenization caused by anthropogenic forces (Qian & Ricklefs, 2006; Rahel, 2002; Rhymer & Simberloff, 1996) (see Box 9.1 for definitions). Although appreciating the important role that hybridization has played in species evolution, plant biologists were among the first to point out the risks to biodiversity posed by increased gene flow, whenever divergent but reproductively compatible species or populations come into contact (see, for example, Levin et al., 1996). Global trade has contributed to the invasion of exotic species, but species invasions and introductions between regions within continents may pose an even greater threat. In this instance, invading or introduced species are often more closely related to regional endemics and thus are more likely to hybridize.

Hybridization and introgression counteract diversification between populations, which can be particularly problematic for a rare species coming into contact with an abundant one. Human activities have facilitated gene flow between taxa through both deliberate and unintentional introductions of species and through failure to consider that habitat modification may bring previously allopatric populations into contact. Introductions may also indirectly disrupt ecological processes in, for example, communities of plants and insects or interactions between host and parasite (see the later section titled "Indirect Effects"). Although many introductions are unintentional (for example, release of ballast water containing nonnative organisms [Ricciardi & MacIsaac, 2000], escape of fish-bait species [Rahel, 2002], and incidental introduction of insect pests and pathogens on ornamental plants [Brasier, 2000]), there are also an alarming number of cases in which problematic introductions were deliberately conducted to fulfill a relatively narrow management goal (Rahel, 2002). As I illustrate in this chapter, unintended consequences of such actions can have serious evolutionary consequences for a wide range of taxa.

Equipped with an understanding of the evolutionary relationships between affected taxa, loss of biodiversity through hybridization with introduced species could often be predicted a priori. This is of more than theoretical interest and, before modifying habitats or introducing species, the onus should be on biologists to assess potential ecological or evolutionary risk of intra- and interspecific hybridization. In fact, models are now available to help predict the outcome of introductions in this regard, and thus management focus needs to be on developing strategies to reduce the impact of human activities and to ensure that biological diversity is protected.

BOX 9.1 Definitions of Terms Used in Chapter 9

admixture	Mixture of genotypes in a population through hybridization of individuals from different parental populations
biotic homogenization	Increase in species similarity in various regions over time that arises through mixing of nonnative and native species
genomic extinction	Loss of monophyletic evolutionary lineages
genetic rescue	Introduction of individuals into small, isolated populations
genetic restoration	To encourage interbreeding and thus increase gene flow and genetic variation among individuals; goal is to eliminate detrimental effects of inbreeding (genetic rescue [Tallmon et al., 2004]), as well as to maintain local adaptation and increase fitness of individuals (genetic restoration [Hedrick, 2006]) in isolated populations
heteroploid	Having a variable number of chromosomes that is not a whole-number multiple of the haploid chromosome number for that species
hybridization	Breeding between individuals from genetically distinct populations, regardless of taxonomic status
hybrid sterility	Hybrid individuals are viable, but are not fertile
hybrid swarm	Population composed entirely or almost entirely of hybrid and backcross individuals
hybrid viability	Hybrids survive, but may or may not be fertile
hybrid vigor (heterosis)	Increase of fitness in hybrid individuals compared with parental taxa
inbreeding depression	Loss of fitness resulting from interbreeding of close relatives
intercross	Crosses between individuals of different "species" as defined by the U. S. Endangered Species Act—in other words, species, subspecies, vertebrate distinct population segment
introgression	Gene flow between distinct populations by hybrid individuals backcrossing to one or both parental taxa
outbreeding depression	Loss of fitness resulting from interbreeding of individuals from genetically divergent populations

SPECIES COLLAPSE VIA HYBRIDIZATION AND INTROGRESSION

Genomic extinction of rare species can result from hybridization with or without introgression (Rhymer & Simberloff, 1996). If premating, reproductive-isolating mechanisms fail, hybrid sterility or hybrid inviability play an important role in preventing future introgression. Divergent lineages remain distinct in these instances, but if one of the hybridizing populations is rare, its reproductive effort is disproportionately wasted and the population may nevertheless be rapidly driven to extinction (Fig. 9.1A). Even if introgression occurs, it may be limited by unidirectional hybridization. For instance, F_1 hybrids may only produce viable offspring if backcrossing with one of the parental taxa but not the other. Other forms of unidirectional introgression include instances when viable offspring are produced only if males of taxon A mate with females of taxon B but not the opposite pairing, or if offspring of only one sex are viable or fertile. If hybrid sterility is unidirectional, hybrid

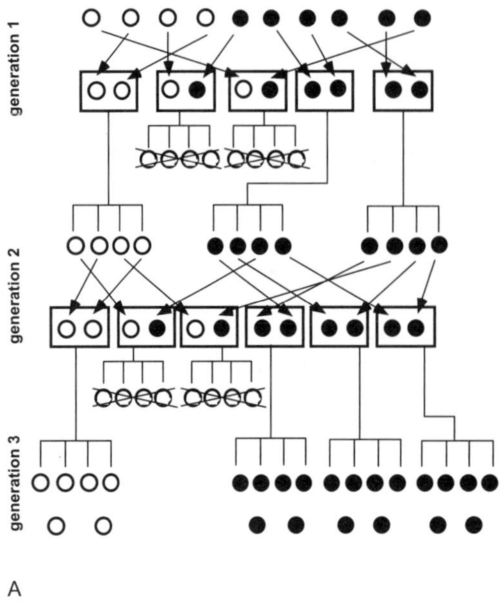

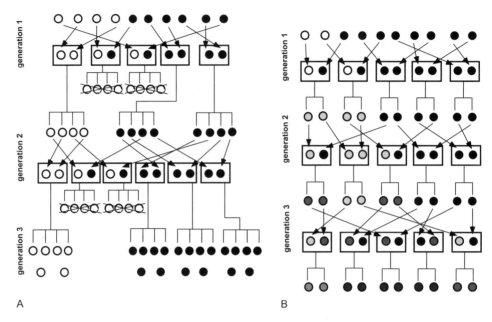

FIGURE 9.1 (A, B) Hybridization can lead to the decline of the rarer species (open circles) regardless of the consequences for offspring fertility. Although the number of individuals has increased after three generations in (A), the offspring are not viable and the frequency of the rarer species has declined. (B) represents a hypothetical scenario in which hybrids are fully viable and fertile and a hybrid swarm results in complete admixture. Partially filled symbols represent F_1, F_2, and backcross offspring; arrows indicate which individuals represent the parents (in boxes) in each generation. (After Levin [2002].)

offspring of the heterogametic sex are usually most negatively affected (Haldane's rule). Hybrid sterility is more common than hybrid inviability in taxa in which males are the heterogametic sex (for example, *Drosophila*, and mammals with males that are XY and females that are XX), but not in taxa with heterogametic females (for example, sterility and inviability in birds and butterflies in which males are ZZ and females are ZW [Coyne & Orr, 2004]).

If fitness of hybrids and backcrosses is similar to that of the hybridizing parental species, a population composed entirely of hybrid and backcross individuals (in other words, a hybrid swarm) can gradually replace parental taxa (Fig. 9.1B). This phenomenon is accelerated if hybrids have higher fitness than parentals (in other words, hybrid vigor). It may require decades, but there are several well-documented cases in which genomic extinction or near extinction occurred in as few as three to seven generations (for example, California cordgrass [Wolf et al., 2001], Pecos pupfish [Rosenfield et al., 2004]). Rapid introgression of the rare Pecos pupfish *Cyprinodon pecosensis* with the ubiquitous sheepshead minnow *C. variegatus*, an accidentally introduced bait fish, is effectively driving the Pecos pupfish to genomic extinction. In this case, the rapid rate of hybridization and introgression is accelerated by sexual selection and ecological superiority of sheepshead minnows and their hybrids (Rosenfield et al., 2004).

PREDICTING THE LIKELIHOOD OF HYBRIDIZATION BETWEEN TAXA

Time since divergence of parental species, the so-called *speciation clock*, should be a strong indicator of whether two taxa are likely to hybridize. Pre- and postmating compatibility are negatively correlated with genetic distance and thus negatively correlated with time since divergence (Mallet, 2005). Most speciation clock studies focus on intrinsic postzygotic isolation, but *incompatibility* clock may be a more accurate term for this phenomenon. This is because other forms of reproductive isolation may be driving speciation prior to evolution of postzygotic isolation (for example, ecological barriers or behavioral mechanisms such as assortative mating [Bolnick & Near, 2005]). Premating reproductive barriers evolve much more rapidly than postzygotic isolation and are likely the major mechanisms maintaining reproductive isolation in animals (Coyne & Orr, 2004). These mechanisms are not absolute, however, as witnessed by the extensive number of cases of hybridization when allopatric species are brought into contact via introductions or habitat modifications.

Degree of genetic divergence between taxa is a reasonable predictor of reproductive compatibility for many animals including butterflies (Mallet, 2005), crayfish, freshwater mussels, and fish (Bolnick & Near, 2005; Perry et al., 2002); and birds and mammals (Fitzpatrick, 2004). However, some taxonomic groups such as butterflies, amphibians, and birds appear to hybridize more readily than mammals, for instance, regardless of time since divergence (Mallet, 2005). Although speciation rates appear to be similar for birds and mammals, the latter lose the ability to form viable hybrids 10 times faster than birds, perhaps because of higher rates of regulatory evolution, immunological interaction between female and fetus, or sex-linked incompatibilities (Fitzpatrick, 2004).

DETECTION OF HYBRIDS

Accurate detection of hybrids has important implications for conservation of rare species, but it is often difficult to identify hybrid individuals unequivocally using morphological characters alone. Hybrids often converge morphologically on the parental phenotypes after one or more generations of backcrossing. However, an integrative approach combining morphological and molecular data has proved valuable in the analysis of potentially cryptic hybridization (Gaubert et al., 2005). New Bayesian methods are available for identifying hybrid individuals based on multilocus molecular data (see, for example, STRUCTURE [Pritchard et al., 2000] and NEWHYBRIDS [Anderson & Thompson, 2002]). Nonetheless, distinguishing between backcrosses, F_1 hybrids, and parentals is difficult and requires use of an extensive number of unlinked marker loci (Vähä & Primmer, 2006).

MIXING OF GENE POOLS

Introductions and Translocations

Intentional introductions for conservation purposes usually have one of three goals: reintroducing species to part of their former range from which they have been extirpated, introducing organisms outside their historical native range, or augmenting declining numbers (Vergeer and colleagues, this volume). If hybridization between one common and one rare taxon endangers the latter, reintroductions to augment the rarer population will not improve its conservation status as long as the potential to hybridize persists (Moritz, 2002). One frequently documented cause of reintroduction failure results from not fully recognizing and removing the original threatening process, such as hybridization. For instance, the red wolf *Canis rufus* was declared extinct in the wild in 1980, and hybridization with the coyote *Canis latrans* is considered one of the factors contributing to decline of the red wolf. Red wolves are closely related to both coyotes and the North American-evolved eastern timber wolf *C. lycaon* (also called the *eastern Canadian wolf* [Wilson et al., 2000, 2003a]). To reintroduce red wolves, captive-bred animals were released onto the Alligator River National Wildlife Refuge in North Carolina in 1987, where hybridization with coyotes is an ongoing problem (Adams et al., 2007). It was predicted that red wolves could disappear in three to six generations (12–24 years) prior to implementation of a rigorous and expensive control program of coyotes and putative hybrids. To survey the area genetically for the presence of hybrids and non-red wolf canids, fecal specimens were collected for genotyping and were used in conjunction with reference genotype data from the experimental red wolf population. Individuals with genotypes not indicative of red wolf stock were then targeted for removal from the population.

Similar concerns pertain in the northeastern United States where wolf reintroductions have been considered to restore their function as top predators in forested ecosystems within their historical range. A 26 million-acre forested area in North America ranging from the Adirondack Mountains of New York east to northern Maine contains suitable gray wolf habitat and an adequate prey base of deer and moose. Apart from political concern regarding the potential depletion of these big game populations by wolves, hybridization with coyotes would be a serious threat to wolf population persistence in the Northeast. Coyotes are ubiquitous in this region and coyote control would be logistically impossible. This underlines one of the problems of invasive species such as coyotes and particularly ones that hybridize, in that they are expensive or impossible to remove and there is little hope of remediation (see also the barred owl example discussed later in "Role of Habitat Modification").

Because wolf taxonomy is controversial, central to the issue of wolf recovery would be determining which species of wolf is the most appropriate to reintroduce in the northeastern United States. Genetic analysis of wolf historical skin samples from the 1800s indicated that the former occupant of this region was not, as previously assumed, the gray wolf *C. lupus*, but rather the eastern timber wolf (Wilson et al., 2003a). The eastern timber wolf is much more likely than the gray wolf to hybridize with coyotes (Hedrick et al., 2002). Currently, the issue of wolf reintroduction is mired in political (rather than biological) controversy, and no introductions of any wolf species seem likely in the near future, unless they colonize on their own from Canada.

Understanding how taxonomic designations define elements of biological diversity has not always been a priority. Worldwide there are 18 named subspecies of the peregrine falcon *Falco peregrinus* that vary in size, shape, color, and migratory behavior. Three of these subspecies occurred historically in North America. A reintroduction program was initiated after their precipitous decline or extirpation in several regions of North America resulting from DDT-induced thinning of their eggshell. Under the guise of genetic rescue (Box 9.1) (Vergeer et al., this volume), seven subspecies from North America and Europe were used as pure- and mixed-pair breeding stock, including several individuals that were a priori known to be hybrids (Tordoff & Redig, 2001). The effort was considered a resounding success because the premise was that having peregrine falcons was the important result, regardless of historical lineages (*elaborated by* Faith, this volume), local adaptations, or any consideration for biotic homogenization. In addition to reduction in local adaptation, the breakdown of distinct lineages and regional distinctiveness could impede future range expansion (García-Ramos & Rodriguez, 2002) or reduce the potential for future diversification because of reduced variability across habitats (Moritz, 2002).

Intentional hybridization efforts like that conducted with the peregrine falcon should be tempered by the possibility that population mixing may induce a loss of fitness through outbreeding depression (Edmands, 2007). Although less well documented than inbreeding depression, there are several cases in which interpopulation hybrids suffer a loss in fitness as a result of disruption of coadapted gene complexes, loss of local adaptation, or genomic incompatibility (as discussed later in "Intraspecific Hybridization") (Vergeer and colleagues, this volume).

As populations of different species continue to decline, one of the strategies for rebuilding depleted populations is to increase recruitment by introducing additional individuals, often from artificial propagation programs (Rahel, 2007). Nowhere is this more prevalent than in fisheries management, where stock enhancements from hatcheries are widely used. Rapid genetic change in hatchery fish has resulted in deliberate and unintentional selection during domestication, as well as from using relatively few breeders (Levin et al., 2001). The result can be complete introgression or displacement of wild fish, with negative effects on performance traits. Escape of farmed commercial salmon has contributed to the extinction of wild salmon through extensive hybridization, leading to loss of genetic diversity and, hence, to loss of capacity to adapt to environmental change (Fleming et al., 2000).

Role of Habitat Modification

Biogeographic barriers are an important historical factor in determining regional composition of species and in promoting diversification among regions (Rahel, 2007). However, anthropogenic influences have reduced isolation imposed by such barriers by increasing connectivity between populations. Habitat modification has contributed to increased probability of hybridization between native populations by fostering mixing of previously isolated taxa. Habitat change may comprise everything from local habitat disturbance and habitat corridors to regional land use changes that allow expansion of one taxon's range into the range of another. In addition to direct anthropogenic effects on the landscape, changes in global climate are predicted to increase the speed of species invasions via range expansion (García-Ramos & Rodriguez, 2002).

When sympatric species lack complete reproductive isolation, habitat disturbance facilitates hybridization between species. Moreover, if fitness of hybrids is enhanced by habitat changes, reversal of speciation can ensue (Seehausen, 2006). Taylor and colleagues (2006) documented the collapse of a sympatric species pair of threespine sticklebacks *Gasterosterus* spp. into a hybrid swarm that arose through loss of habitat heterogeneity coinciding with recent introduction of the American signal crayfish, *Pascifasticus leniscului*. Signal crayfish modified the habitat by destroying aquatic vegetation used for stickleback nesting and by increasing water turbidity, factors that contributed to loss of premating isolation between the species pair (for more details, see Smith and Grether, this volume). Many recently evolved (in other words, postglacial) species pairs are vulnerable to reverse speciation when environments are disturbed (Seehausen, 2006). As a result, we are well on the way to a less diverse world as we transition from natural to biologically homogenized ecosystems.

An interesting example of the effects of large-scale habitat modification involves the barred owl *Strix varia*. During the last century, barred owls dramatically expanded their range from eastern North America across central Canada to British Columbia, and south to northern California (Haig et al., 2004). In the process, they have invaded the range of endangered northern spotted owls (*S. occidentalis caurina*) and are now fully sympatric. Competition and hybridization between species are considered the most serious threats contributing to the decline of spotted owl populations (U.S. Fish and Wildlife Service, 2007). Prevailing hypotheses for the rapid westward invasion of barred owls include an increase in suitable habitat across central Canada that is related to tree planting for farm shelterbelts, and global climate change. There is a controversy raging over the recovery action to cull invading barred owls and barred owl–spotted owl hybrids experimentally, as outlined in the 2007 draft of the northern spotted owl recovery plan (U.S. Fish and Wildlife Service, 2007). Given the extent of the barred owl invasion, such action would have to be maintained in perpetuity. If left unchecked, however, barred owls are projected to drive spotted owls to extinction.

UNINTENDED CONSEQUENCES

Indirect Effects in Ecological Communities

Many examples of the negative effects of hybridization and introgression involve direct effects of interactions among species. There is a growing awareness, however, of indirect effects of hybridization and introgression on ecological communities—for instance, on plant–insect and host–parasite interactions. The geographic range of many host plants has expanded either because the distribution of native plants has changed in response to landscape disturbances or, more directly, through plant introductions. Host plant range expansion has contributed to gene flow between phytophagus insect populations that in turn reduces insect genetic diversity and adaptation to local environmental conditions (Oliver, 2006).

A complex example involves hybridization between two formerly allopatric subspecies of butterflies that occurred when deforestation facilitated spread of their native weed hosts into previously forested areas. Differential infection of butterfly subspecies with a male-killing *Spiroplasma* bacterium causes a skewed sex ratio that limits female choice in one subspecies, thus facilitating interspecific mating and the decline of the male-depauperate taxon (Lushai et al., 2003).

A freshwater mussel in the eastern United States provides an interesting example of how host–parasite interaction contributes to biological homogenization. Intentional introductions of game fish for sport have been implicated in the decline of the yellow lampmussel, *Lampsilis cariosa*, in the Potomac River drainage in West Virginia and Maryland (Kelly, unpublished master's thesis). Larval mussels (glochidia) are obligate parasites on the gills of a fish host for several weeks. After *L. cariosa* transforms to the juvenile stage, the parasites drop off the host fish to the stream benthos, where they mature into reproducing adults. In the Potomac River drainage, most of the remaining populations of yellow lampmussel are composed of hybrids; specifically, hybrid mussels carry the mitochondrial genome of a congener from the midwestern United States that uses introduced game fish as a host. It is hypothesized that nonnative game fish were infested with glochidia when they were introduced into the river drainage, thus facilitating hybridization between native yellow lampmussels and the Midwest species.

Natural and human-mediated hybridization have also been implicated in the ecological niche expansion of pests, pathogens, and disease vectors (Arnold, 2004). The evolution of new virulent pathogens is of concern for cultivated and natural ecosystems, because more plants along with their pathogens are introduced into nonnative areas. One dramatic example is the outbreak of a fungal disease that has destroyed substantial areas of riparian habitat for the European alder *Alnus glutinosa*. Alder is important for stabilizing riparian ecosystems and was not susceptible to *Phytophthora* fungi until recently (Brasier, 2000). The newly invasive alder pathogens are heteroploid hybrids of *Phytophthora* that appear to have evolved from crosses between two introduced fungal species. Introduced fungi co-occur on unaffected ornamental plants in the rose family (*Rubus* spp.) imported for horticultural purposes (see also in this chapter the section titled "Evolution of Invasiveness").

Escape of Transgenes from Genetically Modified Organisms

Are genetically modified crops more invasive than their nongenetically modified counterparts? Based on a summary of several studies, Chapman and Burke (2006) suggest that genetically modified crops are not more invasive. However, there are some interesting examples of crop-to-crop intraspecific gene flow that show how rapidly transgenes can be transmitted over large spatial distances (Ellstrand, 2003a; Marvier, this volume). For instance, hybridization occurred between three varieties of canola (oilseed rape, *Brassica napus*), each with a transgene for a different herbicide. Within 17 months, a triple-resistant volunteer plant was discovered more than 550 m from source crops and, unfortunately, such transgenic volunteers are difficult to control. However, some of the concern regarding the consequences of using genetically modified organisms focuses on the ecological risk of trangene escape via hybridization from genetically modified plants and animals to wild relatives. There is plenty of evidence that introgressive hybridization occurs between a high proportion of crop-by-wild relative hybrids, and genomic extinction of wild relatives has occurred in several cases (Ellstrand, 2003a).

Fitness of natural populations can also be compromised by outbreeding when domesticated traits that are disadvantageous in the wild are inherited by native relatives. Recent studies indicate that transgenes targeting insect herbivores and disease pathogens can function as well in F_1 progeny of a crop-by-wild mating as in the crop itself. For instance, when *B. napus* modified with *Bacillus thuringensis* (*Bt*) transgenes hybridizes with a wild relative, the birdseed rape *B. rapa*, *B. napus,* and F_1 hybrids with the *Bt* transgene have enhanced fitness under high insect herbivore pressure compared with *B. rapa* (Vacher et al., 2004). However, in the absence of herbivores or pathogens, crop transgenes for *Bt* and for pathogen resistance contribute to reduced fitness in hybrids (as discussed in Hails & Morley, 2005). Thus, transgene introgression could exact a cost of resistance to herbivory that is important in regulating spread of transgenes among populations of wild relatives (Chapman & Burke, 2006). Strong genotype-by-environment interactions have also been observed in the fitness of crop–wild hybrids of radish, and risk assessments should include replicated studies across environmental gradients in multiple localities (Campbell et al., 2006).

Does escape of transgenes from crop species via hybridization contribute to the evolution of increasingly invasive wild plant relatives? Although the evidence is more equivocal, introgression of nongenetically modified crops with wild relatives has increased invasiveness, transformed populations into agricultural weeds, or contributed to range expansion of some wild species (summarized in Chapman & Burke, 2006). In fact, crop–wild hybrids might replace parental species in new environments (Hegde et al., 2006).

A concern regarding human disease transmission involves introgression of an insecticide resistance gene from the ancestral S form to the incipient M species of the malaria vector *Anopheles gambiae* s.s. (Weill et al., 2000). Introgression in this case may threaten malaria prevention programs by extending the transmission potential of the M form of *A. gambiae*, a form that is more ecologically adapted to habitats created by human activities than the S form (della Torre et al., 2002). By targeting the propensity of vectors to hybridize in the wild, researchers hope to block malaria transmission by releasing transgenic mosquitoes that inhibit malaria parasite development (Zhong et al., 2006).

HYBRIDIZATION AS AN EVOLUTIONARY FORCE

Evolution of Invasiveness

Intra- and interspecific hybridization can also serve as a stimulus for the evolution of invasiveness in introduced taxa, a process that is accelerated by human-mediated dispersal and disturbance (Lee, 2002). Multiple introductions are often followed by a lag phase between establishment of local populations and their invasive range expansion. In the interim, hybridization between divergent taxa or populations leads to new lineages that spread well beyond the site of introduction (Ellstrand & Schierenbeck, 2000).

Havoc caused by introductions of invasive species can be particularly difficult to predict outside their native range. After its accidental introduction into Europe and North America, the invasive spread of the Dutch elm disease microfungus *Ophiostoma ulmi* was facilitated by interspecific transfer of mating and vegetative incompatibility-type alleles with the much more aggressive *O. novo-ulmi* (Paoletti et al., 2006). Prior to introgression of viral resistance genes from *O. ulmi*, *O. novo-ulmi* was highly susceptible to infection. Introgressed *O. novo-ulmi* is now responsible for the global pandemic of Dutch elm disease, highlighting the potential of interspecies gene transfer for facilitating rapid adaptation of invasive organisms to new environments. In the case of the Dutch elm disease fungus, selection pressure from a fungal virus favored sexual outcrossing and diversity over clonality as a reproductive strategy (Paoletti et al., 2006). Interspecific hybridization in fungi is rare, but is increasing as a result of environmental disturbance (Brasier, 2000).

Although not a case of imperiled native species per se, African honeybees (*Apis mellifera scutellata*) serve as an interesting example of how hybridization and the ecological complexity of species interactions can affect invasions. After dispersing from the original introduction site in South America, African honeybees subsequently invaded feral European honeybee populations in the southern United States. A combination of factors, including a relatively small colonist population of European honeybees, unrelenting invasion of aggressive African honeybees, and susceptibility of European honeybees to an introduced parasitic mite, has accelerated the rate and extent of hybridization between honeybee subspecies (Pinto et al., 2005). Genetic analyses indicate that paternal and maternal bidirectional gene flow has led to the formation of a hybrid swarm of Africanized bees that is on the move. Northward invasion is slowed, because hybrids are less cold tolerant than European honeybees—in other words, hybrids appear to have inherited their cold tolerance from the more tropical African bees. However, a fortuitously limited range of thermal tolerance does not alleviate the dangerous implications of honeybee hybridization for human populations farther south. Modeling approaches are now available to help predict the likelihood of hybrid invasions (see "Future Directions").

Ecological Divergence and Hybrid Speciation

Hybridization can facilitate ecological divergence in wild plants (Rieseberg et al., 2003), but its influence on adaptive evolution and increased fitness of hybrids is generally considered much weaker in animals (Hendry et al., 2007). One exception is the case of homoploid hybrid speciation (in other words, no change in chromosome number) of two native tephritid fruit flies: the blueberry maggot *Rhagoletis mendax* and the snowberry maggot *R. zephria* (Schwarz et al., 2005). The hybrid of these two flies, the so-called *Lonicera* fly, shifted to invasive honeysuckle *Lonicera* spp. that provided a novel plant host and an ecological niche for the hybrid fly. Interestingly, the *Lonicera* host population is also composed of hybrid and backcrossed individuals.

Ecologically driven speciation has also been demonstrated in an aquatic system in which hybridization generated novel adaptations that allowed a hybrid lineage of fish to thrive in a new ecological niche (Nolte et al., 2005). Channeling, dredging, and damming have cut the Rhine River Delta off from the open sea, eliminating tidal influences and creating a novel interconnected system of habitats. Within the past 20 years, a new hybrid lineage of sculpins (*Cottus perifretum* × *C. rhenanus*) has invaded these warmer, more turbid reaches of the Lower Rhine drainage. This represents novel habitat for sculpin, which prefer colder, well-oxygenated water. Locally adapted sculpin populations are relatively resistant to introgression if the environment remains undisturbed, emphasizing again the need to preserve natural habitats and ecosystem function.

CASE STUDIES

Intraspecific Hybridization

To study patterns of dispersal within subtropical rainforest streams in Australia, freshwater shrimp (*Paratya australiensis*, Atyidae) were translocated between two pools in separate subcatchments within the same drainage (Hughes et al., 2003). No thought was given a priori to whether populations in different subcatchments might have diverged genetically over time and adapted to local environmental conditions. However, genetic analysis conducted after translocation revealed a surprisingly high level of divergence (6%) between populations in the two subcatchments. Subsequent sampling only 7 years after experimental translocation indicated that the resident genotype at one site had gone extinct. Further genetic analyses suggested that sexual selection and outbreeding depression had played a role. Analysis of maternally inherited mtDNA indicated that both resident and translocated females mated preferentially with translocated males—in other words, resident females produced hybrid offspring that all had the resident mitochondrial lineage. Unfortunately, hybrid matings resulted in wasted reproductive efforts because hybrid offspring had low viability compared with those of translocated pairs. The resident lineage was lost in fewer than seven generations through hybrid matings. These results support predictions based on simulations (see, for example, Wolf et al., 2001) showing that if a rare population lacks a competitive advantage and reproductive barriers are weak, then hybridization can lead to genomic extinction in as few as five generations.

Similar concerns pertain to translocation of plants to restore native ecosystems and the subsequent effects of intraspecific hybridization between locally adapted plants and translocated individuals (Hufford & Mazer, 2003; Vergeer et al., this volume). Ecological restoration efforts of native plant communities often fail to take into consideration the occurrence of genetically distinct ecotypes within a single species. Thus, ecotypes adapted to different climatic and edaphic conditions elsewhere in the species range are often chosen for translocation to the restoration site. Although there is often an initial optimistic phase of hybrid vigor in the F_1 generation as translocated and local plants begin to breed, it is often followed by outbreeding depression in later generations of hybrids and backcrosses. One unintended consequence is that introgression of maladaptive genotypes may threaten long-term success of ecological restoration efforts. Keeping this in mind, a combination of research and monitoring is essential for successful translocation experiments that ensure population recovery as well as maintenance of its future evolutionary potential (Stockwell et al., 2003).

Interspecific Hybridization

Deliberate as well as unintentional introduction of the mallard duck *Anas platyrhynchos* has been widespread and contributed to the decline of closely related native species in North America, Hawaii, and New Zealand (Rhymer, 2006). In New Zealand, for instance, a once-common species, the gray duck (*A. superciliosa superciliosa*), was hunted until the 1990s, is now listed as endangered, and is on the verge of genomic extinction because of hybridization and introgression with introduced mallards. Hybridization with mallards has also been implicated in the decline of two ducks: the American black duck, *A. rubripes*, in the eastern United States, and the Mexican duck, *A. diazi*, in the southwestern United States. Changing land-use practices across the continent contributed to range expansion of mallards from the central plains, bringing them into contact with these previously isolated species. In the case of black ducks, intentional introductions of mallards for hunting accelerated rates of hybridization and introgression between species. Extensive introgression resulted in Mexican ducks being officially declared conspecific with the mallard (in other words, *A. platyrhynchos diazi*), a political move designed to prevent federal listing of Mexican ducks as endangered. Taxonomic revision was undertaken despite the fact that there are only a few pure populations of Mexican ducks remaining in central Mexico that are as genetically divergent from the common mallard as are the other taxa in the mallard complex in North America (McCracken et al., 2001), each of which is considered a valid species.

Escaped mallards from parks, backyard ponds, and hunting clubs have hybridized with the mottled duck *A. fulvigula* in southern Florida to such an extent that the public is being notified that this could lead to the demise of their indigenous species. In 2005, genetic analysis showed that approximately 11% of putative mottled ducks sampled in Florida were in fact hybrids (Williams et al., 2005a). A more critical example concerning feral mallards involves the endangered Hawaiian duck or *koloa*, *A. wyvilliana*, that has a population of only about 2,500 birds. This duck is so threatened by hybridization that the population on the island of Kauai is now considered the only one remaining that is not a hybrid swarm (Rhymer, 2001).

FUTURE DIRECTIONS

Assessing Biological Risk

Although a widespread management response to hybridization is to cull hybrids, hybrids may contain the last remaining genes of a species on the brink of extinction. Hybrids might also fulfill the ecosystem function of the rare taxon. Moreover, removal of hybrids and reversal of detrimental environmental influences after the fact, even if possible, do not necessarily ensure restoration of rare species integrity. We now have many examples in which interbreeding appears to increase threats to rare taxa and increase rates of extinction (Levin et al., 1996; Rhymer & Simberloff, 1996). Having identified potential threats from hybridization, part of decision making regarding species introductions and habitat modification should be a cost–benefit approach to management that takes into account such risks.

Modeling approaches are available to examine the effects of interbreeding of native taxa with invading species. One of the first of such papers (Huxel, 1999) tested whether hybridization, regardless of introgression, affects the rate of species displacement. This approach involved single-locus, two-allele models with varying levels of inbreeding and fitness of the heterozygote. Epifanio and Philipp (2001) explicitly altered three variables to predict extinction of animal populations: initial proportion of parental taxa, fitness gradients among parental and introgressed taxa, and the strength of assortative mating. Wolf and colleagues (2001) took a different approach to model extinction through hybridization by incorporating factors such as selfing rate and allelic frequencies. Model predictions were in close agreement with the rapid decline of native California cordgrass *Spartina foliosa* observed in the field in San Francisco Bay. More complex multilocus models incorporate these factors plus others such as mutation rate, recombination rate, and probability of outbreeding depression via disruption of intrinsic coadaptation or local adaptation (Edmands & Timmerman, 2003).

More recently, Hall and colleagues (2006) focused on models that incorporate a quantitative genetics approach to predict the potential for hybrid invasions of plants. They assume that a large number of loci affect life history traits and that many genotypes occur in a population. The key variable in their model is the area occupied by individuals of a particular genotype in a given generation, where area occupied is defined as a combination of vegetative growth rate and adult survival and recruitment. Their model could be used to design effective control strategies by predicting the rate of hybrid invasions and the genetic structure of hybrid populations.

The upshot of these theoretical and empirical studies is that hybridization poses an important risk to taxonomic diversity. Management needs to take a conservative approach when considering translocating species or populations into the range of related species or conspecific populations, or altering the landscape such that contact of previously allopatric taxa is facilitated. The potential loss to biodiversity represents a cost that needs to be included in biological risk assessments.

Policy Issues

Indeed, endangered species biologists are eager for guidance. The previously enforced hybrid policy under the Endangered Species Act stated that hybrids of listed species were not protected (Haig & Allendorf, 2006). Although the hybrid policy was abandoned as simplistic and biologically irrelevant in 1990, there is little guidance for biologists struggling with the complexity that hybridization introduces into the management of rare species. A subsequent, more flexible "intercross" policy that would provide for possible protection of hybrids has neither been adopted nor withdrawn (*intercross* was considered less controversial than *hybrid*). Allendorf and colleagues (2001) outlined three hybridization scenarios, each with one specific management recommendation, that help guide conservation priorities when hybridization results from anthropogenic actions: hybridization without introgression (remove nonnative taxa and any viable F_1 hybrids), widespread introgression (ignore hybrids and focus on protecting remaining pure populations), and complete admixture (protect hybrids because they may represent the only genetic legacy of a species suffering from genomic extinction). The authors caution that policy actions must be flexible enough to account for the wide variety of situations that occur in nature.

In 2000, a controlled propagation policy was adopted that actually allows, in rare cases, for controlled intercrosses to effect genetic restoration of an endangered species (Haig & Allendorf, 2006). The policy sanctions rescue attempts such as the one undertaken in 1995 for the Florida panther, *Puma concolor coryi*. Cougars (= panthers) from Texas (*P. c. stanleyana*) were released into Florida to interbreed with the remaining 30 panthers, with the idea that the resulting increased genetic variability would counter inbreeding depression and improve the chances of population recovery. The Florida population of panthers subsequently increased to 87 individuals by 2006, but whether the increase was attributable to genetic restoration (sensu Hedrick, 2006) or demographic rescue remains controversial (Creel, 2006; Tallmon et al., 2004; Vergeer et al., this volume).

There are no easy answers and a key goal is for evolutionary biologists to help elucidate the complexities of taxonomic loss via hybridization and offer sound judgment on how to interpret each situation. Increasingly useful modeling approaches to predict the process will be essential along with practical approaches to incorporating science into policy decisions.

SUGGESTIONS FOR FURTHER READING

Reviews by Levin and colleagues (1996) and Rhymer and Simberloff (1996) are a good introduction to the topic, whereas Rieseberg and colleagues have fully explored the genetic mechanisms and issues related to hybridization in plants (see, for example, Rieseberg et al., 2003). Hall and colleagues (2006) provide an overview of models to assess the potential risk of hybridization and suggest an innovative modeling approach of their own. Papers by Allendorf and coworkers (2001) and Haig and Allendorf (2006) discuss hybridization and introgression in a practical context by addressing the difficult policy issues as they relate to endangered species. The excellent book by Coyne and Orr (2004) on speciation provides a wealth of information on the underlying concepts relevant to species divergence, hybridization, and introgression.

Allendorf, F. W., R. F. Leary, P. Spruell, & J. K. Wenburg. 2001. The problems with hybrids: Setting conservation guidelines. Trends Ecol Evol. 16: 613–622.

Coyne, J. A., & H. A. Orr. 2004. Speciation. Sinauer Associates, Sunderland, Mass.

Haig, S. M., & F. W. Allendorf. 2006. Hybrids and policy (pp. 150–163). In J. M. Scott, D. D. Goble, & F. W. Davis (eds.). The endangered species act at thirty: Conserving biodiversity in human-dominated landscapes. Vol 2. Island Press, Washington, D.C.

Hall, R. J., A. Hastings, & D. R. Ayres. 2006. Explaining the explosion: Modeling hybrid invasions. Proc R Soc B. 273: 1385–1389.

Levin, D. A., J. Francisco-Ortega, & R. K. Jansen. 1996. Hybridization and the extinction of rare plant species. Conserv Biol. 10: 10–16.

Rhymer, J. M., & D. Simberloff. 1996. Extinction by hybridization and introgression. Annu Rev Ecol Syst. 27: 83–109.

Rieseberg, L. H., O. Raymond, D. M. Rosenthal, et al. 2003. Major ecological transitions in wild sunflowers facilitated by hybridization. Science 301: 1211–1216.

III

EVOLUTIONARY RESPONSES TO ENVIRONMENTAL CHANGE

Throughout the history of life on earth, organisms have been challenged by changes in the physical and biotic environments. Contemporary populations continue to face these natural challenges, but at the same time must also cope with anthropogenic influences that may increase intensity or frequency of stressors such as temperature, habitat degradation, and biological invasions, as well as novel challenges in the form of pesticides, agronomic development, and urbanization. Environmental stress is defined as a response to an external force that directly affects fitness by reducing reproductive output or causing increased mortality. Environmental stress can indirectly alter patterns of adaptive evolution by increasing recombination and mutation rates, maintaining genetic variation, and increasing expressed phenotypic variation (Hoffmann & Parsons, 1991). Three major human-enhanced factors that have increased exposure of populations to environmental stress are climate change, species introductions, and habitat destruction. In Part III, the authors explore stresses that ecologists and evolutionary biologists associate with these axes of environmental change and examine how adaptive stress responses (or lack thereof) shape geographic ranges, environmentally sensitive performance, fitness profiles, and demographic dynamics of populations.

GLOBAL WARMING

According to the fourth Intergovernmental Panel on Climate Change (Alley et al., 2007b), both land and ocean temperatures are warming rapidly. The consensus among climatologists is that the principle cause of global warming is the increase in atmospheric CO_2 and other greenhouse gasses resulting from human activities such as the burning of fossil fuels, drastic modifications in land cover (for example, deforestation), and industrial processes. Scientists predict that human activities will continue to elevate atmospheric CO_2, and thus global warming, throughout the 21st century. Models of climate change predict a 1.5 to 4.5 °C increase in mean global temperature during the next century. Ecosystems will be exposed to the warmest conditions prevailing on earth within the last 100,000 years. Not only is temperature increasing, but the *rate* at which it is increasing is predicted to have a dramatic impact on biodiversity (Parmesan, 2006).

The 2001 Intergovernmental Panel on Climate Change (IPCC) report (Houghton et al., 2001) describes "fingerprint" events as those that occur as a direct consequence of global warming. These events include (1) heat waves and periods of record-breaking warmth; (2) ocean warming, coupled with rising sea levels and coastal flooding; (3) high rates of glacier melting; and (4) polar warming. In addition, the IPCC defines "harbinger" events as those that, although not conclusively linked to global warming, are predicted to increase in frequency as warming continues. These events include (1) the spread of diseases as the ranges of insect vectors expand (Altizer & Pederson, this volume), (2) earlier arrival of spring, (3) shifting ranges of plant and animal populations, (4) changes in population structure in response to abiotic stresses, (5) bleaching of coral reefs, (6) increases in precipitation accompanied by heavy rain and snowfalls and flooding, and (7) extended droughts. Historically, many organisms responded to climate change through latitudinal or altitudinal migration (Bartlein & Prentice, 1989). In the contemporary landscape, migration may be greatly impeded by human activities and disturbances. Thus, organisms are forced to rely, to an even greater degree, on evolutionary adaptation (Etterson, this volume; Gilchrist & Folk, this volume).

BIOLOGICAL INVASIONS

Every species on earth today was, at one time, a colonist in an environment filled with potential competitors, predators, parasites, and prey. Biological introductions and invasions are nothing new, but the frequency with which new colonizing populations are introduced has increased dramatically with the scope of global commerce. Elton (1958) first recognized the frequency and importance of human-mediated invasions as major factors shaping ecological communities. Ecologists have rightly tended to focus on the negative effects of invaders on the invaded community (Ruesink et al., 1995; Vitousek et al., 1997). Invaders may increase deleterious competitive effects on native species (Callaway & Ridenour, 2004) or physically alter the landscape, disrupting extant communities (Pollock et al., 1995; Singer et al., 1984). Invaders can also introduce parasites or disease agents that decimate native populations (Daszak et al., 2000). The loss of native species as well as the cost of managing invaders can have direct and severe economic consequences (Wilcove et al., 1998). A purely ecological perspective, however, will ignore important evolutionary changes that accompany biological invasions and may prove important in management (Gilchrist & Folk, this volume).

For evolutionary biologists, invasions create opportunities to view evolution of species and communities firsthand. Joseph Grinnell (1919) argued that the arrival of English house sparrows in Death Valley set up an "experiment in nature" for students of ecology and evolution. Johnson and Selander (1964) followed Grinnell's suggestion and, in their classic study, discovered evidence of rapid adaptive evolution of these invaders across North America. A recent review of contemporary evolution (Stockwell et al., 2003) highlighted several cases spanning diverse life histories where recently introduced species have undergone significant evolutionary change within a few generations of colonization.

Intuitively, one would expect that colonization would often involve a significant genetic bottleneck, resulting in reduced genetic variation and limited potential for rapid evolutionary change (Boulding, this volume; reviewed in Willis & Orr, 1993). Yet such bottlenecks may also change epistatic (Cheverud & Routman, 1996; Goodnight, 1988) and dominance (Willis & Orr, 1993) variance into additive genetic variance. Coupled with changes in direction and intensity of natural selection during invasion (Lee, 2002; Sakai et al., 2001a), perhaps we should not be surprised that evolutionary change often follows introduction and invasion. Wares and colleagues (2005) review the literature and conclude that most invasions do not result in a dramatic loss of genetic diversity, suggesting that adaptation to new biotic and abiotic stresses should not surprise us. Carroll and Watters (this volume) show how changes in genetic variation may interact with phenotypic adaptation to influence population persistence in colonized environments.

Far less is known about the impact of biological invaders on the evolution of native species and their communities (Callaway et al., 2005a; Carroll & Watters, this volume; Strauss et al., 2006a). Allelopathic effects of invasive plants impose natural selection on the invaded community (Callaway et al., 2005b), potentially leading to long-term changes in community structure and function as well as adaptive responses on the part of native species. Evidence of natural selection in two native grass species in response to the invasive forb *Acroptilon repens* was uncovered by amplified fragment length polymorphism (AFLP) analysis of adjacent invaded and uninvaded communities (Mealor & Hild, 2006). A study examining the evolutionary response of the common native plant *Lotus wrangelianus* to sequential invasion by two exotic competitors revealed no evolutionary response until the density of the second invader was experimentally reduced (Lau, 2006). The author suggests that adaptation to multiple invaders may be impossible, given the diversity of selection pressures imposed by two or more exotic species. Interestingly, that scenario is reminiscent of the situation facing invaders that must respond to multiple native competitors.

HABITAT CHANGE: LOSS AND FRAGMENTATION

The IUPN (currently known as the *World Conservation Union*) declares that as of 2004, 15,589 species including vertebrates, invertebrates, plants, and fungi were considered threatened with extinction as a result of global-level habitat degradation and destruction (Baillie, 2004). This assessment was based on studies comprising 38,047 species, or less than 3% of the world's 1.9 million described species. The conservation status of most species is unknown. Threatened species, those falling into the categories of "Critically endangered," "Endangered," and "Vulnerable" include 12% of birds, 23% of mammals, 32% of amphibians, 42% of turtles and tortoises, as well as significant numbers of fish and plant species. Very few invertebrate species have been evaluated. The only insect species that have been closely evaluated are swallowtail butterflies (Order: Lepidoptera), and dragonflies and damselflies (Order: Odonata), many of which are highly threatened. Species extinction rates are higher than background rates by two to four orders of magnitude, thus bringing us to what is considered to be *the sixth great extinction of life on earth*. Most extinctions since 1500 AD have taken place on oceanic islands, yet extinctions on continents are now nearly as common (Baille 2004).

Loss of habitat poses the greatest threat to biodiversity (Baillie, 2004). Habitat loss is coupled with the explosive growth of the human population, which is likely to climb by about 66% in the 21st century to 10 billion people (Cincotta & Engelman, 2000). Linked to human population growth are increased habitat destruction resulting from urbanization and agriculture, increasing energy consumption and increased pollution, and high demand for water, all of which affect natural populations. For example, in California, where water for human consumption is a valuable commodity, a controversial water project threatens tortoises and endangered bighorn sheep in the Mojave National Preserve (ADBSS, 2004; Longshore et al., 2003).

The Food and Agriculture Organization of the United Nations (FAO, 1997) estimates that about 40% of earth's primeval forests have been destroyed through human activities. The loss continues as millions of hectares of natural forest are destroyed annually. In the 1990s alone, approximately 4% of global forests were lost. The habitat degradation and fragmentation that results from deforestation, both within the forest and in adjacent areas, can negatively affect species and further isolate subpopulations. Habitats of many species are now mere fragments of their original area. The degree to which dispersal between fragments is impeded has a direct effect on the genetic structure of the populations inhabiting the fragments (Templeton, 1990). If these fragmented "islands" are genetically isolated, then populations within each island are demographically independent and subject to local extinction, increasing the probability of extinction for the global population. Species may be at high risk for extinction as a result of *mutational meltdown*—that is, a dramatic loss of genetic variability fueled by inbreeding depression as a consequence of habitat fragmentation (Lynch et al., 1995b; Tomimatsu & Ohara, 2006).

GLOBAL CHANGE AND EVOLUTIONARY RESPONSES

Climate change, biological invasions, and habitat destruction thrust all organisms into new physical environments. Chapters in this section examine the ensuing evolutionary responses to these sources of environmental change. Etterson (chapter 10) begins with a review of the concepts of evolutionary genetics that provides a theoretical basis for understanding adaptive responses to environmental perturbation. She reviews her own work on reciprocal transplant experiments of various ecotypes of the annual legume *Chamaecrista fasiculata*. In this case, she finds evidence of local adaptation to climatic variation; however, the genetic correlations among fitness-related traits suggest that northern populations of this species will face severe evolutionary challenges in the face of climate change. She concludes with some important ideas for management of biological reserves and environmentally sensitive communities.

Gilchrist and Folk (chapter 11) review how environmental stress induced by anthropogenically driven change can be a potent evolutionary factor. They provide a general review of insect adaptation to thermal and desiccation stress, noting that climate change at the population level can be a significant consequence of invasions and habitat destruction as well as global warming. Case studies are presented of evidence for genetic adaptation to thermal and desiccation stress in *Drosophila* and of the emerging evolutionary disharmony between unchanging photoperiodic cues and changing local climate for the pitcher-plant mosquito, *Weyomyia smithii*.

Carroll and Watters (chapter 12) consider the interplay of factors that will determine how populations cope with environmental change. These factors include the amount, form, and distribution of genetic variation; developmental and behavioral plasticity; reproductive and developmental performance; and population size. For example, even while favoring adaptation, natural selection in vulnerable populations may also involve mortality costs that can lead to extinction (Gomulkiewicz & Holt, 1995). Carroll and Watters review how plastic responses of individuals may mediate mortality during the selective transition to genetic adaptation. Taking advantage of such plasticity, conservation managers may be able to reduce interspecific competition for resources by cultivating a diversity of individual phenotypes, independent of genetic diversity, and thereby increase population size.

Boulding (chapter 13) reviews how genetic diversity affects adaptive potential and viability of small populations during periods of environmental change. Various kinds of neutral molecular markers are used to estimate population structure, gene flow, and effective population size. These are often contrasted with markers that are potentially under natural selection, including specific candidate genes and QTL. Her review includes consideration of quantitative traits under selection and the ability of populations to sustain directional natural selection. The literature reveals that populations bottlenecked for several generations are likely to have reduced variation in fitness-related quantitative traits, which may limit their resilience in the face of environmental stress.

George W. Gilchrist and Donna G. Folk
Williamsburg, Virginia

10

Evolution in Response to Climate Change

JULIE R. ETTERSON

Global climate change is imposing a natural experiment on earth's biota that will require populations of organisms either to adjust to the changing environment or face extinction (Davis et al., 2005). Adaptation of the planet's biota to such environmental challenges is not unique; indeed, the climatic history of earth is marked by cycles of warming and cooling that occur with different periodicity and are punctuated by periods of rapid change (Zachos et al., 2001). The fossil and genetic records suggest that range shifts (Mitton et al., 2000), adaptation in situ (McGraw & Fetcher, 1992), and simultaneous evolution and range shifts (Cwynar & MacDonald, 1987) are processes that allow populations to persist through climatic cycles. At the same time, however, many plant (Jackson & Weng, 1999) and animal taxa (Martin, 1984) have gone extinct in the face of climate change.

Human-induced climate change is expected to differ from previous changes in several important respects that might increase the evolutionary challenge to natural populations. First, the predicted rate of climate change will exceed by a factor of 10 any period during the last 10,000 years (Alley et al., 2007a). Second, rapid climate change is superimposed upon other anthropogenic factors that already imperil native organisms. For example, fragmented populations embedded in a matrix of altered habitat may have reduced opportunities for range shifts and may be cut off from input of novel genetic variation (Swindell & Bouzat, 2006). For many species, contemporary population sizes are reduced, which may cause genetic diversity to be lost by drift and inbreeding, and may increase susceptibility to extinction by stochastic environmental events (Heschel & Paige, 1995; Reed, this volume). Habitat degradation may also allow invasion of exotic species that compete for resources and compound stress (Strauss et al., 2006a). Furthermore, positive interactions between organisms (for example, between plants and pollinators) may become decoupled as species respond to climate change in different ways (McCarty, 2001). Thus, the persistence of organisms will depend upon a multiplicity of interacting factors.

ARE ORGANISMS ALREADY RESPONDING TO CLIMATE CHANGE?

Movement of many plant and animal species in response to recent warming has already been observed. In a meta-analysis that included 893 species, 80% exhibited range shifts within the last 17 to 1,000 years that match climate change predictions with an overall average rate of 6.1 km per decade toward the poles (Parmesan & Yohe, 2003). However, not all species are equally likely to keep pace with climate change. Animal species associated with specific vegetative communities may migrate at a much slower rate (Hill et al., 2002). Among plants, those with specialized soil requirements or those lacking mechanisms for long-distance seed dispersal (for instance, wind- or animal-mediated dispersal) may expand their ranges more slowly. In contrast, generalists or organisms that favor disturbed habitats, including many invasive species, will be more likely to become established beyond their current range (Dukes & Mooney, 1999). Estimates of postglacial range expansion based on molecular data for North American trees suggest that historical rates of movement are slower than what is necessary to keep pace with 21st-century warming (McLachlan et al., 2005). More information on the dispersal capacity of organisms is necessary to predict movement in response to climate change, given that range shifts are typically driven by long-distance dispersal events at the leading edge of the migratory front (Pearson & Dawson, 2003).

There are few contemporary examples where extinction can be directly attributed to climate change because of interacting effects of other environmental factors. However, a strong case has been presented that implicates climate change as a primary cause of extinction of amphibian species in cloud forests of Costa Rica (Pounds et al., 1999). Similarly, it is difficult to predict precisely future rates of extinction resulting from climate change (Botkin et al., 2007). Nevertheless, "first-pass" estimates for the year 2050 based on species–area relationships and a range of climate and dispersal scenarios are alarming and indicate that anthropogenic climate change could become a major cause of extinction (Thomas et al., 2004). However, these predictions of extinction may either overestimate or underestimate true risk in light of evolutionary aspects of populations that were not taken into account including (1) the extent to which populations across the species range are already locally adapted to climate and (2) the potential for populations to mount adaptive evolutionary responses with ongoing climate change.

Extinction rates may be greater than predicted because species are composed of populations that are adapted to a range of environments that is more narrow than for the species as a whole. The models of Thomas and colleagues (2004) and others that predict migratory responses of organisms to climate change rely upon the assumption that a *species* is characterized by a single set of climate tolerances, or *climate envelope,* across their ranges (Pearson & Dawson, 2003). However, we know that *populations* within a species range are not uniform but differ in morphology, phenology, and physiology in a way that often corresponds to local environment (Gilchrist & Folk, this volume; Linhart & Grant, 1996). Because tolerance limits of a given population are likely to be narrower than for the species as a whole, climate change will exceed population limits more rapidly than would be predicted based on a single climate envelope for the species (Fig. 10.1). Thus, one consequence of local adaptation is that, as climate changes, conditions may deteriorate for populations across the species range rather than just, for example, at species' range margins.

However, extinction rates may be lower than predicted if populations undergo adaptive evolution in response to climate change. Theoretical models suggest that adaptive evolution can enhance population persistence in a changing environment even when range shifts are possible (Bürger & Lynch, 1995). The potential for evolutionary response depends upon genetic structure at the population level, not for the species as a whole. Populations will differ in potential for adaptive evolution because they have been uniquely molded by historical processes such as founder events, genetic drift, gene flow, and selection. Consequently, each population differs in the extent to which it is already adapted to local conditions and each has its own range of environmental tolerance and evolutionary potential. Thus, the ultimate fate of a species depends upon the evolutionary response of these genetically differentiated populations across the species' range.

Adaptive evolution is an almost universal response of the global biota to environmental change. It is becoming increasingly clear that environments that are "local" today will probably not be so in the future (Intergovernmental Panel on Climate Change, 2007). Thus, questions of imminent concern to conservationists should be the following:

1. To what extent are populations indeed locally adapted to the environment they currently occupy?
2. What is the range of conditions that populations can tolerate through adaptive physiological, behavioral, and phenotypic responses?
3. What is the potential for evolution of a different climate optimum and/or a broader range of tolerance?
4. Are there management practices, such as facilitated gene flow, that could mediate the effects of climate change?

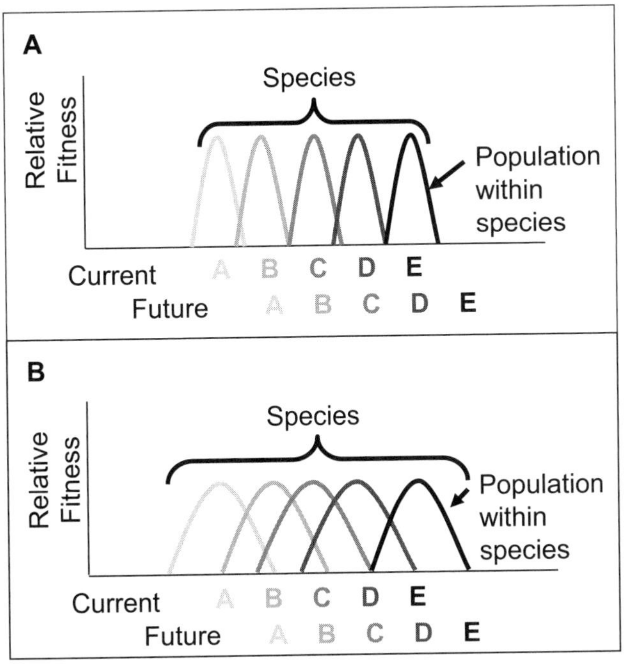

FIGURE 10.1 Hypothetical bioclimatic envelopes for a species compared with populations within a species. Fitness is on the y-axis and an environmental gradient (for example, temperature or length of the growing season) is on the x-axis. (A) Populations are specialized with regard to their position along the climate gradient and have high fitness at their optima but a narrow range of tolerance for conditions around the optima. (B) Populations are locally adapted to their position along the climate gradient but have a wider breadth of tolerance around their fitness optima. If all populations are equally well adapted to conditions throughout the species range, as is implied by the species-level envelope, when the environmental gradient shifts with climate change, the fitness of populations will be reduced only at the trailing edge of the range where the climate is deteriorating. However, if populations are highly specialized, as shown in (A), climate change will expose populations across their range to conditions outside their tolerance limits where persistence is unlikely. In (B), the greater ecological amplitude of populations may allow populations to persist with climate change but with lower fitness. See Figure 11.4 of Gilchrist and Folk (this volume) for a more detailed description of how the shape of these curves may be altered by selection. (Modified from Figure 1 of Davis and colleagues [2005].)

CONCEPTS

What Factors Influence Evolutionary Response to Climate Change?

The probability of ongoing adaptive evolution depends upon a number of genetic and ecological factors, including the pattern of natural selection (Box 10.1) and the genetic architecture of populations (Box 10.2). The *breeder's equation* is a simple expression that illustrates the basic relationship between these factors in determining the rate of evolutionary change. The amount of phenotypic change expected in response to

BOX 10.1 Natural Selection

Natural selection occurs whenever individuals within a population differ in fitness because of the traits they possess. The strength and direction of natural selection on these traits can be statistically estimated using techniques that are rooted in multiple regression and are often referred to as *phenotypic selection analyses* (popularized by Lande and Arnold [1983] and reviewed in Brodie and colleagues [1995]). In these analyses, a measure of relative fitness is regressed onto a number of other measured traits that are putative targets of selection. If there is a significant linear or curvilinear relationship between fitness and trait values, then significant selection is inferred.

A new method of analysis, "aster," allows multiple components of fitness with different underlying distributions (e.g., survival and fecundity) to be considered jointly in a single analysis (Shaw et al., 2008, in press).

To use this approach, fitness correlates (in other words, survival or fecundity) and other traits that are hypothesized to influence fitness (in other words, morphology, physiology, and behavior) are measured on a number of individuals in a population. Selection coefficients are obtained by regressing relative fitness (w) onto other trait data. Partial regression coefficients of these analyses are interpreted as the measure of direct selection, the selection gradient (β_i), on trait i (Box Fig. 10.1A). Covariance between relative fitness and trait i—in other words, the selection differential (S_i)—is interpreted as a composite measure of direct linear selection on trait i and indirect linear selection mediated by phenotypic correlations with other traits. Estimates of the curvature of selection surfaces (stabilizing or disruptive selection), can be obtained from multiple regressions that include as predictors quadratic (β_{ii}) and cross-product functions of the traits (β_{ij}). The β_{ii} reflect curvature in the selection surface and indicate stabilizing or disruptive selection if the peak or valley of fitness corresponds to an intermediate phenotype (Box Fig. 10.1B, C). The β_{ij} reflect selection on trait i that varies depending upon the value of trait j. In other words, selection favors particular combinations of traits (Box Fig. 10.1D).

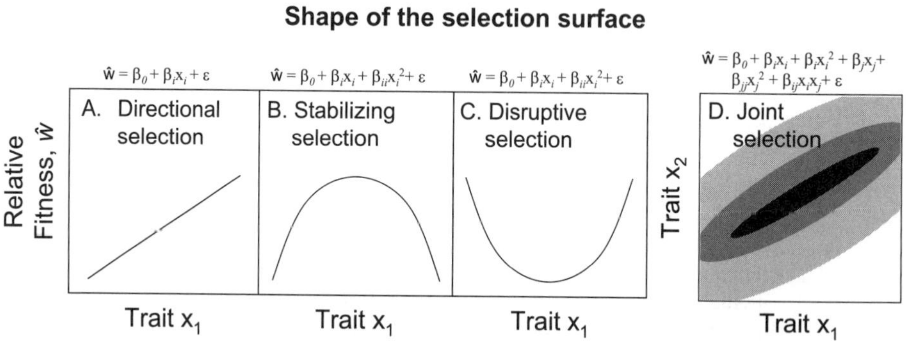

Box Fig 10.1 Possible shapes of selection surfaces estimated by phenotypic selection analysis. (A) A significant coefficient, β_1, indicates that the trait is under selection favoring either larger (as shown) or smaller trait values. (B, C) Significant quadratic terms in the regression model, β_2, indicate stabilizing selection if the partial regression coefficient is negative (B) and indicate disruptive selection if the coefficient is positive (C). (D) Significant joint selection on a pair of traits is indicated by a significant cross-product term in the regression model (β_3).

(continued)

To predict rates of evolutionary change for specific traits, selection coefficients must be expressed in original trait units. However, to compare overall strength of selection among traits, populations, or environments, it is more appropriate to use a standardized measure: selection intensity (i). Selection intensity is estimated by simply standardizing original trait data ($\bar{z} = 0$, standard deviation $= 1$) before estimating covariance between relative fitness and traits as described earlier.

It may be difficult to identify traits that are under selection in studies that are conducted on populations in their native habitat, because traits may already be optimized by historical selection. Power to detect selection on traits in a particular environment can be enhanced by expanding the phenotypic distribution such that more extreme trait values are present than in the native population (Mitchell-Olds & Shaw, 1987). Such broadening of the phenotypic distribution has been accomplished through the use of mutant and transgenic lines, physiological manipulation, hybridization of ecotypes, physical manipulations, artificial selection, or mixing genotypes obtained from different environments (see Etterson, 2004a).

BOX 10.2 Quantitative Genetics Terminology

Genetic architecture broadly refers to the distribution and nature of genetic variation at multiple scales, including variation at the levels of species, region, population, family, and individual. Genetic architecture at the population level is of primary interest for predicting evolutionary change and encompasses genetically based variation that can be transmitted across generations as well as differential expression of variation in different environments (in other words, plasticity) and genetic correlations among traits.

Heritability is an estimate of the fraction of that variance in phenotype that can be attributed to additive genetic variation (V_A) that is inherited directly from parents. Remaining phenotypic variation may be attributable to dominance genetic variation (V_D) that is reestablished each generation, and environmental variation (V_E), which is generally not inherited. I restrict my discussion to these basic effects, although experimental designs are available to estimate variance resulting from epistasis (V_I) and parental effects (V_M) (Lynch & Walsh, 1998).

There are two kinds of heritability estimates that differ in precision: broad-sense heritability (H^2) and narrow-sense heritability (h^2) (Lynch & Walsh, 1998). In both cases, heritability estimates are obtained by measuring individuals in families that are related to some degree. Broad-sense estimates can be obtained from measurements made on replicated genotypes (for example, clones in plants) or from broods obtained from natural matings (for example, nestlings or seed collected from maternal plants). However, estimates based on these kinds of relatives are confounded with effects that are generally not transmitted across generations and therefore will not contribute to evolutionary change. Consequently, broad-sense heritability is a coarser upper-bound estimate:

$$H^2 = V_A + V_D/V_A + V_D + V_E \qquad (10.1)$$

Narrow-sense heritability is a more precise estimate because it partitions additive genetic effects from other effects (compare the numerator of Box Eq. 10.1 and 10.2).

$$h^2 = V_A/V_A + V_D + V_E \qquad (10.2)$$

(continued)

BOX 10.2 Quantitative Genetics Terminology *(cont.)*

Breeding values estimate the genetic value of an individual, accounting only for additive genetic effects. The variance in breeding values is the additive genetic variance V_A. In contrast to a family mean, breeding values are not confounded by other sources of variation (dominance, maternal effects). Thus, the difference between a family mean and a breeding value is analogous to the difference between broad-sense and narrow-sense heritability.

G-matrices describe the pattern of additive genetic variance and covariance for multiple traits or character states in different environments (Fig. 10.3). The G-matrix is symmetrical, with the additive genetic variances on the diagonal and additive genetic covariances off the diagonal as shown here.

$$G = \begin{pmatrix} V_{A_i} & Cov_{A_{ij}} & Cov_{A_{ik}} \\ Cov_{A_{ij}} & V_{A_j} & Cov_{A_{ik}} \\ Cov_{A_{ik}} & Cov_{A_{jk}} & Cov_{A_k} \end{pmatrix} \quad (10.3)$$

Heritability or G-matrices can be obtained in three basic ways: (1) measuring traits in parents and offspring—for instance, in natural populations where parents are known (see, for example, Réale et al., 2003); (2) measuring offspring that were produced from the intentional matings according to a specific experimental design (see, for example, Etterson & Shaw, 2001); or (3) from response to artificial selection from one generation to the next (see, for example, Lenski, 2001). Although narrow-sense heritability estimates are preferable, they are more difficult to obtain because pedigree information for the offspring must be available. Also note that heritability is not a fixed attribute of a population, but may change in different environments (see Etterson, 2004b).

selection per generation for a single trait (R) is a function of heritability (h^2) and the strength of selection as measured by the selection differential (S):

$$R = h^2 S \quad (10.1)$$

In the simplest case, we expect a large evolutionary response to occur if natural selection is consistently and strongly targeting traits that are heritable and are uncorrelated with other traits under selection. The following sections describe how to estimate these coefficients and explore the evolutionary consequences if the conditions stated here are not met.

Natural Selection

To predict evolution in response to climate change, we need to know how patterns of natural selection will change in the future. In other words, what will be the targets and strength of selection with climate change? Selection regimes may simply shift to higher elevations or latitudes with climate warming, or may change in more complex ways because the ecological context, including both abiotic and biotic factors, will be altered. The most direct way to assess trends in the pattern of selection is to conduct phenotypic selection analyses on data collected from long-term studies of field populations. However, there are few data sets that are complete enough to document such temporal changes in selection on wild populations (but see Grant & Grant, 1995). An alternative approach is to compare natural selection in current environments with selection in experimental conditions that mimic those predicted for the future. However, most studies that have manipulated environmental conditions, such as temperature, precipitation, and CO_2, have focused on changes in species composition rather than on patterns of natural selection, although there are a few exceptions (see Totland, 1999). Insight into temporal changes in selection may also be obtained by characterizing spatial changes in selection along environmental gradients that encompass a range of environments similar to those predicted for the future (Etterson, 2004b). However, this approach also may provide an incomplete picture of future selection because native species composition will also be altered by climate change, and invasive species, pests, and diseases may invade new territories and alter patterns of selection. Furthermore, patterns of selection may be more erratic in the future if climates become more prone to extreme events such as drought, heavy precipitation, heat waves, and intense tropical cyclones (Alley et al., 2007).

Genetic Variation

Populations respond to natural selection through changes in allelic frequencies. Thus, the most fundamental requirement for adaptive evolution is that populations harbor diversity at loci that underlie traits that are the targets of selection. Genetic diversity within populations is sometimes inferred from measures of neutral molecular variation (Boulding, this volume). However, molecular variation will not necessarily correspond to genetic variation in complex traits that are a product of many genes that map to locations throughout the genome and interact with each other and with the environment throughout the course of development to influence phenotype (Pfrender et al., 2000). Timing of life history events, dispersal ability, thermal and drought tolerance, and competitive ability are examples of polygenic traits that do not have a simple genetic basis but will nevertheless be likely targets of selection under a changing climate. Although some of these traits may be influenced mainly by just a few chromosomal regions (QTLs), ongoing research has yet to resolve how many genes are represented within these regions and whether these genes are relevant under field conditions (see, for example, Weinig et al., 2002). Quantitative genetics offers alternative measurements of genetic variation for single traits (additive genetic variance and heritability) or multiple traits (G-matrix). Recently, models have been developed that can accommodate genes of major effect into a quantitative genetic framework (Walsh, 2001).

Genetic Correlations

Phenotypic traits are often not inherited independently but are genetically correlated such that a change in one trait results in a concomitant change in another trait. This arises either because single genes affect more than one trait (pleiotropy) or because multiple genes tend to be inherited together as a unit (linkage). Genetic correlations can enhance or reduce evolutionary rates depending upon their relationship to the direction of selection. Two kinds of genetic correlations are relevant in the context of climate change: genetic correlations among traits and genetic correlations across environments.

Additive genetic correlations among traits can enhance response to selection if the direction of the correlation is in accord with the direction of selection. For example, if two traits are positively genetically correlated and selection is favoring high values for both traits, then the joint vector of selection on these two traits matches the direction of the correlation and reinforces evolutionary response (Fig. 10.2A, inset 1). In contrast, if selection is favoring high values for one trait but low values for the other, then the joint vector of selection on these two traits is antagonistic to the direction of the genetic correlation and may thus slow evolutionary response (Fig. 10.2A, inset 2). When genetic correlations among traits are not in accord with the direction of selection, adaptive evolution may be slowed or maladaptive evolution may occur. Furthermore, traits that are not directly under selection may evolve because they are correlated with other traits that are direct targets of selection (see also Box 11.1 on linkage in Gilchrist and Folk, this volume).

Additive genetic correlations across environments can also influence evolutionary rates. Such correlations are especially important if the pattern of selection fluctuates over time, which may occur if climate varies between wetter/drier or warmer/cooler conditions. A single genotype may respond differently to these different environments, a phenomenon called *genotype-by-environment interaction*. These different *character states* of a genotype are often depicted in a reaction norm diagram (Fig. 10.3A, C, E). The reaction norm can

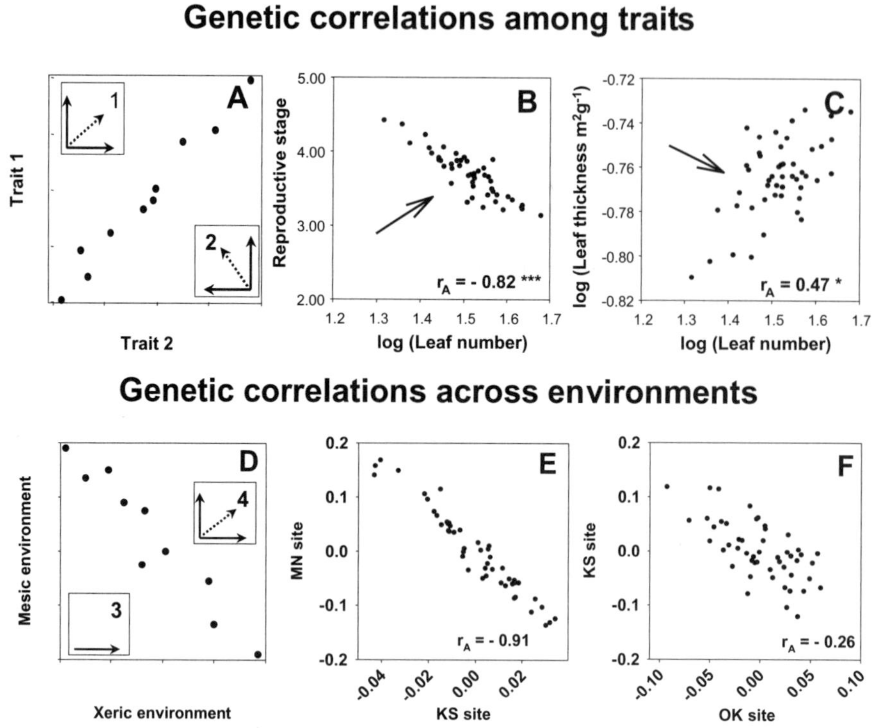

FIGURE 10.2 (A–F) The effect on adaptive evolution of genetic correlations among traits (A–C) and across environments (D–F). (A) This graph shows a hypothetical positive genetic correlation between a pair of traits where points on the graph are family means or *breeding values*. The arrows in inset 1 show selection favoring families that have high values for both traits, resulting in a joint vector of selection (dotted arrow) that is positive and thus in accord with the direction of the genetic correlation. Evolutionary change is reinforced in this case because the direction of the genetic correlation matches the direction of selection acting on the same pair of traits. In other words, some families express the combination of trait values that are favored by selection (upper right). In contrast, the arrows in inset 2 show that selection is favoring families that have high values for trait 1 but low values for trait 2. In this case, evolutionary change is constrained because the direction of the genetic correlation is antagonistic to the joint vector of selection on these traits. No families possess the combination of trait values that are favored by selection (upper left). Positive selection on trait 1, for example, would result in a maladaptive positive response in trait 2. (B) This scatterplot of Minnesota population breeding values for reproductive stage and leaf number shows a negative genetic correlation that is antagonistic to the positive vector of joint selection on these traits. (C) This scatterplot of the Minnesota population breeding values for leaf thickness and leaf number breeding values shows a positive genetic correlation that is antagonistic to the negative vector of joint selection. (D) This graph presents a hypothetical negative genetic correlation among breeding values for fitness across two environments (mesic and xeric), indicating a genetic trade-off (in other words, genes that contribute to high fitness in the mesic environment are associated with low fitness in the xeric environment). In contrast, a positive correlation (near one) would indicate that the evolution of a generalist phenotype with high fitness in each environment is possible (for example, see Fig. 10.3D). Intermediate values of genetic correlation would indicate that independent adaptation to either or both environments is possible (for example, see Fig. 10.3F). If climate is changing directionally, only selection in the xeric environment matters (inset 3). In this case, across-environment genetic correlation is irrelevant because selection will consistently favor genotypes with relatively high fitness under xeric conditions. However, if climate fluctuates within the lifetime of the organism, only families with high fitness in both environments are favored by selection (inset 4). (E, F) Negative genetic correlations for estimated lifetime fecundity are shown for the Minnesota population when raised in its home site and in Kansas (E), and the Kansas population when raised in its home site and in Oklahoma (F). KS, Kansas; MN, Minnesota; OK, Oklahoma. $^{*}P < 0.05$; $^{***}P < 0.001$. (Views A through C are from Figure 1 of Etterson and Shaw [2001]; views D through F are modified from Figure 1 and Table 3 of Etterson [2004b].)

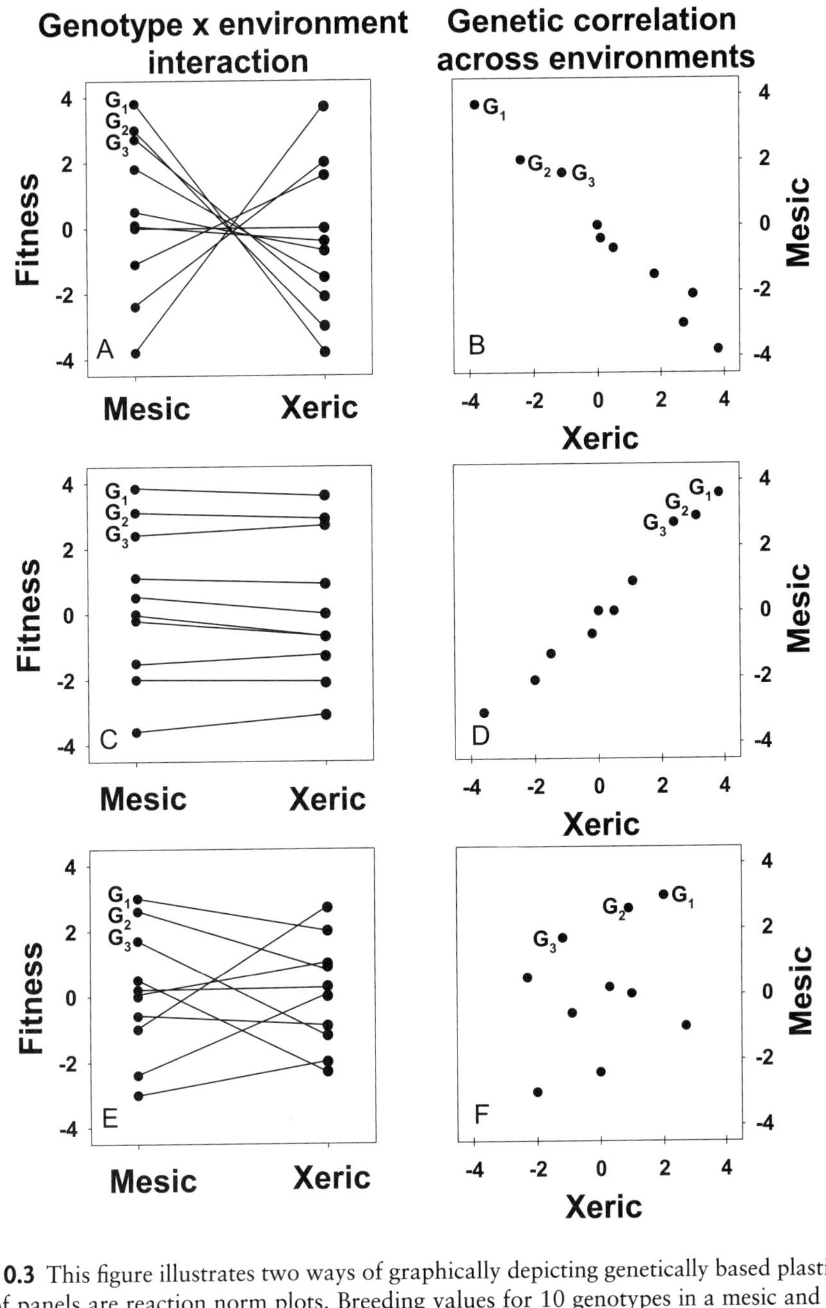

FIGURE 10.3 This figure illustrates two ways of graphically depicting genetically based plasticity. The left column of panels are reaction norm plots. Breeding values for 10 genotypes in a mesic and xeric climate are connected by a line [for example, G_1in (A) has the highest fitness in the mesic environment but the lowest fitness in the xeric environment]. The right column of panels shows the same data but is plotted to show the genetic correlation with the environments represented on the x- and y-axes. Breeding values for a particular genotype are shown as a single point [for example, G_1 in (B) has the highest fitness in the mesic environment but the lowest fitness in the xeric environment, as in (A)]. (A) Genotypes attain high fitness either in the mesic condition or the xeric condition, but not in both. (B) This trade-off is depicted as a negative genetic correlation across environments. (C) Genotypes differ in fitness regardless of their environment. (D) Inherent differences in genotypic quality translate into a positive genetic correlation across environments. (E) There is genetically based variation in the pattern of reaction norms for the 10 genotypes, indicating that selection could mold plastic responses. (F) There is no significant genetic correlation across environments.

be redrawn as a genetic correlation by plotting breeding values for each genotype with the two environments represented on the x- and y-axes (Fig. 10.3B, D, F).

If climate change is more consistent and directional (for example, becoming progressively warmer or drier), then natural selection may favor adaptive evolution via fixed genetic changes that result in, for example, earlier breeding or greater thermal and drought tolerance (Fig. 10.2D, inset 3). However, if natural selection fluctuates (for example, extreme weather events such as droughts), then natural selection may favor adaptive solutions involving plasticity in traits such that fitness remains high across a range of environments (Fig. 10.2D, inset 4).

Genetic correlations among traits or across environments can be incorporated into predictions of adaptive evolution using a multivariate extension of the breeder's equation:

$$\Delta\bar{\mathbf{z}} = G\beta \qquad (10.2)$$

where the predicted change in a vector of traits ($\Delta\bar{z}$) across one generation can be predicted using the additive genetic variance–covariance matrix (G) and the vector of selection gradients (β) (Lande, 1979). However, this method requires taking the product of two estimates, each of which has an associated error as a result of sampling and measurement error. Thus, significance testing is not straightforward. Alternatively, the response to selection can be estimated in a single step as the additive genetic covariance between relative fitness and traits (Price, 1970):

$$\Delta\bar{\mathbf{z}} = \mathrm{Cov}_A[w, \mathbf{z}] \qquad (10.3)$$

where w is individual relative fitness (absolute fitness divided by mean fitness), and $\mathbf{z}$ is the vector of trait values. This approach is advantageous because the response to selection can be directly tested by assessing the significance of the additive genetic covariance using maximum likelihood techniques (Shaw & Shaw, 1994).

The rate of adaptive evolution may be positively or negatively influenced by gene flow between populations. Evolution may be more rapid with gene flow because genetic variation is increased (Swindell & Bouzat, 2006). Alternatively, gene flow may slow evolution if it occurs between populations that are adapted to distinct environments (García-Ramos & Kirkpatrick, 1997). Theoretical models show that if populations occur across gradual environmental gradients, gene flow between adjacent populations does not inhibit local adaptation because populations are similarly adapted. However, if the environmental gradient is steep, the process of local adaptation is swamped by gene flow between populations that are adapted to substantially different environments. Under these conditions, the level of genetic variation harbored by populations will have little effect on local adaptation and population persistence. However, if gene flow is interrupted, evolutionary change of peripheral populations may proceed quickly.

Species that occur across environmental gradients frequently express clinal patterns in phenotype that may suggest underlying local adaptation. However, these patterns do not necessarily reflect genetic differentiation, but may be a product of environmental responses or phenotypic plasticity that may or may not be adaptive. The extent to which clinal patterns in phenotype can be attributed to genetic differentiation, phenotypic plasticity, or some combination of these can be ascertained through common-garden experiments in which populations are sampled along an environmental gradient and then reared in common conditions. Because the populations experience the same environment, differences detected among them in morphology, physiology, and phenology must be genetically based. Furthermore, if this differentiation is related to the gradient from which they were sampled, they will retain clinal patterns in a common environment (see, for example, Etterson, 2004a). Clinal patterns that disappear in a common garden can be attributed to phenotypic plasticity (see, for example, Maherali et al., 2002). The adaptive value of this plasticity cannot be evaluated with this experimental design (see Table 10.1 for an overview of experimental approaches).

A particularly interesting and extensive study of this kind compared patterns of trait variation among 16 California populations of an annual plant species that were sampled along elevation gradients at each of six latitudes (Jonas & Geber, 1999). This study supported some adaptive hypotheses—for example, development time decreased with elevation, and rates of photosynthesis decreased with latitude. However, patterns of variation across elevation and latitude were not necessarily consistent, even though these sampling transects encompassed

TABLE 10.1 Overview of Experimental Approaches That Can Be Used to Detect Genetic Differences among Populations, Infer the Adaptive Value of These Differences, and Predict Potential Evolutionary Responses.

Approach	Observation	Inference	Value
Observations in the field	Differences among populations	Genetic differentiation, phenotypic plasticity, or a combination of these two effects	Develop adaptive hypotheses
	Differences in fitness among individuals within populations that are associated with particular trait values	Phenotypic selection	Identify traits that are the targets of selection; estimate the strength and direction of selection to predict evolutionary response
Common garden in one environment	Differences among populations	Genetic differentiation between populations	Determine that population differences are genetically based
Common garden with an experimental treatment	Differences among populations	Genetic differentiation between populations	Determine that population differences are genetically based
	Differences among treatments	Phenotypic plasticity	Test the adaptive values of plastic responses to the experimental treatment
	Different responses among populations in response to the treatments	Genetic differentiation between populations in phenotypic plasticity	Test the adaptive value of genetic differences among populations in relation to the experimental treatment
Reciprocal transplant across a range of environments	Populations have highest fitness in their home site	Local adaptation	Test the adaptive value of population differences; assess the scale of local adaptation
Pedigreed offspring from one population in one environment	Differences among families	Genetic variation within the population	Estimate heritability and genetic correlations among traits; can be used to predict evolutionary response
Pedigreed offspring in more than one environment	Differences among families	Genetic variation within the population	Estimate heritability and genetic correlations among traits; can be used to predict evolutionary response in each environment
	Differences among treatments	Phenotypic plasticity	Test the adaptive values of plastic responses to the experimental treatment
	Differences among families in response to the treatments	Genetic variation within the population in phenotypic plasticity	Estimate genetic variation within populations for phenotypic plasticity and across-environment genetic correlations to predict further evolution of plasticity

Shown in order from the least informative to the most informative.

similar clines in temperature and precipitation. Hence, this study provides a cautionary note and indicates that selection along different geographical clines used as surrogates for future change may not generate consistent patterns.

Information gained from common-garden experiments can be enhanced by exposing populations to different environmental treatments such as temperature, precipitation, CO_2, and competition. If a specific environmental factor (for example, temperature) has been important in trait divergence, then populations should respond according to the conditions in their native habitats (for example, populations from hotter environments perform better in the warmer treatment). The adaptive value of plastic responses can also be evaluated with this experimental design. For example, adaptive plasticity would be inferred if a plant maintained fitness by changing the angle of its leaves to reduce incident light, thereby preventing thermal damage to photosynthetic enzymes. Overall, this experimental approach can provide valuable insight regarding specific attributes of the environment that are responsible for patterns of genetic differentiation (in other words, the agents of selection), in addition to providing evidence of inherent genetic divergence among populations.

Local adaptation is most commonly tested with reciprocal-transplant, or *provenance,* experiments that compare fitness of populations in their native environment with their fitness in alternative environments. Local adaptation is inferred if populations obtain higher fitness in their native site. These experiments are particularly powerful in the context of climate change if populations have been moved in a direction that is concordant with predictions of climate change. Such experiments provide insight into the extent and scale of local adaptation as well as decrements in fitness that may occur with climate change assuming no range shifts or adaptive evolution. Data from large-scale provenance experiments of commercially important tree species that have been established for as long as 50 years are now being reexamined in the context of climate change. These valuable data show that populations have evolved different climate optima and breadth of tolerance, and suggest that these populations will have substantially reduced survival and growth over the long term if climate changes as predicted (Rehfeldt et al., 1999).

Are Populations Evolving as Climate Changes?

Many organisms are already responding to climate change as evidenced by changes in plant and animal phenology, physiology, and behavior. In a meta-analysis, Paremsan and Yohe (2003) examined 677 species with published records and found that 62% showed trends toward earlier timing of life history events. Changes in life history were manifest as earlier frog breeding, earlier bird nesting, earlier date of first flowering and tree budburst, and earlier arrival of migrant birds and butterflies. It is rarely known whether these changes reflect underlying genetic change or phenotypic responses to an altered environment. However, evidence from artificial selection experiments suggests that it is at least plausible that adaptive evolution is contributing to these patterns. For example, a native herbaceous species flowered 13 days earlier after three generations of artificial selection (Burgess et al., 2007) which is larger than the 3- to 8-day change observed for plant species in nature. Likewise, artificial selection for more than 2,000 generations in the bacterium *Escherichia coli* showed that thermal tolerance can undergo adaptive evolution, even under conditions of fluctuating selection (Lenski, 2001).

Evolutionary response to climate change has already been documented for a few wild organisms for which long-term study of field populations have been conducted. For example, during the past 30 years, mosquitoes that breed in pitcher plants in the eastern United States have evolved different genetically based photoperiodic cues for breaking dormancy that correspond to increases in the length of the growing season in recent decades (Bradshaw & Holzapfel, 2006; Gilchrist and Folk, this volume). In Australia, 20 years of temporal sampling of fruit fly populations has demonstrated latitudinal shifts of clinal variation of genetic traits (Umina et al., 2005). Rapid evolution of clinal variation has also been observed for several invasive species (see, for example, Huey et al., 2000). Evolution of the timing of breeding has also been shown for a population of red squirrels in Canada. During the past decade, spruce trees have been producing their cones progressively earlier (Réale et al., 2003), which has in turn favored earlier-breeding squirrels. By tracing inheritance of timing of breeding from parents to offspring for several generations, these researchers were able to distinguish genetic change from phenotypic plasticity and ultimately attributed observed changes to both factors. Evolution of timing of life history events may be a more important mechanism of adaptation to climate change in temperate regions than the evolution of thermal tolerance per se (Bradshaw & Holzapfel, 2006).

How much genetic change must occur in order for populations to keep pace with climate change? The answer to this question depends upon the extent to which populations are already locally adapted and the extent to which they can adapt through plastic responses without genetic change. Populations that are highly locally adapted and that have specific climate requirements will experience changes in selection immediately and, assuming that traits under selection are genetically based, evolve. In contrast, populations that are less specialized and can maintain fitness across a wider range of environmental conditions through physiological and morphological plasticity will not experience strong changes in selection until the margin of their ecological tolerance has been exceeded. Thus, the rate of adaptive evolution may be slower for more plastic organisms because they possess greater ecological amplitude whereas the rate of evolution may be faster for less plastic organisms because their range of tolerance may be more rapidly exceeded. Plasticity itself may be a genetically based trait that may be under selection and undergo evolutionary change. Changes in selection resulting from global warming are likely to target genetic variation at both of these levels because climate is predicted to change not only directionally but also become more variable with an increased frequency of extreme weather events such as drought. Such a pattern of climate variation may favor the evolution of plasticity.

Adaptive Evolution in Response to Climate Change: A Case Study from the Great Plains

Little information is available regarding how much adaptive evolution will be required to keep pace with climate change and whether it can occur fast enough to rescue populations from extinction. Yet, this information is crucial for understanding the impact of climate change on native populations and for developing appropriate management practices. To obtain a better understanding of these issues, I conducted a series of experiments that were designed to address the following questions:

1. How far will climate change shift populations away from their optimal environments?
2. What traits need to evolve to close the fitness gap that results from this shift?
3. How much genetic variation is there within populations for traits that are the targets of selection?
4. Will genetic correlations between the targets of selection slow adaptive evolutionary responses?
5. What impact will the biotic community have on selection response?
6. Can we facilitate adaptive evolution by human-mediated gene flow in the direction of climate change predictions?

The focal species for this work was an annual legume, *Chamaecrista fasciculata* (Fabaceae), that is native to the Great Plains of the United States. This species is primarily outcrossing and is pollinated by bumblebees. In addition to more generalist herbivores, the developing pods harbor a specialist seed herbivore, *Sennius cruentatus* (Bruchidae), whose larvae develop inside maturing fruit and consume one to three seeds (personal observation).

Across the study area, temperature increases from north to south and precipitation decreases from east to west, producing an aridity gradient (Fig. 10.4). The underlying premise of this work was that this spatial gradient in climate could serve as a proxy for the temporal gradient expected with climate change. For example, plants from Minnesota may experience a climate more similar to the current climate of Kansas or Oklahoma in the future, depending upon the rate of climate change and the time frame of interest.

Are Populations Already Genetically Differentiated According to Climate?

My first observation was that *C. fasciculata* populations varied clinally in the field—for example, leaves become smaller and thicker as one moves from north to south along the cline (unpublished data). To determine whether clinal variation in leaf morphology was genetically based and associated with the cline in water availability, I collected seed from five populations located between southern Oklahoma (35°N) and southern Minnesota (44°N) and reared these plants in a common-garden environment in the greenhouse. Plants from both populations were then exposed to one of two watering treatments: saturated and restricted (Etterson, 2001). This experiment showed clear patterns of clinal variation in phenotype that corresponded to the latitude of population origin (Fig. 10.5A). For example, northern plants exhibited early rapid growth rates, were shorter, had more branches and more leaves, had smaller flowers and flower parts, and tended to fold their leaves to a greater extent in response to midday water stress. Furthermore, northern plants responded positively to increased

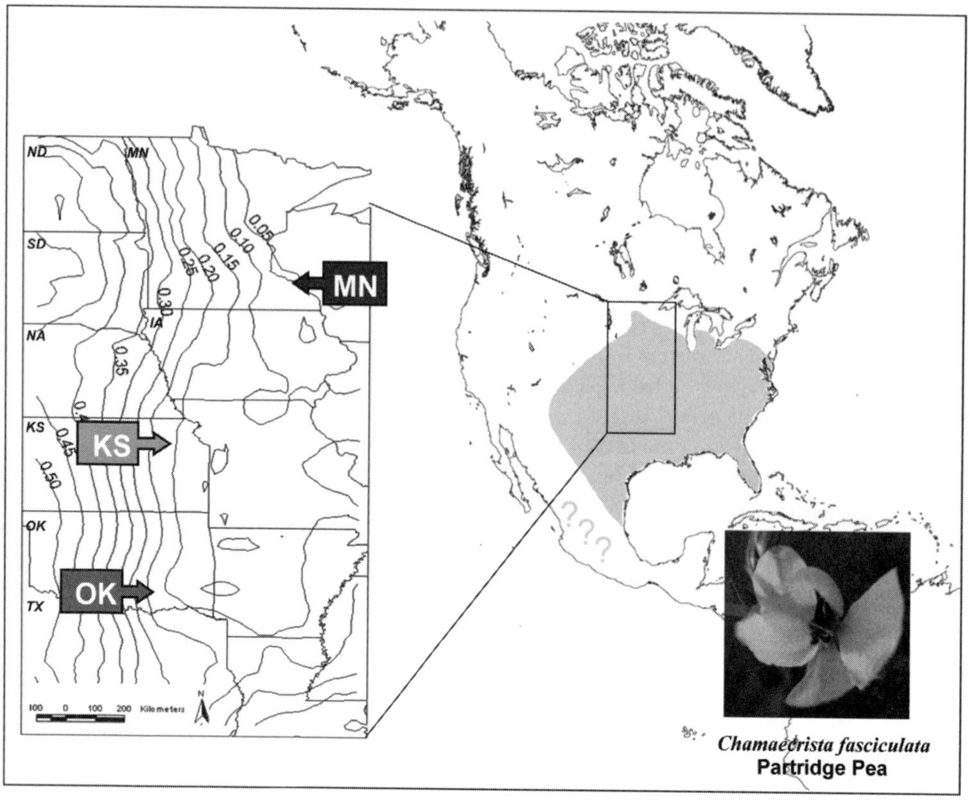

FIGURE 10.4 The study area in the Great Plains is boxed within the known distribution of *Chamaecrista fasciculata* in North America. Temperature increases from north to south and precipitation decreases from east to west produce an aridity gradient across the Great Plains of the United States. Shown are isoclines of α, an integrated measure of seasonal growth-limiting drought stress on plants that takes into account temperature, precipitation, and soil texture (α for evergreen trees, 1951–1980 [Thompson et al., 2000]). Arrows point to the locations of focal populations for quantitative genetics study. (Modified from Figure 1 of Etterson [2004a].)

water availability for vegetative traits but not for fitness as measured by flower number (Fig. 10.5B). In contrast, southern plants showed no difference in vegetative traits but produced significantly more flowers when exposed to drought. Overall, this experiment showed genetic differences between C. *fasciculata* populations that vary according to latitude, and suggested that water availability may be an important factor that has contributed to this differentiation.

However, for these traits to continue to evolve, there must be genetic variation within populations. To get a preliminary estimate of genetic diversity for traits associated with drought tolerance, I designed a second common-garden experiment in which full-sibling families from the northernmost and the southernmost populations, and one centrally located population, were each raised in the greenhouse and then subjected to short-term drought (Etterson, 2001). As in the previous experiment, there was a strong pattern of clinal variation among *C. fasciculata* populations for both morphological and physiological traits. Furthermore, populations exhibited different physiological responses to drought for some traits (Fig. 10.5C, D), suggesting that southern populations were more drought adapted. Broad-sense heritability could be inferred for about half the morphological traits, but only weakly for physiological traits.

Based on these studies, I developed the following three adaptive hypotheses. In more arid climates,

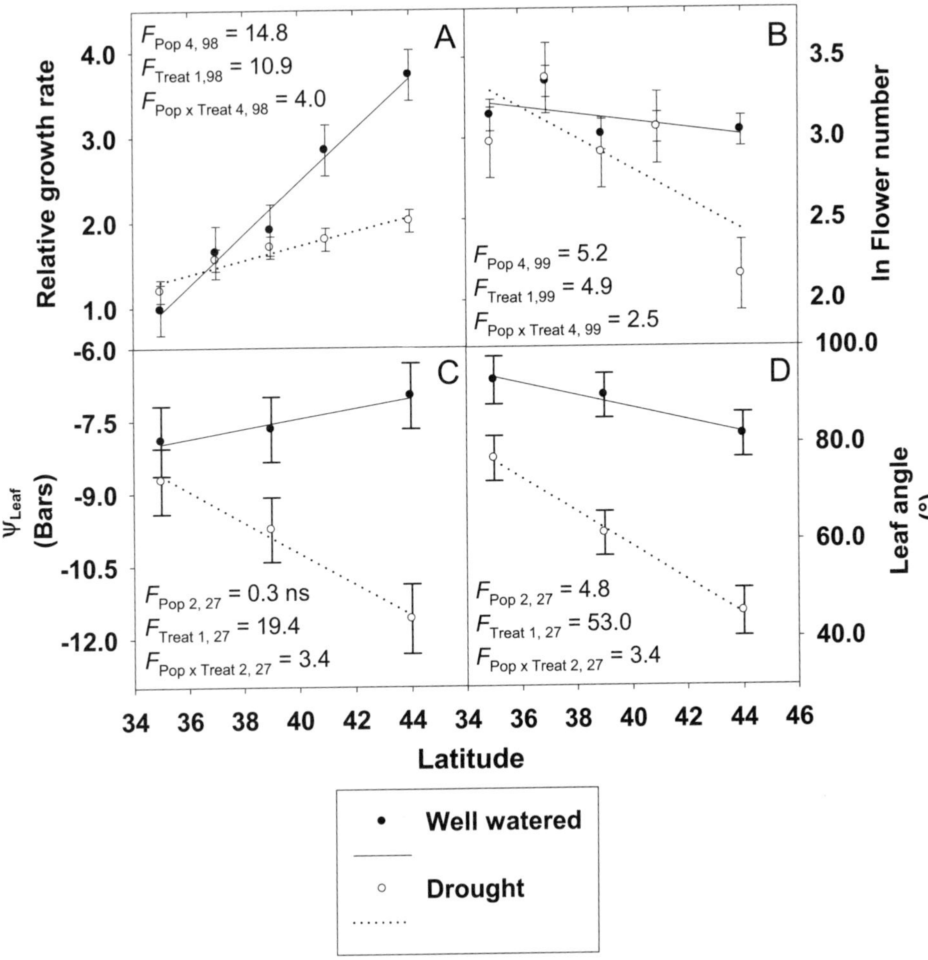

FIGURE 10.5 (A–D) Data from two separate common-garden studies in which populations were sampled from the aridity gradient in the Great Plains and subjected to different watering treatments. Long-term drought experiment with five populations located in Minnesota, Iowa, Kansas, northern Oklahoma, and southern Oklahoma: (A) early relative growth rate calculated as ln(leaf number_{t2}) − ln(leaf number_{t1})]/$t_2 - t_1$, where t = 1 month; (B) the sum of 13 weekly flower counts. Short-term drought experiment (3 days) with three populations located in Minnesota, Kansas, and southern Oklahoma: (C) leaf–water potential and (D) leaf angle (0 deg means that leaves are fully folded and 180 deg means that leaves are fully unfolded) (Modified from Etterson [2001].)

selection should favor (1) rapid progression through the life cycle during the short growing season of the north but should be weaker or reversed in sites further south, (2) smaller plant size in the south because of greater transpiration–lost water suffered by plants with more leaf surface area, and (3) thicker leaves in the more southern sites because smaller but thicker leaves are less subject to water loss but can maintain rates of photosynthesis.

Can Populations Evolve in Response to Future Climate Change?

My paramount goal was to predict whether adaptive evolution of these traits could occur rapidly enough to keep pace with climate change. To this end, I used a formal breeding design so that estimates of narrow-sense heritability and among-trait and across-environment genetic correlations could be obtained. Pedigreed *C. fasciculata* families were reciprocally planted into each of three common-garden plots located, respectively, in a northern site, a central site, and a southern site (Etterson, 2004b).

A strong pattern of local adaptation was evident in this experiment, with native populations having the highest relative fitness in each of their home sites as measured by estimated lifetime seed production (Fig. 10.6A). At the central site, reproductive fitness was strongly influenced by the presence of the bruchid seed herbivore *S. cruentatus*. Plants from all populations were attacked, but percentage of infested plants was greatest for each of the local populations (percentage of total plants attacked: Minnesota, 26%; Kansas, 36%; Oklahoma, 21%; $\chi^2 = 83.1, P < .0001$, N = 2754) (Etterson, unpublished data). Despite losses to herbivores, local populations still had the greatest seed production relative to the nonlocal populations. When northern populations were reared in more southern sites, they experienced severe reductions in fitness. The Minnesota population produced 31% and 67% less seed than the local populations in Kansas and Oklahoma, respectively, and the Kansas population produced 11% less seed than the local population in Oklahoma. This pattern of local adaptation was further supported by phenotypic–selection analyses that showed natural selection favored trait values of the native population (Fig. 10.6B–D) (Etterson, 2004a).

All tested *C. fasciculata* populations exhibited substantial phenotypic plasticity that could in some cases be adaptive because it was in the direction of trait values favored by selection (Fig. 10.6B–D). Even though plasticity was in an adaptive direction, it was not adequate to maintain fitness across the broad increments of climate tested here (Fig. 10.6A).

Populations had significant heritabilities for most traits under selection, although the magnitude of the estimates varied widely and were generally lower for the northern population at the periphery of the species range (Etterson, 2004b). Despite appreciable selection and genetic variance, when among-trait correlations alone were taken into account using multivariate methods (Eq. 10.3), predictions of evolutionary response were more often less in absolute magnitude than univariate predictions (Eq. 10.1), and in many cases one half or less (Etterson & Shaw, 2001). Slow rates of evolutionary response are attributed to genetic correlations among traits that are antagonistic to the direction of selection under a changed climate (Fig. 10.2B, C). For example, selection favored plants with many thick leaves in the southern environment but when northern plants were grown in this site they produced either many thin leaves or few thick leaves, indicating a fundamental genetic trade-off. Overall, these analyses suggested that adverse genetic correlations among traits could retard evolutionary rates in response to climate change.

Four substantial negative across-environment genetic correlations were estimated for *C. fasciculata* fecundity, the closest proxy of fitness in this study (Fig. 10.2E, F). The extent to which these correlations will pose further obstacles to adaptive evolution depends upon the constancy of climate change that ensues. If future climates are more variable as has been predicted (Alley et al., 2007), antagonistic pleiotropy across environments may also impede evolutionary change.

Some climate models predict conditions in Minnesota similar to current-day Kansas within 25 to 35 years (for example, Canadian Global Coupled Model 1, [Canadian Climate Center, 1999]). Making the simplistic assumptions of constant genetic variation and selection coefficients, the number of generations required before the trait means of the Minnesota population are expected to match those of the native Kansas population generally exceeds the time predicted for this climate change (rate of phenological development, 21 generations; leaf number, 42; leaf thickness, 79). The Minnesota population is predicted to achieve the local population means in Oklahoma in fewer generations because of stronger selection and higher expression of additive genetic variance. However, these are likely underestimates of the number of generations required because strong selection over as few as 10 generations can substantially deplete genetic variation. Moreover, selection coefficients are not likely to remain constant. Thus, even though there is significant genetic variation for most traits, the rate of multivariate evolution is expected to be slower than the rate of climate change. Even if adaptive evolution gradually occurs, severe reductions in seed production suffered by the Minnesota population may increase the risk of extinction because of demographic instability. Theoretical models that couple evolutionary and demographic dynamics of closed populations have shown that even if adaptive responses are possible, densities may fall

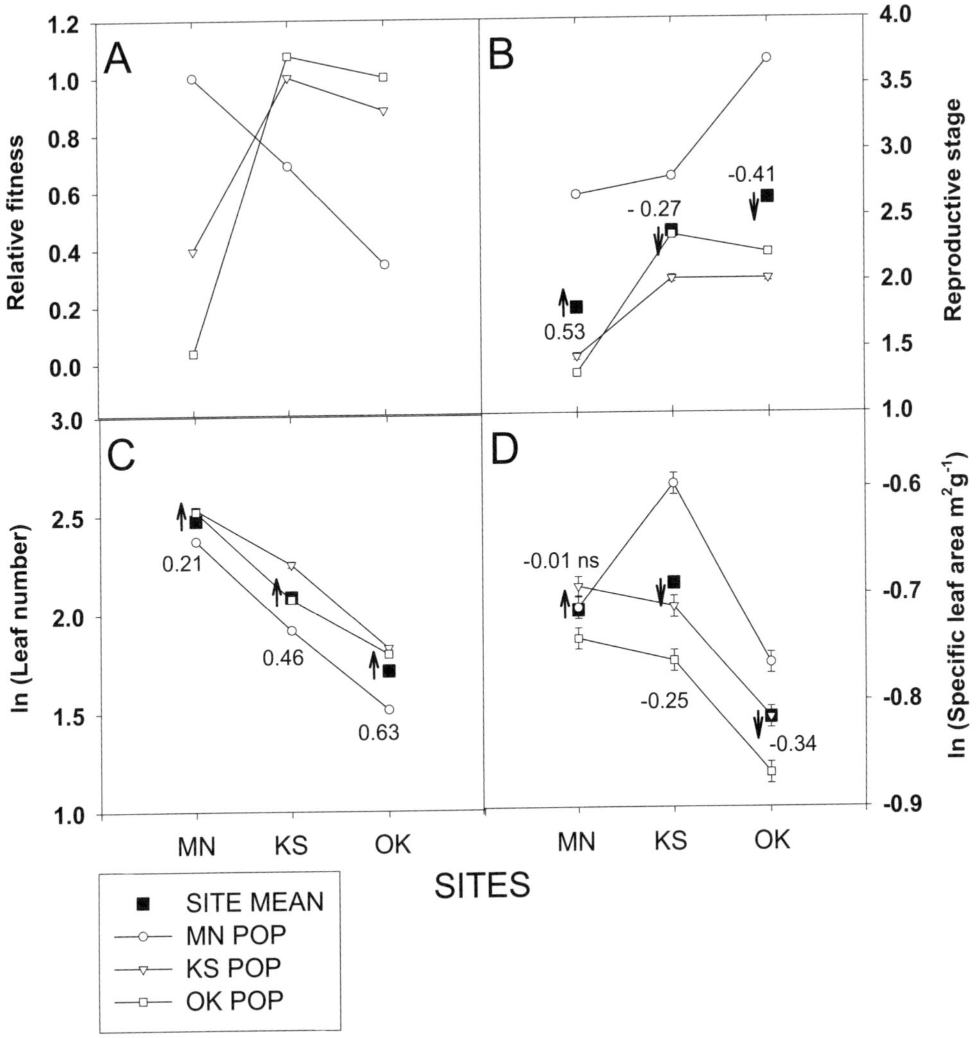

FIGURE 10.6 Data from three populations of *Chamaecrista fasciculata* reciprocally transplanted into three sites (Minnesota [MN], circles; Kansas [KS], triangles; Oklahoma [OK], squares). (A) Relative fitness calculated for each site as population mean fecundity divided by maximum population mean fecundity. (B–D) Least-squares means and standard errors (very small) of reproductive stage (B), leaf number (C), and specific leaf area (D) measured in three populations. Black squares show the mean for one site averaged across populations. Arrows indicating the direction of selection are shown with an estimate of S, the selection differential at each site. All S are significant at P < .001 except one marked NS. (nonsignificant). (Modified from Etterson [2004a], Fig. 2 and Table 2.)

below some critical level such that populations are susceptible to extinction by stochastic processes (Bürger & Lynch, 1995; Reed, this volume). In contrast, populations that are only slightly maladapted to changed conditions and have large populations are expected to be rescued by evolutionary change.

Although climate change is predicted to occur at a rapid rate, the abrupt stepwise change modeled in this experiment is not realistic. It is difficult to predict how the genetic architecture of these populations would differ if exposed to climate change gradually over many generations. However, the conditions of this experiment likely provide a conservative test of evolutionary potential because more gradual climate change would allow greater opportunity, relative to the magnitude of selection, for mutation and recombination to break up antagonistic trait correlations. The extent to which these populations can adapt to changes in selection may depend upon the stability of genetic correlations among traits, which is largely unknown (Cheverud & Routman, 1996).

Overall, this study suggests that if climate changes as predicted, northern populations of *C. fasciculata* will face a severe evolutionary challenge in the future. In particular, constraints on evolutionary change such as modest heritabilities, among-trait and across-environment genetic correlations antagonistic to selection, and demographic instability resulting from lower seed production in a hotter and drier climate may inhibit adaptation to novel environments.

Can Populations Be Managed to Mitigate Effects of Climate Change?

If species cannot respond to selection fast enough to keep pace with climate change, it may be necessary to manage populations to maintain fitness as climate changes. In principal, this could be accomplished by moving organisms with the band of climate to which they are adapted. However, this kind of human-mediated movement is not without risk. Introduced genotypes from lower latitudes or elevations are likely to interbreed with members of remnant native populations. This could benefit native populations if genetic variants that are well adapted to the new climate are introduced or if overall levels of genetic variance are increased. However, if populations are substantially diverged, genetic mixing could lead to reductions in fitness for subsequent generations of both transplanted and remnant native populations (Rhymer, this volume). Fitness losses among interpopulation crosses are referred to as *outbreeding depression* and can be the result of disruption in local adaptation, underdominance, and the breakup of positive epistatic gene interactions (reviewed in Edmands, 1999). Persistence of these negative genetic effects is generally not known (but see Erickson & Fenster, 2006), but will likely vary depending upon the extent of divergence among populations.

To evaluate the benefits and risks of human-mediated gene flow in this system, I produced interpopulation hybrids between Minnesota and Oklahoma plants. Seedlings from parental, F_1, F_2, and F_3 generations were reciprocally planted into the Minnesota and Oklahoma site (unpublished data). As in previous experiments, a strong pattern of local adaptation was evident (Fig. 10.7). In their native sites, Minnesota produced 20 times more seed than Oklahoma, and Oklahoma produced three times more seed than Minnesota. Hybrids had intermediate fitness in all cases. In the context of climate change, these data suggest that facilitated gene flow from Oklahoma to Minnesota today would reduce population fitness because hybrids produce less seed than native Minnesota plants in the current climate of Minnesota. However, if climate in Minnesota became more similar to Oklahoma, then facilitated gene flow could have an overall positive effect because hybrids exposed to the Oklahoma climate produce more seeds than Minnesota plants. Although human-mediated gene flow appears to be a promising management strategy for these populations, more studies of this kind should be conducted because outbreeding depression among interpopulation hybrids is often reported.

FUTURE DIRECTIONS

More experimentation is required to increase our understanding of the role that adaptive evolution will play as climate changes. It is essential to obtain more detailed genetic information for traits that will be the targets of selection for more species with different life history patterns and geographic scales of differentiation. Controlled experiments should be conducted using populations from a broad range of environments and in experimental conditions that may actually materialize in a changed climate. Estimates of survival and reproductive fitness should be taken whenever possible to evaluate whether morphological, phenological, and physiological responses are adaptive. Reciprocal transplant experiments of both long- and short-lived organisms should continue across multiple environmental clines and into the direction of predicted climate change. These experiments will be improved if other members of the biotic community are also included, because they can have a strong influence on population fitness. However, it is not sufficient simply to transplant into intact native communities, because these associations may not persist into the

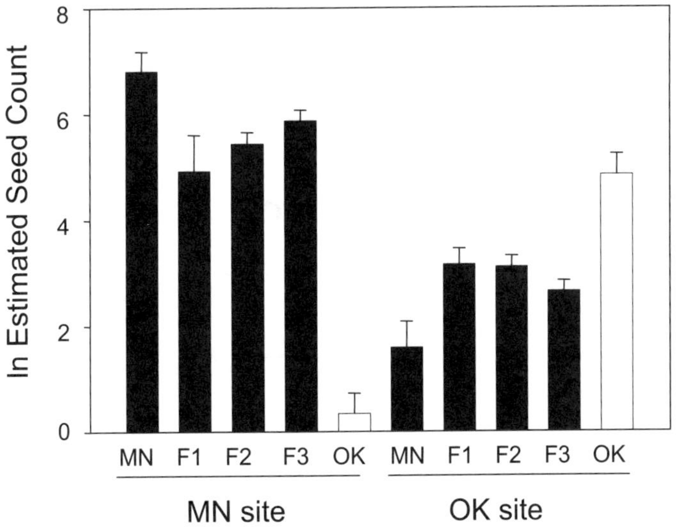

FIGURE 10.7 Estimated seed count for Minnesota (MN) and Kansas (KS) populations and their interpopulation F_1, F_2, and F_3 hybrids reciprocally planted into the Minnesota and Kansas site. (J. Etterson & R. Shaw, unpublished data.)

future as some organisms migrate, some evolve, and some go extinct. Lastly, collaboration should be fostered between quantitative geneticists, conservation biologists, ecologists, and physiologists to gain the greatest advantage from these experiments that tend to be long term and laborious.

SUGGESTIONS FOR FURTHER READING

Conservation biologists will find a cogent review of genetic inferences that can be made in natural populations using molecular genetic and quantitative genetics approaches in Pfrender and colleagues (2000). Barton and Turelli (1989) provide an entry into the literature on evolutionary insights gained through quantitative genetics. The vulnerability of populations to extinction with climate change is explored in theoretical models of Bürger and Lynch (1995) that simultaneously account for the rate and fluctuation in natural selection, the demographics of populations, and the genetics underlying quantitative traits under selection. An excellent empirical guide to conducting phenotypic selection analysis can be found in the review of Brodie and colleagues (1995). Davis and colleagues (2005) provide a more complete review of studies of evolution in response to climate change both in the past and the present.

Barton, N. H., & M. Turelli. 1989. Evolutionary quantitative genetics: How little do we know? Annu Rev Genet. 23: 337–370.

Brodie, E. D., III, A. J. Moore, & F. J. Janzen. 1995. Visualizing and quantifying natural selection. Trends Ecol Evol. 10: 313–318.

Bürger, R., & M. Lynch. 1995. Evolution and extinction in a changing environment: A quantitative-genetic analysis. Evolution 49: 151–163.

Davis, M. B., R. G. Shaw, & J. R. Etterson. 2005. Evolutionary responses to climate change. Ecology 86: 1704–1714.

Pfrender, M. E., K. Spitze, J. Hicks, et al. 2000. Lack of concordance between genetic diversity estimates at the molecular and quantitative-trait levels. Conserv Genet. 1: 263–269.

11

Evolutionary Dynamics of Adaptation to Environmental Stress

GEORGE W. GILCHRIST
DONNA G. FOLK

Organisms have always lived in changing environments; however, human influences have dramatically increased the frequency and intensity of environmental change. The forces of environmental change, including habitat destruction, biological invasions, and climate change, may each involve changes in the physical environment and thus expose organisms to environmental stress—that is, cause fitness-reducing organismal responses to external factors. In particular, habitat destruction may induce surviving members of a population to emigrate to new, potentially less favored habitats where new challenges must be faced by a genetically and demographically depleted population (Travis, 2003). Similarly, biological invasions directly thrust the invading organism into novel and potentially stressful physical habitats, wherein they may engage in biotic interactions with the native community. Establishment of nonnative species and its ultimate impact on the community may depend on its ability to adapt to these changed physical conditions (Lee et al., 2003). Ultimately, climate change acts both in situ, by changing the characteristics of the ancestral habitat in ways that may increase stress (Pörtner & Knust, 2007), and ex situ, by forcing species to emigrate to habitats that may have a less stressful climate (Parmesan, 2006). Climate change may be driven locally by habitat alteration, such as deforestation or increasing pavement area, or globally by increasing greenhouse gases.

In each of these examples of environmental change, altered abiotic factors such as temperature cycles, humidity profiles, photoperiods, and other chemical and climatic factors can induce stress, which may in turn have direct and indirect demographic consequences that ultimately drive evolutionary change (Bijlsma & Loeschcke, 2005; Hoffmann & Hercus, 2000). The potential for adaptation to stress imposed by altered climatic conditions, be they driven by increasing greenhouse gases, local habitat change, or population relocation, is the subject of this review.

We focus on an unlikely group of organisms for conservation biologists: fruit flies in the genus *Drosophila* (but see Frankham, 2005c). Although not generally viewed as an ecological model, probably more is known about the ecological genetics of adaptation to climate in drosophilids than any other group of organisms. Drosophilids are found on every continent, and many species occupy a wide range of latitudes where they have undergone adaptive evolution of heat and cold tolerance (reviewed in Hoffmann et al., 2003c), desiccation tolerance (Da Lage et al., 1990; Hoffmann et al., 2001), and body size (Gilchrist et al., 2004; James et al., 1995). Clinal variation (graded variation across an environmental gradient, such as latitude or temperature) in tolerance of environmental stress provides a means for assessing the evolutionary potential of species to adapt to climate change over time.

Many drosophilids, including the lab rat of arthropods, *D. melanogaster,* were until recently, isolated endemic species. *Drosophila melanogaster* was probably confined to tropical Africa until the last glaciation (David & Capy, 1988). The fly colonized the New World within the last few centuries. Widespread dispersal of cosmopolitan drosophilids within continents has been associated with the exploitation of breeding and feeding sites provided by human activities (Dobzhansky, 1965), setting up many replicated "experiments in nature" across broad latitudinal gradients. Despite humankind's role in providing food resources, the influence of humans on the introduction of some species into new regions may be relatively inconsequential. For example, *D. melanogaster* arrived in Australia about 100 year ago, and genetic evidence suggests that the contemporary continental population may be descended from a single release event in northern Australia (Hoffmann & Weeks, 2007).

Comparison of multiple-introduced drosophilids can provide insight into the qualities that make a species a good invader. Important determinants of successful dispersal and colonization may include general adaptability and resistance to stressful conditions (Carson, 1965; Dobzhansky, 1965). Resistance to desiccation at high temperatures and to cold stress was compared among six of the tropical members of the *D. melanogaster* species subgroup, of which only two (*D. melanogaster* and *D. simulans*) have dispersed into temperate regions. The highest stress resistance was shown by cosmopolitan *D. melanogaster; D. simulans* was only moderately desiccation resistant yet had high cold tolerance relative to the endemics. These findings reinforce the notion that resistance to environmental stresses may be a requirement for widespread invasions.

In this chapter we focus on how drosophilids and some close relatives adapt to spatial and temporal variation in abiotic stress. Our discussion focuses on responses to climate change; however, as outlined earlier, both invasions and habitat destruction can force an organism into a new climatic habitat where its adaptability will be tested. Although biotic interactions are likely to be important in many cases, virtually all will take place within a physical arena shaped by climate. For both endotherms and ectotherms, characteristics such as the ability to escape from a predator or to infect a potential host may well hinge on an energetic capacity that is a function of temperature, humidity, and insolation. Although our understanding of how the distribution of climatic variation may constrain demographic performance of populations is in its infancy, at least one study has implicated the rapid decline of a population as a result of a shrinking arena of favorable temperature conditions (Pörtner & Knust, 2007).

ABIOTIC FACTORS LIMITING SPECIES DISTRIBUTIONS

Temperature

Environmental temperature affects nearly every aspect of organismal biology. Thus, temperature changes produced by local habitat modification or global climate change may significantly alter species ranges and patterns of seasonal timing. Temperature, sometimes in concert with other abiotic cues, mediates a variety of physiological processes that have direct demographic influences on populations (Angilletta et al., 2002). In ectotherms, increasing environmental temperature directly increases growth and metabolic rates up to a critical thermal maximum (Cossins & Bowler, 1987). In endotherms, local warming can reduce energy expenditures when ambient temperature is comfortably below regulated body temperature.

As temperatures change, suitability of the local habitat for a given species may also change. Studies indicate that the ecological ranges of many species are changing rapidly (Parmesan, 2006) and, furthermore, that some regions will no longer be suitable if climate changes continue at the current rate. Suitable ranges may shift in latitude or altitude just as ranges of many extinct taxa seem to have shifted in response to climate change that occurred in deeper geologic time (Bush & Hooghiemstra, 2005; Huntley, 2005). What differs today (in addition to the rapid rate of climate change) is widespread fragmentation and exploitation of habitat to satisfy human needs. These barriers and regions of altered habitat mean that the ranges of many wild species may shrink or even disappear. If species cannot track suitable habitat in space, then they will have to remain within the current range and adapt to changing conditions. It is unknown how many species will not be able to adapt fast enough to keep up with the changing thermal environment. Species that are capable of shifting their range may do so at variable rates, thus disrupting community structures. It is predicted that 20% to 30% of species will go extinct if temperatures increase by 1.5 to 2.5°C (Adger et al., 2007).

Both summer heat and winter cold can affect organismal fitness. High temperatures generally increase rates of metabolism and water loss in insects and other ectothermic animals, and are thus energetically costly to endure. Elevated temperature causes stress, resulting in the induction of heat-shock proteins (Hsps) and other elements of the stress response (Feder & Hofmann, 1999). Although induction of these proteins aids survival and helps cells return to normal functioning after stress, expression of Hsp70 reduces fecundity and has other negative impacts on fitness (Krebs et al., 1998; Silbermann & Tatar, 2000). Exposure of insects and other ectotherms to cold generally reduces metabolic rate, with a corresponding decrease in activity levels and a slowing of foraging, reproduction, and growth. Endotherms must either allow body temperatures to decrease and enter torpor, or ramp up metabolism to maintain body temperature above ambient levels. In each of these cases, less energy becomes available to support growth and reproduction.

Many *Drosophila* species exhibit a limited range of thermal adaptations to overheating. Most species can plastically increase their upper thermal limits by acclimation. Short-term exposure to high or low temperatures produces a rapid acclimation response known as *hardening* (Loeschcke & Sorensen, 2005), which has been widely studied in drosophilids. Exposure to high-temperature stress in the laboratory induces synthesis of various Hsps that increase survival and aid return to normal cellular function. The precise role of this response in nature is only partially understood (Krebs & Feder, 1997; Roberts et al., 2003). Hardening to cool temperatures (Sinclair & Roberts, 2005) can also increase cold tolerance.

Changes in seasonality appear to be driving shifts in behavior, environmental sensitivity, and migration patterns, with concomitant fitness effects for natural populations (Bradshaw & Holzapfel, 2006). For example, within the past 25 to 50 years, some populations have experienced an earlier-than-average onset of ambient temperatures sufficient to initiate onset of spring reproductive behavior (Parmesan, 2006). Interacting species may differ in sensitivity to the onset of spring warming, resulting in changes in synchrony between predators and their prey or hosts and their parasites. For example, Netherlands populations of the great tit, *Parus major*, have evolved a timing of egg laying that coincides with the historical peak of spring caterpillar abundance (Visser et al., 1998). During the past 25 years, however, the peak of caterpillar abundance has shifted approximately 20 days earlier in the year, a change in moth life history that is correlated with warmer springs and an earlier budbreak of oak hosts (Visser & Holleman, 2001). The result has been a decrease in average breeding success of great tits; individuals with greater plasticity in breeding time are selectively favored over those with more rigid breeding schedules. Whether or not there will be an evolutionary response to this selection is not presently known.

Humidity and Precipitation

Drought-caused collapse of plant communities is responsible for significant levels of environmental degradation. Drought has various definitions, but is generally characterized by unusually low precipitation, resulting in a water shortage, abnormal dryness, and decreased relative humidity for one season or longer. Drought occurs as a normal feature of climate, but the frequency and duration of drought conditions vary with climatic zones. Drought should be assessed within a region relative to a long-term balance between precipitation and evapotranspiration. Factors such as global warming and human activity can increase the likelihood of drought. Climate models predict that although global precipitation could increase by 7% to 15%, evapotranspiration could increase by 5% to 10% (Alley et al., 2007). Precipitation is predicted to increase at high latitudes and decrease at low and middle latitudes, creating the potential for more severe droughts in these regions. Additionally, with increased drought comes increased risk of fire. Wildfire activity has increased since the mid 1980s, with more frequent fires of longer durations and over a longer wildfire season (Westerling et al., 2006). Moreover, the spatial distribution of fires is correlated with increased spring and summer temperatures and an earlier spring snowmelt, factors that interact to extend and intensify the dry season. Fire clearly shapes natural plant and reptile communities and is likely to have significant effects on organisms in other taxa (see, for example, Bond & Keeley, 2005; Wilgers & Horne, 2006).

CONCEPTS: ADAPTATION TO ENVIRONMENTAL STRESS

Genetic Variation in Stress Response Is Necessary for Evolution

Researchers have generally assumed that stress responses are polygenic traits best modeled by quantitative genetics. The phenotypic variance (V_P) of a trait is expressed as $V_P = V_A + V_D + V_I + V_E$, where V_A is the additive genetic variance; V_D, the dominance variance; V_I, the variance resulting from genetic epistatic interactions; and V_E, environmental variance. Contributions of V_D and V_I to phenotypic variance are generally considered to be negligible, although this is not always the case. Narrow-sense heritability of a trait (h^2) is the proportion of the total phenotypic variance that is the result of the additive genetic variance (V_A/V_P). It is this component of the total phenotypic variation upon which natural selection acts; hence, its magnitude is viewed as an indicator of the "evolvability" of a trait. Changes in the heritability (h^2) of a trait can be caused by fluctuations in V_A among different environments, as well as changes in V_E. Understanding how environmental conditions affect trait heritability is vital in predicting the ability of populations to respond to environmental changes (Hoffmann & Parsons, 1991).

Stressful environmental conditions affect heritable variation (Hoffmann & Merila, 1999). Additive genetic variation changes in response to environmental conditions, but can we predict the direction in which V_A will change relative to a specific environment? Expression of genetic variation, and thus heritability, may increase under stressful conditions (in other words, increase in V_E). For example, a combination of stresses led to increased genetic variability and heritability for fecundity in *D. melanogaster* (Sgrò & Hoffman, 1998). Yet, increasing V_E can also lead to reduced expression of genetic variation and thus reduced heritability. For example, *D. melanogaster* shows a significant decrease in genetic variability and heritability for development time when exposed to high culture temperatures (Imasheva, 1998). Hoffmann & Merila (1999) suggest that further work is needed to study the effect of novel environmental conditions on the expression of genetic variation.

Genetic Variation in Thermotolerance and Evolved Responses

Two approaches are used in analyzing genetic variation for traits of interest in *Drosophila* species: (1) comparative common-garden studies among and between species from different geographic locations and (2) laboratory selection studies that allow estimation of heritabilities and selection responses for traits. Common-garden studies entail bringing individuals from the field into the lab, where they are bred for one or more generations under conditions that standardize environmental effects on the phenotype before testing. Selection studies provide a direct assay of how traits change under a specific selection regime. In both cases, ecological realism is sacrificed for a direct assay of genetic potential.

Extensive research has linked expression of Hsps to differences in thermotolerance. For example, higher levels of Hsp70 increase the survivorship of larvae, pupae, and adults in both natural and laboratory populations (see for example, Bettencourt et al., 2002; Dahlgaard et al., 1998; Kreb, 1999; Krebs & Feder 1997, 1998). Other members of the Hsp family have similarly been implicated in changes in thermotolerance, including Hsp68 (McColl et al., 1996), *hsr-omega* Anderson et al., 2003; McColl & McKechnie, 1999), and Hsc70 (Folk et al., 2006). In general, induction of Hsps with a pretreatment exposure to sublethal heat increases survivorship under severe heat stress (Feder, 1996). Much of the effect is the result of the role of Hsps as molecular chaperones that help prevent aggregation of damaged proteins during stress and assist in refolding denatured proteins when conditions improve (Parsell & Lindquist, 1993). This, however, is not the primary mechanism of adaptation to warm climates by species and populations. Equatorial populations of *D. melanogaster* have a limited heat-shock response (Zatsepina et al., 2001) at temperatures more than 36°C compared with standard laboratory strains. In this case, reduced expression of Hsp70 is correlated with insertion of two transposable elements: one in the intergenic region of the 87A7 *hsp70* gene cluster and one in the promoter region of the *hsp70Ba* gene. A limited response at 36 to 39°C is also observed in desert-dwelling *D. mojavensis* (Krebs, 1999); however, the mechanism that reduces gene expression is not known. In both cases, flies adapted to a high-temperature lifestyle have greatly elevated thermotolerance despite the delayed and diminished heat-shock response. These experimental results suggest that mechanisms other than induction of Hsp70 must exist for increasing tolerance of high temperature.

Two studies have identified several QTL for temperature tolerance in *D. melanogaster*. Norry and colleagues (2004) identified regions on chromosomes 2 and 3 associated with high-temperature tolerance. All the regions of chromosome 3 showed dominant effects and were linked to various heat-shock loci. Candidate genes identified on chromosome 2, however, are of less obvious functional significance. This study also provided preliminary evidence of regions on the X chromosome that contributed to elevated thermal tolerance. A second study by Morgan and Mackay (2006) used recombinant inbred lines to identify seven autosomal QTL for thermotolerance. Three were linked to variation in cold tolerance and four to variation in heat tolerance. Nearly all showed sex-specific effects. Four of the QTL include Hsps among the candidate genes. These studies and many others (Sorensen et al., 2003) suggest that Hsps play an important role in environmental tolerance in plants, invertebrates, and vertebrates. The role of other physiological mechanisms affecting thermal sensitivity, such as changes in membrane properties (Overgaard et al., 2006) or ion channels (Rosenzweig et al., 2005), are less well explored, but such mechanisms are also likely to be important.

In a warming environment, selection for increased thermotolerance is to be expected. Studies that examine geographic variation in thermotolerance among populations of biological invaders suggest that selection has changed patterns of thermal adaptation throughout recent history. For example, *D. melanogaster* was introduced into Australia about 100 years ago. Both physiological and genetic evidence show adaptive evolution of heat and cold tolerance in these populations (Hoffmann & Weeks, 2007). Moreover, clinal patterns in several genetic markers associated with adaptation to local climate conditions have shifted toward a more equatorial composition during the last 20 years in Australian *D. melanogaster* (Umina et al., 2005), indicating a rapid adaptive response to climate warming (see the case studies presented later in the chapter).

Polytene chromosomes are multiple chromatids that remain synapsed together and occur in the salivary glands of *Drosophila*. These chromosomes have been used in studies of geographic variation since the mid 1900s, and many species have visible molecular markers in the form of chromosomal

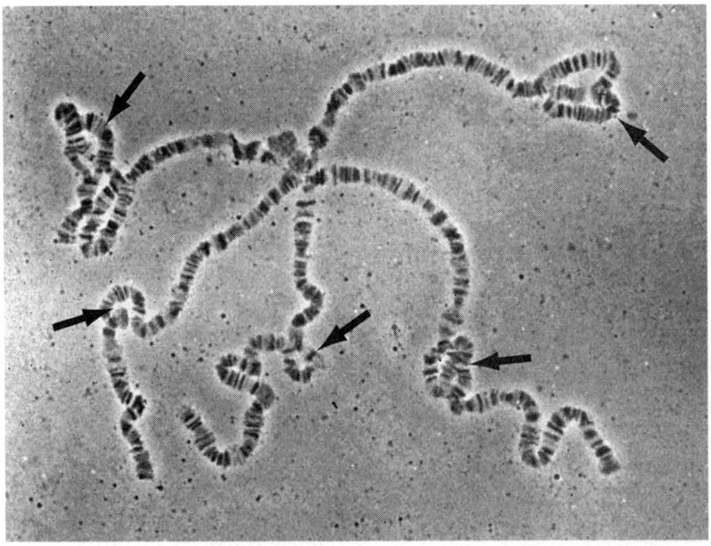

FIGURE 11.1 The five polytene chromosomes of *Drosophila subobscura*. The inversion loops (see arrows) on each chromosome represent regions that are heterozygous for specific inversions. (Image courtesy of Dr. Joan Balanyá, University of Barcelona.)

BOX 11.1 The Language of Linkage and Selection

The genes within a chromosome are physically linked to each other. In *genetic hitchhiking,* selection on alleles at one locus can change the allelic frequencies at adjacent, tightly linked loci. Pairs of genes, each having two or more alleles that differ from their joint Hardy-Weinberg expectation are said to be in *linkage disequilibrium* (or LD). Over time, LD will diminish toward zero as a result of recombination. One potential way of maintaining LD is through *inversion* of a chromosome segment. As the name implies, an inversion arises when a chromosome breaks in two places and the intervening sequence of DNA is flipped around. Heterozygotes for an inverted sequence undergo reduced recombination because crossing over within the inversion results in various kinds of damage to the recombinant chromosomes. High levels of LD can also be maintained between alleles of two loci by natural selection. For example, allele *A* at locus 1 might work better with allele *B* on locus 2 than if locus 2 carries allele *b*. This association could maintain LD for *AB* chromosomes. This is also an example of *epistasis,* defined as an interaction between alleles at two or more loci. Genomics studies are beginning to reveal extensive networks of epistatic interactions for fitness-related traits (Sanjuan & Elena, 2006).

arrangements of inversions (Krimbas & Powell, 1992) that provided the first assays of molecular variation among populations (Fig. 11.1, Box 11.1). Frequency of inversion loops and banding patterns recorded decades ago provide a historical molecular snapshot of populations that can be compared with contemporary patterns. Moreover, many species exhibit predictable clinal variation (Balanyà et al., 2003; Dobzhansky, 1947) and seasonal variation (Dobzhansky, 1948; Rodriguez-Trelles et al., 1996) in chromosomal arrangement frequencies, suggesting that these markers have adaptive significance related to climatic stress. Several studies have found changes in arrangements within populations during the last several decades that are suggestive of global warming (Anderson et al., 2005; Etges et al., 2006; Rodriguez-Trelles & Rodriguez, 1998). Balanyá and colleagues (2006) examined shifts in clinal patterns of environmental temperature and chromosome arrangements of flies from 26 sites spanning three continents and most of the latitudinal range of *D. subobscura* (Fig. 11.2). In 21 sites both the thermal index and the chromosome index for a given site has become more "equatorial"—that is, realized an average shift in the composition of both indices equivalent to about a 1° latitudinal shift toward the equator during the past 20 to 40 years. In most cases, however, we do not know how a particular gene arrangement contributes to thermal adaptation (but see Anderson et al., 2005). Nevertheless, these data suggest that genetic variation is available for adaptation to changing conditions.

Genetic Variation in Desiccation Tolerance and Evolved Responses

As in the study of thermotolerance, both common-garden comparisons and laboratory-based selection studies have been applied to assay genetic variation in desiccation tolerance. Comparative studies of desiccation tolerance in *Drosophila* species generally indicate that species from drier habitats have greater desiccation tolerance (Eckstrand & Richardson, 1980; Gibbs & Matzkin, 2001; Hoffmann & Parsons; 1991; Karan & Parkash; 1998). In the lab, response to selection in *D. melanogaster* for enhanced desiccation tolerance is generally substantial (realized heritabilities are approximately 0.60), indicating a high level of genetic variation for this trait (Hoffmann & Harshman, 1999; Rose et al., 1992). However, responses to selection for increased desiccation tolerance in other species of *Drosophila* have been mixed in terms of both direction and magnitude of response. In this section we take a closer look at these conflicting results.

Adaptive constraints may derive from lack of genetic variation, which may be the result of slow mutation rates, genetic drift, inbreeding, and past selection regimes. Some drosophilids lack sufficient genetic variability to respond adaptively

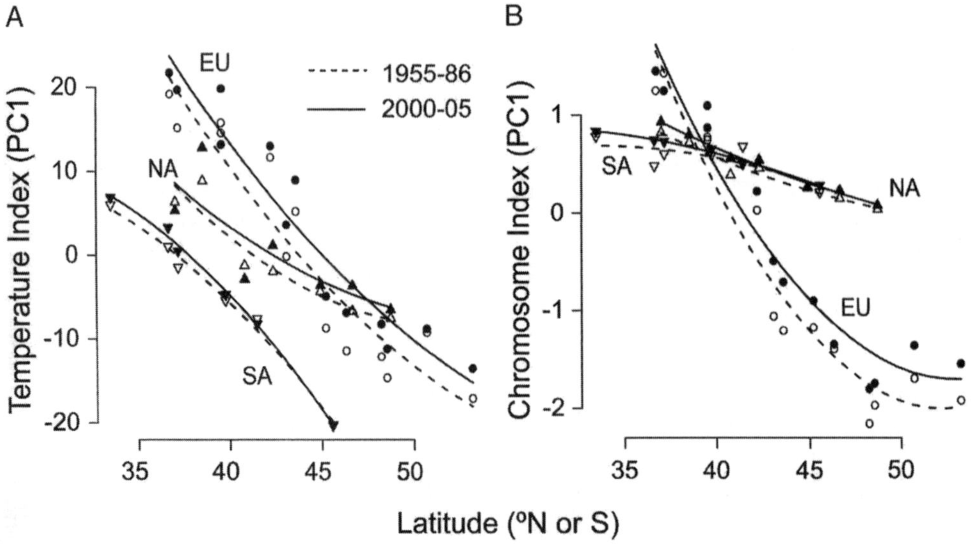

FIGURE 11.2 (A, B) Parallel shifts toward more equatorial clinal patterns in an analysis of the principal components of temperature (A) and chromosome arrangement frequencies (B) during the past 20 to 40 years in *Drosophila subobscura.* EU, European populations (circles); NA, North American populations (upward triangle); SA, South American populations (downward triangle). (After Figure 1 in Balanyá and colleagues [2006].)

to desiccation stress. Hoffmann and colleagues (2003a) examined genetic variation for desiccation tolerance in *D. birchii,* a rainforest species found on the northeastern coast of Australia. Populations of *D. birchii* distributed across nearly 7° latitude showed clinal, albeit weak, variation in desiccation tolerance, suggesting that heritable genetic variation for desiccation tolerance occurs among populations. Interestingly, the most resistant populations (in other words, those from the southern region) did not respond to laboratory selection for enhanced desiccation tolerance, even after strong selection for more than 30 generations in the lab. Despite having low heritable variation for desiccation tolerance, these same populations had high genetic variation in morphology and microsatellite loci. This study highlights the low potential for adaptive response to desiccation stress in *D. birchii.* The authors conclude that these findings emphasize "the importance of assessing evolutionary potential in targeted ecological traits and species from threatened habitats" (Hoffman et al., 2003a).

Populations found in border (in other words, ecologically marginal) habitats provide an opportunity to study constraints on the distribution of organisms (Hoffmann & Blows, 1994). To uncover why some species may be unable to extend their ranges beyond borders, the genetic composition of marginal and centrally located populations has been examined. Blows and Hoffmann (1993) hypothesized that the inability to move beyond the borders of marginal environments may be the result of limited genetic variation. Little is known about either pattern or mechanism; however, reduced genetic variation at species boundaries might arise from strong directional selection for tolerance to local environmental stresses. Which traits, then, lack sufficient genetic variation to allow dispersal beyond marginal habitats? Blows and Hoffmann (1993) performed a selection experiment to study genetic variation for desiccation tolerance (and that of other stresses) using four populations of *D. serrata* collected along the Australian eastern coastline: two from central locations and two from marginal, more southerly ones. Desiccation tolerance would be considered a limiting factor in the distribution of marginal populations if (1) marginal populations showed low heritability for the trait and (2) the mean for the trait was higher in marginal populations relative to more centrally located ones. Rainfall along the coastline varies dramatically, with the northern coast tropical and wet and the southern coast temperate and dry. Thus Blows and Hoffman

(1993) predicted that desiccation tolerance should be higher in the south.

Replicate lines from each population were established and selected for desiccation tolerance. Lines generated from marginal populations had lower realized heritabilities, indicating lower genetic variation. Even after 14 generations of selection, marginal populations did not show an increase in desiccation tolerance. (The second prediction for marginal populations, that of higher desiccation tolerance, was not observed.) These findings are consistent with the prediction that low genetic variation for desiccation tolerance limits the distribution of marginal (southern) populations of *D. serrata.*

No evidence of latitudinal clinal variation in desiccation tolerance was observed in Australian *D. serrata* (Hallas et al., 2002) or among 18 populations of Australian *D. melanogaster* (Hoffmann et al., 2001). These results suggest that most of the variation for desiccation tolerance is found within and not between populations—at least in Australia. Low levels of differentiation were found among geographic locations, but this was not correlated with selective factors that varied along the cline. Alternatively, the lack of clinal variation and population differentiation may indicate that desiccation is not a significant stress affecting *D. serrata* or *D. melanogaster* along the eastern coast of Australia.

These findings contrast with those of Karan and colleagues (1998), who examined desiccation tolerance along a latitudinal cline for three drosophilid species in India where seasonal variation increases with latitude. Southern regions are humid and thermally stable throughout the year whereas summers in the northern regions become progressively warmer and drier. For all three species, desiccation tolerance increased significantly with latitude as well as with altitude. A particularly intriguing finding involves *D. ananassae,* which is exceptionally desiccation sensitive, yet widespread and numerically dominant in India. The authors suggest that the success of this species, despite its desiccation sensitivity, is the result of its commensal lifestyle. Essentially, domestic habitats provide refugia and thus protection against abiotic stress.

Theory of Performance Curve/ Tolerance Curve Evolution

Thermal sensitivity of a fitness-related trait (or fitness itself) can be described by a performance curve

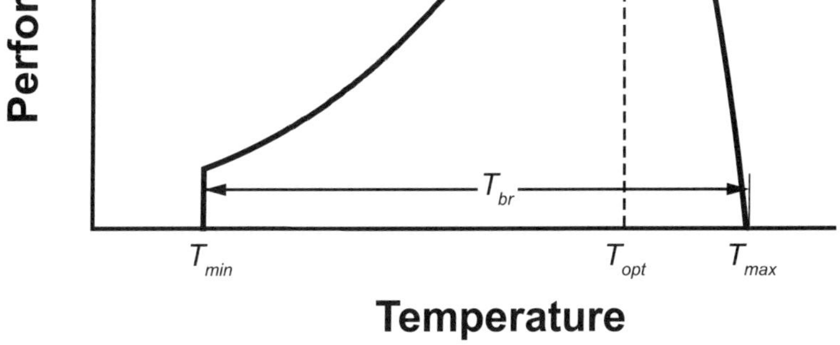

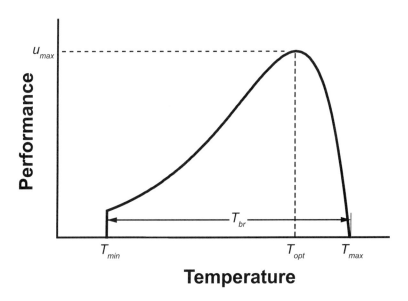

FIGURE 11.3 A hypothetical performance curve, showing key features (see text for detailed explanation). T_{br}, performance breadth; T_{max}, maximum limit of performance; T_{min}, minimum limit of performance; T_{opt}, optimal temperature; u_{max}, temperature at which performance realizes a maximum rate.

in which performance level is plotted as a function of environmental temperature (Fig. 11.3). Key features of the performance curve include the optimal temperature (T_{opt}), which is the temperature at which performance realizes a maximum rate u_{max}, and performance breadth (T_{br}), which reflects the degree of thermal specialization. The variables T_{min} and T_{max} define the limits of performance. Thus a performance curve provides an abstract representation of the thermal sensitivity of fitness-related performance. For example, in many insects, flight ability is directly related to fitness-enhancing behaviors such as foraging, mate location, and oviposition. Flight ability is also strongly related to temperature; insects have a critical lower temperature limit below which flight is not possible. Flight ability increases with temperature to a point where high temperature degrades neuromuscular coordination and flight again becomes impossible. The range of temperature within these limits is the performance breadth: the range of conditions permitting fitness-related activity.

Lifetime fitness of an individual, assuming that fitness is proportional to the level of performance, is estimated by the summation (over the lifetime temperature profile) of the amount of time at each temperature multiplied by performance at that temperature (Gilchrist, 1995, 2000). Global and local temperature changes can mean a change in the distribution of time periods at a given temperature over the lifetime of an organism. This change can impose selection on the shape and position of the thermal performance curve. The evolutionary response to such selection will depend on available genetic variation in specific directions (Izem & Kingsolver, 2005).

Beginning with Richard Levins (1968), several authors have argued from first principles that the area underneath a performance curve would be constant on some scale. The result is a classic specialist–generalist trade-off: Specialists have high performance over a narrow range of conditions, whereas generalists have somewhat lower performance over a broader range of conditions (Fields, 2001; Gilchrist, 1995, 1996; Lynch & Gabriel, 1987). If selection generally favors increased performance, then it seems reasonable to assume that the observed performance curve represents a near optimum, given genetic and physical constraints and the distribution of environmental conditions encountered. For example, enzymes with relatively low energies of activation and greater conformational flexibility tend to allow more rapid reaction rates; however, such enzymes are also more liable to heat damage. Increasing stability of an enzyme to allow it to retain function at higher temperatures may increase the energy of activation and reduce flexibility (Fields, 2001). If selection favors increased performance at higher temperatures and the evolution of the performance curve is constrained by a specialist–generalist trade-off, then selection imposed by increasing temperatures could result in a horizontal shift (Fig. 11.4A) of the curve to the right. A second possibility, still assuming a specialist–generalist trade-off, is that T_{max}, the upper limit on performance, might shift to the right with no corresponding change in T_{min} (Fig. 11.4B). Because of the specialist–generalist constraint, maximal performance would decline. In some species, the upper end of the performance curve might be constrained by an absolute upper thermal limit (see, for example, Gilchrist & Huey, 1999) as well as the specialist–generalist curve. In this case, one might see a narrowing of the curve with increased performance at higher temperatures as the population evolves to specialize in warmer temperatures (Fig. 11.4C). If there is no specialist–generalist constraint, then one might see performance curves increasing in T_{max} without a corresponding change in T_{min} (Fig. 11.4D).

Relatively few studies have examined specialist–generalist trade-offs in performance curves (reviewed in Angilletta et al., 2003). Gilchrist (1996) found a negative genetic correlation between performance breadth and peak performance, consistent with a generalist–specialist trade-off. A recent statistical innovation (Izem & Kingsolver, 2005) allowed partitioning of genetic variation in performance curves of *Pieris* caterpillars into specialist–generalist (Fig. 11.4B, C), horizontal shift (Fig. 11.4A), and vertical shift (relaxing the constant area constraint) components. The model explained 67% of observed genetic variation in the performance curve. Most of the variation explained reflects constraint on the area under the curve (38% specialist–generalist plus 16% horizontal shift) as hypothesized by specialist–generalist trade-offs. Only 13% of variance was the result of vertical shift.

In each scenario of performance curve evolution (Fig. 11.4), there may or may not be a correlated shift in thermal preference (indicated by an arrow). Thermal preference is the temperature that the organism chooses; in lizards, thermal preferences are positively, but not completely, correlated with T_{max} and T_{opt} of sprint speed (Huey & Bennett, 1987). Correlations are further weakened when phylogenetic relationships are considered (Garland et al., 1991). If selection acts directly on performance curves, then

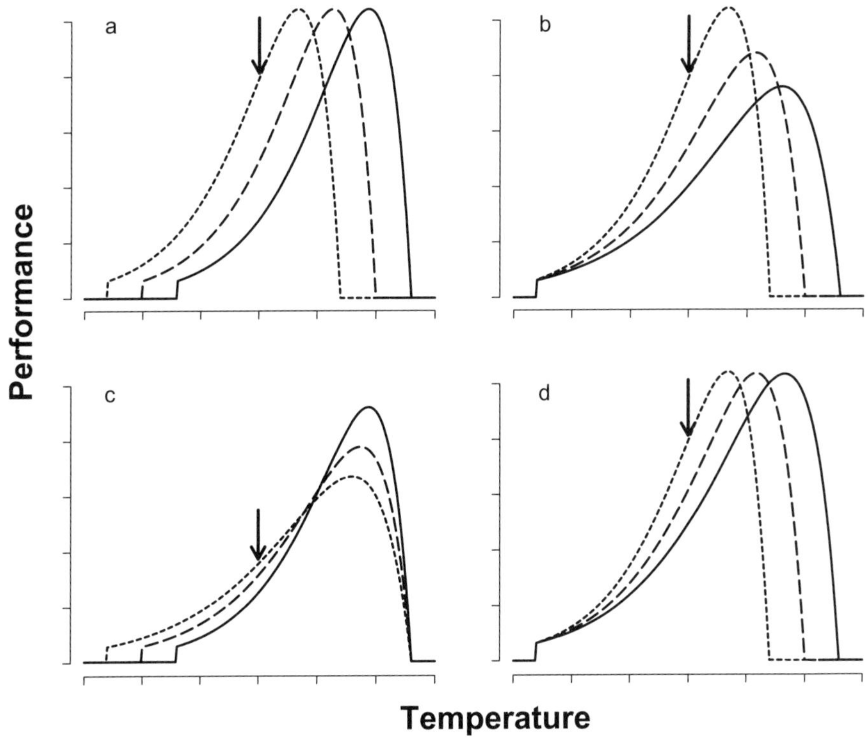

FIGURE 11.4 Some scenarios for performance curve evolution in response to environmentally induced stresses such as increasing temperature or increasing aridity (direction of increase indicated by dotted to dashed to solid lines). In (A) through (C), we assume a constant area constraint on performance curve evolution. (A) A horizontal shift in the performance curve. (B) Evolution of T_{max}, with T_{min} constrained by other factors. (C) Evolution of T_{min}, with T_{max} constrained by other factors. (D) Evolution of T_{max}, relaxing the assumption of constant area under the curve. The arrows show the point of preference prior to selection. Note that selection generally decreases performance at the preference point.

weak correlation suggests that preferences may lag behind the evolution of performance. This may be because of thermal constraints on other aspects of fitness-related activity or because plasticity implied by the performance curve reduces intensity of selection on preference. In any case, inspection of the evolving curve suggests a general reduction in rate of performance (and hence fitness) at preselection thermal preference in nearly every scenario of adaptive change in performance curves in response to warming. Thus, populations undergoing directional selection resulting from climate change or a habitat shift may have a lower rate of reproduction or reduced foraging ability if thermal preference remains unchanged, leading to lower rates of population growth and reduced fitness.

One possible alternative to evolution of the performance curve itself is simply switching to a different thermal preference (Fig. 11.5). This may or may not represent a genetic change; however, shifting thermal preferences to a higher temperature generally places an organism closer to the right-hand plummet of the performance curve. Asymmetry in thermal performance curves is generally agreed to reflect the nature of changes in enzyme structure and function with temperature. Colder temperatures tend to slow down reactions and immobilize organisms whereas hotter temperatures tend to alter the shape of proteins and slow or end metabolism. Thus, choosing to abide at a higher temperature is a risky game and probably not a long-term option.

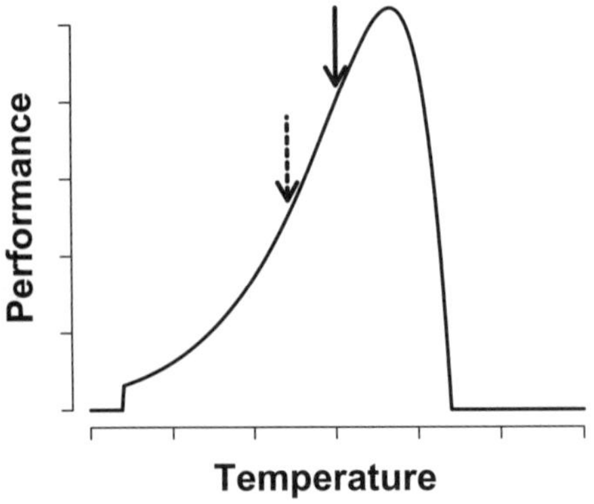

FIGURE 11.5 Shifts in preference rather than shifts in the performance curve. A shift in preference may require no genetic change; however, it increases risk of thermal (or desiccation) damage (shift occurs from dashed to solid arrow).

CASE STUDIES

The Evolution of Thermotolerance in Australian *Drosophila melanogaster*

Drosophila melanogaster is derived from an African endemic. Approximately 100 years ago, a propagule of flies colonized eastern Australia (Hoffmann & Weeks, 2007) where they now range from tropical Queensland to temperate Tasmania. Microsatellite evidence from populations on the eastern coast of Australia suggests a single colonization event with a large pool of genetic variation and relatively little population structure. Despite minimal differentiation at neutral loci, Australian *D. melanogaster* exhibits highly significant clinal variation in various traits and in genes associated with variation in thermotolerance. Adaptive variation in both heat and cold tolerance was found in multiple isofemale lines spanning approximately 28° of latitude (Fig. 11.6). Knockdown temperature (a measure of an individual's ability to maintain locomotory control at high temperatures) decreased from low to high latitudes. In contrast, cold tolerance (measured by chill coma recovery, an index based on the time required to recover from a severe cold stress [Gibert et al., 2001]) increased with latitude.

Recent studies have begun to unravel some of the genetic changes that influence thermotolerance in these flies. Both heat and cold tolerance appear to be at least partially linked to an inverted region, *In(3R)Payne,* on the right arm of chromosome III. This inversion is nearly fixed in low-latitude populations and almost absent at higher latitudes, and has been linked to clinal variation in body size (Calboli et al., 2003; Weeks et al., 2002). The Hsp locus *hsr-omega* lies within this region and shows correlations in allelic frequencies with heat and cold tolerance (Anderson et al., 2003; McColl & McKechnie, 1999; McKechnie et al., 1998). Dobzhansky (1970) hypothesized that chromosomal inversions reduce rates of recombination within the inverted regions and thus promote maintenance of sets of coadapted alleles. Linkage is strongest near inversion breakpoints (Andolfatto et al., 2001; Hasson & Eanes, 1996), suggesting that some loci near the middle of the inversion (where *hsr-omega* is located) may be able to recombine more or less freely as a result of double crossover within the inverted region. Are clines in linked heat-shock genes the result of directional selection, or are clines the result of genetic hitchhiking? If allelic variants of a locus are, in fact, under directional selection, then clinal variation in those loci ought to be apparent within any one of the arrangements of the inverted region. No clinal variation for *hsr-omega* alleles was found within flies carrying the *In(3R)Payne* inversion (Anderson et al., 2005), a pattern more consistent with hitchhiking as part of a coadapted gene complex within the inversion than with direct selection on *hsr-omega.*

In the most exacting test yet of Dobzhansky's hypothesis, Kennington and colleagues (2006)

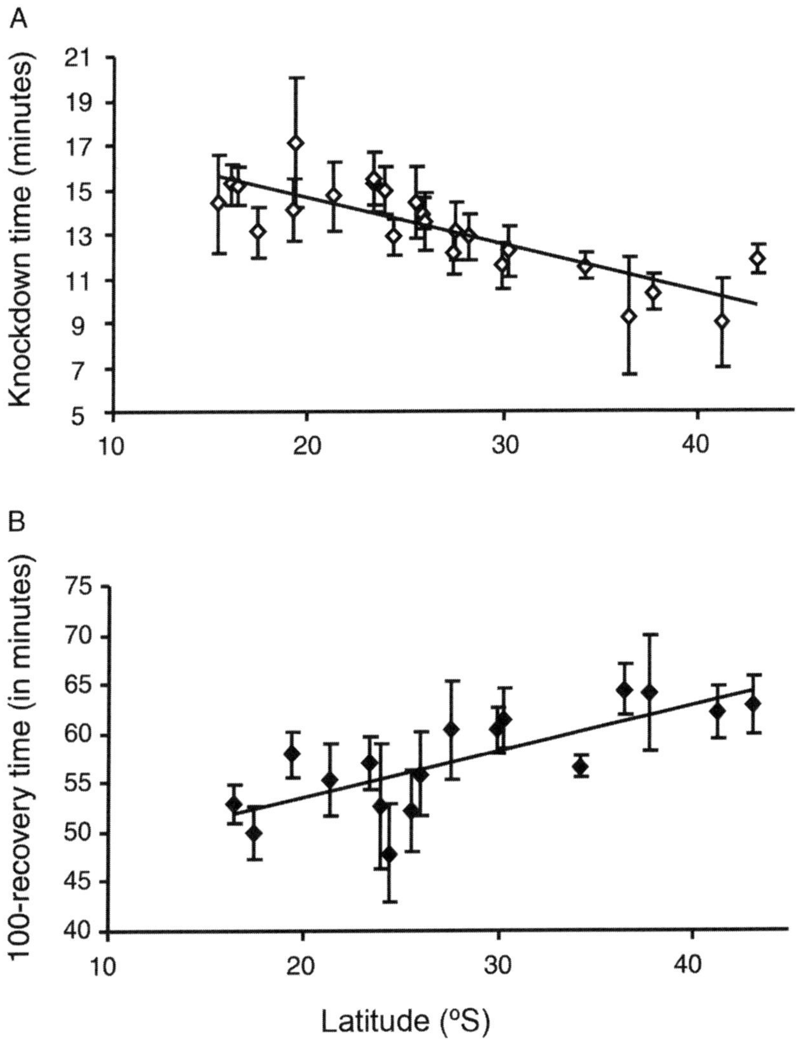

FIGURE 11.6 (A, B) Clinal change in high (A) and low (B) temperature tolerance for multiple populations from the eastern coast of Australia. (From Hoffmann and colleagues [2002].)

examined a set of 24 polymorphic markers across the *In(3R)Payne* inversion. They found extensive patterns of LD that could not be explained simply by reduced recombination near the breakpoints. Some regions within the inversion showed extremely high LD, whereas regions immediately adjacent showed no LD. The most likely explanation is epistatic selection among specific loci within the region. The adaptive value of the *In(3R)Payne* inversion itself is that it helps to maintain specific groupings of coadapted alleles. Strong latitudinal clines in chromosome inversions previously linked to climatic variation exist in several other drosophilids. Studies of *D. subobscura* (Munte et al., 2005) and *D. pseudoobscura* (Schaeffer et al., 2003) support this pattern of high LD maintained by epistatic selection. These results suggest that some or all clinal variation in thermotolerance genes within *In(3R)Payne* represents adaptive epistatic variation that has evolved by natural selection within the past 100 years. Selection is apparently acting to maintain specific combinations of alleles across multiple loci rather than the "optimal" allele at each single gene.

In an apparent response to climate change, latitudinal clines of alcohol dehydrogenase (*adh*) (Umina et al., 2005) and *In(3R)Payne* (Anderson et al., 2005) have shifted toward a higher temperature composition during the past 20 years. The cline in *adh* is one of the classic stories in evolutionary genetics (Berry & Kreitman, 1993), with higher frequencies of the slow allele at low latitudes on three continents. Molecular and physiological studies suggest that the slow allele has higher thermal stability than the fast allele. The *adh* locus is just outside of an inversion on the left arm of chromosome II, *In(2L)t,* which also shows clinal variation (Umina et al., 2005). Despite close linkage, however, it is only the *adh* cline that has shifted in clinal elevation (in other words, the y-intercept) but not the slope between 1979 through 1982 and 2002 through 2004. The *In(2L)t* inversion has retained the same clinal pattern throughout this period, suggesting that directional selection has been acting directly on the *adh* locus. Observed changes in allelic frequencies are equivalent to an average migration of 3.9° latitude toward the South Pole, an equivalent distance of some 400 km. A similar shift in clinal elevation with no change in slope has occurred in the *In(3R)Payne* inversion, with an equivalent average migration of 7.3° latitude (Umina et al., 2005), a distance of more than 800 km. During this same time interval, temperatures have increased by 0.1 to 0.3°C and annual rainfall has decreased 10 to 70 mm per year along the eastern coast of Australia. The temperature increase, however, is only equivalent to a 1.3° latitudinal shift (Umina et al., 2005), suggesting an interaction among changing climatic factors. Clearly these studies show that genotypes associated with thermotolerance and possibly with desiccation tolerance are changing in natural populations in response to a changing climate.

Changing Photoperiodic Sensitivity in Pitcher-plant Mosquitoes

Wyeomyia smithii is a small, weak-flying mosquito with a distribution from Florida to Maine in North America. These animals lay eggs in the water-filled cavities of pitcher plants where *W. smithii* larvae feed on small insects that fall into the plants. Reproduction cycles of this species vary from several generations per year at low latitudes to a single generation per year at high latitudes. One of the extraordinary features of this system is the ability to collect pitcher plants in the field and transplant them to controlled environmental chambers that have cycling thermal and photoperiod environments. Thus, controlled larval habitats in this system are far more natural than the usually constant temperature, vial-and-milk-bottle-rearing habitat provided by most *Drosophila* labs (although the larval *W. smithii* diet of guinea pig chow and freeze-dried brine shrimp is only slightly more natural than the agar, sugar, and yeast concoctions fed to captive fruit flies).

Populations of *W. smithii* from different latitudes vary in their ability to withstand transient heat or cold stress. Zani and colleagues (2005) tested heat tolerance of mosquitoes from three latitudes (with four populations sampled per latitude) that were maintained on long-day (18 hr light [L]/6 hr dark [D]) photoperiods and sinusoidal temperature cycles ranging from 15 to 30°C each day. Heat stress was applied to replicate cultures from each line by increasing the temperature from 30 to 41°C over 4 hours each afternoon for 4 days of the larval life cycle. Zani and colleagues (2005) measured survivorship to adulthood and R_0 (the per-capita replacement rate [Reed, this volume]). Although heat stress dramatically reduced survivorship, no statistically significant effect of geographic location was observed.

Cold tolerance was also studied in the same experiment. Here, mosquito larvae were held on diapause-inducing short days (11 hr L/13 hr D) and sinusoidal temperature cycles from 10 to 27°C. Temperature was gradually reduced over 3 months to a 0 to 9°C cycle. Feeding ceased at daily maximum temperatures less than 12°C. These larvae were then transferred to a constant control temperature of 5°C or to a cold stress of −3°C in which the water froze around the larvae for 4 days. Both groups were then held at 5°C for 5 days, then 10°C, 10 hr L/14 hr D for 4 weeks, then switched to diapause-breaking long days (18 hr L/6 hr D) and "normal" thermal periods (sinusoidal, 15–30°C). Eclosing adults were scored for survivorship and R_0 for heat stress. A significant regional effect on survivorship was apparent for cold tolerance (Zani et al., 2005). Despite relatively high mortality in both stress treatments, R_0 was similar in both control and experimental stress treatments, suggesting that the most stress-tolerant genotypes had relatively high fitness regardless of the presence or absence of stress. When only controls were considered, significant clinal variation for R_0 was observed: Low-latitude populations had higher heat tolerance than high-latitude populations, and high-latitude populations had higher cold tolerance. It should be noted, however, that one weakness of this study is that temperature perturbation is confounded with differences in photoperiod treatment.

Similar results showing clinal adaptation to heat and cold were observed in a study of *W. smithii* that considered fitness over an entire thermal year (Bradshaw et al., 2004). In this study, one treatment group was exposed to a uniform southern photoperiod that allowed multivoltine demography, whereas another treatment group was exposed to a northern photoperiod that enforced a univoltine, diapausing demography. Mosquitoes were placed on low-, middle-, or high-latitude thermal periods, and an index of fitness including survivorship and reproduction was estimated over a year. Fitness decreased with latitude in the low-latitude thermal period treatment and increased with latitude under high-latitude thermal periods, with similar relative patterns under both northern and southern photoperiodic treatments. Interestingly, this genetic variation was not observed in a single generational study (Bradshaw et al., 2000).

Despite these genetic differences in thermal sensitivity in *W. smithii,* much of the adaptation to thermal stress, including stress imposed by climate change, has been accomplished indirectly by adjustment of seasonality through photoperiod sensitivity rather than by directly changing thermotolerance. To test this idea, Bradshaw and colleagues (2004) altered photoperiodic patterns independently of thermal periodic patterns and assessed fitness over an entire year. The question is: What happens to fitness when a locally adapted population must endure photic and thermal cycles that do not coincide? All the following experiments used a common, benign, mid-latitude thermal year, with temperatures that oscillated both daily and seasonally. The minimum winter temperature was approximately −3°C and the maximum summer temperature was approximately 35°C.

Wyeomyia smithii from multiple high-, low-, and mid-latitude populations were tested during a control mid-latitude photic year programmed to resemble the natural annual cycle of day lengths at a mid-latitude photic and thermal environment that occurs at about 40° latitude. Performance of the high- and low-latitude populations in this control environment was then compared with that of the same populations in their "proper" photic environment mismatched with the same benign mid-latitude thermal year. Specifically, cohorts from high-latitude populations were exposed to a high-latitude photic year: approximately 8 months of short days (10 hr L/14 hr D) followed by 4 months of long days (18 hr L/6 hr D). Cohorts from low-latitude populations were exposed to a low-latitude photic year: approximately 4 months of long day, then 4 months of short day, then 4 months of long day. Fitness consequences of having a mismatch between thermal period and photoperiod were severe for both high- and low-latitude populations. High-latitude populations in their "natural" high latitude natural photoperiod but at "benign" mid-latitude thermoperiod suffered an 88% fitness reduction relative to the same populations in the mid latitude thermo- and photoperiod. Fitness reduction here was caused by a mixture of low fall abundance, low spring survivorship, and low female fecundity. Despite mild temperatures, high-latitude populations spent much of the summer in diapause, having to survive the remainder of the year on stored resources at temperatures that generally support high metabolic expenditures. In contrast, low-latitude populations suffered a 74% fitness reduction in the low-latitude photoperiod as a result of fewer larvae entering diapause in the fall, thus producing pupae and adults that were unable to survive and reproduce during winter conditions.

The dramatic impact of changing the cyclic relationship between temperature and photoperiod suggests that selection on seasonal timing ought to be a particularly potent source of evolutionary change. Indeed, Bradshaw & Holzapfel (2001) have documented a genetic shift in critical photoperiod between 1972 and 1998 (Fig. 11.7 [Bradshaw & Holzapfel, 2001]). As implied earlier, *W. smithii* from higher latitudes have a longer critical photoperiod triggering entry to diapause than do lower-latitude mosquitoes. During the past 20 years or more, these mosquitoes have evolved a shorter critical photoperiod, with the greatest effect at higher latitudes. Thus, high-latitude populations today wait to enter diapause some 9 days later in the late 1990s than they did in the 1970s.

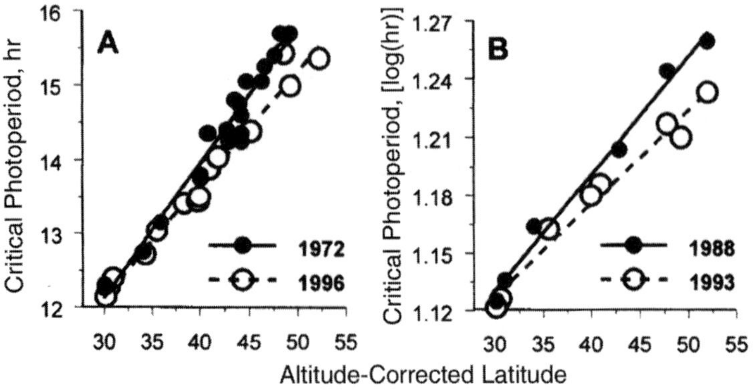

FIGURE 11.7 Critical photoperiods of *Wyeomyia smithii* collected during the overwintering generation from 1972 to 1996, determined from static (1972, 1996) or changing (1988, 1993) photoperiods. Comparison of slopes shows significantly shallower slopes in the later year in each comparison, suggesting that high-latitude populations are evolving more southern phenotypes. (From Figure 2 in Bradshaw and Holzapfel [2001], ©2001 National Academy of Sciences, USA.)

FUTURE DIRECTIONS

How Do Abiotic Stresses Interact with Demographic Challenges?

One of the challenges for conservation biologists and evolutionary biologists is to understand better the demographic consequences of natural selection. Wallace (1968) characterized natural selection as a continuum between soft selection and hard selection. Under soft selection, a population is filled to carrying capacity with the most fit genotypes available. For example, the abundance of a particular food resource determines population size, and competition for those food items determines which genotypes will be represented. It is ecology, not natural selection, that determines population size. In contrast, hard selection directly affects the number of individuals that survive and reproduce; selective mortality can reduce the population below carrying capacity. For example, consider time under a certain range of permissive climatic conditions to be the critical resource. Here, individuals that have the physiological capacity to mature and reproduce within the permissive amount of time will be the only survivors. For example, if successful development and reproduction for most genotypes requires a certain number of hours within the range of 15 to 29°C, and the climate warms, then individuals that can "use" time at 30°C will increase their fitness, whereas those that cannot may suffer. In this case, there may be direct, demographic consequences of selection: Individuals who are harmed by exposure to a 30°C temperature will fail to grow or reproduce. If most of the population is harmed by this stress, then population size will decrease. Stressful environmental conditions that increase mortality and reproductive failure always exert a selective pressure for increased tolerance of the stress. However, populations may not be able to evolve due to a lack of genetic variation (Blows & Hoffmann, 2005). Alternatively, hard selection can decrease population size to the point that random forces associated with demographic stochasticity may dominate population dynamics (Reed, this volume), and thus adaptation ceases to be a likely outcome (Lande & Shannon, 1996).

Of particular importance is the interaction of selection and demography at the limits of species ranges. The theory of source–sink dynamics in harsh environments suggests that a balance of migration and selection at the edge of a species range may prevent local adaptation, even if genetic variation in the necessary direction is present (reviewed in Holt & Gomulkiewicz, 1997). Such constraints on adaptation may be relaxed somewhat by temporal variation that creates a temporary refuge from selection (Holt et al., 2004). It seems likely that anthropogenically driven changes that result in increased global temperature and aridity are likely to increase mortality and reproductive failure in many populations in the near future. Currently, elevated rates of population extinction are found at the more equatorial edge of many species of plants and animals (Parmesan, 2006), a pattern correlated with increased warming or changes in precipitation patterns. Simultaneously, many of these species' ranges are edging the earth's poles. Conservation biologists should be concerned about the effects of environmental stress on local demographics at the ends of species ranges. In particular, a better understanding of the relative roles of hard versus soft selection in stressful environments might help guide management of bounded populations facing change resulting from climate or competition with invasive species.

Fitness and Life History Consequences of Stress Tolerance: Costs and Benefits

Relatively little is known about the true demographic costs of evolving stress tolerance. Most of the studies that have examined costs focus on a single generation or single component of fitness. For example, Schmidt (2005a) documented geographic variation in the incidence of ovarian diapause in *D. melanogaster*. This variation has allowed Schmidt to examine demographic costs to populations possessing this trait. Under nondiapause-inducing conditions, populations with a high incidence of diapause have lower fecundity early during reproductive life but higher survivorship of cold and starvation stresses (Schmidt et al., 2005b). Similar trade-offs between increased stress tolerance and decreased early fecundity have been found in response to artificial selection in the laboratory for desiccation tolerance (Albers & Bradley, 2006), and in outdoor cage experiments comparing overwintering ability and spring fecundity of flies from different geographic regions (Hoffmann et al., 2003b). Thus it appears that a reduction in early fecundity is commonly associated with a variety of stress-resistant traits.

Demographically, reduced early fecundity is expected to lower the intrinsic rate of reproduction of a genotype, but the effect in a single generation often seems small. The study of *W. smithii* by Bradshaw and colleagues (2004) is one of the few to look at the fitness consequences of adaptation to environmental stress over multiple generations. What they found was that many small, often insignificant positive and negative contributions to fitness can, throughout the course of a few generations, cause very dramatic differences in population size. Their results show that having the appropriate diapause strategy makes a crucial contribution to fitness. To date, no comparable study of the costs and benefits of thermal or desiccation tolerance over multiple generations has been carried out in the field or in a realistic lab environment (but see Junge-Berberovic, 1996).

Conservation biologists routinely use population viability analyses to forecast the fate of populations and have incorporated methods to include climatic covariates to improve predictive power (Ellner et al., 2002). These models have been modified to allow stochastic variation of demographic parameters; however, we are not aware of any attempts to incorporate potential demographic effects of natural selection imposed by climate change (but see Maschinski et al., 2006). Clearly, evolutionary time and ecological time act on similar timescales, and models addressing the demographic consequences of natural selection (Hairston et al., 2005) are needed to forecast population responses to climate change and range expansions.

SUGGESTIONS FOR FUTURE READING

We are in the midst of a global experiment on the impact of climate shifts on biological processes. Parmesan (2006) provides an excellent and timely review of the ecological and evolutionary shifts that have already been observed in a wide range of biological systems. Reviews by Angilletta and colleagues (2002) and Huey and Kingsolver (1993) provide insight into how physiology of thermal stress can drive demographic patterns and fitness. These ideas are illustrated with an important study by Pörtner and Knust (2007), showing that increasing ocean temperatures result in a mismatch between oxygen availability and oxygen demand in marine fish that is strongly correlated with decreased growth rates and population declines. Hoffmann (2003c) and Weeks and colleagues (2002) provide reviews of the relationship between genes and thermal sensitivity for drosophilids. Bradshaw and colleagues (2004) highlight the importance of photoperiodic adaptation to seasonality and the challenges that many organisms will face as thermal periods and photoperiods become out of sync as a result of climate change, population emigration, or biological introductions.

Angilletta, M. J., Jr., P. H. Niewiarowski, & C. A. Navas. 2002. The evolution of thermal physiology in ectotherms. J Therm Biol. 27: 249–268.

Bradshaw, W. E., P. A. Zani, & C. M. Holzapfel. 2004. Adaptation to temperate climates. Evolution 58: 1748–1762.

Hoffmann, A. A., J. G. Sørensen, & V. Loeschcke. 2003c. Adaptation of *Drosophila* to temperature extremes: Bringing together quantitative and molecular approaches. J Therm Biol. 28: 175–216.

Huey, R. B., & J. G. Kingsolver. 1993. Evolution of resistance to high temperature in ectotherms. Am Nat. 142: S21–S46.

Parmesan, C. 2006. Ecological and evolutionary responses to recent climate change. Annu Rev Ecol Evol Syst. 37: 637–669.

Pörtner, H. O., & R. Knust. 2007. Climate change affects marine fishes through the oxygen limitation of thermal tolerance. Science 315: 95–97.

Weeks, A. R., S. W. McKechnie, & A. A. Hoffmann. 2002. Dissecting adaptive clinal variation: Markers, inversions and size/stress associations in *Drosophila melanogaster* from a central field population. Ecol Lett. 5: 756–763.

Acknowledgments We thank Chuck Fox and Scott Carroll for inviting us to contribute to this volume. Michael Angilletta and an anonymous reviewer offered comments that improved this chapter. We acknowledge support by National Science Foundation DEB-0344273 and EF-0328594.

12

Managing Phenotypic Variability with Genetic and Environmental Heterogeneity: Adaptation as a First Principle of Conservation Practice

SCOTT P. CARROLL
JASON V. WATTERS

Seasons pass, succession proceeds, landscapes rise and erode. Species invade and disappear. All life experiences and is part of continual environmental flux. Environmental variation, coupled with the changing states of organisms themselves, generates tension between the design and performance of individuals. Across populations, we witness remarkable patterns of adaptation that emerge on brief timescales (see, for example, Benkman, this volume), yet no individual member of a population shows the best response to all circumstances. Many factors limit the degree and form of adaptation, including history, genes, environment, development, experience, decisions, and fortune. It follows then that a combination of variation—within individuals, among individuals, and among subpopulations—will determine a population's adaptability and resilience to environmental change (Boulding, this volume; Reed, this volume).

In contrast to predictably cyclic phenomena such as daily or seasonal rhythms, substantial and persistent shifts in environmental regimes have been comparatively rare (Vermeij, this volume). We are now in an epoch in which human-induced change is a force of extraordinary and enduring influence (Palumbi, 2001b). Altered climates, overharvesting, and introductions of exotic species have stressed many populations irretrievably. Population declines as well as local and global extinctions are sure signs of a failure to adapt (Gomulkiewicz & Holt, 1995), a consequence of which is the biodiversity crisis.

Recent discoveries in biology suggest that evolution is not best viewed as a strictly historical process, but rather as ongoing. Thus, evolution in response to global change phenomena, such as shifting seasonality and species introductions, changes the character of many organisms. Several well-known cases of "rapid evolution" during the middle 20th century, such as industrial melanism in peppered moths and heavy-metal tolerance in plants on mine tailings, were originally regarded as exceptional results of unnaturally high selection pressures. Although this interpretation may have been fair at the time, we now know that these cases were harbingers of the present. Such adaptive evolutionary events are now commonplace, occurring on timescales traditionally thought of as *ecological*. As examples accumulate, it remains clear that many cases of rapid evolution involve some degree of anthropogenic perturbation. Indeed, analyses by Hendry et al. (2008) indicate that human influences are especially potent agents of phenotypic change in contemporary animal populations.

The unanticipated fact of contemporary evolution in disturbed populations raises both interesting dilemmas and opportunities for biological conservation. Historically, protection and restoration of habitats and populations have been paramount and practiced in isolation from evolutionary concerns. Now, however, more and more species no longer genetically "match" their recent ancestral, less-disturbed habitats. Just as conservation biology reluctantly began to include human populations in the analysis of sustainable biotic communities, it must also face the prospect that permanently altered ecological conditions will result in permanently altered species. It is important to consider that the best conservation practices for evolving species may not be those derived from historical analyses under different or more stable conditions (Schlaepfer et al., 2005).

In addition, for that increasing portion of remnant species restricted to managed populations, it is critical to design ways to control evolutionary processes to promote sustainable conservation practices. Paradoxically, adaptation to environmental challenges will not always promote long-term population persistence, particularly in cases in which intraspecific competition or frequency-dependent selection are strong (Ferriére et al., 2004). Applied evolutionary methods to protect populations may include the cultivation of suites of phenotypes that use resources more efficiently, the augmentation of breeding population size, and selection for special traits (for example, pathogen resistance, sedentary behavior). Applied methods may also include retardation of specialization or runaway processes that would otherwise increase the probability of extinction. Thus, as organisms face environmental change, the challenge for biologists is to devise alternative methods of incorporating evolutionary concerns into sound conservation practice.

This chapter thus focuses on the relatively unexplored interplay of contemporary evolution and conservation (Stockwell et al., 2003). The pertinent concepts span the range from genetic variation through gene expression, to development and plasticity of phenotypes, and ecological communities and the niche. Given this conceptual breadth, we focus more on outlining the concepts and domain of the problem rather than on specific prescriptive advice. Modified environments will affect species by altering niche structure, which in turn affects reproductive success of phenotypes and fitness of genotypes. Although in many species change limits or eliminates resources, in others it will result in species proliferation. New conditions may be abiotic or biotic, including such factors as community and social structure. Change may also have direct effects on the manner in which individuals interact with their environment, or indirect effects through an influence on other community members. Under altered conditions, behavioral and physical adaptations to food, activity levels, or climate, and the organization of the life cycle itself may not have the same fitness benefits as in the recent past. Although population decline and ultimate extinction is one response in the face of pervasive change, what is happening in species that manage to persist?

CONCEPTS

Adapting to Change: Homeostasis, Plasticity, and Selection

Probably very few species have neutral relationships with global change (see, for example, Parmesan & Yohe, 2003), and many effects will be deleterious. Populations that persist in the face of enduring or accelerating environmental change will be those that are resilient to it, or that benefit from it (at least in the short term). Benefits may accrue for species that lose predators and competitors, or that gain new habitats or food resources (see, for example, Carroll, 2007a; Mittelbach et al., 1995). Whether persistence stems from tolerating habitat decline or from exploiting niche improvements, adaptive phenotypic and genetic responses will likely be involved (Rice & Emery, 2003, Stockwell et al., 2003, Strauss et al., 2006a).

To the extent that metabolic, enzymatic, and osmotic functions basic to life are optimized in a limited range of conditions, organisms stressed by environmental change may respond plastically to maintain homeostatic internal conditions. Choosing a better thermal site, generating metabolic heat, searching for water, digesting fat reserves—all of these are short term responses that stabilize other elements of the phenotype. Direct measures of behavioral or physiological changes can reveal stress symptomology, but we may often be blind to any response if our observations are of homeostatically stabilized characters. When plasticity is overt, it may provide ready clues to the problems individuals face and their possible solutions. However, organisms in unprecedented environments will be prone to making mistakes. When there is a relative absence of plasticity, selection may ensue among individuals based on differences in trait values. Likewise, variation among individuals in their plastic responses may cause selection among alternative forms of plasticity. Homeostasis, adaptive and maladaptive plasticity, and evolution will thus tend to interact as a population is faced with environmental change. The following section illustrates important aspects of those interactions.

Responding to Change: Taking First Steps

An example of tolerating negative anthropogenic change without apparent adaptation occurs in the Monarch butterfly, *Danaus plexippus*. These large, brightly colored butterflies usher in summer throughout North America, where their milkweed host plants in the genus *Asclepias* occurs. Novel biotic interactions are perturbing Monarch habitat use in what is likely a common fashion. In New England, the European black swallow-wort *Vincetoxicum nigrum*, a plant in the milkweed family, escaped from gardens in the 1800s and has now invaded early successional habitats (Tewksbury et al., 2002). While searching for larval food plants on which to oviposit, gravid female Monarchs are attracted to the exotic host and lay substantial numbers of eggs on its leaves and stems. Proximate cues that Monarchs use to identify native hosts have led them to include this new plant, an apparent flexibility that might expand the Monarch resource base in the northeastern United States. However, larval hatchlings on black swallow-wort do not survive to maturity, most likely because of growth-inhibiting compounds not found in the native host (Tewksbury et al., 2002). What at first seems like an increase in the availability of a larval resource in the butterfly's habitat (in other words, a *quantitative* change) is instead a *qualitative* change in which a poison masquerades as a resource. Whether Monarchs escape this *evolutionary trap* (Schlaepfer et al., 2005) (Box 12.1) remains to be seen, but the continued availability of suitable native host species should contain the threat of the introduced plant to the butterfly population. Similar cases have been reported in a number of insect species.

In contrast, pied flycatchers, *Ficedula hypoleuca*, in northern Europe are adapting in the face of global change, but with a twist. These small birds have responded to climatic warming during the past two decades by advancing their breeding season by about 10 days. This is functionally important because nesting success depends strongly on synchronizing nestling development with peak emergence of locally ephemeral insect prey that now emerge earlier in spring. Indeed, earlier-laying pairs do better than later pairs, but selection favoring early laying has nonetheless become progressively stronger during the past 20 years. This is because even though the birds are responding to advanced local phenology by breeding sooner, they are not actually arriving from tropical Africa to occupy their breeding territories any earlier in the season (Both et al., 2006). So although showing adaptive plasticity in one phase of the life cycle, pied flycatchers are currently constrained by now-obsolete responses to photoperiodic migration cues. The climate has warmed, but daylength periodicity remains the same. The birds may be near their physiological maximum in terms of being able to build nests and lay eggs more quickly. Only if and when variant individuals appear that arrive earlier will the species adapt more effectively to the new timing of insect emergence in northern Europe. Indeed, those flycatcher populations in areas with the earliest-emerging insect communities have declined by close to 90% throughout the study period, whereas those in areas with the latest-emerging peaks have fared much better, declining only about 10% on average (Both et al., 2006). Whether earlier breeding stems from adaptive plasticity that is induced by advanced conditions experienced on arrival, and proximately caused by either experience (for example, learning across years) or from genetic change remains to be discovered.

Reaction Norms: Evolution and Plasticity

The case of the pied flycatcher shows that as environmental perturbations alter the suitability of particular life histories, organisms themselves may change, and it is the interaction of these changes that becomes important. Although some traits such as human eye color are much more determined by genes than environment, others (such as surface color of some arctic birds and mammals) respond more strongly to the environment. Each trait has a *reaction norm* (Box 12.1) characterized by its expression in different environments. Traits that vary little across environmental conditions are *canalized,* whereas those that vary are labile or *plastic.* All organisms possess remarkable adaptive plasticity in at least some of their attributes. In contrast to humans, for example, mantis shrimp (a stomatopod crustacean) alter the pigment composition of their compound eyes to improve vision at different depths beneath the sea

BOX 12.1 Terms and Definitions

ecological and evolutionary traps	An ecological trap exists when a poor-quality habitat maintains evolutionarily relevant cues. The trap occurs when animals bypass a higher-quality habitat that lacks specific cues to settle in the lower-quality patches. An evolutionary trap occurs when animals make behavioral or life history decisions based on specific cues that were historically, but no longer, correlated with higher fitness. The opposite of a trap is a release.
key trait	A key trait is a trait that is a primary determinant of fitness in a given condition, and one whose presence or value is a principal determinant of fitness in a given environment.
phenotypic and evolutionary potential	Phenotypic potential is the full range of phenotypes that a population is capable of expressing across all possible environments. Some phenotypes may be adaptive for a given environment, others may not. Evolutionary potential is the genetic variation maintained in a population that will allow for future evolutionary change. Evolutionary change suggests an adaptive response over time to a new or modified environment.
reaction norm	The reaction norm is the pattern of phenotypes produced by a genotype under different environmental conditions.
strategy or syndrome	A strategy or syndrome is a set of possible phenotypes that serve a given function. Variant phenotypes that comprise a strategy are called *tactics*. A strategy may also be called a *syndrome* (a set of functionally, coadapted traits). Alternative phenotypes of a polymorphic syndrome may be called *types*.

surface (Cronin et al., 2000). Phenotypic plasticity that is adaptive under ordinary circumstances may likewise be adaptive under extraordinary circumstances (for example, rapidly rising sea levels). Biotic interactions have especially profound selective influences (Carroll, this volume), and mutually plastic interactions may catalyze rapid adaptation (Agrawal, 2001). However, not all individuals will react similarly to different environments, and their constituent traits will likely not all react in parallel, leading to a diversity of phenotypes with different evolutionary fates.

Interaction between plasticity and selection is also fundamental to the adaptive process, yet it has received comparatively little attention. Rapidly induced, nongenetic adaptation realized through physiological and behavioral adjustments alters selection by buffering stress and mortality, by permitting exploitation of altered conditions, and by promoting use of new niches (Ghalambor et al., 2007). Nonetheless, many new environments present organisms with unprecedented, potentially stressful challenges. Despite the power of integrated homeostatic mechanisms and adaptive phenotypic plasticity, ultimate limitations of *phenotypic potential* (Box 12.1) mean that genetic responses to directional selection may also be important in population-level response to change.

For several reasons, however, plasticity still may have a role in subsequent adaptation. First, if selection in new environments is in the direction of phenotypic shifts expressed in existing plastic reaction norms, and if there is genetic variation for the reaction norm, directional selection may favor assimilation (in other words, canalization) of induced phenotypic extremes that are best adapted to new circumstances. With phenotypes teetering on a performance edge, plasticity may thus span what would otherwise be the "valley of extinction" in the altered fitness landscape, providing raw material for selection to continue adaptive evolution (Price et al., 2003). Second, in stressful, changed circumstances, reaction norms may produce a range of novel phenotypes, most of which lack chance adaptations to the extraordinary conditions (maladaptive plasticity), yet the increased phenotypic variation may also produce novel, adaptive phenotypes (Badyaev, 2005; Badyaev et al., 2005). Lastly, stress of environmental change can also increase genetic diversity by increasing rates of mutation and recombination (Hoffmann & Parsons, 1997).

Expression of novel phenotypes in new circumstances is an interesting developmental phenomenon with potentially important consequences for population-level resilience to anthropogenic change. Adaptive reaction norms, evolved to evolutionary equilibrium in previously more stable environments, will tend to accumulate neutral, unexpressed genetic variation. Now those genes are in new environments, capable of manifesting unprecedented genotype-by-environment interactions. Current selection has had little chance yet to winnow novel reaction norms into functional strategies, or new genotypes into functional phenotypes. As a result, most novel variants will not perform well. However, among those chance forms that do survive the stresses of change, directional selection may then result in evolution toward a new adaptive state (Badyaev, 2005; Badyaev et al., 2005; Price et al., 2003).

There is empirical evidence that in changing environments, adaptive phenotypic and genetic responses may occur in tandem. For example, Losos and colleagues (2004) introduced a predatory lizard species into a population of prey lizards. They observed adaptive plasticity in avoidance behavior by prey, but went on to show that plasticity in response was insufficient to shelter the prey population from strong selection on limb length. Elongated limbs, which are related to running speed, evolved quickly despite behaviorally mediated attenuation of selection. Similarly, Réale and colleagues (2003) conducted a decade-long study of the response of a subarctic squirrel population to global warming. They observed concurrent plastic *and* genetic shifts in breeding phenology that were adaptive in relation to changes in timing and abundance of seed crops. If evolution is a common response to environmental change even in adaptively plastic populations, what are the implications for biodiversity and its management over different timescales?

Working to understand such complexity is a basic challenge of applied evolutionary biology. Can we predict phenotypic and genetic responses to environmental change, and how those responses will interact with demography and community ecology? Considered broadly, many factors suggest that evolution is predictable. Phenotypic convergence between unrelated taxa in similar environments is a strong indicator that predictions based on knowledge of natural selection have the potential to be robust. Repeated and recent evolution of benthic near-shore marine stickleback fishes (*Gasterosteus aculeatus*) into lacustrian benthic and pelagic forms (Rundle et al., 2000) is a clear indicator that certain sequelae are more likely than others in the face of change. Likewise, detailed knowledge of trait function has led to repeatable matches between predicted and observed evolution in a number of studies in which selection on those traits was altered experimentally or accidentally (Hendry & Kinnison, 1999). Moreover, in the absence of specialized knowledge, known boundaries of physiological or morphological function may permit predictions based on biophysical first principals (for example, constraints of energetics, temperature, nutrition, hydration, or physical scaling). One generalization is that adaptive evolution is more likely to occur in expanding populations (Boulding, this volume; Reed, this volume)—for example, when habitat change involves addition of resources or subtraction of risk. Some reviews suggest that evolutionary response to anthropogenic influences is so commonplace that evolution in enduring populations may be an appropriate first assumption for almost any focal taxon, from microbes to trees (Palumbi, 2001b; Strauss et al., 2006a).

If phenotypes and populations are indeed malleable by the environment and selection, then the range of potential responses to changing environments is so wide that predicting evolutionary responses may be a challenge. In particular, if plasticity has a complex, multifaceted interaction with selection and if plasticity simultaneously evolves, then the evolution of population-level traits may be quite complicated. For example, across a gradient of variability in sexual selection, Carroll and Corneli (1999) found that male soapberry bugs *Jadera haematoloma* in the southwestern United States made adaptively plastic (in other words,

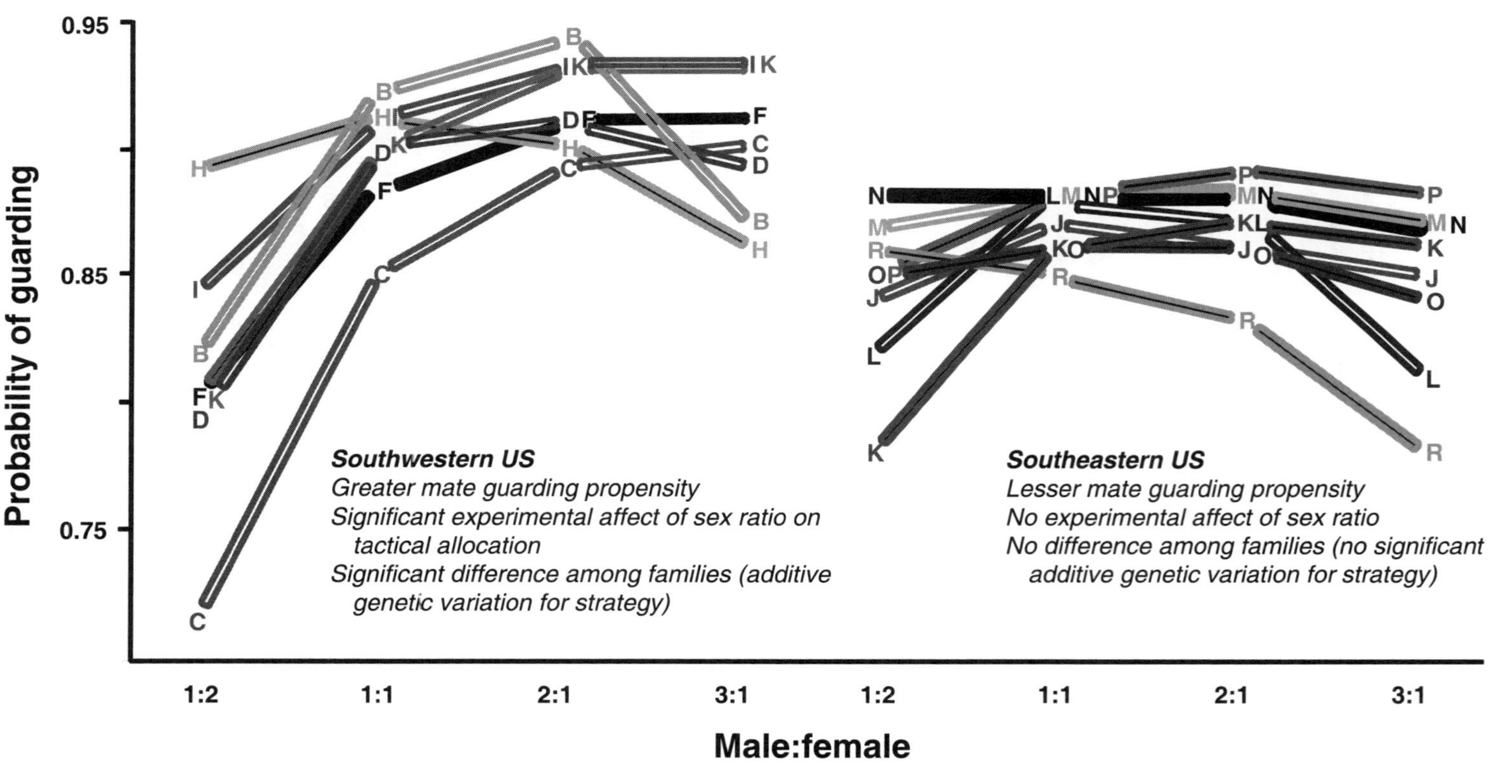

FIGURE 12.1 The experimental influence of sex ratio on behavior of male soapberry bugs in each two geographic regions. Shown is allocation to serially monogamous mate guarding versus promiscuous nonguarding for families of half-sib brothers distributed across a series of experimental sex ratios. Letters distinguish families. Their position shows the mean probability (in four replicates) that a male continued to guard across observations made at 3-hour intervals for 8 days. Adult sex ratios in nature are much more male biased and variable in the southwestern United States. Declines in guarding probability at the highest sex ratio, observed in a subset of families, probably results from vulnerability to increased harassment by unpaired males. (Figure modified from Carroll and Corneli [1999].)

conditional) mating decisions and also had significant additive genetic variation for mating strategy. This contrasted with an adaptively nonplastic southeastern U.S. population that showed little genetically based variation (Fig. 12.1). These populations thus differ in phenotypic and genetic variation and their interaction, and so may have very different evolutionary potential for mating strategy and related traits. Predicting evolutionary responses will depend on having information about the adaptive developmental, behavioral, and genetic resources a population possesses. Understanding the limits of adaptability is as important as understanding the types of environmental challenges a population faces.

Limitations in the adaptive responses of individuals to environmental change often result from constraints imposed by previous selection pressure. Such constraints are usually called *trade-offs* to emphasize that a performance increment in one context or function is likely to come at a cost to performance in another (for example, the number of offspring vs. the quality of care or provisioning for each, time spent foraging vs. time spent being vigilant for danger or other opportunities). Each individual's set of plastic responses is like a tool kit for coping with environmental change. However, performance of plasticity depends on anteceding phenomena: the ability to assess the environment accurately, the quality of cues that are detected, and the capacity to respond appropriately to information that can never be more than partial (see, for example, Schlaepfer et al., 2005). Adaptive variation observed in the magnitude of phenotypic plasticity between populations may indicate trade-offs between plasticity and other critical functions (Carroll & Corneli, 1999). Overall, the evidence suggests that there is substantial inter- and intrapopulation variation in the responses of individual traits to environmental change, such that adaptive flexibility in some traits may pair with substantial performance deficits in others as environments change (Carroll, 2007b; Schlaepfer et al., 2005).

CHARACTERIZING CHANGE AND PREDICTING RESPONSE

One way to refine the calculus of forecasting adaptation is to characterize environmental change and then predict classes of response at individual and population levels. For example, environmental change occurs at varying timescales. It may be comparatively slow, resulting in an environment that is only slightly different for each consecutive generation, or it may be rapid, such that the environment experienced by one generation is very different from that experienced by the next. Moreover, as suggested earlier, altered environments may be quantitatively or qualitatively different from the original environment. Examples of quantitative change include long-term shifts in climatic conditions that produce slight differences in ecological opportunities from one generation to the next. Quantitative changes also occur when organisms are transferred to new environments similar in most respects to their former environment. For example, a large proportion of species investigated show altered phenologies or range shifts in response to climate warming; in only a few cases has the mode of adaptation been described (Parmesan & Yohe, 2003). Species that take advantage of similarities between their ancestral habitat and human-altered habitats also experience this type of quantitative change (for example, urbanized populations of the peregrine falcon *Falco peregrinus* adopt skyscrapers as nesting habitat that are similar to natural high-cliff aeries (Fig. 12.2). In environmental shifts of this type, the selective environment experienced by one generation is qualitatively similar to that experienced by the next, although it may vary in the strength of specific selective pressures.

Qualitative environmental change occurs when the contemporary environment no longer retains the same set of key, evolutionarily relevant selection pressures. Such change can occur as a result of gradual modifications that bring about the end of similarities between evolutionarily relevant and contemporary environments, or as a result of rapid environmental modification. For example, species evolve means of detecting and avoiding familiar predators, but introduced predators may be different enough to represent a qualitatively different challenge. In the 60 years since introduction of the bullfrog, *Rana catesbeiana,* to western North America, larvae of native red-legged frogs, *R. aurora,* have evolved adaptive bullfrog-avoidance behavior (Kiesecker & Blaustein, 1997). Plastic avoidance behaviors are inducible simply by exposing red-legged frog tadpoles to water flowing first over bullfrog tadpoles (or adults) in a different chamber. This adaptive reaction norm has not evolved in *R. aurora* populations where bullfrogs are still absent. Kiesecker and Blaustein

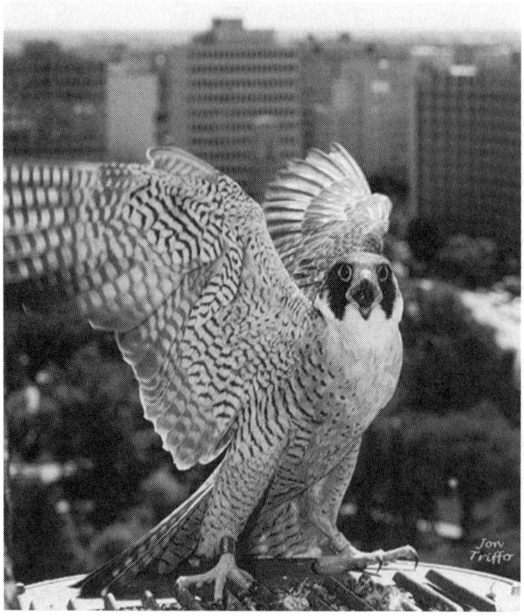

FIGURE 12.2 Like their rock dove prey, peregrine falcons now nest within niches on clifflike urban buildings. (Photo © Jon Triffo.)

(1997, p. 1758) conclude that "knowledge of how individuals vary in their response to introduced organisms may be extremely important in understanding the dynamics of biological invasions."

To explore possible routes toward better predicting adaptive responses (or their absence), it may be useful to formalize further the discussion of qualitative versus quantitative environmental change. Although both types of change can be substantial enough to stress and eliminate individuals, differences between them influence the predictability of adaptive responses. The key difference between qualitative and quantitative environmental change is the manner in which the selective environment changes. Qualitative change represents a switch to a fundamentally different selection environment, rendering previous knowledge of phenotypic potential less useful. As a result, predicting if and how populations tolerate qualitative changes is difficult. Quantitative change, on the other hand, is a shift in the strength of selection pressures that are already relevant. Thus, previous knowledge of phenotypic potential will help predict whether a particular quantitative change will be met with adaptation, whether via development, selection, or their interaction. Although it is likely that any environmental change will have both qualitative and quantitative components, and qualitative changes may exert their influence through quantitative channels, dichotomizing the types and understanding which is more prevalent should be helpful in determining a course of action.

In Figure 12.3 we present a scheme for classifying traits and attributes of populations, individuals, and genomes that underlie adaptive responses to environmental change and are hence key to the management of evolving populations. Whether a population responds to environmental change successfully will depend on the nature and rate of the change, the genetic and phenotypic form of individual response, and the population demographic context (Boulding, this volume; Reed, this volume). Although natural selection is the prime creative force behind biological diversity, there is no necessary expectation that this process will rescue threatened populations. Instead, interaction among genetic, phenotypic, and demographic responses depicted in Figure 12.3 will determine how populations react to environmental change.

Small populations have the twin vulnerabilities of higher extinction risk (whether from gradual

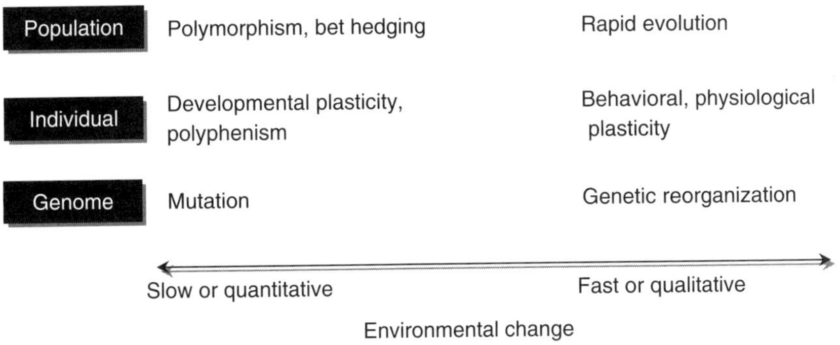

FIGURE 12.3 Sources of variation significant to adaptation and conservation in the face of environmental change. Some characters span levels—for example, polyphenism is a product of individual experience that is assessed with regard to among-individual variation.

decline or catastrophe) and lower genetic diversity (Brook, this volume; Sambatti et al., this volume). When environments change significantly, populations are faced with altered selection pressures. During the lag between environmental change and adaptation, selective mortality may reduce population size before mean fitness improves. As a result, even while molding adaptation, strong selection in vulnerable populations may itself prove catastrophic and lead to extinction (Gomulkiewicz & Holt, 1995). Plastic responses of population members to rapid environmental change will often be key mediators of the strength of selection in the changing environment. Thus, traits we observe in populations are products of past selective environments, genetic and phenotypic correlations, trade-offs, gene flow, and the current developmental environment (see, for example, Etterson, this volume).

The suite of factors that may convert an extinction trajectory into one of adaptation that permits population persistence or growth is thus manifold. This suite includes the amount, form, and distribution of genetic variation through the population; adaptive developmental and behavioral plasticity; reproductive and developmental performance (including generation time) under the altered conditions; and, of course, population size. Achieving and maintaining populations of sufficient size is far from certain for many large or rare organisms, particularly those in fragmented habitats, and in those for which inbreeding depression is a likely problem.

In contrast, other models examining *evolutionary potential* (Box 12.1) of catastrophically reduced populations suggest an unexpected consequence of the interaction of demography and genetics. Founder–flush models describe how population bottlenecks cause inadvertent genetic reorganization that can actually enhance additive genetic variation and thus promote evolutionary response to selection (Willis & Orr, 1993). Overall genetic configuration of remnant populations may differ substantially in allelic frequencies from their larger, ancestral populations (Brook, this volume; Reed, this volume). By reorganizing genes in novel genetic backgrounds, new sources of genetic diversity may appear (Fig. 12.3). Such reorganization could result in the appearance of new genotypes, and potentially very different phenotypes resulting from novel gene-by-gene (in other words, epistatic) interactions (Goodnight, 2000). Population disruptions that generate founder–flush scenarios may underlie cases of unexpectedly rapid adaptive evolution (Carroll, 2007b; Regan et al., 2003).

Modalities that buffer environmental stress, or bolster adaptive potential, are dichotomized for illustrative purposes in Figure 12.3. However, the implication remains that management of phenotypes and genotypes for conservation may take a variety of forms. The first alternative is doing nothing, a cost-effective tactic that in certain circumstances may be sound practice. For example, populations shifting from natural to anthropogenic food sources (for example, refuse reliance in dumpster-diving black bears, *Ursus americanus*) may experience altered selection regimes yet persist more successfully than without such novel resources (Beckmann & Berger, 2003). Likewise, propagules or gene flow from remnant natives may in some cases be sufficient to reconstitute decimated populations with genetically diverse descendants (Rice & Emery, 2003). At the other extreme, intense management of phenotypes may be merited when doing so will enhance effective population size and thereby protect viability and genetic variation. A further possible management tactic is manipulation of selection regimes to enhance adaptation to changing environments. Such tactics could be aimed at, for example, promoting the persistence of threatened taxa or customizing organisms to improve degraded habitats.

In the following sections, we focus on management possibilities that emphasize adaptive processes including adaptive decision making, phenotypic plasticity, and polyphenism and polymorphism expressed as discontinuous phenotypes. We focus on how individuals respond to changing environments, how their responses are distributed phenotypically within populations, and how those responses may be manipulated to achieve particular conservation goals.

TRAITS, PHENOTYPES, AND ADAPTATION

Individuals that express different values of specific traits vary in their capacity to respond to environmental stressors. For example, all else being equal, individuals with longer legs are likely to run faster than short-legged individuals and are thus more adept at eluding predators (see, for example, Losos et al., 2004). In this case, having long legs is a key adaptive advantage in terms of predator avoidance. Here, short-legged individuals will likely be lost from the population, unless short-legged individuals fare better than long-legged ones in other situations (for example, if short-legged individuals are more efficient foragers when no predators are present). Although both continuous and discrete phenotypic variation mean that individuals differ in their responses to environmental change, little consideration has been given to such differences in most genetic or ecological approaches to conservation. Instead, it is often assumed that behavioral rates or ecological needs of the average individual are sufficient to describe a population's state. Nonetheless, variation among individuals in *key traits* (Box 12.1) is likely of considerable importance to population viability.

Whether in discrete or continuously varying morphs, discerning the context-specific functions of key traits is likely critical to understanding why some taxa fare better than others when environments change. As a result, it should be useful to develop a framework by which key traits can be recognized. Doing so will inform predictions about which species are most at risk from human-induced change and which are most capable of persisting. Furthermore, understanding variation in key traits within populations will increase our ability to promote adaptive response to environmental change.

When individuals first encounter a new or altered environment, adaptive response is mediated by their ability to detect change, recognize challenges or opportunities presented, and react appropriately. Thus, the sequence of events that occur when individuals interface with a modified environment provides insight into the mechanisms at work in both adaptive and maladaptive responses to novelty. At each step in the sequence there is a key developmental, perceptual, or behavioral response. For example, animals may not perceive novel predators as threats (Schlaepfer et al., 2005). The key response here is recognition of that threat; thus individuals that fail at this step may be eliminated quickly. Further along in the sequence, animals may detect new predators and perceive them to be a threat, but nevertheless respond with inappropriate behavior. Failure to respond to novel predators with appropriate antipredator behavior has been observed in the Tasmanian quoll (Jones et al., 2004). In contrast to their responses when exposed to acoustic signals of native predatory owls and mammals, acoustic signals from novel predators (for example, introduced foxes and cats) elicited behaviors that would likely increase vulnerability of quoll to such predators.

The sequence of adaptive response—detect, recognize, respond—to environmental change suggests particular areas of study to determine how populations will fare. When individuals respond to cues that do not provide substantive information about their environment, they may fall into ecological or evolutionary traps. *Ecological traps* (Box 12.1) occur when individuals use environmental cues to choose suboptimal habitats (Schlaepfer et al., 2005). The result is a fitness decline for individuals who use the trap habitat. Generally, the cues used to make inappropriate habitat choices are similar to cues used historically, but there is now a disassociation between the cue and habitat quality. For example, sage sparrows, *Amphispiza belli*, living in a small reserve surrounded by suburban housing and modified open space, preferred to settle in areas where fitness returns were lower than they would have been had the sparrows settled in less-preferred areas (Misenhelter & Rotenberry, 2000).

The concept of the ecological trap falls within the more general notion of the evolutionary trap. In an evolutionary trap, individuals make behavioral or life history decisions based on cues that appear to have evolutionary relevance but are in fact superfluous and evoke decisions that are deleterious to survival or reproduction (Schlaepfer et al., 2005). For example, brown glass beer bottles resemble the dorsal surface of unusually large, female buprestid beetles of the species *Julidomorpha bakewelli*. In the throes of misdirected matings, males attempt to copulate with the bottles, the result being losses in time, energy, and mating opportunities (Gwynne & Rentz, 1983), and possibly even death should inebriated ants swarm from the bottle and consume misguided males.

Because evolutionary traps are driven by perception of specific cues, investigations of sensory ecology will offer insight regarding the types of mistakes that individuals are likely to make. To determine the role of evolutionary traps in population risk, the forms of change should be broadly considered in relation to the relevant cues used by the population of concern. For example, if the primary change has been climatic warming, one should consider the decisions that individuals make based on temperature, such as reproductive phenology, microhabitat use, migration, thermoregulation (Huey et al., 2003; Réale et al., 2003), and the interaction of temperature with photoperiod.

Notably, although environmental change challenges many populations, it represents opportunity for others. Thus while some species are caught in evolutionary traps, others may experience *ecological* and *evolutionary release* (Box 12.1). For example, *Jadera* and *Leptocoris* soapberry bugs have undergone rapid diversifying evolution because of new opportunities brought about by introduction of nonnative host plants. Novel host plants have massive, uncontested seed resources that soapberry bugs are preadapted to exploit for growth and reproduction. In as little as 30 years (fewer than 100 generations) soapberry bugs have evolved remarkably different values of host preference, proboscis length, survival, fecundity, and development time on the new host (Carroll, 2007b). In some cases, estimated fitness on novel hosts may have increased as much as sevenfold relative to fitness on native hosts (Carroll, unpublished analysis).

Such human-caused ecological and evolutionary changes may increase population sizes and thus lower probabilities of local extinction. In addition, these changes may increase intrapopulation diversity and lead toward natural processes like speciation. Both such outcomes favor biological diversity, and so may be construed as positive from a conservation standpoint. Indeed, there is no a priori reason to expect that anthropogenic changes will not improve survival and reproduction of some species through sudden shifts in niche structure that provide benefits or reduce risk. Populations that inadvertently benefit from change may become increasingly prominent, experience ecological and evolutionary release, and have disproportionate impacts on other community members. Thus, anthropogenic influences may result in reconfigured communities. It follows that beneficiaries of reconfiguration should be emphasized in conservation biology studies, even though such taxa may themselves be secure.

CASE STUDIES: INTRASPECIFIC BIODIVERSITY AND POPULATION PERSISTENCE

Polymorphism and Polyphenism

Although it is critical to study traits key to adaptive response to environmental change, it is also important to consider performance of individuals expressing discrete, alternative phenotypes within a species. Alternative phenotypes are relevant to conservation management because such variation may allow individuals to exploit resources very different from one another. For example, niche partitioning by individuals who interact with their environment in fundamentally different ways can lead to increased population stability (Bolnick et al., 2003).

Polyphenisms are extreme forms of adaptive phenotypic plasticity in which environmental variation results in expression of discrete phenotypic variants rather than a continuous distribution of variants. Depending on the nature of environmental cues, a single genotype may develop into one or another alternative morphs like that seen in winged versus wingless insects, terrestrial versus aquatic newts, horned versus hornless beetles and mammals, and male versus female fish. Polymorphisms may be phenotypically indistinguishable from polyphenisms, but instead result from genetically determined development pathways that are independent of environmental control. Both spatial and temporal variation in selection can theoretically contribute to the persistence of genetic polymorphisms within populations (Roff, 1994). In reality, however, morphs may commonly be determined by a combination of genetic and environmental factors, but the degree of environmental influence may vary from slight to substantial. Environmental change may also result in the rapid evolution of polyphenic thresholds (Moczek & Nijhout, 2003).

Although polyphenic and polymorphic forms are usually defined with morphological criteria, they may also include attributes of life history, physiology, and behavior that collectively comprise adaptive *syndromes* (Sih et al., 2004). Both polymorphisms and polyphenisms are widespread sources of phenotypic variation and are regarded as important in niche diversification and speciation (Ghalambor et al., 2007).

Discrete polyphenic syndromes are *threshold* phenomena in which a large change in the probability of developing into an alternative morph may occur over a small range of environmental variation. Although cues that trigger alternative developmental trajectories are often continuously distributed, environments in which organisms must function may be distinctly dichotomous (for example, absence vs. presence of predators, terrestrial vs. aquatic habitat, and the probability that a resource patch will persist over a given span of time) (Roff, 1994). Individual condition (phenotypic state) may be the major determinant of the probability of success in intraspecific resource competition in cases such as male–male competition in dung beetles (Fig. 12.4) and salmon. Discontinuous alternative morphs imply that selection is very different under various circumstances of phenotype-by-environment interaction, and that very different, functionally incompatible traits perform better in the respective circumstances.

Alternative phenotypes within a syndrome may show fundamentally different responses to the same environmental challenge, and may be important to conservation in at least four ways that are relevant here. First, functional differences between morphs in key traits may reveal what resources are important and how resources are distributed. Second, changes in morph frequencies may indicate how environments are changing. Third, multiple morphs within populations may partition resources and permit larger effective population sizes to persist. And lastly, discrete morphs may be easier to manipulate through developmental and genetic management for conservation purposes than more continuous morphs (Watters et al., 2003).

In some circumstances, the different responses of alternative phenotypes result in similar fitness payoffs, whereas in others, differential response results in unequal fitness returns. For example, males of the butterfly *Heliconius charitonia* express two alternative mating phenotypes that also exhibit a foraging polymorphism (Mendoza-Cuenca & Macias-Ordonez, 2005). Large males mate with females that are still in the pupal stage whereas small males search for and mate with fully emerged, virgin adult females. Pupal mating means that adult virgin females are often rare, and small males spend nearly twice as much time foraging as large males, and make pollen-collecting visits to nearly twice as many flowers as large males. Adult male size may result largely from the quality of his developmental environment, and smaller males may be "making the best of a bad job." Searching for flying virgins may be more costly energetically, but it is an alternative to direct physical encounters with large males attending pupae. In addition, it leads to small males being more active pollinators, giving their behavioral differences from large males potentially ecological significance as well.

Alternative Male Phenotypes in Salmon

In another example of alternative mating types, males of the coho salmon *Oncorhynchus kisutch* express two life histories that differ in time to maturity and mating tactics (Watters et al., 2003). This species spends its first year in freshwater and then

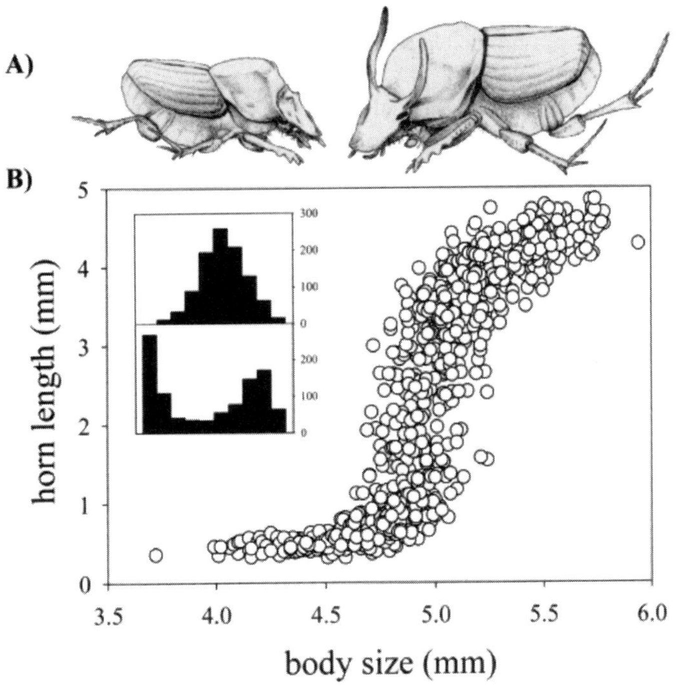

FIGURE 12.4 Occurrence of horns is a key trait in male dung beetles. (A) Hornless and horned males of the polyphenic *Onthophagus taurus*. (B) The probability of being a horned male increases steeply for males that metamorphose at an adult body length above 4.6 mm. Adult body length reflects environmental conditions experienced during larval development. Horned males use their horns to block access to tunnels, sheltering their mates and the dung balls (the larval food source) that they have created together. In contrast, hornless males attempt to burrow beneath defenders and inseminate females underground before they complete oviposition. A sigmoidal scaling relationship between body size and a specific character is typical of polyphenism. Inset histograms show frequency distributions for body size (above) and horn length (below). The apparent disruptive selection has led to rapid divergence in the threshold body size in the 40 years after the introduction of *O. taurus* into new habitats. (After Moczek and Nijhout [2003].)

moves downstream to the ocean to mature, finally returning to freshwater to spawn and die. One male type—hooknose—matures after 1.5 years spent in the ocean and returns to freshwater after attaining a relatively large size. The other male type—jack—spends only about 6 months in the ocean and returns to spawn at a relatively small size. These types result from juvenile growth differences that are influenced by success in gaining access to prime foraging habitats. Juveniles that acquire territories in food-rich riffles of natal streams are able to grow faster because territoriality reduces energetic costs of maintaining position in the stream. Juvenile males that grow quickly mature as jacks, whereas slow-growing males take an additional year to mature, ultimately becoming much larger hooknoses. Thus, differences in resource acquisition, in this case space- and food-use by juveniles, is a proximate trigger of discrete developmental differences in adult phenotypes.

Such ecological differences between alternative reproductive phenotypes are not often investigated but may be quite common. Indeed, discrete phenotypic differences may facilitate population persistence. For instance, *O. kisutch* populations with both types of males may be more resilient against environmental change than populations with only a single male type. Because jack and hooknose from within the same generation of offspring spend different amounts of time in the ocean, deleterious conditions in the marine environment are less likely to eliminate an entire cohort. Similarly, nonoverlapping return times of males within a cohort may increase chances that some individuals reproduce in the event that spawning conditions are poor in a given year.

Alternative phenotypes may also have effects on population dynamics and genetics. For instance, by using resources differently in space and time, a polymorphic population may attain a larger size than a monomorphic population. Differences between male types in development time also means that genes mix between generations, and that the breeding population size varies less between years than it would in a monomorphic species (Nunney, 1993). Reducing variation in the number of breeding individuals (effective population size) reduces the probability that a breeding population will drop to zero. "Bet-hedging" life histories such as these may be favored over monomorphic strategies in variable environments. In the case of salmon, conditional male dimorphism may both reduce variation in population size and increase annual productivity. Because different male life histories are likely to lead to population persistence when the environment changes, Watters and colleagues (2003) suggested that restoration of salmon streams should focus on supplying a diversity of juvenile habitats. In this way it is possible for individuals to develop along more than one alternative, environmentally determined trajectory

Behavioral Types

When there are consistent differences among individuals in response to the same stimuli, maintaining variation in these responses promotes population persistence. A growing body of literature demonstrates that, in many species, individuals exhibit consistent behavioral types across contexts (Sih et al., 2004). An upshot of such consistency is that a disposition that serves an individual well in some circumstances may serve it poorly in others. Individuals may be classified as *shy* or *bold,* for example. Bolder individuals tend to be more active than shy individuals and retain that tendency even in the presence of predators (Bell & Stamps, 2004). Bolder individuals may simultaneously be more likely to acquire food and risk predation. To the extent that individual behavior and the differences among individuals are structured by such types, the ability of individuals to perform well in a diversity of circumstances is undoubtedly limited.

Of particular relevance to the problem of managing populations to cope with contemporary environmental change is the manner in which different behavioral types deal with novelty. Given that behavioral types respond differently to the same stimuli, circumstances that challenge some types may be environmental opportunities for others. There are many examples of how different behavioral types act in novel situations. Behavioral types in the great tit *Parus major* are referred to as *fast explorers* and *slow explorers* based on their readiness to explore new areas and approach novel items. Fast explorers are more likely to sample novel foods (Verbeek et al., 1996) and are seemingly self-motivated (van Oers et al., 2005). Slow explorers, in contrast, pay more attention to what other birds are doing (van Oers et al., 2005) and may avoid the mistakes that others make. In addition, these types appear to vary in their survival after different types of environmental change (Dingemanse et al., 2004). When resources were rich, for example, fast-exploring males fared better than slow-exploring males. For females, though, the trend was opposite. Populations composed of both types may be more persistent than monotypic populations would be.

Similarly, in a review of *coping styles* in a variety of species, Koolhaas and colleagues (1999) considered how different types interact with their environment. *Reactive* individuals are shy, and monitor their surroundings and modify their behavior to meet new challenges. In contrast, *proactive* individuals live more by routine. They do well in predictable situations in which they can maintain efficient function. Additionally, proactive individuals are more aggressive than reactive ones. Along with these behavioral differences, the types also differ physiologically and have different risks of acquiring disease (Koolhaas et al., 1999).

In social species, then, it may be important to maintain a mix of behavioral types. Social groups composed of variant behavioral types may, on average, fare better than groups comprised of individuals that respond similarly to the same stimuli. The niche has many domains, and in those in which individuals are critically interdependent, fitness of an individual expressing one behavioral type depends strongly on how many others in the group express the same or other behavioral types (Dugatkin & Reeve, 1998). Thus, there are many contexts in which, and functions for which, individual differences may permit exploitation of alternative resources or underused portions of the niche. Through such diversifying mechanisms, populations may more thoroughly exploit resources and achieve larger effective population sizes (Watters et al., 2003). For both canalized and plastic traits, population-level impacts of environmental challenges are likely to depend heavily on the phenotypic distribution among individuals. Although for slight changes, most individuals may perform well, response to greater changes will be limited by phenotypic variance. When some individuals cannot respond effectively, populations that express greater phenotypic diversity are more likely to persist in the face of change.

That such phenotypic variation promotes population persistence is the basic tenet of the phenotype management approach to habitat restoration and population conservation (Watters & Meehan, 2007; Watters et al., 2003). If developmental circumstances influence variation, even without considerable heritability for important traits, populations can be tailored to be more resilient to environmental change. As shown in the previous examples, resource availability influences individual development, as does the manner by which resources are acquired and used. Because competition may favor individuals that use alternatives means of acquiring specific, preferred resources, providing a diversity of alternative resources may thus be effective in promoting expression of multiple phenotypes in a population. Captive populations may likewise be driven to produce multiple phenotypes through variation in rearing environments and training opportunities. Watters and Meehan (2007) suggest that groups of animals that are planned for reintroduction may fare better if they are composed of individuals that vary in their behavioral responses to environmental challenge. Managers should pay particular attention to specific behavioral and morph differences among individuals when planning reintroductions.

FUTURE DIRECTIONS

There is ample evidence that populations and species evolve in response to human influences on nature, and such adaptive change is probably more common and widespread than is currently recognized. In the face of pervasive environmental change, efforts to prevent adaptive phenotypic or genetic change with the goal of preserving more ancestral or "pristine" states are at best difficult and at worst misguided. For populations that do not disappear, evolution is likely an inevitable consequence, if not a prerequisite, for persistence (Carroll, 2007a). Only when there is a plan to return or transfer individuals to former conditions should efforts to retard or reverse evolution be considered. Allowing evolution to proceed should in general facilitate conservation goals while potentially preserving genetic variation in taxa of interest. If wildlands are, in the long run, restored and expanded, we can likely count on future evolution and evolutionarily enlightened management to assist in the process of adapting old populations to new environments.

This is not to argue that management schemes that reduce stress and risk cannot also buffer the intensity of selection and thus permit plastic responses to be the main path of adaptation. For example, after initiating a risk-averse tactic of sheltering the endangered tammar wallaby, *Macropus eugenii,* on predator-free islands off Australia, restoration of mainland populations began only after managers taught wallabies to avoid introduced predatory foxes. Once entrained in a pilot population, there is evidence that predator avoidance may be copied by conspecifics (Griffin & Evans, 2003), potentially generating a self-replicating, adaptive antipredator meme of conservation value. If effective, such learned, culturally transmitted adaptive responses could strongly limit both predation and evolution on the mainland.

Nonetheless, populations are changing in important ways beyond simply the number of surviving individuals, suggesting that monitoring adaptive responses to environmental change should be a first principle of conservation biology. Clearly, any long-term, sustainable approaches would have to assay evolution in managed populations, and the current rates of change are pushing the problem to the forefront. The challenge now is to design ways to use awareness of inevitable adaptive responses to avoid problems stemming from mistaken assumptions of stasis, as well as to devise ways to enhance how adaptive processes may serve conservation goals. Evolutionary conservation biology is a new discipline in its intellectual infancy. The notion of species as fixed entities pervades endeavors ranging from global change models to PVAs. Moreover, standard conservation genetics approaches use molecular markers that are presumed to be neutral. Rapid adaptation is often associated with the same factors that are causing extinctions (Stockwell et al., 2003), and thus conservationists are well positioned to work with species change, provided they remain alert to such processes.

If we accept the likelihood of ongoing adaptation to anthropogenic change, the next challenge from a management perspective is to predict and deal with its outcome, and, if possible, to manipulate its expression. Like other problems in biology, solutions must generally come from detailed knowledge of natural history. Yet the degree and form of environmental change may make some adaptive responses more likely than others (Fig. 12.3). Relatively slow or qualitative responses may be met with individual developmental responses—including polyphenisms and polymorphisms like bet hedging—that depend on the pace and spatial structure of change. Standing genetic variation, reassortment, and sexual recombination, enhanced by occasional beneficial mutations, may be sufficient to keep pace with slow or relatively inconsequential changes.

In contrast, rapid or qualitative changes may tax populations much more severely than slow or quantitative changes, resulting in more dramatic biotic reorganizations. New genotype-by-environment interactions may result in unprecedented phenotypes with scant but nontrivial chances of success and ultimate genetic assimilation (Grether, 2005; Price et al., 2003). Stress may increase rates of mutation (Hoffmann & Parsons, 1997), and population crashes may reorganize genomes in ways that likewise increase phenotypic diversity (Goodnight, 2000). Through such seemingly arcane avenues, stressed populations, although in danger of extinction, may nonetheless be granted a disproportionate number of different "lottery tickets," offering a chance at persistence in the altered environment. Rapidly inducing behavioral and physiological plasticity may be important in buffering mortality and morbidity threats, permitting adaptive evolution with the opportunity to proceed and perhaps rebuild genetically and phenotypically altered populations.

In cases in which such self-rescue appears unlikely, conservation intervention may promote population persistence—in ways analogous to endogenous variation and plasticity—by feeding, sheltering, moving, or altering species at risk. For example, through *phyloclimatic* modeling of the ancestral niche of plants in the ant-dispersed genus *Cyclamen*, Yesson and Culham (2006) predicted that many of the species are threatened by the northward shift in the Mediterranean habitat to which they are adapted. Specialization of ant-dependent dispersal is a key trait in *Cyclamen* that we can predict will limit its capacity to respond to abrupt geographic shifts in physiological jeopardy and opportunity. Humans might consider moving plants and attendant ants northward and perhaps faster than they can travel themselves.

Different environmental challenges will influence different individual traits, and not all affected traits of a species will respond similarly to change. The goal of evolutionarily enlightened management is to choose traits that can be manipulated and that are key to building resistance and sustainability into populations. The phenotype management approach (Watters & Meehan, 2007; Watters et al., 2003) points to individual tactical variation as a source of information and leverage for conservation biologists. By managing the genetic and environmental bases of phenotypic expression, it may be possible to promote population persistence in several ways. Methods may include increasing the diversity of resources used, decreasing intraspecific competition, increasing effective population size, reducing risk of catastrophic extinction, and enhancing the diversity of genotype-by-environment interactions that may lead to long-term adaptive changes. Important progress will likely be made in the next decade in implementing tests of these ideas.

Historically, evolution has been studied in native, multispecies communities in a variety of contexts, including character displacement from competition, interactions of hosts and prey with their enemies, and mutualisms. Human activities are influencing the selective outcomes of such interactions as well as adding novel selection pressures (Stockwell et al., 2003; Strauss et al., 2006a), and it will ultimately be important to consider how evolution in multiple traits, among multiple interacting species, will play out and affect conservation and other practical agendas. We know that populations can quickly evolve with respect to how individuals detect and respond to changes in the distribution of risks and opportunities through time and in the landscape, and that there may be evolution of behavior, physiology, morphology, and life history. Indeed, organisms from microbes to trees have changed in the past century (Strauss et al., 2006a).

An evolutionary approach to conservation requires that we consider multigenerational patterns among as well as within species. Little is known about how, for example, short-term changes in interpopulation diversity will play out in longer term influences on a focal taxon or other species with which it interacts. The geographic mosaic model of community structure (Thompson, 1999) predicts that traits of species will reflect shifting community composition and characteristics across regions. A variety of natural systems appear to show such patterns (for example, seed-feeding insects, crossbills, and squirrels each select for different trait combinations in lodgepole pine cones [Benkman et al., this volume]). Within communities, even local variation in species composition may result in heterogeneous evolution (Rudgers & Strauss, 2004). Human impacts, in addition to altering community composition, are changing the characteristics of constituent species, and there is reason to expect that those changing characters will likewise affect community processes. Whether a particular taxon is threatened by or benefits from change, networks of new direct and indirect effects that result in novel or accelerating coevolutionary relationships will alter basic ecological outcomes from trophic cascades (Loeuille & Loreau, 2004; Strauss et al., 2006a) to speciation (Hendry et al., 2006).

Lastly, there is the prospect that widespread, deep, or irretrievable alteration of native landscapes will invite assembly of artificial communities. Such communities may perhaps consist of modified organisms that preserve key types of biological function and interaction or ameliorate particular environmental problems. Until recently, such was the general indifference to contemporary evolution that, for example, in 1972, the U.S. National Academy of Sciences concluded that lead had no toxic effects on plants because plants could grow on contaminated soils! An informed perspective has meant that newly evolved toxin-tolerant strains of plants and microbes are increasingly used for remediation and reclamation of polluted sites. Other forms of degradation resulting from the loss of particular ecosystem processes such as pollination, decomposition, nitrogen fixation, water retention, population regulation of aliens or natives, and evolvability itself may lead to analogous interventions in which communities are artificially assembled to solve specific problems and to sustain ecosystem functions. Organisms used for "triage" for threatened systems may increasingly be genetically designed. Clearly, predicting and managing the dynamics of novel biotic communities, whether they result from human recklessness or from community engineering, will be one of the principal challenges of incorporating evolutionary biology into conservation.

SUGGESTIONS FOR FURTHER READING

The challenges of integrating adaptation's multifarious modalities with the pressing practical needs of conservation are daunting. Ferriére and colleagues (2004) assist this effort by presenting a concise and coherent conceptual introduction to the interaction of genes, populations, and environments in the management of biodiversity. Gomulkiewicz and Holt (1995) provide illuminating theory about the costs and benefits of adaptive evolution in populations threatened by extinction. Ghalambor and colleagues (2007) offer an in-depth review of the role of phenotypic plasticity in contemporary evolution, and Schlaepfer and colleagues (2005) introduce the important concept of evolutionary traps and their conservation significance.

Ferriére, R., U. Dieckmann, & D. Couvet. 2004. Introduction (pp. 1–14). In R. Ferriere, U. Dieckmann, & D. Couvet (eds.). Evolutionary conservation biology. Cambridge University Press, Cambridge, UK.
Ghalambor, C. K., J. K. McKay, S. P. Carroll, & D. N. Reznick. 2007. Adaptive versus

non-adaptive phenotypic plasticity and the potential for adaptation to new environments. Funct Ecol. 21: 394–407.
Gomulkiewicz, R., & R. D. Holt. 1995. When does evolution by natural selection prevent extinction? Evolution 49: 201–207.
Schlaepfer, M. A., P. W. Sherman, B. Blossey, & M. C. Runge. 2005. Introduced species as evolutionary traps. Ecol Lett. 8: 241–246.

Acknowledgments We thank M. Singer, M. Loeb, and an anonymous reviewer for providing helpful suggestions; A. Sih for useful discussion; J. Triffo for allowing us to use the image of the peregrine falcon free of charge; and The Institute for Contemporary Evolution and the Australian-American Fulbright Commission for financial support of S. Carroll.

13

Genetic Diversity, Adaptive Potential, and Population Viability in Changing Environments

ELIZABETH GRACE BOULDING

Heterozygosity is the most widely used estimator of genetic diversity. The enduring appeal of heterozygosity may stem from a widely held belief that a correlation between heterozygosity and fitness reflects genomewide benefits of genetic diversity (reviewed by Dewoody & Dewoody, 2005). Heterozygosity can be thought of as the probability that an individual will have more than one allele at a particular genetic locus and is usually reported as expected heterozygosity, H_E (Box 13.1). Lasting popularity of H_E with conservation biologists is rather surprising because it has long been known that this estimator is relatively insensitive to population bottlenecks of short duration (Hedgecock & Sly, 1990). Other measures of within-population genetic diversity include individual multilocus heterozygosity, proportion of polymorphic loci, squared size difference between two alleles at a locus, and the effective number of alleles per locus (Box 13.1).

In addition to estimating genetic diversity within populations, conservation biologists are often interested in estimating the proportion of the total genetic diversity that is partitioned among geographically separated populations. This is useful when deciding where to locate protected areas and whether to translocate individuals between populations. The most popular method of estimating population structure is with Wright's hierarchical *F* statistics (Wright, 1965), which are derived from Wright's inbreeding coefficient (Box 13.1) and can be calculated from the deficiency of heterozygotes at nuclear molecular markers. His most famous statistic, the fixation index F_{ST}, is the genetic variation within subpopulations relative to that within the total population. At equilibrium, F_{ST} is inversely related to the product of the migration rate among the subpopulations and their effective population size. Another statistic that is useful, especially when the pedigree is unknown, is Wright's within-individual inbreeding coefficient F_{IS}, which partitions the amount of heterozygosity in individuals relative to that within the subpopulation. His final statistic, F_{IT}, partitions the amount of heterozygosity in individuals relative to that within the total population. Excoffier and colleagues (1992) describe a computer program (Arlequin) that computes F_{IS}, F_{ST}, and F_{IT} from molecular markers and can test for structure among predefined groupings of the subpopulations.

Inbreeding depression has been broadly defined as a decline in fitness that arises from decreasing heterozygosity (or increasing homozygosity) across the genome (Reed, this volume). An increase in homozygosity caused by small population size can decrease population viability through three mechanisms. First, matings between close relatives becomes more common. This is problematic because close relatives are likely to have inherited the same recessive deleterious allele at a given locus, which increases the frequency of expression in the population. This is inbreeding in the strict sense,

BOX 13.1 Introduction to Neutral Molecular Markers for Ecologists

Routine nonlethal collection of small tissue samples for later genotyping with molecular markers is becoming an increasingly common component of ecological studies. Small tissue samples can be preserved for later extraction drying them, or by putting small pieces in 95% ethanol or commercial salt solutions. Commercial kits are now available that can reliably extract DNA from tissue, hair, blood, fin clips, or feces. Most molecular markers in common use today (Box Table 13.1) require only small amounts of crude DNA extract because they use PCR to amplify the target DNA region exponentially until there is enough to genotype. After PCR, one of three genotyping steps usually takes place: (1) the PCR fragment is sent to be sequenced by a local automated DNA sequencing facility, (2) the PCR fragment is subjected to electrophoresis in a capillary tube filled with acrylamide gel to determine its size, or (3) a PCR fragment containing an SNP is labeled with allele-specific primers or probes, which are genotyped automatically in a specialized machine.

Neutral Molecular Markers Glossary

allozymes	Different alleles at an enzyme locus produce proteins that differ in net charge, which affects their relative migration rate on a starch gel.
effective number of alleles per locus, A_E	The number of alleles it would take to obtain a given level of the expected heterozygosity if all alleles were equally frequent. $A_E = 1/(1 - H_E)$
expected heterozygosity, H_E	Often called genetic diversity, D. For a single locus it is estimated using the formula $H_E = 1 - \Sigma p_i^2$, where p_i is the frequency of the ith allele.
expressed sequence tag (EST) markers	Complementary DNA (cDNA) libraries are created when messenger RNAs (mRNAs) are harvested from a particular tissue and converted to DNA using the enzyme reverse transcriptase before being cloned into a plasmid vector. These "complementary" or cDNA libraries are meant to have a complete copy of every mRNA in that tissue. cDNA libraries contain different alleles for a particular gene either because the organism was heterozygous at that locus or because the library was made from multiple individuals. The forward or reverse sequences of each clone in a cDNA library are often publicly available in EST databases and can be used by ecologists to discover SNPs or microsatellites that are within or immediately adjacent to a particular protein-coding region.
F_{IS}	Wright's (1965) within-individual inbreeding coefficient. This measures the amount of nonrandom mating within subpopulations that results in excessive homozygosity at the individual level.
F_{ST}	Wright's (1965) fixation index. The average number of pairwise differences among subpopulations relative to the average number within populations. This is the most used of Wright's *F* statistics that estimate the hierarchical partitioning of genetic variance within and among populations.

continued

inbreeding coefficient	The probability that two alleles for a particular gene are identical by descent as determined from a known pedigree containing a shared ancestor.
individual multilocus heterozygosity	The proportion of heterozygous loci within an individual (Coltman & Slate, 2003).
microsatellites	A genetic locus with alleles that consist of a variable number of copies (usually 10–30) of simple tandem repeat (STRs) of 2 to 4 bp of DNA sequence. Different alleles have different PCR fragment sizes and can be distinguished by their relative migration rates on an acrylamide gel.
neutral molecular marker	A base pair substitution, insertion, or deletion in a DNA region that is close to neutral with respect to the fitness of the organism in that it does not change its viability or fertility. For example, third codon positions in protein-coding genes are synonymous substitutions.
nonsynonymous substitution	A single base pair substitution in a protein-coding DNA region that does result in an amino acid substitution and therefore could experience a different selection coefficient.
observed heterozygosity, H_O	Number of individuals that are heterozygous at a locus over the total number sampled.
PCR	Polymerase chain reaction. This is a modern method of making millions of copies of a DNA sequence of 100 to 5,000 bp pairs that is surrounded at each end by a known "primer" sequence of 20 to 30 base pairs. This is done by putting some dilute DNA extract, the forward primer (usually labeled with a fluorescent dye), the reverse primer, extra bases of all four types, and heat-resistant DNA polymerase, into an eppendorf tube and then using a thermocycler machine to cycle the reaction between three temperatures that are typically: 95°C, 50°C, and 72°C. This has largely replaced molecular cloning as a method of obtaining sufficient quantities of a particular gene fragment to allow genotyping in conservation genetics laboratories.
proportion of polymorphic loci	Proportion of the loci that have more than one allele that has a frequency greater than 5%.
single nucleotide polymorphism (SNP)	DNA sequence variation at a single base pair in a nuclear or mtDNA fragment. The differences in sequence were traditionally detected by DNA sequencing, but high-throughput genotyping methods that use hybridization to oligonucleotide chips or primer extension reactions with labeled nucleotides are becoming more common.
squared size difference between two alleles at a locus	This is a useful measure of genetic diversity at a microsatellite locus where the absolute size differences among alleles is known precisely.
synonymous substitutions	A single base pair substitution in a protein-coding DNA region that does not change the amino acid.

From Hartl and Clark (2007)

continued

BOX 13.1 Introduction to Neutral Molecular Markers for Ecologists *(cont.)*

BOX TABLE 13.1 Popular Neutral Molecular Markers (Box 13.1) Used to Estimate Genetic Diversity in Multicellular Animals, Their Characteristics, and Assumptions, and Best Methodology for Genotyping

Marker	Type	Neutral?	Genotyping	Allelic Diversity	Reproducible among Labs?	Comments
Allozyme (protein)	Biparental nuclear, codominant	Often	Starch gel electrophoresis	Low per locus	Somewhat if relative allele mobility used	Historical data sets abundant
SNP (nDNA exons single-copy genes)	Biparental nuclear, codominant	Third codon position	Primer extension, SSCP or direct sequencing of PCR product	Low per site but one SNP per 1,000 bp	Highly if sequenced	Direct sequencing if many alleles
SNP (nDNA introns)	Biparental nuclear, codominant	Usually	Primer extension, SSCP; can sequence homozygotes	Moderate	Highly if sequenced	Length polymorphism common
SNP (ITS, NTS ribosomal genes)	Biparental nuclear, codominant	Usually	SSCP	Moderate and several copies	Highly if sequenced	Concerted evolution of gene copies
SNP (mtDNA)	Maternal haploid, codominant	Third codon position	Primer extension, SSCP or direct sequencing	Moderate	Highly if sequenced	Nuclear pseudogenes
Microsatellite = STR = SSR	Biparental nuclear, codominant	Usually	Capillary DNA automated sequencer best	High per locus	Yes, if same machine with test samples	Often too many alleles, if sample size is small
AFLP	Biparental nuclear, dominant	Most	Fragment length analysis	Moderate but many loci per gel	Somewhat	Useful for little-studied genomes
RAPD	Biparental nuclear, dominant	Often	Fragment length analysis	± only, but many loci per gel	No	Historical interest
MHC	Biparental nuclear, codominant	Balancing selection	SNP, SSCP	High	Highly if sequenced	High gametic phase disequilibrium

See also Avise, 2004.

which has been defined as the proportion of alleles identical by descent (Brook, this volume). Second, genetic drift results in fixation of deleterious and slightly deleterious alleles. Accumulation of mutations through drift can cause extinction by mutation meltdown (Reed, this volume). Third, strong genetic drift causes rare alleles to be lost and thus reduces a population's ability to adapt to new environments. I will show that reduction in evolutionary potential (Soulé, 1987) can also cause extinction.

Most morphological, behavioral, and life history traits are quantitative and thus show continuous variation that cannot easily be separated into discrete phenotypic classes (Falconer & MacKay, 1996). Estimating the genetic diversity of quantitative, or complex, traits is challenging, because phenotypes in this case are determined by multiple genetic loci and environment. Furthermore, unlike single-locus Mendelian traits such as eye color, loci responsible for complex traits are only now being identified (Hill, 2005). The most common within-population measure of genetic diversity used for complex traits is still narrow-sense heritability (Charmantier & Garant, 2005), which is defined as the proportion of the phenotypic variance that is directly heritable (Box 13.1). There is also a between-population measure of genetic diversity for quantitative traits, Q_{ST} (Merilä & Crnokrak, 2001), which is analogous to F_{ST}.

NEUTRAL VERSUS NONNEUTRAL MARKERS

Neutral theory (Kimura, 1968; Kimura & Ohta, 1971) proposed that the majority of base substitutions that become fixed in populations are nearly neutral with respect to selection. Theory also predicts that fixation rate of neutral substitutions υ is equivalent the mutation rate per gene per successful gamete. Neutral theory has been supported by empirical population genetics studies, beginning in the 1960s with allozyme protein electrophoresis (Avise, this volume). Allozymes are variants for a particular enzyme that have an amino acid substitution that affects protein net electrical charge and enables alleles to be distinguished using their relative migration on starch gels (Avise, 2004a). It is now recognized that although substitution of alleles segregating at most allozyme loci are neutral with no detectable effect on fitness, substitutions at other loci are not. Correlation between the frequency of particular allozyme alleles and particular environments have been observed and are likely the result of habitat-dependent selection (see, for example, Johannesson & Tatarenkov, 1997). As discussed later, modern PCR-based markers can also be nearly neutral or nonneutral, depending on linkage to other genes under differential selection in different environments.

IS GENETIC DIVERSITY RELEVANT IN A CONSERVATION CONTEXT?

One paradigm of modern conservation biology is that management of small populations entails estimates of molecular genetic diversity (Avise, 2004a). It is unclear, however, whether molecular markers yield an accurate estimate of population viability. Lynch (1996) reviewed the usefulness of molecular markers in conservation biology and questioned whether they provided much insight into (1) the loss of potential to adapt to future environmental change, (2) accumulation of deleterious mutations, and (3) the fate of individuals translocated from a large and genetically diverse population to a small and inbred one. In contrast, O'Brien and colleagues (1996) described two examples in which reduced variation at molecular markers in populations of large cats was correlated with elevated sperm abnormalities and reduced testosterone levels. Thus, some evidence suggests that reduced variation at molecular markers can signal a severe loss of mean population fitness. The question remains, however: Are molecular markers useful for estimating the potential of a population to adapt to future environmental change?

Stockwell and colleagues (2003) argue that conservation biologists thus far have been mostly concerned with maximizing genetic diversity rather than considering trade-offs arising from contemporary (or rapid) evolution. For example, they point out that strong selection for adaptation to current environments erodes genetic variation that might be needed to adapt to future environmental change. As a second example they remind us that translocation into an inbred population will reduce the level of inbreeding but will decrease its local adaptation.

In this chapter I discuss the effect of genetic diversity on the adaptive potential and viability of small populations experiencing environmental change.

I review the most widely used types of molecular markers and discuss methods of using such markers to estimate N_e. I specifically investigate (1) the extent to which genetic diversity at neutral molecular markers predicts population viability, (2) nascent technical and statistical solutions that address the problem of estimating evolutionary potential, and (3) the strength of the relationship between heritability and the adaptive potential of a complex trait in a series of populations connected by migration.

VARIATION AT MOLECULAR MARKERS

"Neutral" Loci

Heterozygosity and N_e

The level of inbreeding per generation and the loss of alleles from genetic drift both increase when the effective population size, N_e, is small (Brook, this volume). Indeed, if we ignore mutation and migration, then the rate of loss of heterozygosity for a single locus, at generation t + 1 as a function of the heterozygosity at generation t, is

$$H_{t+1} = H_t\left(1 - \frac{1}{2Ne}\right) \quad (13.1)$$

where N_e is the effective population size (Freeman & Herron, 2006). The decline in heterozygosity with N_e is somewhat more complex to model for a quantitative trait because the phenotype is determined by a few to hundreds of loci that combine additively, by epistasis, by dominance, and by the maternal and rearing environments (Falconer & Mackay, 1996). A past meta-analysis of 19 studies by Reed and Frankham (2001) found no correlation between heritabilities and genetic diversity at molecular markers. However, Reed (this volume) now argues that levels of genetic variation, whether measured as heritabilities, allelic diversity, or heterozygosities, correlate well with theoretical and empirical evolutionary potential.

Commonly Used "Neutral" Molecular Markers

Allozyme electrophoresis has been almost entirely replaced in conservation genetics by other molecular markers that use PCR to amplify specific DNA fragments (Avise, this volume). The main advantages of PCR-based DNA markers are that they (1) allow genotyping of very small tissue samples, (2) can amplify DNA fragments from impure genomic DNA extracts from ethanol or salt-preserved tissues, (3) can use noninvasively obtained tissues such as hair or fecal samples, and (4) vary in their rates of molecular evolution, allowing the most appropriate one to be chosen. This has made possible studies that were never possible before. Tissues used for allozyme electrophoresis had to be fresh or stored at temperatures less than −80°C to avoid denaturation of the enzyme because it was needed to catalyze the reaction that stained the gel (Avise, 2004a).

Currently, the two most commonly used markers in conservation genetics studies of animal populations are SNPs, both in mitochondrial and single-copy nuclear genes, and length polymorphisms in microsatellites (Box Table 13.1, Box 13.1). Microsatellites, which are also known as *simple tandem repeats* (STRs) or *simple sequence repeats* (SSRs), are particularly useful for paternity analysis because they typically have a large number of alleles per locus. Both SNPs and microsatellites are popular because they are PCR-based and because, with care and automation, the genotyping results are reproducible among different laboratories (Box Table 13.1). However, several new methods of genotyping a large number of SNP loci for a large number of individuals have been developed that use multiplexing of allele-specific extension primers or probes (Morin et al., 2004). These allele-specific methods emit distinctive fluorescent signals that are interpreted by automated allele-scoring software and need only a little human correction. Therefore, studies using SNP markers are usually lower in cost and higher in throughput than microsatellite markers, even after the lower polymorphism of SNP markers is taken into account. Also, large numbers of SNP loci are becoming available for many model organisms because of the availability of short DNA sequences of expressed sequence tags (ESTs) in public databases (Box 13.1) and because of new, more efficient methods of resequencing previously published genomes (Morin et al., 2004). If a large number of markers is needed for a nonmodel organism, then AFLP is still widely used, even though heterozygotes usually cannot be distinguished from homozygotes and reproducibility is only moderate (Box Table 13.1).

Variation at "Adaptive" Molecular Markers

Major Histocompatability Complex Markers

Neutral molecular markers may not resolve fine-scale population structure nor be able to assign individuals to closely spaced populations. This problem has encouraged a number of laboratories to develop "selected" or "adaptive" molecular markers. If local adaptation causes some alleles to become substantially more common at some sites than others, then that will enable the assignment of individuals to particular populations. For example, major histocompatibility complex (MHC) loci are currently a popular type of "selected" molecular marker for jawed vertebrates and have been used successfully to look at the genetic diversity of bottlenecked populations in the Felidae (Yuhki & O'Brien, 1990). Major histocompatibility complex genes are part of the immune response to foreign antigens, and their high mutation and recombination rates are thought to be an adaptive strategy for rapid evolution in response to disease pathogens.The antigen recognition site part of the protein is unusual in showing more nonsynonymous substitutions among individuals than synonymous substitutions (Box 13.1), suggesting positive selection is taking place (Hughes & Nei, 1988, 1989). This positive Darwinian selection is hypothesized to promote higher charge profile diversity in the antigen-binding cleft of class I MHC molecules and thus high substitution rates (Hughes et al., 1990). Major histocompatibility complex loci were useful for fine-scale population differentiation in Sacramento River chinook salmon (Kim et al., 1999). They were also found to give higher F_{ST} values than microsatellite markers, allowing finer scale detection of among-population differentiation in sockeye salmon (Miller et al., 2001).

Markers that Are Linked to Loci Under Selection

A low percentage of the loci for any type of molecular marker can be classified as "selected" or "adaptive" molecular markers because it is tightly linked physically to QTL that experience selection for different optima in different habitats. For example, 5% of the 306 AFLP loci genotyped in a marine snail population showed different frequencies in the upper and lower shore ecotypes that were divergent in morphological and life history characteristics (Wilding et al., 2001). An average of 2% to 3% of the AFLP loci showed different frequencies for normal and dwarf ecotypes of whitefish living in the same lake (Campbell & Bernatchez, 2004). In both these cases, the neutral AFLP loci showed little or no significant genetic differentiation between the two ecotypes, likely because gene flow between them was too high. A promising new type of marker to use to study adaptive genetic divergence is SNPs in ESTs (Culling et al., in review). Vasemägi and colleagues (2005) found correlations between 5 of 75 EST-associated microsatellites, and the salinity and latitude of the habitat of Atlantic salmon populations.

ADAPTIVE POTENTIAL

The importance of maintaining the evolutionary potential of a population was first mentioned by Soulé (1987). Franklin (1980) used mutation rates for neutral quantitative traits in *Drosophila* to propose that populations had to have an effective size, N_e, of at least 50 in the short term to avoid inbreeding and of 500 in the long term to maintain their evolutionary potential. Lynch and Lande (1998) do not agree that $N_e = 500$ is large enough to maintain the evolutionary potential, and argue that effective population sizes of 1,000 to 5,000 are needed, especially for single-locus traits because of their lower mutation rates. This is a concern because populations targeted for conservation are likely to have effective population sizes much smaller than 5,000.

Complex Traits Determined by Quantitative Trait Loci

There has been considerable effort recently to locate the QTLs that determine complex morphological, life history, and behavior traits that are of most interest to conservation biologists. Only 10 to 100 neutral molecular markers are typically available for most nonmodel organisms. However, fine-scale genetic linkage maps with thousands of markers now exist for model organisms used in agriculture, and these can allow identification of the actual DNA sequence that comprises the QTL (Hill, 2005).

Detection of QTLs is often achieved by genotyping backcrosses or F_2 crosses between divergent populations with neutral molecular markers and looking for statistical associations with phenotypic traits. With family sizes of 300, only QTLs that contribute 5% or more to the additive genetic variance can typically be detected (see, for example, Tao & Boulding, 2003). Two lines of corn that had been divergently selected for 100 generations were used to discover 440 SNP markers that enabled detection of QTLs that contributed less than 1% to the genetic variance for oil content (Hill, 2005). The results from the corn QTL analysis and results from a recent QTL analysis on body size in a line of poultry support the hypothesis that many quantitative traits are determined by many segregating genes (about 50 in corn and 13 in poultry), each with small effects that combine additively (Hill, 2005). However, in many other lines of poultry, body size is determined by mostly a few segregating loci, each with a large effect (Hill, 2005), perhaps because other loci that could potentially contribute have become homozygous. The number of segregating loci cannot be determined by just observing at the phenotypic distribution of the trait. Even only two loci may give a continuous phenotypic distribution after epistasis, dominance, maternal, and environmental effects are included (Lynch & Walsh, 1998).

Variation at Quantitative Trait Loci Currently "Neutral" and N_e

The amount of allelic variation at a neutral QTL in an isolated population depends on the equilibrium between mutation and genetic drift (Lynch & Walsh, 1998). The magnitude of mutation rate depends on the DNA region (Hartl & Clark, 2007) whereas the amount of genetic drift depends inversely on N_e, as shown in Eq. 13.1. The situation for a quantitative trait like body size is more complex than for a single-locus Mendelian trait because if the effects at different loci are additive, then they can cancel each other out. Imagine that each locus has only two possible alleles, one "plus" allele that increases body size and a second "minus" allele that decreases body size. Then at some QTLs, the plus allele will drift toward a higher frequency and at others it will drift toward a lower frequency so that small changes in allelic frequencies will cancel out, to some extent, and body size will remain constant. Evolution of quantitative traits is often modeled using the infinitesimal model, which assumes that the phenotype, z, depends on an infinitely large number of genes, each having an infinitely small additive effect (Bulmer, 1985). The large number of loci means that that the allelic frequencies at each QTL and the additive genetic variance at linkage equilibrium remain approximately constant over time and that the breeding values (Box 13.2) of each individual in the population are normally distributed (Bulmer, 1985).

Variation at Quantitative Trait Loci Currently under Measurable Selection

The amount of allelic variation at a QTL in an isolated population depends on the equilibrium between mutation, selection, and genetic drift (Lynch & Walsh, 1998). Strong selection for a prolonged period will reduce the additive genetic variances, and therefore the heritabilities for traits important in local adaptation will decrease (Falconer & MacKay, 1996). If directional selection is strong, as when there are human-induced effects on a population, then the breeding values will no longer be normally distributed and the loci may show positive gametic phase disequilibrium (Box 13.2) and may deviate from the traditional infinitesimal model.

After the population mean has reached a new optimum and is under pure stabilizing selection, then excessive genetic variance will actually decrease the short-term fitness of the population. Lande and Shannon (1996) point out that conservation biologists often assume that a higher additive genetic variance is always better, but that this is not true if the optimum is fixed, because the environment is constant. They argue that when there is strong stabilizing section toward a fixed optimum, a larger genetic variance increases the average distance of the individual's phenotype from the optimum and reduces the mean fitness of the population. In contrast, as I discuss later, higher heritabilities are beneficial if discrete or continuous environmental change is occurring (Lande & Shannon, 1996).

Geographic Variation at Quantitative Traits and Q_{ST}

The geographic distribution of a species at risk is often more fragmented than it was historically, and managers must often decide whether to translocate individuals among locations. Although translocation will reduce inbreeding, it might cause the population to be maladapted if, for example, species in

BOX 13.2 Basic Quantitative Genetics for Ecologists

Quantitative genetics is the study of the underlying genetics of continuous traits, meristic (or discrete) traits, and threshold traits, which includes most traits of interest to ecologists. Such complex traits are usually determined by multiple genetic loci and by the maternal and external environments. The theory of quantitative genetics was developed by animal and plant breeders who traditionally used parametric statistics to compare the phenotypic values of related to unrelated individuals for a particular trait. This allowed estimation of the heritability that, when multiplied by the intensity of artificial selection, successfully predicted the evolution of the phenotypic mean in the next generation (Falconer & Mackay, 1996). Modern applied quantitative genetics still relies on statistical analysis of the phenotypic measurements of a pedigreed population of animals rather than knowledge of what is occurring at the molecular level (Falconer & Mackay, 1996). Indeed it is only recently that quantitative trait lociQTLs that contribute to such complex traits have actually been identified (Hill, 2005). However, location of QTLs has been achieved by creating crosses that result in gametic phase disequilibrium between molecular markers and QTLs (Lynch & Walsh, 1998). Higher values of a particular trait will be statistically associated with a particular allele at a molecular marker locus (Lynch & Walsh, 1998). If the genome is saturated with enough molecular markers, then it is possible to clone and sequence the DNA region that contributes to the trait. This is being attempted for the California condor to locate the mutation for chondrodystrophy (Zoological Society of San Diego, 2008) so that likely carriers of this disease can be identified.

Quantitative Genetics Glossary

additive genetic variance	V_a, proportion of phenotypic variance that is directly heritable and results from the sum of the allelic effects at all the loci that affect a particular trait.
animal model	A maximum likelihood method of estimating heritabilities and genetic correlations that is particularly precise when there are measurements of traits for a population over several generations, a an estimate of the pedigree from observation or paternity analysis with molecular markers.
breeding values	The heritable genetic value of an individual as judged by the average value of a particular trait in its offspring. When mated at random, its breeding value will be twice the average deviation of its progeny from the population mean.
dominance genetic variance	V_d, variance in the phenotype caused by inheritance of alleles at a particular heterozygous quantitative locus that cause a single copy of the plus allele to contribute to the phenotype as if the locus was homozygous for the plus allele.
environmental variance	V_e, variance in the phenotype caused by variation in the environment.
epistatic variance	V_i, variance in the phenotype caused by nonadditive interactions between particular alleles at different loci that contribute to a quantitative trait.
gametic phase disequilibrium (or linkage disequilibrium)	A nonrandom association of alleles at multiple loci created by physical linkage, nonrandom mating, or mixing of two populations under different selection regimes. Positive gametic phase disequilibrium occurs when particular gametes have more plus alleles at the multiple loci that contribute to a trait than would be expected given the average frequency of plus alleles in the population.

continued

genetic correlation	Degree to which two traits respond to selection independently; can be caused by two genes being determined by some of the same genetic loci or by gametic phase disequilibrium.
genetic variance	The sum of the additive genetic variance, the dominance variance, and the epistatic variance.
genetic variance/covariance matrix	Matrix with additive genetic variances on main diagonal and genetic covariances on the subdiagonals.
heritability	h^2 or H^2, proportion of phenotypic variance that is heritable. Narrow-sense heritability (h^2) is ratio of additive variance to phenotypic variance. Broad-sense heritability (H^2) is ratio of total genetic variance to phenotypic variance. Both are estimated by a breeding design that compares similarity of a trait in relatives and nonrelatives that is due only to their genes. The most common way to estimate narrow-sense heritability is from a pedigree that includes half siblings.
linear selection differential	$S = \mu_s - \mu$, where μ is the original population mean and μ_s is the mean of those that survive to breed. S measures direct and indirect selection caused by selection on traits that are highly phenotypically correlated with the trait of interest. S is equivalent and can be converted to β, the univariate linear selection gradient (Freeman & Herron, 2006).
maternal effects	Variance in the phenotype caused by genetic and environmental variation in the maternal environment. Offspring are more similar to their mother than expected from the additive effect of the genes that they have inherited from her.
multivariate linear selection gradient	Statistical methodology that separates direct selection on a trait from indirect selection on the trait; traditionally obtained by using the standardized trait values as the independent variables in a multiple linear regression with relative fitness as the dependent variable (Lande & Arnold, 1983).
nonadditive genetic variance	The sum of the dominance variance and the epistatic variance.
phenotype	The value of the traits that you observe on the organism.
phenotypic correlation	Pearson's correlation between two different traits.
phenotypic variance	V_P, sample variance of trait from actual measurements on the organism.
Q_{ST}	Amount of among heritable population variation in a quantitative trait relative to the amount of additive genetic variation within populations.
response to directional linear selection	$R = \mu' - \mu$, where μ is the original population mean and μ' is the mean of the next generation. The response is the change in the mean value of the trait in the next generation relative to their parent's generation.

From Falconer and Mackay (1996)

different populations are locally adapted to different climate optima along a latitudinal gradient. One way to assess whether maladaptation of translocated individuals will be a problem is to compare the amount of genetic differentiation among populations shown by neutral molecular markers with that shown by quantitative traits.

Spitze (1993) defined Q_{ST} to be the among-population variation in a quantitative trait relative to the within-population variance. He pointed out that if Q_{ST} was larger than the F_{ST} value, then the among-population variance in the trait (in his case, body size in the water flea *Daphnia*) must be the result of local adaptation rather than neutral phenotypic evolution.

Palo and colleagues (2003) compared Q_{ST} from a maternal half-sib design with F_{ST} from eight microsatellite loci for six frog populations along a latitudinal gradient and found that Q_{ST} exceeded F_{ST} for all three life history traits, suggesting considerable local adaptation. However, they also found there was no correlation between pairwise estimates of Q_{ST} and F_{ST} for any populations. This lack of correlation supports the hypothesis that knowledge of F_{ST} from neutral molecular markers is not helpful in predicting geographic differentiation in the quantitative traits under strong selection (Stockwell et al., 2003), and therefore other methods such as estimation of Q_{ST} must be used.

Linkage between Neutral Molecular Markers and Quantitative Trait Loci

Estimation of Q_{ST} may be difficult or impossible for small populations of a species at risk because precise estimates require at least five families in a controlled breeding design per population and at least 20 populations (O'Hara & Merilä, 2005). One possible alternative is to use the $\cong$ 2% to 5% of neutral molecular markers that are tightly and physically linked to QTLs and to calculate F_{ST} instead of Q_{ST}. Calculation of F_{ST} from only those molecular markers showing exceptionally high values was done for the marine snail (Wilding et al., 2001), the whitefish (Campbell & Bernatchez, 2004), and the salmon populations (Vasemägi & Primmer, 2005) discussed earlier. Thus, we may be able to estimate geographic differentiation at QTLs by looking at the F_{ST} values of physically linked molecular markers. It would be optimal to use high-throughput markers such as SNPs so that a large number of closely-spaced markers for a large number of individuals from each population could be genotyped.

Heritability and Fisher's Equation

The breeder's equation states that the rate of evolution in response to new selective pressures is proportional to the heritability (Etterson, this volume; Falconer & Mackay, 1986):

$$R = h^2 S \qquad (13.2)$$

where h^2 is the narrow-sense heritability and S is the selection differential (Box 13.2). Therefore, to maintain the future ability to adapt to changes in the environment, managers of populations must maximize or at least maintain heritability. Unfortunately, narrow-sense heritabilities are difficult to assess quickly for animals because their estimation usually involves rearing offspring to adulthood in a paternal half-sib design to avoid maternal effects (see, for example, Boulding & Hay, 1993). Rearing the offspring in several environments is also important because the heritability of traits not closely tied to fitness, such as morphometric traits, has been shown to increase when environmental conditions are favorable (reviewed by Charmantier & Garant, 2005). Furthermore, estimation of heritabilities involves estimating variance components and, because of this, heritabilities typically have very large 95% confidence limits unless more than 500 half-sib families are used (Falconer & MacKay, 1996; Lynch & Walsh, 1998). This makes it difficult to detect whether a prolonged population bottleneck has resulted in a significant reduction in the additive genetic variance. For these reasons, the best method of determining whether a prolonged population bottleneck has resulted in a small effective population size might be with neutral molecular markers, because such markers are quicker and less invasive than traditional heritability estimates.

Genetic Variance/Covariance and Lande's Multivariate Equation

In many cases, local adaptation of populations along a gradient can involve several traits that are correlated phenotypically with one another (see, for example, Palo et al., 2003). If estimates of the multivariate linear selection gradient and genetic variance/covariance matrix are both available, then one can predict the change in the mean of each trait using Lande's 1979 matrix equation for multitrait evolution:

$$\Delta z = G\beta \quad (13.3)$$

where $\Delta \mathbf{z}$ is a vector of the response to selection, G is the genetic variance/covariance matrix estimated by measuring traits on related and unrelated individuals, and β is a vector of multivariate linear selection gradient coefficients (Etterson, this volume; Lande & Arnold, 1983) (Box 13.2). This matrix equation was a major step forward because it incorporates both the direct and the correlated response to selection that previously had to be calculated separately (Falconer & Mackay, 1986). The effect on evolutionary potential of genetic constraints caused by genetic correlations among multiple traits has rarely been considered by conservation geneticists (Etterson, this volume; Etterson & Shaw, 2001) but could further increase the minimum population size required to maintain evolutionary potential.

POPULATION VIABILITY

Do Neutral Molecular Markers Estimate Population Viability?

Viability after a Recent and Prolonged Population Bottleneck

Lynch (1996) reviewed the literature on the relative mutation rates for molecular markers ($10^{-8} - 10^{-5}$ per locus per year) and quantitative traits ($10^{-3} - 10^{-2}$ per trait per generation) and concluded that the latter would recover their levels of additive genetic variance after a population bottleneck more quickly because they have higher mutation rates. However, this difference in mutation rates is not relevant when only very recent population bottlenecks are of interest, as is typically the case for endangered populations. If the bottleneck was recent, then not enough time will have passed for mutation to regenerate the genetic variance at either QTLs or the marker loci. Furthermore, many modern molecular markers, such as adenosine and cystine dinucleotide microsatellite markers have per-gamete mutation rates averaging 10^{-4} to 10^{-3} per year, depending on their length (Whittaker et al., 2003). This rate is more similar to the mutation rates that Lynch (1996) cites for quantitative traits.

Equation 13.1 showed that the rate of loss of heterozygosity at neutral molecular markers was inversely related to effective population size. Equation 13.4 shows that the same is true for the rate of loss of variation at neutral quantitative traits. At equilibrium between mutation and genetic drift, the broad-sense heritability h_b^2 for a neutral trait is (Falconer & Mackay, 1996, p. 351)

$$h_b^2 = \frac{2N_eV_m}{2N_eV_m + V_E} \quad (13.4)$$

where V_m is all the mutational genetic variance and V_E is the environmental variance. If $V_m = 10^{-3}V_E$, which has been estimated from mutation rates for neutral characters in *Drosophila*, then at mutation–drift equilibrium the equation predicts that if $N_e = 100$ then $h^2 = 0.17$ and if $N_e = 10,000$ then $h^2 = 0.95$ (Falconer & Mackay, 1996, p. 351). However, most heritabilities for large populations are considerably lower than 0.95, suggesting that natural selection must be important in reducing them. Unfortunately, the effects of selection on additive genetic variance are only understood for major components of fitness in which most mutations are deleterious (Falconer & Mackay, 1996).

If there is no selection on a trait and the nonadditive genetic variance is assumed to be zero, then the reduction of heritability after t generations of inbreeding is

$$h_t^2 = \frac{h_{t-1}^2(1 - F_t)}{1 - h_{t-1}^2F_t} \quad (13.5)$$

where h_t^2 is the heritability of the population and F_t is the inbreeding coefficient (Box 13.1) at time t, and h_{t-1}^2 is the heritability at time $t - 1$ (after Eq. 15.1 in Falconer & Mackay, 1986). The increase in the inbreeding coefficient per generation is a function of the effective population size, N_e (Eq. 3.6 in Falconer & Mackay, 1986):

$$F_t = \frac{1}{2Ne} + \left(1 - \frac{1}{2Ne}\right)F_{t-1} \quad (13.6)$$

If we assume that in the base population at time $t-1$ that $F_{t-1} = 0$, then the second term in Eq. 13.6

disappears. Substituting the simplified version of Eq. 13.6 into Eq. 13.5 gives us the reduction in the heritability of the population after one generation at a small effective size:

$$h_t^2 = \frac{h_{t-1}^2\left(1 - \frac{1}{2Ne}\right)}{1 - h_{t-1}^2 \frac{1}{2Ne}} \qquad (13.7)$$

However, such a reduction would be impossible to detect after only one generation. For example, if we assume that $N_e = 5$ and that initially $h^2 = 0.50$, then Eq. 13.7 says that after one generation, the heritability will have been reduced to 0.43, which is 86% of its initial value. Similarly Eq. 13.1 shows that after one generation at $N_e = 5$ the heterozygosity at a single neutral molecular marker locus will be reduced to 90% of its initial value. It would be impossible to detect statistically such a small reduction in the heritability using the methodology typical for wild animal populations. It would also be impossible to detect the reduction in heterozygosity at any known molecular marker locus after one generation if the census population size was also only five animals.

Estimation of N_e with Molecular Markers

Considerable recent effort has been expended toward using molecular markers to estimate N_e, and two methods are currently popular. The first method uses the observation that populations that have recently experienced a very small effective population size often show a reduction in the number of alleles before they show a reduction in their observed heterozygosity, because rare alleles are lost first (see, for example, Hedgecock & Sly, 1990). Cornuet & Lucas (1996) devised a method using allelic frequencies from nuclear genes to determine whether a significant number of loci exhibit a larger heterozygosity than that expected from the observed number of alleles when it is assumed that the locus is at mutation–drift equilibrium (their program, BOTTLENECK 1.2.02, is freeware available at http://www.montpellier.inra.fr/URLB/bottleneck/bottleneck.html). Luikart and Cornuet (1998) took this approach with 56 allozyme and 37 microsatellite data sets from real populations, and under the assumptions of the stepwise mutation model, found a significant heterozygote excess in the bottlenecked populations about half the time. Beebee & Rowe (2001) likewise used BOTTLENECK on data comprising eight microsatellite loci genotyped for 20 to 40 tadpoles from each of 50 endangered toad populations to confirm that they could identify populations for which the census data documented a severe population bottleneck (down to tens of toads) within the past 20 to 30 years. Luikart and Cornuet (1998) postulate that such bottlenecks should be detectable for 0.2 to4 N_e generations after the bottleneck, where their N_e is the bottleneck population size.

A second method for precisely estimating N_e uses temporal changes in the allelic frequencies at molecular marker loci and assumes that selection, mutation, and migration can be ignored (Waples, 1989; Williamson & Slatkin, 1999). Anderson (2005) describes an efficient Monte Carlo method for estimating N_e from temporally spaced samples using a coalescent-based likelihood and even provides a computer program to do the calculations. Wang and Whitlock (2003) describe a new methodology for estimating effective population size (N_e) and migration rates (m) simultaneously from changes in the allelic frequencies of neutral molecular markers between two or more samples that are separated by periods of time that are short enough so that mutation can be ignored, but they do not provide a computer program.

It is reassuring that molecular markers such as microsatellites can detect a recent population bottleneck if the population has been recently reduced to a few tens of individuals, but the question is whether a decline of that magnitude will result in a reduction in the additive genetic variance.

It would seem likely that prolonged bottlenecks less than $N_e = 50$ would be detectable in wild populations using molecular markers. If bottlenecks are detected, then heritabilities could be estimated using modern statistical methods to determine whether they had declined (Falconer & Mackay, 1986; Lynch & Walsh, 1998). However, not everyone agrees that even severe bottlenecks can be detected. After their meta-analysis, Coltman and Slate (2003) conclude that, because of the low correlation between multilocus heterozygosity and phenotypic variability, at least 600 individuals might have to be genotyped for microsatellite markers before there is sufficient statistical power to detect inbreeding depression. Their conclusions might have differed, however, if they had analyzed correlations between multilocus heterozygosity and the additive genetic variance instead of correlations between multilocus heterozygosity and the phenotypic variance.

Does Evolutionary Potential Decrease after Bottlenecks?

There is some evidence that reduced variation at molecular markers is correlated with a reduced response to selection, at least over the long term. Unfortunately, most available data are from founder–flush experiments where the bottleneck typically only lasts for one generation and is followed by exponential population growth up to a large carrying capacity. Briggs and Goldman (2004) found that genetic variation in AFLP markers and long-term selection response was reduced in laboratory plant populations (*Brassica rapa*) that had experienced a recent bottleneck of two individuals relative to control populations. In contrast, they found that heritability and short-term selection response actually increased for two of the three bottlenecked replicates for the first three generations after the bottleneck. They used the CoNe software (Anderson, 2005) to estimate N_e from the AFLP markers and got underestimates ($N_e = 9-19$) of the true N_e, which was known from the experimental design to be 25.

Founder–flush bottleneck experiments using laboratory populations of houseflies have also found that genetic variation at molecular markers declined, but heritabilities can actually increase after a single population bottleneck. Flies that had experienced a single population bottleneck of 1, 4, or 16 mated pairs were compared with a larger control population (Bryant et al., 1986). For five of the eight traits, the highest heritability was for four pairs, and this was interpreted as conversion of some of the nonadditive genetic variance into additive genetic variation (Bryant et al., 1986). The average rate of fixation of several allozyme loci for these lines were 7%, 13%, and 38%, respectively, and the viability of the one- or four-pair lines was significantly lower than that of the 16-pair lines (Bryant et al., 1986). Allozyme analysis of these bottlenecked populations documented that only 39.1%, 75.6%, and 80.1% of the alleles remained for the one-, four-, and 16-pair treatments, respectively, after five separate bottleneck founder–flush episodes. In a second experiment the additive genetic variance increased in the one-pair treatment after the first bottleneck but returned to the same level as the control after the fifth bottleneck, suggesting that by that point all the nonadditive genetic variance had been converted to additive genetic variance (Bryant & Meffert, 1995).

A recent meta-analysis suggests that severe inbreeding may cause a linear decrease in the levels of genetic variation, as predicted in Eq. 13.5, for quantitative traits that are not closely associated with fitness. Van Buskirk and Willi (2006) reviewed 22 published studies, mostly on insects, in which they compared the level of genetic variation in experimentally inbred populations (measured by V_A or h^2) with that in outbred control populations. They found the expected linear decrease in variation with inbreeding coefficient (Eq. 13.5) only for morphological or behavioral traits such as bristle number, wing length, and mating speed. For life history traits, the heritability actually increased after a bottleneck up to an inbreeding coefficient of 0.4, because of increases in the dominance and epistatic variance (Box 13.2). Despite this observation, they questioned whether bottleneck-induced variation actually increased the viability of the populations because of the inbreeding depression that often accompanies it.

Gilligan and colleagues (2005) estimated genetic variation for two neutral traits—abdominal and sternopleural bristle numbers—and allozyme heterozygosity in 23 populations of *Drosophila melanogaster* that they had maintained at effective population sizes of 25, 50, 100, 250, or 500 for 50 generations. They found that quantitative genetic variation was being lost at a similar rate to variation at molecular markers. However, both rates were significantly slower than those predicted by neutral theory. They attributed their success at detecting this significant relationship to a large range of inbreeding values present in the different treatments of their experiment.

Reed and associates (2003a) compared the number of offspring produced by outbred and by fully inbred population of *D. melanogaster* maintained in either single-stress, variable-stress, or benign conditions for seven generations before being transferred to an environment with novel stresses for seven generations. They used the addition of either copper sulfate or methanol to the food medium as the single or variable stress and then used the absence of sugar in the medium as the novel stress. They found that outbred populations adapted to the absence of sugar significantly better than inbred populations, and that populations that had been maintained in a stressful environment were better at adapting to a novel stressful environment. They attributed the slower adaptation by flies from a benign environment where selection was relaxed to two possible factors: the loss of costly stress-resistant traits or mutation accumulation.

Estimating Heritabilities in the Field Using Molecular Pedigrees

To avoid complications arising from estimating evolutionary potential using molecular markers, many researchers prefer to estimate heritability directly. However, heritability estimation for a population of a species at risk is challenging because the number of individuals that are available is limited and because breeding is not controlled. Without controlled breeding it is not possible to apply a traditional half-sibling breeding design for estimating genetic parameters (Lynch & Walsh, 1998). One solution to this is to collect trait measurements over several generations and to use an animal model (Box 13.2) to estimate the heritabilities (Kruuk, 2004). Estimation of heritabilities and additive genetic variances in nonpedigreed wild populations is becoming more feasible because of the development of highly polymorphic molecular markers that estimate more accurately the relationships among the individuals on which the quantitative traits are measured (Wilson & Ferguson, 2002 Garant & Kruuk, 2005;). Microsatellite markers have been used successfully to estimate genetic parameters in wild populations of deer (Kruuk et al., 2000) and trout (Wilson et al., 2003b). However, estimation of parental genotypes with molecular markers is not trivial, even if at least five offspring from each full-sib family are genotyped for 5 to 10 microsatellite loci (Lemay and Boulding, unpublished data; Smith et al., 2001). Furthermore, heritabilities estimated in natural populations, where the breeding design cannot be controlled, will be inflated if related individuals share a more similar environment than unrelated individuals (Lynch & Walsh, 1998). On the positive side, the environmental variance exhibited in the field will be at a natural high level rather than at the reduced level typical of a controlled laboratory environment. On the other hand, laboratory estimates of heritabilities and genetic correlations allow efficient breeding designs with known parents and can be improved by rearing offspring in more than one environment, which allows estimation of genotype-by-environment interactions (see, for example, Boulding & Hay, 1993; Mahaney et al., 1999). A multivariate maximum likelihood analysis of three different cholesterol subfractions on 942 pedigreed baboons fed one of two diets found that the genetic correlations between three subfractions differed between diets because of epistatic interactions (Mahaney et al., 1999).

ADAPTATION TO CHANGING ENVIRONMENTS

After a change in the environment, populations will be initially maladapted, their fertility and viability will decrease, and their population growth rates may become negative (Lynch & Lande, 1993). Unless the populations evolve fast enough to adapt to the changing environment, they may go extinct. Fast rates of evolution are only possible if the populations are large enough so that they remain genetically diverse.

Discrete Environmental Change

One Population: r, K, and h^2

The question of whether populations with higher heritabilities are more likely to survive a discrete, or steplike, change in the environment was first addressed by Gomulkiewicz and Holt (1995). They presented an analytical model of the evolution of a quantitative trait after a discrete change in the environment caused the population to decline. In their model, the population did not go extinct if the trait evolved rapidly enough to adapt to the new optimum before the population dropped below an unspecified critical population size. They showed that heritability reduced the amount of time that a population spent below this critical, but undefined, population size and concluded that the effect of heritability was only important for large populations experiencing small discrete changes in the environment.

Boulding and Hay (2001) disagreed with this conclusion and used the results from an individual-based finite-locus model to argue that heritability was important in preventing extinction after environmental change. They showed that populations that had higher heritabilities recovered faster from a discrete environmental change than those with lower heritabilities, particularly when the initial population size was large (Fig. 13.1), and

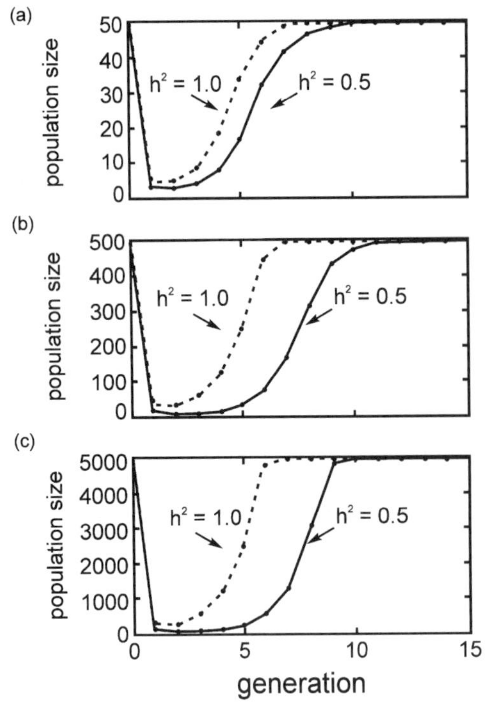

FIGURE 13.1 (A–C) Effect of heritability (h^2) on the number of generations required for an isolated population to recover to carrying capacity after a discrete shift in its optimal phenotype. Initial population size is 50 (A), 500 (B), and 5,000 (C). The percentage of populations that avoided extinction was dependent on initial population size when $N_i = 50$: 16% for $h^2 = 0.5$, 65% for $h^2 = 1.0$; when $N_i = 500$: 89% for $h^2 = 0.5$, 100% for $h^2 = 1.0$; when $N_i = 5{,}000$: 100% for $h^2 = 0.5$, 100% for $h^2 = 1.0$. Increasing the heritability and the initial population size increased population viability by increasing the minimum population size reached before the population began to increase again. (After Figure 1 in Boulding and Hay [2001].)

were consequently less likely to go extinct. Indeed, Figure 13.2 shows that for large populations (N = 10,000), increasing the heritability from almost zero to one increased the size of the shift in the optimum that could be tolerated by three phenotypic standard deviations (PSDs). They also found that the size of the shift in the optimum to which the organisms could adapt increased linearly with the $\log_{10}$ of the fecundity ($r^2 = 0.97$, Fig. 13.3). Thus, species with high fecundities should be able to adapt to larger shifts in the optima than species with low fecundities. The results of their model support the hypothesis that higher heritabilities, larger population size, and higher population growth rates increase population viability (Reed, this volume).

One Population: Migration

Migration is important to population viability because it can increase the heritability of traits that are under different local selection pressures in adjacent populations. A large difference between the genotypic means of the immigrants and that of the local population will generate strong, positive gametic phase disequilibrium, which will inflate the additive genetic variance and increase the response to selection until it is gradually broken down by recombination during reproduction (Tufto, 2001). Therefore, if migrants are prevented from reaching a population because of habitat fragmentation, there may be a large decrease in the heritability even if the population is moderately large.

In quantitative genetic models without demography, gene flow prevents local adaptation unless stabilizing selection is strong (Boulding, 1990; Bulmer, 1985). Similarly, in quantitative genetic models with demography, recurrent gene flow from a population adapted to an environment where the optimum is very different will result in local maladaptation and thereby increases the probability of extinction (Boulding & Hay, 2001; Tufto, 2001). In contrast, Holt and Gomulkiewicz (2004) present a density-dependent, individual-based quantitative model that shows that a somewhat higher number of migrants from a source population can facilitate slow adaptation by a low-density sink population to a novel environment over long periods of time. However, they found that migrants were no longer beneficial after local adaptation had taken place.

Linear Series of Populations Connected by Migration

The evolutionary potential of a population affects its ability to adapt to the abiotic and biotic effects of climate change. Populations at either end of a linear environmental gradient have an increased risk of extinction because unbalanced gene flow from the center of a cline results in poor local adaptation to abiotic factors such as temperature (Reviewed by Brindle and Vine, 2007). Recent analytical quantitative genetic models have shown that

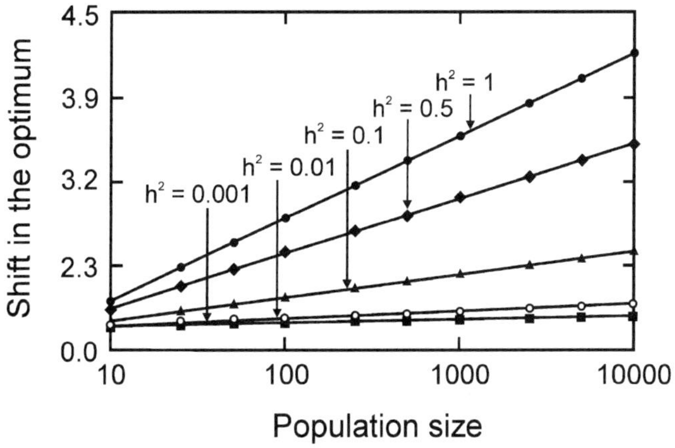

FIGURE 13.2 Effect of heritability on the maximum shift in the optimum that 95% of the populations tolerated for different initial population sizes. Increasing the heritability substantially increased the amount of discrete environmental change that a single, isolated population could withstand. (After Figure 2a in Boulding and Hay [2001].)

gene flow can prevent populations on the periphery of a species' range from adapting to their local environment (García-Ramos & Kirkpatrick, 1997), causing them to become demographic sinks (Kirkpatrick & Barton, 1997).

Climate change can also change selection from the biotic environment. Global warming after the last ice age resulted in species that were good dispersers moving poleward at faster rates than those that were poor dispersers (Gates, 1993). This suggests that predators may move poleward ahead of their usual prey species and begin to prey on indigenous prey. Subtropical invertebrate predators often have feeding appendages that are more specialized

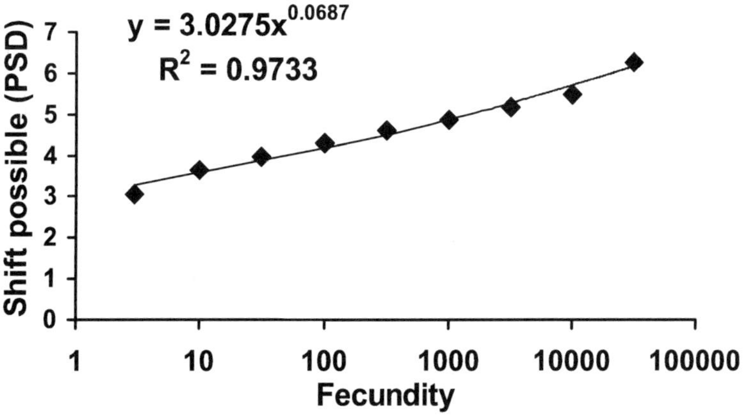

FIGURE 13.3 Effect of fecundity on the maximum shift in the optimum that 95% of the populations tolerated. The amount of discrete environmental change that an isolated population could withstand increased linearly with the $\log_{10}$ of the fecundity. PSD, phenotypic standard deviation. (Data from Boulding and Hay [2001].)

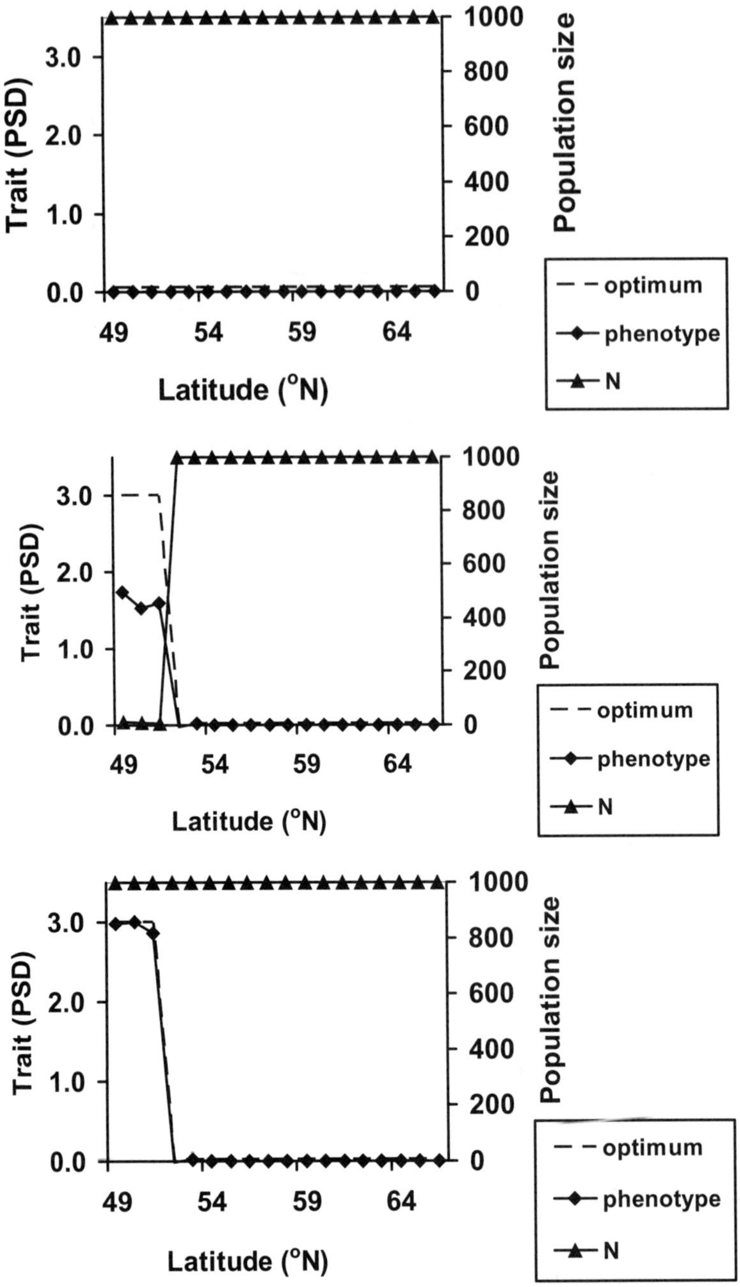

FIGURE 13.4 Evolutionary response to a discrete change in the optimal phenotype. (A) Before shift in optimal phenotype (all cline simulations). (B) Population sizes and evolution of the mean phenotype in the cline three generations after optimum in three southern populations was shifted 3 PSDs. Note that the populations in three end populations have dropped well below carrying capacity. Selection intensity = 0.1, heritability = 0.5, migration rate from adjacent uninvaded population = 0.025. (C) Mean phenotype and population size along the cline 17 generations after the shift in the optimum by 3 PSDs in the southern three populations. Note that the populations at 49°N and 50°N have completely adapted to the new optimum, but the population at 51°N still has not. Heritability = 0.5, migration rate from adjacent uninvaded population = 0.025. PSD, phenotypic standard deviation.

than those of comparable temperate predators likely because predator–prey coevolution is more escalated in the tropics (Vermeij, 1977, 1987). Range extensions by subtropical predator species are interesting because they may cause extinction of temperate prey species, which could otherwise tolerate the stress of warmer temperatures.

Local adaptation to a discrete change in the environment, such as one caused by an invasion by a new predator, will be less likely if there are large numbers of maladapted immigrants from adjacent uninvaded populations. Boulding and colleagues (2007) describe an extension of an individual-based finite-locus model of the phenotypic evolution of a quantitative trait (Boulding & Hay, 2001) to a linear series of populations connected by migration. The new data in Figure 13.4 show the evolutionary response to a discrete change in the optimal phenotype that occurs only in the three most southern populations. In generation 1 (Fig. 13.4A), the optimum in the three most southern of 18 populations is shifted by 3 PSDs, but the width of the fitness function is kept constant at 0.75 PSD. In this run, the population size three generations after the shift in the optimum declined from the carrying capacity of 1,000 to 12, nine, and five individuals at 49°N, 50°N, and 51°N latitude, respectively (Fig. 13.4B). The population at 49°N was buffered from gene flow from sites where the optimum was unchanged and recovered up to the carrying capacity by generation 9. However, the population at 50°N did not recover until generation 11, and the population at 51°N went temporarily extinct and did not recover until generation 15. For this run with a migration rate of m = 0.025 (adjacent populations exchanged 2.5% of their individuals), the three invaded populations eventually showed complete adaptation to their new optimum. For a higher migration rate of 0.25 (adjacent populations exchanged 25% of their individuals), adaptation was complete in the populations at 49°N and 50°N, but was not complete in the population at 51°N (91%), which directly received immigrants from its uninvaded neighboring population at 52°N.

Figure 13.5 shows that the initial heritability also had an effect on population viability on the most southern of the three populations. When the heritability was intermediate ($h^2 = 0.5$), as shown in the previous example, the rate of adaptation to the new optimum was relatively rapid (Fig. 13.5). When heritability was the maximum value possible ($h^2 = 1.0$), adaptation to the new optimum occurred more quickly so that the minimum population size was larger and the population returned to the carrying capacity sooner (Fig. 13.5). In contrast, when the heritability was low ($h^2 = 0.1$), the population went extinct and did not recover within 180 generations (Fig. 13.5). Thus, higher heritabilities reduce the chance that a population would go extinct as a result of a discrete change in the local optimum that occurs in only part of a species' geographic range.

Continuous Environmental Change

Heritability and Continuous Linear Rates of Change

All analytical models published so far demonstrate that population persistence is more likely when the heritability of the quantitative trait is high. Pease and colleagues (1989) considered migration and adaptation in response to a moving linear environmental gradient and found that higher heritabilities increased the probability that the species would persist in the face of continuous environmental change. Lynch and Lande (1993) found that populations experiencing continuous environmental change were more likely to maintain positive rates of population growth if the heritability was high. They proposed that the critical rate of environmental change for a low-fecundity population that requires 50% of the individuals to survive and reproduce would be only 1.6 $N_e \times 10^{-3}$ PSD per generation. Bürger and Lynch (1995) present an analytical model that explicitly includes demography, and a simulation model of adaptation to continuous environmental change as well. They found that populations in environments with larger carrying capacities could tolerate higher rates of environmental change. This was because when N_e was larger, more additive genetic variance was maintained in the population at equilibrium. They concluded that the maximum rate of continuous environmental change that can be tolerated could be as little as 0.1 PSD per generation when the carrying capacity was $K = 32$, but increased to 0.7 PSD per generation when $K = 512$.

Demography versus Genetics

Lande (1988) has argued that demographic processes are more important than genetic factors in small populations; however, genetic processes become important when there is strong selection

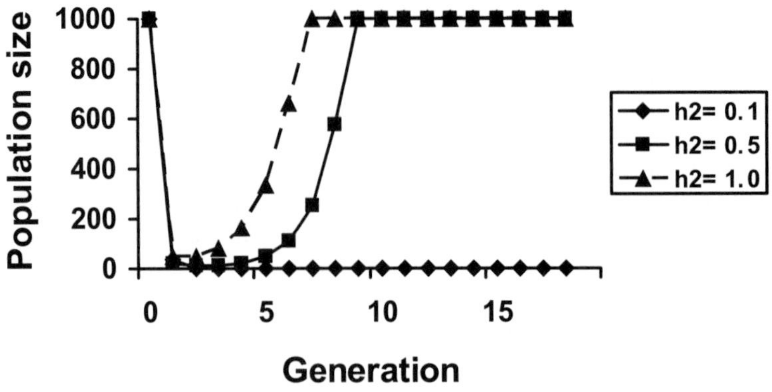

FIGURE 13.5 Effect of heritability on population dynamics of the population at 49°N, which is experiencing directional selection toward a larger optimum because of a predator invasion of the southern three populations of the cline. Migration rate from adjacent population = 0.025. Higher heritabilities increased population viability.

for new optima as a result of climate change. Brooks (this volume), as well as the new data from our individual-based model (Boulding et al. 2007), shows that both demographic and genetic parameters facilitate adaptation by populations to changes in the environment. Higher heritabilities increase the likelihood that a population will adapt to discrete changes in the environments. Given sufficient time for the population to reach equilibrium between mutation, migration, and drift, the heritability of neutral traits will be larger if the effective population size is larger (Eq. 13.4), and the effective population size will be larger if the carrying capacity is larger (Freeman & Heron, 2006). However, under nonequilibrium conditions, heritability can vary independently of population size. For discrete environmental change, a large initial population size gives the population more time to adapt before it goes extinct. Furthermore, when stabilizing selection was strong and the shift in the optimum large, adaptation in our model often happened within 10 generations, which made the initial heritability more important than the mutation rate (Boulding & Hay, 2001).

GENERAL CONCLUSIONS

I conclude that neutral molecular markers can often be used to identify populations that have recently experienced a prolonged population bottleneck. Populations with a reduced number of alleles at neutral molecular markers will be those that have had a small effective population size for several generations. The literature suggests that bottlenecked populations are also likely to have low heritabilities for quantitative traits at least for those traits that are closely related to fitness.

I presented new results from a previously described individual-based model of a single quantitative trait (Boulding & Hay, 2001; Boulding et al. 2007) that demonstrate that the viability of a population after a discrete change in its environment depended on its heritability, as well as the number and source of the migrants from adjacent populations. The model showed that a decrease in the heritability resulted in reduced evolutionary potential of populations along the cline to adapt to environmental change. Populations that were not able to adapt to changes in their environment had a higher probability of going extinct.

FUTURE DIRECTIONS

My review of the literature leads me to conclude that the most urgent need is for better statistical methodology for detecting reductions in heritable genetic variation using neutral molecular markers. This would be facilitated by more laboratory experiments (see, for example, Gilligan et al., 2005) that simulate small population size and simultaneously look at the genetic diversity of neutral molecular markers and the heritabilities of quantitative traits that are neutral as well as those closely related to fitness. Because Q_{ST} is so difficult to estimate, it would also be useful to have more population studies that estimate F_{ST} with a large set of selected markers linked to QTLs, and then compare those values with the amount of genetic differentiation at neutral markers. Most ecological studies that have been published so far have compared F_{ST} at less than 10 selected markers with F_{ST} of a much larger number of neutral markers. This is important because estimation of F_{ST} with selected markers might be an alternative to estimation of Q_{ST} when less than 20 populations are available. It would also be useful to have theoretical models of genetics and demography for multiple quantitative traits rather than just for single quantitative traits. These are needed to assess the role of genetic architecture in constraining the potential of populations to adapt to environmental change. For example, antagonistic genetic correlations among traits could reduce the rate at which prairie plants adapt to global warming (Etterson, this volume; Etterson & Shaw, 2001).

SUGGESTIONS FOR FURTHER READING

For a gentle but complete introduction to molecular markers, see Avise's book (2004a). For a solid understanding of introductory quantitative genetics, see the classic by Falconer and Mackay (1996). Hartl and Clark (2007) give a detailed but clear introduction to population genetics, ecological quantitative genetics, and even genomics in their new edition. Holt and Gomulkiewicz (2004) give a nonmathematical review of quantitative genetic models that include demography and review their own work, which emphasizes evolution over long timescales.

Avise, J. 2004. Molecular markers, natural history and evolution. 2nd ed. Chapman & Hall, New York.

Falconer, D. S., & T. Mackay. 1996. Introduction to quantitative genetics. 4th ed. Longman, New York.

Hartl, D. L., & A. G. Clark. 2007. Principles of population genetics. 4th ed. Sinauer Associates, Sunderland, Mass.

Holt, R. D., & R. Gomulkiewicz. 2004. Conservation implications of niche conservatism and evolution in heterogeneous environments (pp. 244–264). In R. Ferrière, U. Dieckmann, & D. Couvet (eds.). Evolutionary conservation biology. Cambridge University Press, Cambridge, UK.

Acknowledgments I thank C. Fox, H. Freamo, E. N. Hay, H. J. Lee, M. Lemay, D. Reed, and an anonymous reviewer for their suggestions for improving the manuscript; and I thank I. Smith for redrawing Figures 13.1 and 13.2. Financial assistance was provided by Natural Sciences and Engineering Research Council of Canada Discovery and Strategic Project grants.

IV

CONSERVATION OF THE COEVOLVING WEB OF LIFE

The web of interactions among species—Darwin's entangled bank—is as much a product of evolution as the species themselves. The relationships between predators and prey, parasites and hosts, competitors and mutualists undergo constant change as species coevolve, colonize new habitats, and confront changing environments. We now know that interspecific interactions can sometimes evolve so rapidly that the changes can be observed within the timescale of our own lifetimes, with human activities sometimes driving these evolutionary changes (Altizer et al., 2003; Palumbi, 2001a; Thompson, 1998). The web of life is therefore never exactly the same in any two places on earth, because a combination of ecological, evolutionary, coevolutionary, and now societal, processes constantly fuels changes in interactions among species.

There is, then, no difference between ecological and evolutionary time. There is a wide range of timescales at which ecological and evolutionary processes shape populations, species, and interactions, and these timescales overlap almost completely within biological communities. Parasites and pathogens have been shown to evolve sometimes in less than a few years (Altizer & Pederson, this volume); some insects have evolved, within only a few decades, new traits that allow them to attack introduced species (Carroll & Fox, this volume); some long-lived plants and birds have coevolved in different ways in different populations since near the end of the Pleistocene (Benkman et al., this volume); and longer term cycles of extinctions and invasions have fostered the ongoing diversification of life over millions of years (Vermeij, this volume). If we could catalog today the genetic structure of all species in almost any biological community and then repeat that study in 100 years, we would inevitably find evolutionary change in multiple species and in their interactions with other species.

Most of this ongoing evolutionary change does not lead to long-term directional change in the traits of species. Instead, most evolutionary change shifts populations first one way and then another as the surrounding abiotic and biotic environment itself continues to change. The resulting evolutionary and coevolutionary meanderings of populations, however, are not trivial processes. These continual changes allow species to stay in the evolutionary game, as populations adapt and readapt locally in a constantly changing world. This is where evolution and ecology meet as processes that sustain the web of life. The number of studies showing local adaptation of populations to other species within their communities continues to grow year after year and now spans a wide range of taxa, habitats, and spatial scales (Laine, 2005; Lively & Dybdahl, 2000; Sotka, 2005; Thrall et al., 2002).

Hence, community ecology and conservation biology are untenable as predictive sciences unless they continue to work toward a perspective that allows for rapid and ongoing evolution. Consider, for example, the most famous example of evolution during the past 100 years—the evolution of Darwin's finches on Daphne Major in the Galapagos. Try to understand the population dynamics and web of interactions of those finches during the past decades without the evolutionary perspective that Peter and Rosemary Grant and their colleagues used to evaluate the ecological significance of available seed sizes for these birds, the rapid changes of bill sizes across years relative to changes in available seed sizes, and the implications of the invasion of other, related finches from neighboring islands (Grant et al., 2004). Without an evolutionary perspective, the short-term ecological dynamics of the traits of these birds, their population dynamics, and their interactions with other species would simply not be interpretable.

There is no reason to believe that these results are unique. They appear unusual only because there are still few biologists worldwide who incorporate rapid evolution and coevolution as one of their working hypotheses in evaluating the dynamics of populations, food webs, and the structure of regional assemblages of species. Yet, incorporation of an evolutionary perspective is crucial to all major questions in community ecology and conservation biology at a time when the composition of major communities is changing so quickly worldwide. The rapid changes we are imposing on the web of life have the potential to impose tremendously strong selection pressures on the ecology of species and interactions.

As natural selection acts locally on species interactions, it creates a constantly changing geographic mosaic of coevolutionary change that connects webs of life across ecosystems (Benkman et al., this volume). Traits for particular defenses or counterdefenses are favored in some environments but not in others. Genetic polymorphisms appear and disappear at the local level, but are retained at the regional and species level as populations exchange genes. Some local mini experiments in coevolution are short-term failures, but others are not. There is a now a solid body of theory and empirical work that shows evidence of selection mosaics, coevolutionary hotspots, and coevolutionary coldspots in interactions among species (see, for example, Benkman et al., this volume; Berenbaum & Zangerl, 2006; Brodie et al., 2002; Burdon et al., 2002; Nuismer, 2006; Thompson, 2005; Toju & Sota, 2006).

These studies suggest that coevolution is an inherently geographic process that begins with local adaptation but gathers coevolutionary complexity as each interaction diverges among ecosystems. The long-term conservation of the web of life probably depends upon conserving this geographic mosaic of coevolution. At a time of wholesale loss of species (Dirzo & Raven, 2003), fragmentation of habitats, the spread of legions of invasive species, and the formation of new communities (Carroll & Fox, this volume), we therefore have a great and current need to understand whether some interactions are more likely than others to collapse in the absence of the potential for an ongoing geographic mosaic of coevolutionary change.

Hence, one of the most pressing challenges facing conservation biology is the development of clear and practical guidelines on how we can best conserve the evolutionary and coevolutionary processes that shape the web of life. As Paul Ehrlich (2001) has argued, we need to understand better how we can best avoid foreclosing the evolutionary options for species. The four following chapters highlight how far we have come in recent years in understanding the evolutionary processes that drive and shape the ever-changing web of life. At the same time, these chapters illustrate that we have huge gaps in our understanding of how these processes are linked over different temporal and spatial scales. We have recently realized, for example, that networks of mutualistic species are often organized in fundamentally different ways from the classic food webs of predators and prey. Mutualistic networks of free-living species (for example, pollinators and plants) are often much less compartmentalized than antagonistic interactions and show strong asymmetries in the number of links among species (Bascompte et al., 2003, 2006). This means that the loss of species, or the invasion of new species, is likely to have very different ecological and evolutionary implications for webs based on different forms of interaction. Yet, no ecosystem has been compared both for emerging changes in mutualistic as well as antagonistic interaction webs.

There is then much to do as we move forward toward a science of applied coevolutionary biology aimed at conserving the web of life.

The goal is to sustain the processes that maintain Darwin's entangled bank amid continual environmental change. The chapters in this section on conservation and evolution in biotic interactions help point the way to a science of conservation biology that considers the evolving links among species to be as important as the species themselves.

John N. Thompson
Santa Cruz, California

14

The Geographic Mosaic of Coevolution and Its Conservation Significance

CRAIG W. BENKMAN
THOMAS L. PARCHMAN
ADAM M. SIEPIELSKI

In 1963, the Newfoundland Wildlife Service introduced the red squirrel, *Tamiasciurus hudsonicus,* in an effort to increase the prey base for the overtrapped Newfoundland marten, *Martes americana atrata.* Like many such interventions, this one had good intentions but was not guided by an understanding of evolutionary interactions within this ecological community. Because martens eat mostly voles in the boreal forests, it is not surprising that introduction of red squirrels has failed to help marten recovery. Thus, the Newfoundland marten remains endangered and, like many such introductions, red squirrel introduction has had unforeseen consequences (Parchman & Benkman, 2002). In this case, there were consequences for species interactions that evolved long prior to the release of the red squirrels.

After the retreat of glaciers on Newfoundland, the black spruce, *Picea mariana,* evolved for about 9,000 years in the absence of red squirrels, which are otherwise the dominant seed predator and selective agent on conifer cones in much of North America. As a consequence of isolation from squirrels, black spruce had lost its seed defenses against squirrel predation. In place of squirrels, crossbills, *Loxia* became the dominant seed predators, so that as black spruce lost squirrel defenses it evolved crossbill defenses. Crossbills in turn evolved larger bills to counter black spruce defenses, and black spruce and crossbills thus coevolved in a predator–prey arms race ultimately leading to the evolution of the endemic Newfoundland crossbill, *L. curvirostra percna.* After the squirrels were introduced, they spread rapidly across the island, feasting on black spruce seeds that were poorly defended and causing the formerly common Newfoundland crossbill to plummet in abundance (Fig. 14.1). The last documented Newfoundland crossbills were a pair photographed in 1988 on an islet that squirrels colonized the same year.

The case of Newfoundland crossbills highlights how geographic variation in the distributions of interacting species affects both the form of coevolution between species and geographic variation within species (Box 14.1). It also provides practical knowledge important to conservation biologists; for example, Newfoundland crossbills demonstrate why introducing nonnative taxa such as squirrels can have negative repercussions. Because coevolution is a widespread and ongoing process (Thompson, 1994, 2005), we believe that the situation faced by Newfoundland crossbills is likely to be common. Here, we discuss an emerging view of coevolution and then further highlight our research and other recent studies to suggest how this view can contribute to conservation biology.

Coevolution, the evolution of reciprocal adaptations in response to reciprocal natural selection, is a key generator of biodiversity. Despite widespread acceptance of coevolution as a central process in such major events in the history of life as the origin of mitochondria and chloroplasts, and in

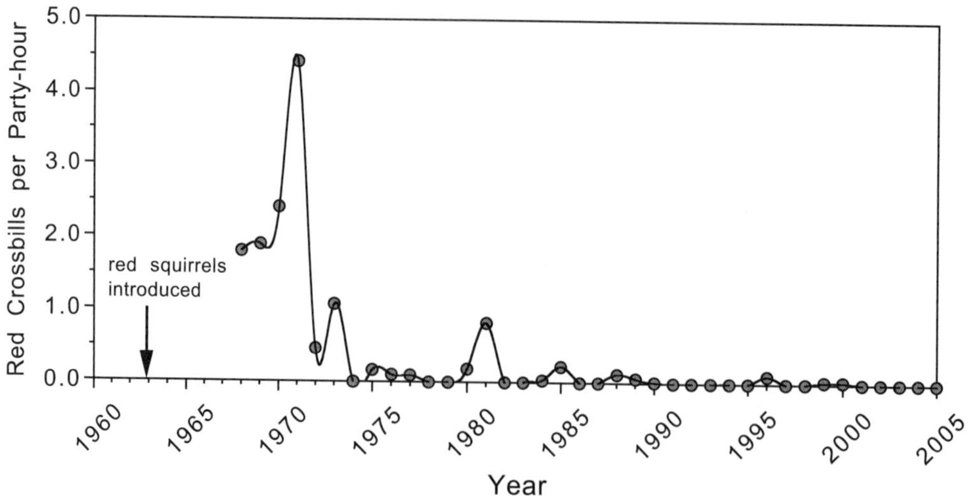

FIGURE 14.1 The number of red crossbills observed per "party-hour" during annual Christmas Bird Counts in Terra Nova National Park, Newfoundland, from 1968 to 2005. A party-hour is one hour of observation by one group of observers. Many of the crossbills reported after 1980 were likely nomadic crossbills from the mainland.

BOX 14.1 Coevolution: The Evolution of Reciprocal Adaptations

Coevolution occurs when two species exert selection pressures on each other that result in reciprocal adaptations. For example, crossbills preferentially forage on trees with thin-scaled cones, thus causing selection among trees in favor of thick-scaled cones. Over time, the conifer population evolves cones with thicker scales, which in turn favors larger billed crossbills that are more efficient at foraging on thick-scaled cones. Such reciprocal selection and adaptation continues to favor increases in scale thickness and bill size until the costs of further increases in these traits in at least one species are outweighed by their benefits. Another visually appealing example of reciprocal adaptation in response to reciprocal selection is the evolution of traits mediating the interaction between the Japanese camellia *Camellia japonica* (Box Fig. 14.1A), and its obligate seed predator, the Japanese weevil *Curculio camellia* (Box Fig. 14.1B) (Toju & Sota, 2006). Japanese weevils have long rostra for drilling holes into the pericarp of fruit where weevils oviposit. In response, the camellia has evolved a thick woody pericarp to protect its seeds from the weevil. Laboratory experiments demonstrate that weevil rostrum length must exceed pericarp thickness for successful oviposition, and field studies show that geographic variation in rostrum length is positively correlated with variation in pericarp thickness, presumably reflecting coevolution (Toju & Sota, 2006). Moreover, Toju and Sota (2006) show that the strength of selection exerted by weevils on pericarp thickness is correlated with geographic variation in pericarp thickness and weevil rostrum length. However, why such geographic variation arises is unknown. In other plant taxa, for instance, there is an increase in frequency of armaments against seed predation at lower latitudes (Toju & Sota, 2006).

(*continued*)

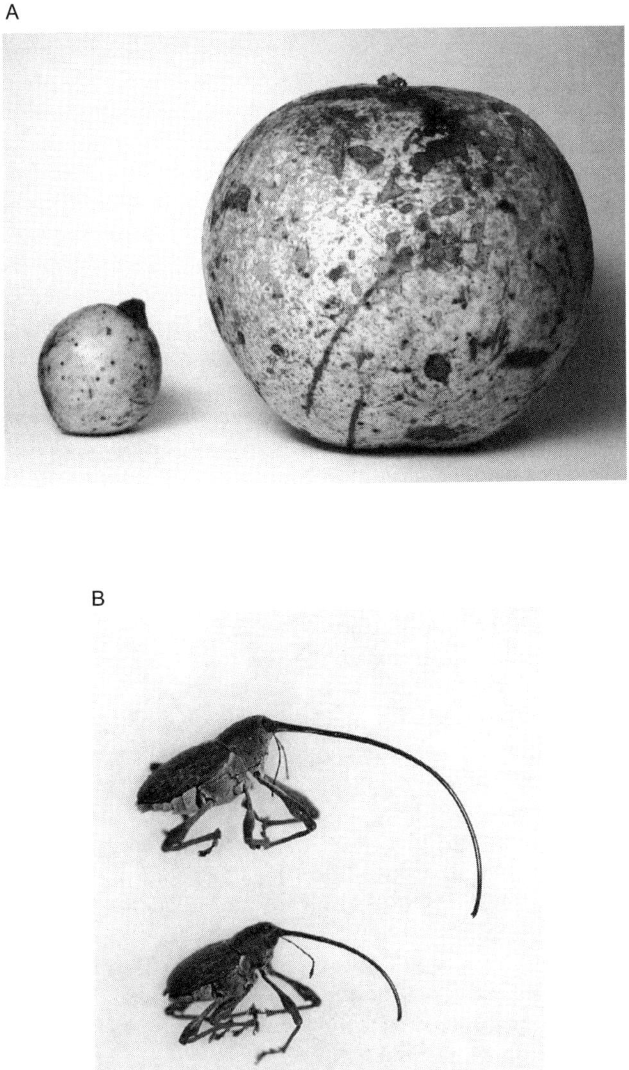

Box Fig 14.1 Geographic variation in the sizes of (A) Japanese camellia fruits and (B) Japanese weevil rostrums. (Photographs © Hiro Toju, provided by the author and used with his permission.)

influencing rates and patterns of phylogenetic diversification, a systematic consideration of coevolution as an important contemporary ecological and evolutionary process has only emerged within the past decade (Thompson, 1994, 2005). A major problem with early studies of coevolution was that investigators looked for evidence of coevolution at the species level and usually based studies of reciprocal selection and adaptation on single populations (reviewed in Thompson, 1994). This approach provided an unrealistic caricature of the ecological and evolutionary dynamics of coevolution, because coevolutionary processes vary among populations and because traits important to coevolution are rarely fixed at the species level.

Advances in three areas during the past decade spurred development of a geographic view of the coevolutionary process. First, empirical studies addressing patterns of genetic variation within species have repeatedly demonstrated that most species are collections of genetically differentiated populations (Avise, 2000). Consequently, microevolutionary processes are often most appropriately investigated at the level of local populations where they may have unique evolutionary trajectories in response to variation in local selection pressures. Second, interacting species rarely have completely coincident ranges or experience the same community context and abiotic conditions throughout their ranges. Thus, species interactions often vary in their form (for example, the nature of interactions span from antagonism to mutualism and differ in evolutionary outcome across populations) (Thompson & Cunningham, 2002). Lastly, an increasing body of literature has laid to rest the notion that adaptation resulting from natural selection is a process too slow to be readily observed in nature. Rapid evolution has been demonstrated in a diverse range of taxa (Carroll & Watters, this volume; Rice & Emery, 2003; Stockwell et al., 2003). Moreover, contemporary evolution and coevolution can occur rapidly enough to function as an ecological process, and may regularly affect ecological dynamics (Thompson, 1998). For example, evolution of a green alga in response to its predator (rotifers) affects the dynamics of their predator–prey cycle (Yoshida et al., 2003). In turn, change in dynamics of the cycle could affect the evolution of predators (although there was no evidence of such in these experiments) and further alter predator population dynamics. Indeed, many coevolutionary interactions have been observed to evolve rapidly during the past several centuries (see, for example, Burdon & Thrall, 1999; Zangerl & Berenbaum, 2005).

Understanding how coevolution influences patterns of adaptation, speciation, and extinction will become increasingly important as natural landscapes are restructured by humans. Conservation biologists have been mostly concerned with the way in which habitat fragmentation affects extinction rates, loss of genetic diversity, ecosystem services, and other ecological processes, with a particular focus on managing endangered species (Meffe & Carroll, 1994). However, a solely ecological viewpoint ignores rapid adaptation and microevolutionary change as processes that affect ecological dynamics and allow species to avoid extinction in the face of habitat loss and fragmentation, climate change, range shifts, and species introductions (Rice & Emery, 2003; Stockwell et al., 2003). Examples of rapid adaptive responses to human-induced ecological change include the evolution of heavy-metal tolerance in plants inhabiting mine tailings (Antonovics et al., 1971) and shifts in morphology in two species of Australian snakes that reduce vulnerability to an introduced and toxic prey, the cane toad *Bufo marinus* (Phillips & Shine, 2004). Conservation of biodiversity, and its future production and maintenance, will require a focus on how these anthropogenically induced changes affect microevolutionary processes relating to organization, structure, and evolutionary outcome of species interactions. The geographic view of the coevolutionary process adopted by the geographic mosaic theory provides a framework for guiding conservation practices in light of ongoing coevolutionary processes.

KEY CONCEPTS: THE GEOGRAPHIC MOSAIC THEORY OF COEVOLUTION

Geographic variation is inherent in the coevolutionary process. Thus, the geographic mosaic theory of coevolution recognizes geographic structure in the form and evolutionary outcomes of species interactions, and in the geographic and population genetic background upon which species interactions evolve. The theory provides a view of how geographic structure combines with a variety of evolutionary processes to determine how the coevolutionary process shapes trait evolution from the population level to the species level. Moreover, the geographic mosaic theory provides a framework for investigating geographic dynamics of coevolution, and for understanding how coevolution shapes diversity within and among species. This theoretical framework is a valuable guide for further development of applied coevolutionary biology (Thompson, 2005).

The geographic mosaic theory of coevolution has three major components. The first component is the observation that species interactions vary among populations. Across the geographic distribution of interacting taxa, variation in species interactions in turn creates a mosaic of divergent selection pressures and evolutionary outcomes. For most well-studied, wide-ranging species, variability in the form and outcome of natural selection stemming from species interactions across multiple populations has been documented, suggesting that geographic selection mosaics are a common if not ubiquitous feature of species interactions (Thompson, 2005). Selection mosaics may arise as a result of differences in community context or variation in abiotic factors over geographic space. For example, the nature of interaction between *Greya* moths and *Lithophragma* host plants upon which moths oviposit and pollinate varies across the shared geographic range from mutualism to commensalism to antagoni... popul...

t ... s ... c ... ti ... wh ... ily ... rec ... (Fig ... tion ... spec... range of outcomes other than coevolution (Fig. 14.2 [Thompson, 2005]). In hotspots, by contrast, both interacting species may be under strong selection that is mediated directly by interspecific interaction.

Traits causing and responding to reciprocal natural selection have been shown to exhibit clinal or mosaic patterns of variation in several systems. Examples include the seed predators and plants mentioned earlier (Benkman et al., 2001; Toju & Sota, 2006) (Box 14.1), interactions involving pollinators or floral parasites and plants (Thompson & Cunningham, 2002), predatory snake and salamander prey (Brodie et al., 2002), and parasitic rusts and plant hosts (Burdon & Thrall, 2000). One result is that natural selection rarely favors the same traits across all populations and that coevolved traits fixed at the species level should be relatively rare. Such examples of geographic mosaics of coevolutionary hotspots and coldspots will presumably become more common as biologists increasingly account for geographic variation in the traits shaped by species interactions.

The third component of the geographic mosaic theory of coevolution is the premise that trait remixing between coevolutionary hotspots and coldspots affects coevolutionary dynamics across the mosaic. The dynamics of coevolution in local populations and across species' ranges are affected by gene flow, mutation, genetic drift, extinction, and recolonization of local populations. Each of these processes can alter the geographic distribution of coevolving traits, and further affect the distribution of genetic variation across geographic ranges of interacting species (Thompson, 2005). Unlike the first two components, the third component and its predictions have mostly been evaluated using theoretical mod... (but see Forde et al., 2004). These models suggest that maladaptation is a likely consequence of geographically structured coevolution (Gomulkiewicz ...l., 2000; Nuismer et al., 1999), and that anal... of geographic variation in the form of natural ...tion alone may oftentimes not be enough to pre... the overall coevolutionary outcomes of these ...ctions. The overall spatial structure of mal...ation (or adaptation) likely depends on the ...of coevolutionary interaction, the underly...ength of divergent selection between pop...s, and the magnitude of gene flow across ...dscape. For example, high gene flow can ...cal adaptation by diluting the frequency of ...ficial alleles, whereas low gene flow can increase genetic variation and facilitate rapid local adaptation (Forde et al., 2004; Gomulkiewicz et al., 2000). Empirical examples of trait mismatches reflecting such maladaptation in interacting species come from studies of predator–prey (Storfer & Sih, 1998), seed disperser–plant (Garrido et al., 2002), and plant–herbivore interactions (Zangerl & Berenbaum, 2003).

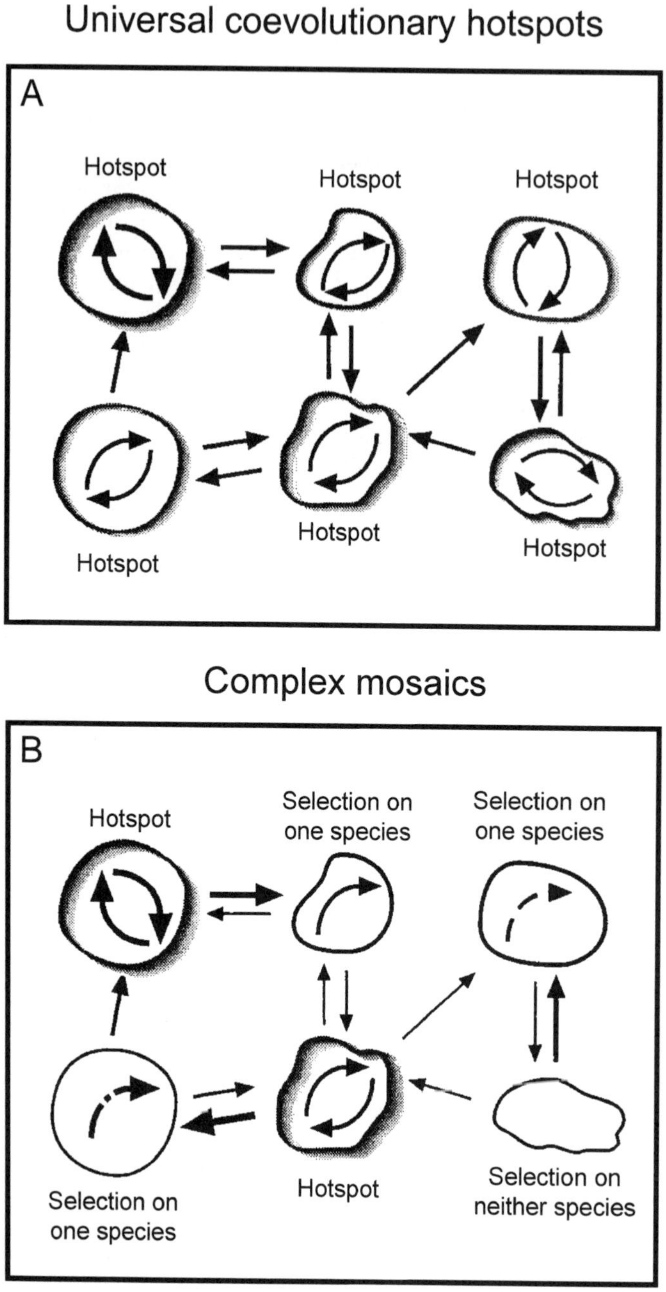

FIGURE 14.2 Schematic of alternative forms and outcomes of reciprocal natural selection for a geographically variable interaction between two species. Arrows within circles represent natural selection caused by one species on another in each community, and differences in arrows represent differences in the form of natural selection. Arrows between communities represent gene flow, with arrow thickness proportional to the amount of gene flow. (A) Reciprocal selection and reciprocal adaptations occur in all communities, although the interaction coevolves differently in each community. (B) Reciprocal selection and adaptation occur only in some communities, so that coevolutionary hotspots are embedded in a matrix of coevolutionary coldspots (From Thompson [2005]. Figure provided by J. N. Thompson, used with permission.)

We conclude this section by noting the general predictions of the geographic mosaic theory: (1) populations should show variation in the traits shaped by interactions, (2) traits of interacting species will be well matched only in some local communities, and (3) coevolved traits will rarely be spread across all populations to become fixed at the species level (Thompson, 2005). This means that investigations of coevolutionary interactions require multidisciplinary approaches aimed at quantifying the form and outcome of reciprocal natural selection across multiple populations, as well as quantifying patterns of genetic variation across geographic ranges of the taxa involved in the interaction.

CASE STUDIES: GEOGRAPHIC MOSAICS OF COEVOLUTION BETWEEN CROSSBILLS AND CONIFERS

Coevolution between red crossbills in the *L. curvirostra* complex and the cones of conifers upon which these birds specialize illustrates the principles of the geographic mosaic theory of coevolution and its relevance to biological conservation. Crossbills are highly specialized birds that have evolved crossed mandibles for spreading apart the overlapping scales of conifer cones to reach the underlying seeds. Because of variation in the structure of conifer cones and seeds, crossbills have diversified into an array of resource specialists on several conifer species (Benkman, 1993, 2003). In North America, nine morphologically and vocally differentiated *call types* exist, with at least six of these known to be specialized on particular conifers. Studies in the laboratory and in the wild suggest that divergent natural selection has caused crossbill diversification (Benkman, 1993, 2003). In addition to the Newfoundland crossbill mentioned earlier, there are call types specialized on Rocky Mountain ponderosa pine, *Pinus ponderosa scopulorum;* Rocky Mountain lodgepole pine, *Pinus contorta latifolia;* Douglas-fir, *Psuedotsuga menziesii menziesii;* and western hemlock, *Tsuga heterophylla* (Benkman, 1993, 2003). Call types also exhibit statistically significant, although subtle, levels of genetic differentiation among one another that are consistent with recent divergence and low levels of ongoing hybridization (Parchman et al., 2006). These results, in addition to the finding that some of the call types exhibit considerable premating reproductive isolation (Smith & Benkman, 2007), are consistent with some of the call types representing incipient species.

Not only does variation in cone structure select for morphological specialization in crossbills, but in some regions selection favors cones that are defended against crossbill predation. This sets the stage for coevolutionary arms races to arise between crossbills and conifers (Box 14.1). However, crossbills are not the only seed predators and selective agents on conifers. Indeed, much more prominent conifer seed predators and selective agents on cone structure in North America are two species of *Tamiasciurus* pine squirrels that harvest and cache vast numbers of cones soon after the seeds mature but before cones begin to open. The consequences of this preemptive seed predation and selection by pine squirrels are quite evident in lodgepole pine forests in the Rocky Mountains (Fig. 14.3). Here, lodgepole pine has evolved defenses in response to selection exerted by pine squirrels, including a decrease in both the ratio of seed to cone mass and the number of seeds per cone. Because of preemptive competition from pine squirrels, crossbills are uncommon in the Rocky Mountains, and any selection on cone and seed traits by crossbill predation is presumably swamped by stronger selection from pine squirrels (Benkman, 1999; Benkman et al., 2001; Siepielski & Benkman, 2005). Crossbills in turn have bills adapted to the average squirrel-defended lodgepole pine cone (Fig. 14.3). Consequently, this region represents a coevolutionary coldspot for crossbills and lodgepole pine, but a coevolutionary hotspot for pine squirrels and lodgepole pine.

Largely as a result of patterns of glacial retreat (Benkman et al., 2001), pine squirrels are absent from several isolated mountain ranges east (for example, Cypress Hills) and west (for example, South Hills) of the main belt of the Rockies (Fig. 14.3). In the absence of pine squirrels, lodgepole pine has lost defenses against squirrels, and crossbill population densities are up to 20 times greater in these areas than in those where pine squirrels are present (Benkman, 1999; Benkman et al., 2001; Siepielski & Benkman, 2005). Lodgepole pine has diverged further in cone traits in these areas as a result of increased selection by crossbills (Benkman, 1999; Benkman et al., 2001). In particular, cones here have thicker distal scales that make it harder for crossbills to reach underlying seeds (Fig. 14.3). As in the crossbills of Newfoundland, crossbills

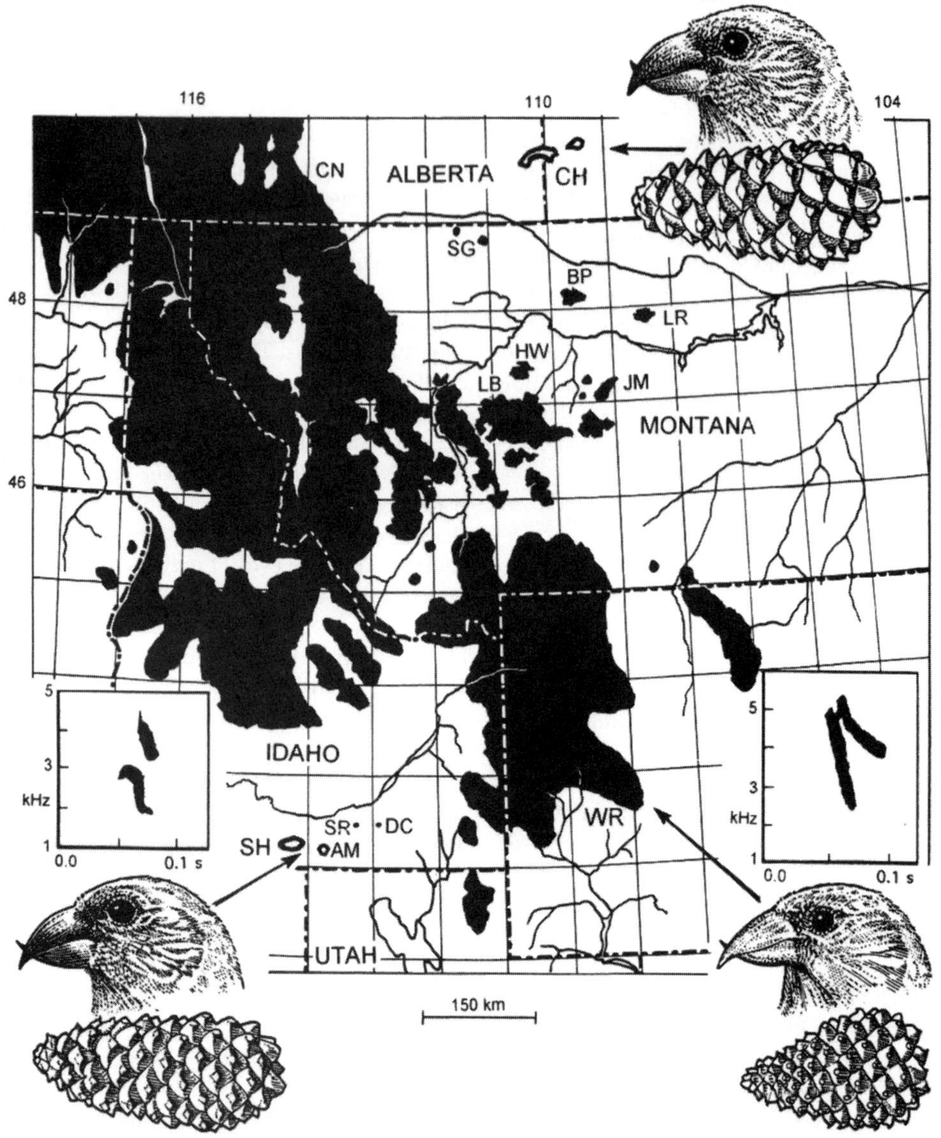

FIGURE 14.3 The distribution of Rocky Mountain lodgepole pine (black) and crossbills and cones in the Rocky Mountains (lower right), Cypress Hills (upper right), and South Hills and Albion Mountains (lower left). Representative sonograms of flight calls are shown for the South Hills crossbill (lower left) and the Rocky Mountain lodgepole pine crossbill (lower right). *Tamiasciurus* pine squirrels are found throughout the range of lodgepole pine, except in some isolated mountains, including the Cypress Hills (CH), Sweetgrass Hills (SG), Bears Paw Mountains (BP), Little Rocky Mountains (LR), South Hills (SH), and Albion Mountains (AM). Pine squirrels were absent from the Cypress Hills until they were introduced in 1950. (From Benkman [1999].)

in mountain ranges that lack pine squirrels have evolved deeper and more "decurved" bills that provide greater access to seeds. (Fig. 14.3). Thus, in the absence of pine squirrels, crossbills have coevolved with lodgepole pine in a predator–prey arms race.

These studies suggest that the presence and absence of a dominant preemptive competitor, the pine squirrel, determines the geographic mosaic of coevolutionary hotspots and coldspots for crossbills and lodgepole pine (Benkman et al., 2001). In addition, parallel and replicated patterns of coevolution between crossbills and their pine and spruce prey (Parchman & Benkman, 2002) (Box 14.2) suggest that geographic mosaics of coevolution may be an important source of divergent selection underlying adaptive variation in conifers and crossbills. Because the mosaics discussed here result in diversifying coevolution between crossbills in areas with and without pine squirrels, mosaics are thus capable of driving ecological speciation in crossbills as a result of divergent selection for resource specialization (Smith & Benkman, 2007). Crossbills in one of these coevolutionary hotspots (South Hills) has diverged in bill morphology and vocalizations from other crossbills specialized on lodgepole pine (Fig. 14.3). Furthermore, a study of breeding crossbills in the South Hills revealed that nearly 99% of the birds paired assortatively in the presence of other nonresident call types (Smith & Benkman, 2007). These studies indicate that coevolutionary selection is causing ecological speciation in the South Hills crossbill and that similar processes may have been driving ecological speciation in the Newfoundland and Cypress Hills crossbills before extinction in response to squirrel introductions (pine squirrels were introduced to the Cypress Hills in 1950 [Benkman, 1999]).

Here we have emphasized the consequences for crossbills of interaction between pine squirrels, crossbills, and conifers. However, these interactions can also have effects that ramify throughout entire ecosystems. Specifically, we want to point out the possibility of what we refer to as *keystone selective agents* whose interactions with other species cause selection and evolutionary changes that alter community structure and even ecosystem processes (Benkman & Siepielski, 2004). For example, pine squirrel predation on lodgepole pine cones reduces the benefits of serotiny (in other words, long-term retention of seeds in woody fruits). Thus variation in the intensity of pine squirrel predation maintains substantial variation in the frequency of serotiny in lodgepole pine (Benkman & Siepielski, 2004).

Pine squirrel predation can have considerable effects on a whole ecosystem because the frequency of serotiny determines the density of seedlings after stand-eliminating fires. Seedling density in turn may have major affects on plant and animal

BOX 14.2 Replicated Evolution

Repeated and independent patterns of evolution imply adaptive evolution. Examples of repeated or replicate evolution include numerous radiations of fishes into benthic and limnetic species pairs after postglacial colonization of northern lakes, and the adaptive radiations of *Anolis* lizards in the Greater Antilles (Schluter, 2000). If coevolution, which results in reciprocal adaptations, is a strong process, then repeated or replicated patterns of coevolution are expected. Replicated patterns of coevolution between crossbills and lodgepole pine east and west of the Rocky Mountains (Fig. 14.3), and the replication of these patterns of trait evolution in crossbills coevolving with black spruce (Parchman & Benkman, 2002) provide evidence that coevolution is sufficiently strong to overwhelm historical contingencies. It also indicates that in some situations we should be able to predict how certain actions may affect the geographic mosaic of coevolution. For example, it would clearly be unwise to introduce tree squirrels to Hispañola, where the endemic, large-billed crossbill *Loxia megaplaga* has coevolved with *Pinus occidentalis* (Parchman et al., 2007). Given the strong and diverse impacts of pine squirrels, introducing them into any area is ill advised.

communities, and biogeochemistry during succession (Benkman & Siepielski, 2004). Thus, not only is it important to protect large mammals that act as keystone predators in communities (Terborgh et al., 1999), it is also imperative to consider evolutionary effects that are not limited to large predators. Perhaps a focus of conservation should be on strongly interacting species that are likely to have strong ecological (Soulé et al., 2003) and evolutionary impacts. Because mammals tend to occur at much higher densities than birds (Brown, 1995), we suspect that the ecological and evolutionary effects of mammals may often dominate those of birds. Moreover, variation in the occurrence of mammals, especially nonvolant mammals with distributions that are more likely to be limited by historical factors, is likely to have large impacts on selection mosaics. We are hesitant to compare other taxa, but we suspect and hope readers will make similar comparisons and predictions for other taxa.

RELEVANCE OF THE GEOGRAPHIC MOSAIC THEORY OF COEVOLUTION TO CONSERVATION BIOLOGY

Humans have become an extraordinary force affecting contemporary ecological and evolutionary processes. Moreover, the same suite of factors commonly cited as affecting these processes can also be expected to influence the coevolutionary process. This includes, but is not limited to, species introductions, habitat loss and fragmentation, and the overall reorganization of the landscape. An important point emerging from the geographic mosaic view of coevolution is that conservation practices cannot be expected to be successful if they only aim to conserve species per se (Thompson, 2005). Rather, this view suggests the need to conserve the full range of species interactions, because the core of biodiversity and its ongoing production resides in the geographic structure of interactions. The importance of *interaction biodiversity,* although long acknowledged, has only become prominent during the past decade as conservation efforts have integrated landscape views with processes organizing biodiversity (Thompson, 1996). However, most of these efforts have focused only on potential ecological effects. It is also important to consider evolutionary history and the potential for contemporary evolution, because changed ecological conditions affect not only the diversity and form of species interactions but also their evolutionary outcomes. Especially for coevolutionary interactions, adaptive evolution is an alternative to extinction and should be taken more seriously now that contemporary microevolution has been widely documented (Rice & Emery, 2003; Stockwell et al., 2003). As anthropogenic forces drastically change the geographic mosaic of coevolution, preservation of wilderness may be especially important to protect ongoing diversification and speciation as well as for understanding the role of coevolution in the origination and maintenance of diversity (Thompson, 2005). Next we discuss several key areas or future directions where the geographic mosaic view of coevolution can be integrated with conservation biology.

Diversifying Coevolution and the Production and Maintenance of Biodiversity

When gene flow is limited among coevolutionary hotspots and coldspots, diversifying coevolution as a result of geographic mosaics may be a major process driving diversification and speciation (Benkman, 1999; Smith & Benkman, 2007). In addition, coevolution has the potential to drive divergence and speciation more rapidly than adaptation to physical environments alone, because genetic feedbacks of the coevolutionary process can produce sustained changes in traits related to selection (for example, in predator–prey arms races). Indeed, because populations are more likely to adapt to large changes that occur gradually rather than suddenly (Holt & Gomulkiewicz, 2004), sustained but gradual changes in the environment as a result of coevolution may be especially conducive for adaptive evolution. Earlier we discussed such diversifying coevolution between crossbills and conifers. Another example comes from laboratory evolution studies with the bacteria *Pseudomonas flourescens* and a bacteriophage. This study demonstrates that geographic structure is capable of causing bacteria and phage to diversify rapidly through antagonistic coevolution (Buckling & Rainey, 2002). Many plant species are also likely to experience selection mosaics that could promote speciation because plant abundance and local plant species diversity, which vary in space, influence selection for pollinator specialization (Sargent & Otto, 2006; Vamosi et al., 2006). And lastly, because ecological divergence is an engine of speciation (Funk et al., 2006), geographic selection mosaics are likely an important fuel of speciation and perhaps adaptive radiations.

The massive scale of species introductions during the past several hundred years has had, for several reasons, major consequences for the dynamics of geographically structured coevolution. First, if divergent selection between populations is the result of different community compositions, then introductions that homogenize habitats in terms of biotic composition will essentially eliminate selection mosaics. For example, introduction of pine squirrels causes conifer populations to experience a common set of selection pressures that may, over time, eliminate differences in cone and seed traits across the geographic mosaic. More immediately, introduced pine squirrels have responded to reduced seed defenses that were lost in their absence by increasing to exceptionally high densities. High squirrel densities in turn depleted cone crops and caused extinction of both Newfoundland and Cypress Hills crossbills. Other conifers located on islands where squirrels are absent—including two pine species in the Mediterranean basin (*P. halepensis* and *P. nigra*), and western hemlock and Douglas-fir in the Pacific Northwest—have also evolved similarly in response to relaxation of squirrel predation (Mezquida & Benkman, 2005; Parchman & Benkman, unpublished data). Indeed, loss of antipredator defenses is a common phenomenon in isolated island organisms. Such evolutionary changes, however, are not just restricted to oceanic islands and are likely a common phenomenon wherever selection varies geographically. This shows the need for understanding community-level interactions and informing conservation priorities from evolutionary and coevolutionary perspectives rather than from merely ecological perspectives.

Habitat Loss and Fragmentation

Habitat loss and fragmentation are well known for their impacts on the ecological dynamics of species. Many breeding bird species, for example, have minimum habitat area requirements for population persistence (Johnson & Igl, 2001). Little attention has been devoted, however, toward understanding the potential consequences of habitat loss and fragmentation on evolutionary and coevolutionary processes (Holt & Gomulkiewicz, 2004). And yet, understanding the effects of habitat degradation, and instigating management practices aimed at preservation and restoration of habitat, are at the very center of conservation biology. Although commonly grouped together in discussions on biodiversity, habitat loss and fragmentation potentially differ in their effects (see Fahrig, 2003). Nevertheless, both factors may be especially relevant to coevolutionary processes.

To our knowledge, the only studies to have considered effects of habitat area on coevolutionary processes are each of two investigations of crossbills and lodgepole pine (Benkman, 1999; Siepielski & Benkman, 2005). In the forest islands of lodgepole pine east and west of the Rockies, where pine squirrels are absent (Fig. 14.3), crossbill abundance increases logarithmically with increasing area of lodgepole pine habitat. Correlation between crossbill density and area of available lodgepole pine in turn increases the strength of selection by crossbills on lodgepole pine, causing increasing seed defenses as habitat area increases. This leads to increasingly larger billed crossbills on larger islands in response to elevated defenses in lodgepole pine cones. Because divergent selection on crossbills as a result of coevolution increases with increasing island area, crossbills on the largest of these islands (the South Hills) have diverged the most from other populations.

We also suspect that habitat area reduces the likelihood of population extinction, leading to populations of both crossbills and lodgepole pine on larger islands that persist long enough to coevolve (Siepielski & Benkman, 2005). These results thus suggest that there may exist some minimum habitat area, varying among species, that is needed to maintain evolutionarily viable populations. This is not surprising given that larger populations are more likely to be able to adapt to environmental change (Holt & Gomulkiewicz, 2004). It is also worth noting that many of the examples of contemporary evolution mentioned here occurred in large populations moving into favorable habitats, allowing populations to increase further—conditions conducive for adaptive evolution (Reznick et al., 2004). In contrast, rapid and sustained evolution is much less likely in small or declining populations, which are also the ones of greatest conservation concern (Reznick et al., 2004). Thus, because large populations are more likely to adapt to environmental change, it is imperative that sufficient habitat is available to support large populations.

Much less is known about the potential effects of habitat fragmentation on the coevolutionary process. However, the geographic mosaic theory itself is predicated on the idea that there is an inherently fragmented geographic structure to the ecology and evolution of species (Fig. 14.2). There are numerous direct and indirect ways that habitat fragmentation can possibly affect contemporary coevolution. Habitat fragmentation leads to a landscape in which suitable habitat is replaced with new habitat types that may harbor an entirely different suite of interacting species, thus increasing community complexity and affecting the origin and nature of reciprocal interaction. Habitat fragmentation can also reduce species connectedness, possibly by breaking up species interactions entirely, or by reducing the strength of interaction. In fact, there is little doubt that historical habitat fragmentation led to geographic selection mosaics for many species (see, for example, Benkman, 1999).

Habitat fragmentation may also influence coevolutionary dynamics by affecting the amount of genetic variation that results from gene flow within and among geographic mosaics. Although increased gene flow can limit local adaptation or lead to local maladaptation, gene flow may also increase local adaptation in some interactions by elevating levels of genetic variation available for selection to act upon. Because the geographic mosaic of coevolution is important in the long-term maintenance of genetic variation within species (Thompson, 2005), erosion of the mosaic resulting from habitat loss and fragmentation may influence the loss of genetic diversity in species by destroying one process that maintains genetic variation in natural populations. However, long-term maintenance of genetic diversity in such interactions is further influenced by gene flow and variable selection across the geographic mosaic of coevolution.

Recent models of coevolutionary dynamics suggest that geographic selection mosaics, coevolutionary hotspots, and gene flow all increase the likelihood that polymorphisms are maintained for parasite–host interactions (Burdon & Thrall, 2000). By restricting the movements of individuals among populations, habitat fragmentation in general can be expected to decrease the contribution of gene flow to the geographic mosaic of coevolution. For some host–parasite interactions, decreased gene flow may result in a rapid loss of genetic variation for defense against pathogens and other enemies, and decreases in host population densities (Nuismer & Kirkpatrick, 2003). Thus, by limiting gene flow, habitat fragmentation may constrain input of genetic variation that is important to adaptation and the maintenance of polymorphisms, thereby increasing the likelihood of extinction on one side of an interaction. Preservation of the full range of interactions and their outcomes across geographic mosaics of coevolution may act to deter the loss of genetic variation and extinctions in species experiencing coevolutionary mosaics. In contrast, by limiting gene flow that can swamp local adaptation, habitat fragmentation can also facilitate local adaptation. However, the potential benefits and costs of gene flow among populations for conservation biology have been hard to resolve (Stockwell et al., 2003), and predicting how this will affect the dynamics of geographically structured coevolution is even more challenging.

Conservation biologists have sought to curb the effects of habitat loss and degradation by restoring species in habitats where anthropogenic forces have caused population decline or extinction. Translocation of organisms during restoration of native ecosystems has provoked serious concern about the genetic characterization of native and introduced organisms in the context of local adaptation and hybridization among native and translocated organisms; thus, the emergence of the field of restoration genetics (Hufford & Mazer, 2003). Restoration geneticists should also turn their attention to the genotype-by-genotype interactions among translocated organisms and species with which they coevolve. The geographic mosaic view of coevolution may provide a useful guide for restoration geneticists in cases in which coevolutionary interactions are concerned. For example, if genetic swamping occurs among native and introduced genotypes, then introducing a plant species in one area that has evolved in response to a different suite of herbivores, pathogens, or pollinators may have drastic consequences for the remaining native populations. Future decisions regarding the genetic identity of translocated organisms should consider selection mosaics and the geographic mosaic of coevolution, particularly the characterization of coevolutionary hotspots and coldspots.

Changes in Species Ranges

Climate change is widely acknowledged to have pronounced affects on ecological and evolutionary processes. The effects of climate change are complicated by the fact that climate change imposes alterations in abiotic factors (for example, temperature, precipitation, elevated CO_2) with cascading effects on biotic processes. For example, elevated levels of CO_2 can alter timing and duration of insect development and plant flowering, which in turn may cause temporal decoupling of pollinators and plants (Harrington et al., 1999).

Shifts in the geographic distribution of species are perhaps the most obvious consequence of climate change. Geographic shifts can affect species interactions and the geographic mosaic of coevolution; however, because taxa within a local community may not change in the same way, such a perturbation has the potential to reorganize communities with considerable affects on selection mosaics. Thomas and colleagues (2001), for example, found that during a 20-year period in Britain, some butterfly taxa increased the total number of habitat types that they occupy. During this same time period, some cricket populations increased in dispersal tendency, which has caused a three- to 15-fold increase in rates of population expansion. Thus some cricket populations are able to cross major habitat barriers, with concomitant reorganization of local communities and species interactions. Even though most local species assemblages have not been constant during past climatic changes (Williams et al., 2004), future climate changes and potential shifts in species distributions are likely to be exceptionally rapid and large in scale, with adaptation to such changes difficult. Furthermore, the ability of some species, particularly plants, to undergo range expansions in response to climate change may be limited by habitat fragmentation. In a recent study of Belgium forest plants, 85% of species had a low probability of successfully colonizing suitable though spatially fragmented habitat, whereas the probability of successful colonization was higher in landscapes with higher connectivity (Honnay et al., 2002).

FUTURE DIRECTIONS

Ongoing reorganization of the landscape and perturbation of the global network of species interactions suggest that conservation biologists should carefully consider evolutionary and coevolutionary consequences of species interactions. In particular, there is a need to understand the geographic structure of coevolutionary interactions, and furthermore to incorporate concern for coevolutionary hotspots and coldspots into conservation and management practice. Conservation programs imbued with an appreciation for coevolution may better guide restoration efforts, help avoid or abate negative consequences of species introductions, and offer a conceptual framework for preserving the very processes that give rise to and maintain diversity. In addition, preserving the diversity of interactions and their evolutionary significance requires that the conservation of geographic selection mosaics and coevolutionary hotspots and coldspots be deemed equally important. Moreover, it is important to maintain the integrity of hotspots and coldspots to avoid deleterious effects of homogenizing gene flow or homogenizing selection. A necessary corollary is that conservation of large areas of habitat, and especially the little remaining wilderness, should be a crucial goal of conservation efforts.

Lastly, the geographic mosaic view of the coevolutionary process suggests that there is a need to appreciate more fully the dynamic nature of species interactions across the landscape and the fact that contemporary evolution and coevolution occur rapidly enough to function as ecological processes. Given the predictions of current climate change models, impending and ongoing anthropogenic change, and the dynamics of populations and community assembly, it is unlikely that the ecological or conservation value of a particular site will remain the same indefinitely.

Conservation biology ultimately rests on our ability to understand and maintain the processes responsible for the production of biodiversity. Conservation biology as a discipline will need to shift focus from conservation of individual species to recognition of the importance of maintaining diversity of interactions among species. This is no simple task. It will require departure from a paradigm that views species as largely adapting to differences in their environments, to one that views species as adapting to and depending on other species (Thompson, 2005). Recognition of geographic variation in these processes is a necessary starting point given that species introductions are occurring rapidly, climate change is imposing shifts in the distributions of species, and habitat fragmentation and loss are all imposing potentially dramatic shifts in the ways that coevolutionary interactions are structured in nature. Incorporating this understanding into conservation practices will require increasing collaboration and cooperation between ecologists, evolutionary biologists, geneticists, conservation biologists and land managers.

SUGGESTIONS FOR FURTHER READING

Thompson (1999) provides an introduction to an issue of *The American Naturalist* dedicated to the geographic mosaic theory of coevolution, and reviews the theory's ecological and evolutionary hypotheses and predictions. An outstanding and more recent and comprehensive review of outstanding comprehensive, and recent review of the major theoretical and empirical studies pertaining to coevolution and the geographic dynamics of coevolution can be found in Thompson's *The Geographic Mosaic of Coevolution* (2005). This book is astonishing in its breadth of coverage. However, relatively little has been written about the implications of the geographic mosaic theory to conservation. The last chapter of Thompson's book (2005) considers how a geographic view of coevolution can be developed into *applied coevolutionary biology*, but Thompson's focus is rather different than ours here. Elsewhere, Benkman and colleagues (2008) discuss further, in the context of the geographic mosaic theory of coevolution, why introductions of strongly interacting species should be avoided, and why *rewilding*, which emphasizes introduction of strongly interacting species that have been extinct for thousands of years, should be used with caution.

Benkman, C. W., A. M. Siepielski, & T. L. Parchman. 2008. The local introduction of strongly interacting species and the loss of geographic variation in species and species interactions. Mol Ecol. 17: 395–404.

Thompson, J. N. 1999. Specific hypotheses on the geographic mosaic of coevolution. Am Nat. 153: S1–S14.

Thompson, J. N. 2005. The geographic mosaic of coevolution. University of Chicago Press, Chicago, Ill.

Acknowledgments We thank J. Thompson and P. Thrall for providing helpful suggestions. An Environmental Protection Agency GRO Fellowship to T. L. Parchman and National Science Foundation grants (DEB-0455705 and DEB-0515735) to C. W. Benkman provided financial support during the writing of this chapter.

15

The Next Communities: Evolution and Integration of Invasive Species

SCOTT P. CARROLL
CHARLES W. FOX

On a global basis . . . the two great destroyers of biodiversity are, first, habitat destruction and, second, invasion by exotic species.

E. O. Wilson (1997) in *Strangers in Paradise*

Among the global perspectives gained by Darwin during his 5-year circumnavigation aboard the *Beagle,* those afforded by witnessing the prevalence and impact of introduced plants and animals were probably seminal to the transformation of his worldview. Exploring the continents of the southern hemisphere and the remote outposts of midoceanic islands, he saw biota replaced, natural economies disrupted, and species deeply altered from their original states. The first volume of Sir Charles Lyell's new *Principles of Geology* series (1830), an embarkation gift from Fitzroy, the *Beagle's* captain, had convinced Darwin that the natural forces he observed linked directly back to those that shaped the world prehistorically. And now he viewed landscapes that appeared as if painted over and repopulated with the familiar characters of the European countryside. Such cataclysms of biotic replacement, following closely on the modest actions of agrarian settlers, could be likened to the first colonizations of new lands, in which the fitting of life forms to the environment, and the assembly of living communities, could be closely inferred.

Since then, human-altered environments have yielded some of evolution's key lessons. The cases of industrial melanism in moths (Kettlewell, 1956), adaptations to toxic waste (Antonovics et al., 1971) and fertilization treatment (Snaydon, 1970; reviewed in Silvertown et al., 2006) in plants, responses to resource extinctions in birds (Smith et al., 1995), the appearance of antibiotic and pesticide resistance (Palumbi, 2001a), and introduced pathogens, pests, and hosts in numerous systems have all helped biologists comprehend the environmental, economic, and social importance of contemporary evolution—evolution occurring on *ecological* timescales of days to years rather than on timescales of centuries to millennia (Carroll et al., 2007). We now realize that contemporary evolution is commonplace (see, for example, Palumbi, 2001a) and that it can have substantial ecological consequences and conservation implications (Carroll et al., 2007).

Among the human agents altering earth's habitats, species introductions offer particularly informative accidental experiments because they mimic natural events important in structuring natural communities (Vermeij, this volume). The movements of species and the peregrinations of continents have shaped the earth's terrestrial and marine biogeography throughout the history of life. During the past few centuries in particular, human transport has augmented rates of biotic exchange among the earth's realms far beyond preindustrial norms (Elton, 1958). As we similarly alter other planet-level processes, including climate and nutrient cycling, the resulting disruption of established patterns of community dynamics may create greater ecological opportunities for invaders (Dukes & Mooney, 1999) while simultaneous recasting the adaptive landscape for all species (see, for example, Réale et al., 2003; Ward et al., 2000). For example, disturbance events should favor stress-tolerant individuals, and Kneitel and Perrault (2006) suggested that if stress tolerance trades off with competitive ability, a prevalence of less competitive phenotypes may render communities more vulnerable to invasion during postdisturbance conditions.

Invasive species provide a tool to study the ecological and evolutionary processes that produce communities (Strauss et al., 2006b). A biotic community is an aggregation of different species living and interacting within an abiotic realm. Constituent members may be largely independent and their associations happenstance, or they may be closely interdependent over vast periods of time, their histories inextricably linked. Each species has characters that are independent of their grouping, and others that depend directly on it and that will change if different species are assembled or if abiotic conditions change. These sorts of variation mean that what we call a *community* is somewhat arbitrary, and that its meaning will differ among taxa as well as within taxa at different places and times. Many of the factors that broadly determine community assemblage, including range expansions, occur naturally and have been common throughout earth's history. However, most of the species movements responsible for generating current patterns of species and community diversity were prehistoric. Thus, the process producing current diversity can only be inferred, largely from modern patterns. To the extent that anthropogenic species introductions are accidental experiments in community ecology, they may offer insights into how communities assemble and function.

Species invasions are unique among the anthropogenic disturbances in that they are naturally dynamic without continued disturbance; organisms interact, but these interactions evolve. In terms of their ecological consequences, this makes species introductions a particularly unpredictable form of environmental perturbation (Pimentel et al., 2000). What will biotic communities look like in 1,000 years or in 10,000 years, and how will they function? What is the role of genetic change in the transition from the colonization of a new habitat to the biotic integration of a population into a new community, and do the effects of an invader change over time? In the face of ongoing evolution, what would creating *sustainable* conservation management involve? These questions address temporal change on scales from seconds to millennia. Although it is common to think of biotic invasions mainly in ecological or management terms, only with an evolutionary perspective do we have the potential of ultimately linking such questions into an integrated framework.

The Consequences of Species Invasions

> *According to the principles so well laid down by Mr. Lyell, few countries have undergone more remarkable changes, since the year 1535, when the first colonist of La Plata landed with seventy-two horses.... The countless herds of horses, cattle, and sheep, not only have altered the whole aspect of the vegetation, but they have almost banished the guanaco, deer, and ostrich [rhea]. Numberless other changes must likewise have taken place; the wild pig in some parts probably replaces the pecarri; packs of wild dogs may be heard howling on the wooded banks of the less frequented streams; and the common cat, altered into a large and fierce animal, inhabits rocky hills.*
>
> Charles Darwin (1860, p. 120) reflecting on, the landscapes of southern South America that he encountered in 1832, three centuries after Spanish settlement

Not all introduced, alien, species become invasive when introduced to new environments. Indeed, most alien species fair poorly, or at best maintain small population sizes in their new communities, and some would quickly go locally extinct without continued human intervention. Although these species may be ecologically significant in certain contexts, it is the invasive species—those with populations that grow rapidly, spread geographically, and integrate into and frequently dominate native communities—that are of major conservation significance and are thus the focus of this chapter.

By definition, invasive species are new actors within biotic communities, and necessarily play many roles in ecological webs—as predators, pathogens, parasites, competitors, mutualists, or hosts (Mitchell et al., 2006). They have important effects on native biodiversity, and annually cause hundreds of billions of dollars in economic losses, many of which are incurred from our efforts to avert the impacts of invasive species in agricultural and natural environments (Mack et al., 2000; Mooney et al., 2005). Despite this focus on control, we are just beginning to appreciate the diversity of ways in which invasive species may alter interactions within their new, anthropogenically modified communities, and how quickly they may do so. Changes in the biology of invasive and native species are often difficult to predict, leading to unexpected outcomes that make invasive (or native) species difficult to control (Carroll, 2007a). Most directly, invasive species may affect population growth of native species. Frequently, invasive species compete with native species for resources, which leads to declines in population sizes of native species. However, because of trophic links within invaded communities, strong indirect effects may lead to the opposite affect on some taxa, with economic or even public health consequences. For example, insects introduced to control invasive spotted knapweed (*Centaurea maculosa*) in North America have fueled population growth in insectivorous native mice, resulting in an increase in their infection rate with hantavirus, the cause of an illness often lethal in humans (Pearson & Callaway, 2006).

Species involved in invasions—both the invaders and those affected by the invaders—may change phenotypically after invasion. This may be the result of phenotypic plasticity, evolved genetic change, or both. A phenotype (morphology, behavior, life history, or any other trait) is plastic when it varies depending on the environment in which individuals express that trait. Because alien species are (by definition) in new environments, it is not surprising that they commonly exhibit different phenotypes in these new environments than in their ancestral ranges. For example, the behavior of the invasive species may change in its new community, often in unpredictable ways and sometimes with devastating results for the native flora or fauna. The nocturnal, ornithophilic brown tree snake (*Boiga regularis*) has altered its habits substantially since it was introduced to Guam around 1960 and eliminated several endemic bird species. The Guam population has become more diurnal and more terrestrial, now feeding primarily on day-active skinks that sleep in relatively sheltered locations at night (Fritts & Rodda, 1998). It is unclear whether this change in behavior reflects adaptive behavioral plasticity, evolved adaptive genetic change, or both. It is likely that the initial change is a plastic response to prey availability, but that the substantial change in prey diversity on Guam will eventually lead to evolutionary (in other words, genetically based) change in snake behavior.

Likewise, because invasive species interact with native species, and thus change the environment of the native species, it is not surprising native species may be different in the presence of invasive species. For example, predatory bullfrogs *(Rana catesbiana)*, introduced from eastern into western North America, have recently led to the evolution of avoidance behavior in native western red-legged frog (*R. aurora*) populations. The behavior is plastic; red-legged frog tadpoles increase refuge use and decrease activity when exposed to bullfrog allelochemicals. However, the response is absent in red-legged frog populations still free from the invasive bullfrogs. We can thus infer that the ability to respond behaviorally to bullfrogs has evolved (and is genetically based), probably in response to bullfrog predation (Kiesecker & Blaustien, 1997).

Many reactions of natives to invasives, and vice versa, are preadapted plastic responses (Carroll & Watters, this volume). For example, native plants have evolutionary histories with their local competitors and should be preadapted to respond adaptively to alien competitors that are similar (for example, congeneric) to native competitors. Similarly, induced defensive responses mounted by plants against native herbivores may be effective against some, but not other, invasive herbivores. Thus, plastic responses that have evolved in response to native competitors or predators may mediate interactions between native and alien species, affecting which species can invade (discussed later) and the response of natives to the invaders (and vice versa). However, selection for further genetically based adaptations is probably inevitable when invasive species reach ecologically significant population sizes. Understanding these evolutionary responses is paramount for understanding and predicting long-term ecological dynamics in the invaded communities. A particularly interesting example of ecoevolutionary complexity has arisen from the interaction of the alien species involved in the effort to control invasive rabbits by introducing the *Mxyoma* virus into Australia. Initial mortality in the rabbits was extreme. A perspicacious experimental design permitted the documentation of rapid evolution of resistance in rabbits and evolution of avirulence in the virus (Fenner & Fantini 1999). After the continent had become populated with derived races of avirulent virus and resistant rabbits, efforts to develop a lethal replacement virus (calicivirus) ultimately led to its (apparently) inadvertent release and irruption (Kovaliski, 1998). Yet, in an instance of preadaptive plasticity, some rabbits were protected by prior exposure to a related but previously unknown native virus common in moist habitats (Cooke et al., 2002). This induced cross-resistance may have facilitated the evolution of genetically based resistance to calicivirus in rabbits. Subsequent evolution of resistance to the new virus was reported in 2007 (Anonymous 2007).

It is important to remember that invasiveness, like any performance measure, is inherently context dependent. Anthropogenic reductions in habitat complexity may reduce the diversity of environments in which introduced species must act. When habitat simplification occurs through the elimination of other species, the challenges faced by introduced species may be substantially diminished. Impacts of prior invaders may reduce the biotic resistance to subsequent invaders, leading to a scenario termed *invasional meltdown* (Simberloff, 2006). In her review of the ecological genetics of invasive species, Lee (2002) concluded that the success of invaders depends more on evolvability than on phenotypic tolerance or plasticity. In biotically homogenized realms, it is possible that the comparative virtues of plasticity may be further diminished, and competition will favor a diminishing subset of species with subpopulations that are evolved *invasion specialists*. On the other hand, as illustrated in the next section, when introduced species enhance resource diversity for some natives, plasticity may continue to play a powerful adaptive role.

An Evolutionary Approach to Invasion Biology

Evolutionary investigations of species invasions, then, have at least a twofold value: What they reveal about the role of ongoing evolution in determining ecological dynamics may in turn be used to predict and manage the impacts of such change in threatened communities (Carroll 2007a; Strauss et al., 2006a; Vermeij, this volume). Adaptive change in both native and alien taxa may determine how communities reconfigure after invasion (Gilchrist & Lee, 2007; Lambrinos, 2004; Strauss et al., 2006a). This chapter focuses on three aspects of an evolutionary approach to biological invasions. First, to the extent that closely related species have similar needs and abilities, how does phylogeny inform us about which species are likely to invade? Second, beyond homologous preadaptation, to what extent does invasion success rely on additional, contemporary, adaptive evolution? Third, how commonly will the biotic changes brought about by invasive species be great enough to select for evolutionary responses in native taxa, and what are the implications for biotic interactions and community structure? After addressing these topics, we then conclude by considering how such changes may alter the dynamics, predictability, and management requirements of biotic invasions. We emphasize the importance of assessing invasions on carefully described time lines beginning at or near colonization.

CONCEPTS

Who Invades?

> *As the species of the same genus usually have . . . much similarity in habits and constitution, and always in structure, the struggle will generally be more severe between them.*
>
> Darwin (1859, p. 60)

Only a small proportion of introduced species become invasive (Williamson & Fitter, 1996). However, because of the environmental and economic impacts, and forecasts of increasing rates of species introductions (Levine & D'Antonio, 2003), biologists have expended considerable effort to identify the attributes of species or habitats that predict invasiveness, in the hope of anticipating and thereby perhaps preventing invasions. Generalization has proved difficult, however (Levine et al., 2003); some invaders do have traits in common, but such lists are generally applicable to only a small group of species, and there are many exceptions (Rejmanek & Richardson, 1996). It is clear that both the properties of the species and the community to which it is introduced must be considered simultaneously—in other words, the key to understanding invasiveness emerges from the match between the invader and its new community (Facon et al., 2006; Ricciardi & Atkinson, 2004).

Information about phylogeny may aid in predicting the opportunities and risks faced by colonists. Much of the discussion has focused on plant invasions. In an early expression of the integrated perspective, Darwin (1859) proposed that species introduced to a region for which they are suited to abiotic conditions would more likely naturalize if the extant community lacked their close relatives. If congeners were present, he reasoned, they would likely use the same resources, leaving little niche opportunity for the new arrival (called *phylogenetic repulsion* by Strauss and colleagues [2006b]).

A complementary conjecture to Darwin's *naturalization hypothesis* is the *enemy-release hypothesis* (reviewed by Colautti et al., 2004), which similarly uses phylogenetic inference in predicting that colonists will more likely succeed if they are not attractive or susceptible to the specialized pests of the native inhabitants. Both hypotheses propose that a colonist will experience reduced *biotic resistance* if it is not too closely related to members of the native community. Contrasting with these hypotheses, Duncan and Williams (2002) suggested that closely related species might naturalize more readily because of conserved traits that render the alien preadapted to the new environment (*phylogenetic attraction* [Strauss et al., 2006b]). This preadaptation could be to abiotic conditions or to any aspects of biotic interactions that may be common to the types of communities they inhabit.

Application of Evolutionary Hypotheses

Attempts to assess the respective merits of these hypotheses have reached mixed conclusions (see, for example, Colautti, 2004; Strauss et al., 2006b), perhaps in part because of the complexities involved in making community-level predictions (see, for example, Urban, 2006). Nonetheless, these and several other recent studies have shed light on basic aspects of the puzzle. For example, Strauss and colleagues (2006b) found that highly invasive grasses in California are, on average, less closely related to native grasses than are established but noninvasive alien grasses. Similarly, Ricciardi and Atkinson (2004) reported that phylogenetic "distinctiveness" augments the impact of invaders in aquatic systems: The highest impact invaders are most often those that belong to genera not already present in the community. Thus, the structure of niche opportunity may be especially important in facilitating the transition from established colonist to competitively superior invader. Parker and associates (2006b) emphasized that such results demonstrate the power of using an evolutionary approach in invasion biology, but also illustrate the need for a better understanding of the mechanisms by which phylogeny influences interactions between invaders and the native species they encounter.

A few studies have examined the mechanisms by which phylogenetic relatedness affects fitness of invasive species in their new habitats. For example, in a phylogenetically controlled experiment, Agrawal and Kotanen (2003) compared the impact of native herbivores on native versus introduced congeneric herbaceous plants in northeastern North America. They found that the alien species suffered significantly greater herbivory than did natives, contrary to the predictions of the enemy-release hypothesis. However, the authors noted that the introduced plants have been in the region for at least 200 years, such that the relationships observed currently may not reflect those that attended the original invasion (for example, substantial evolution may have taken place since introduction; discussed in Carroll and coworkers [2005]). Furthermore, it is not known whether the same herbivores were attacking native versus invasive congeners, or even whether the herbivores were largely native or introduced insects.

With a meta-analysis of published herbivory and plant survival data, Parker and colleagues (2006a) assessed relative risk with respect to the origins of both the plants and the herbivores. They found that native herbivores tended to suppress introduced plants, whereas introduced herbivores attacked primarily native plants and thus promoted abundance and diversity of invasive plants. They concluded that colonizing plants in general are at risk from novel *generalist* herbivores to which they lack specific adaptation. Thus, although native herbivores may provide biotic resistance to plant invasions, the ongoing accumulation of alien herbivores substantially compromises indigenous biological control, probably through a variety of direct and indirect effects on native herbivore and plant populations.

Ricciardi and Ward (2006) further analyzed the data compiled by Parker and colleagues (2006a) and showed that the suppression of alien plants by native herbivores was greatly reduced in aliens with close relatives in the native plant community. The implication is that they share resistance traits with their relatives. This is not inconsistent with the conclusions of Parker and colleagues (2006a), but suggests that in contemporary communities the interactions within new mixes of producers and consumers simultaneously generate several types of strong and significant effects. Moreover, both analyses suggest that the enemy-release hypothesis may be inadequate because it emphasizes the importance of escape from coadapted herbivores, whereas generalist herbivores may be more important sources of herbivory and mortality.

Phylogenetic analyses of the patterns and probabilities have recently been extended further up the scales of evolutionary time and biotic organization. Mattson and coworkers (2007) suggested that difference in the geological/climatic histories in different regions of the Holarctic realm may have created a global gradient in the susceptibility of different continents to invasion by forest. Far fewer North American natives have invaded European forests than the reverse. The authors argue that this may be because more frequent cycles of natural disturbance in Europe, related mainly to climate history, have diminished biotic heterogeneity in ways that simultaneously decrease niche opportunity for colonists and increase biotic resistance. This conclusion is consistent with the theoretical prediction of Melbourne and colleagues (2007) that environmental heterogeneity and "invasibility" should be positively correlated.

The Genetics of Invasiveness

Traversing the scale of time and biotic events to the other extreme, pioneering workers are attempting to discover the genetic bases of invasiveness through quantitative and molecular genetic analyses (Weinig et al., 2007). Such a pursuit is likely to suffer from the same limitations as efforts to define invasive phenotypes, and researchers do not anticipate the discovery of "invasiveness genes" common to numerous taxa. Instead, it is likely that a wide diversity of genes underlies invasiveness, and that the effect of specific genes on invasiveness will be context specific—in other words, which genes influence invasiveness will depend on the hurdles that a species needs to overcome in its new community. Nonetheless, some advances have been made in understanding the genetics of invasiveness. For example, the major genes affecting vegetative versus sexual reproduction and dispersal, a trait that affects weediness, have been localized in sorghum (Paterson et al., 1995). Likewise, multiple studies of *Arabidopsis* have identified specific genes that affect responses to competitors, and competitive ability (Weinig et al., 2007). However, the degree to which any of these genes affect potential for invasiveness is unclear. Also, a particular genes *has* been implicated in promoting invasiveness in fire ants (*Solenopsis invicta*) in the United States (described in the next section), and new analyses will likely identify similar genetic elements in other taxonomic groups.

Hybridization of native and introduced species may be particularly important in allowing taxa to share attributes that result in the development of *superinvaders* (Ellstrand & Schierenbeck, 2006; Rhymer, this volume). Hybridization generates novel genotypes. Although the large majority of the new genotypes produced may be poorly adapted to the environment, "a minority of them may represent better adaptations to certain environments than do any of the genotypes present in the parental species populations" (Stebbins, 1969, p. 26). Hybridization also increases heterozygosity, which can lead to heterosis (hybrid vigor). Mechanisms that fix heterotic genotypes (such as allopolyploidy, permanent translocation heterozygosity, asexual production of embryos [agamospermy], or clonal reproduction) can preserve heterotic genotypes in a population (Ellstrand & Schierenbeck, 2006). In their survey of the relationship between hybridization and invasiveness, Ellstrand and Schierenbeck (2006) found that the majority of invasive species known to be derived from hybridization were capable of fixing heterotic genotypes.

Understanding the genetics underlying invasiveness—both the role of specific genes and the creation of new genetic variation via hybridization—in the context of specific evolutionary hypotheses may assist in the long-term management and control of deleterious traits in evolving populations.

Time Lags on the Path to Invasion

> *Lighten any check, mitigate the destruction ever so little, and the number of the species will almost instantaneously increase to any amount.... striking is the evidence from our domestic animals of many kinds which have run wild in several parts of the world.... The obvious explanation is that the conditions of life have been very favourable, and that there has consequently been less destruction of the old and young, and that nearly all the young have been enabled to breed. In such cases the geometrical ratio of increase, the result of which never fails to be surprising, simply*

> *explains the extraordinarily rapid increase and wide diffusion of naturalised productions in their new homes.*
>
> Darwin (1859, p. 64–66)

The progression from immigrant to invader often involves an initial lag, eventually followed by a period of rapid increase until the species ultimately reaches the bounds of its new range. For example, the Brazilian pepper tree *(Schinus terebinthifolius)* was introduced to Florida about 100 years ago, but it did not become widely apparent in the flora until the 1960s. It now inhabits almost 300,000 ha in South Florida, often in stands so dense as to exclude all other vegetation (Williams et al., 2007). Kowarik (1995) reviewed lag times for invasive plants in Europe and found the average "sleeper" period to be more than 150 years! Although there are many examples of such lags preceding invasions (see, for example, Mack et al., 2000), few studies have shown what factors underlie the transition.

Delays between introduction and invasion may have multiple causes, the simplest being the stochastic vagaries of multiplication for small initial populations of colonists. Such stochastic suppression will only be worsened if strong selection winnows all but the most adapted individuals from the breeding population. When colonization is by a small or closely interrelated founding population, genetic variance may be greatly reduced from that of the parental population (see, for example, Gilchrist & Lee, 2007), limiting prospects for evolution by natural selection. Even though population bottlenecks may also result in genetic reorganization that expresses new genetic variation, much of that variation may not be adaptive (Carroll & Watters, this volume). Even when some of the variation facilitates persistence, refining a new course of adaptation may require generations of further genetic compensation (sensu Grether, 2005) before selection among colonizing genotypes produces traits that vault a population out of its suppressed state by, for example, resolving a particular source of mortality at a critical phase in the life cycle.

The frequent failure of even intentional species introductions suggests a cost of maladaptation—most alien species are unsuited to their new environments or communities and either go extinct or remain at low population sizes. It is therefore likely that among successful introductions are populations that persist only through their abilities to respond adaptively to their new circumstances. Both phenotypic plasticity and genetically based evolution may be important in responses to sudden environmental changes. In particular, phenotypic plasticity may permit populations to persist long enough for novel genetic variation to arise (for example, via mutation or recombination) or until selection can sort among the variation already present in the population. Phenotypic plasticity may also allow production of a broad enough range of phenotypes that populations can bridge *adaptive valleys* (evolutionary intermediates of low fitness) that would, in the absence of plasticity, prevent local adaptation of alien species and thus prevent invasion (see for example, Carroll, 2008; Carroll & Watters, this volume; Ghalambor et al., 2007; Price et al., 2003).

Regardless of why alien species often show time lags before becoming invasive, time delays before invasion have several important implications. First, predicting which immigrants will become invasive and which will remain rare, or will simply disappear, will be difficult to judge based on demographic measures taken over any brief period of time. Second, attempting to measure directly the factors that catalyze the transition to invasiveness in nature will require luck, patience, or a resort to indirect methods. Third, throughout the decades during which a recently resident population is comparatively quiescent, a great many ecological and evolutionary changes may take place. Because of this, the deme that ultimately invades may differ substantially in both its constitution, and the environmental challenges and opportunities it meets, from its colonizing ancestors. This possibility reduces certainty about some of the conclusions we might reach when we attempt to analyze the transition to invasiveness. Knowing the phylogenetic history, genetics, and the environmental history will all be useful.

Colonization as an Evolutionary Event

Although reduced genetic variation in founding populations may often reduce the rate at which colonists respond to selection, colonizations may in themselves be evolutionary events that promote invasiveness. In such cases, lags on the path to invasiveness may be brief, not as a result of preadaptive functionality of traits also favored in the natal environment, but as a result of accidental genetic changes occurring during colonization. The two most environmentally and economically important invasive ants in North America, the Argentine ant (*Linepthema humile*) and the imported red fire ant, exemplify this. Like some invasive plants, these ants have reduced or eliminated many native species as they have spread since their initial introductions to the southern United States from South America about 100 years ago. Both invasions have proceeded as a result of genetically based changes in polygyny (reviewed by Tsutsui & Suarez, 2003). Ants have eusocial breeding systems, and in their case polygyny means having multiple queens per colony. Many invasive ant species form polygynous *supercolonies*. Polygyny is a derived condition in the invasive North American populations of these ants that may have facilitated their invasion, but it appears to have arisen in very different ways in the two species.

In the case of the Argentine ant, population bottlenecks and founder effects at introduction have reduced genetic diversity and increased the genetic similarity of descendant populations. A single, genetically homogenous supercolony of Argentine ants occupies virtually the entire Californian range. This supercolony has only about 50% of the alleles and one third the expected heterozygosity of populations in the native range, where populations have a genetic structure over tens to hundreds of meters, attended by substantial intercolony aggression (Tsutsui & Suarez, 2003). Inherent to their colonization of North America, then, is that genetic similarity and relatedness became decoupled. Descended from a genetically segregated condition in which relatives and nonrelatives were closely discriminated, cooperative behaviors are now displayed toward individuals who are genetically similar but distantly related. Extreme unicoloniality appears to have arisen during or shortly after introduction. Experiments suggest that the loss of intraspecific aggression in introduced populations, resulting in the "endless colony," underlies the ability of Argentine ants to displace native ants via numerical superiority (Tsutsui & Suarez, 2003).

It is important to remember that the genetic change behind the social transition that facilitated invasion was not the product of selection, but instead was the consequence of loss of variation associated with initial colonization. The phenotypic response to that change, although maladaptive under the species' former circumstances, is the adaptation that appears key to invasion success. This adaptation is a manifestation of phenotypic plasticity, by chance beneficial, induced by the founder-effect evolution of diminished genetic variation.

The imported red fire ant (*S. invicta*) also lost genetic diversity during introduction. Although not more homozygous on average, introduced populations have only 50% of the alleles present in native populations. Both monogyne (single queen) and polygyne forms occur in the native and introduced ranges. In the United States, polygyny either arose secondarily from the monogyne or is the result of another introduction about 20 years subsequent to the first. Whatever its origin, the polygynous form is more ecologically destructive than the monogynous form, displacing both native ant species and the monogynous form (Tsutsui & Suarez, 2003). Because of its high heterozygosity, the ability to distinguish relative relatedness should not be an issue for fire ant workers. However, queens from the two forms (monogyne and polygyne) typically possess different genotypes at the general protein-9 locus. This locus (or perhaps loci in close proximity) appears to govern faculties that discriminate relatedness. The North American polygynous genotype fails to discriminate against nonrelatives in a manner analogous to that observed in North American Argentine ants (Tsutsui & Suarez, 2003). As long as the numerical superiority of large colony size is selectively advantageous, outweighing any cost of intraspecific competition, the "cooperative allele" that facilitates invasion should prevail. These conditions will probably persist as long as the population continues to expand into underexploited habitats.

The Natives are Restless: Evolution in Response to Invasion

> *What havoc the introduction of any new beast of prey must cause in a country, before the instincts of the indigenous inhabitants have become adapted to the stranger's craft or power.*
>
> Darwin (1860), speculating based on his observations of the remarkable tameness of Galapagos vertebrates

In the preceding sections we discussed how introduced taxa overcome novel challenges through adaptive evolution. This adaptive evolution frequently takes substantial time, likely contributing to observed time lags between colonization and invasion of alien species. Sometimes, however, adaptive evolution occurs remarkably quickly (Lee, 2002). Whether invasive species evolve significantly before or during expansion into their new habitat, their success imposes a cost on many native species and provides benefits to others. The interactions may be direct—such as an invasive species preying on a native—or indirect, such as by altering the outcome of competition for resources. Invasive species may broadly alter ecosystem properties such as biogeochemical cycles and hydrology, changing conditions throughout ecosystems (Strayer et al., 2006). Changes in the community and ecosystem wrought by introduced taxa may be a potent source of selection on native species (Carroll 2007a, b, 2008; reviewed by Strauss, 2006a). Some of these responses will be largely demographic, but many will be evolutionary (Strauss et al., 2006a). When an alien taxon is sufficiently established to exert selective force, mutual selective shaping of existing phenotypic variation can occur among interacting taxa. Hence, it is possible that, during the next few decades, escalating habitat alteration by alien organisms will result in species that, although remaining in their native locales, evolve into organisms quite different from their current states.

Strauss and colleagues (2006a) identified more than 30 published cases of adaptive evolution in response to the ecological effects of introduced species. For example, many studies have now documented the evolution of competitive ability in native animals and plants living in communities invaded by aliens (see, for example, Calloway et al., 2005b; reviewed by Strauss et al., 2006a), although others have failed to detect evolution in natives (Lau, 2006). However, the best-studied cases are those in which native phytophagous insects have colonized alien hosts with known introduction times. These studies provide some of the most completely documented evidence of recent and ongoing evolution in response to invasion (and, similarly, agriculture). When it is possible to compare directly populations on a new host plant with those remaining on the original hosts, we can then test hypotheses about the direction, rate, and sometimes the genetic basis of adaptive evolution. During the past two centuries, and in some cases the past few decades (tens to hundreds of generations), host shifts have led to the evolution of functionally distinct ecotypes, subspecies, and even species (reviewed by Strauss et al., 2006a).

For example, to look at performance evolution in response to a new host, we have cross-reared and hybridized races of Florida soapberry bugs (*Jadera haematoloma*) that occur on the native balloon vine (*Cardiospermum corindum*) and the phylogenetically related Asian flamegold (or "goldenrain") tree (*Koelreuteria elegans*), the latter being an ornamental commonly planted beginning about 50 years ago (Carroll, 2007a; Carroll & Boyd, 1992; Carroll et al., 1997, 1998, 2001). Flamegold differs from the native in fruit size, seed nutritional quality, and seed availability. Adults of the contemporary balloon vine race closely resemble museum specimens of bugs collected prior to the introduction of *K. elegans* (Carroll & Boyd, 1992), suggesting that they retain the ancestral condition. Flamegold trees are potentially serious environmental weeds in Florida. From the bugs' standpoint, their seeds are an abundant new resource, and we predicted that the plant's differences from the native host would favor changes in a several of the insects' traits that relate to host utilization.

In rearing contemporary bugs from the native host on seeds of the introduced plant, our idea is to recreate how early colonists responded to the new host five decades ago, providing a baseline for comparing how much contemporary bugs in the derived population have changed over about 100 generations (or fewer, depending on how early most of the change has occurred). During this period, lifetime fecundity has nearly doubled, the bugs mature 25% faster, and they are 20% more likely to survive the juvenile period. The length of the mouthparts has evolved from an average of 9.3 mm to 6.9 mm in response to the smaller fruit of the invasive host, which is now preferred almost two to one in choice tests. The population frequencies of flying and flightless morphs have changed a great deal, as has the genetic control underlying the flight polymorphism. The transformation in beak length in these (and other) populations is evidenced by historical series of museum specimens (Carroll & Boyd, 1992; Carroll et al., 1997, 1998, 2003a, b, 2005). Although some of this adaptation has been facilitated by adaptive phenotypic plasticity, the majority has depended on evolved, genetic change (Carroll, 2007a). At the same time, pleiotropic loss of performance on native hosts has evolved with similar speed and often in a symmetrical manner (Carroll et al., 2001).

In a telling turnabout, New World balloon vine has become a serious invasive species in eastern

Australia during the past 80 years. A native soapberry bug on that continent has colonized it as a new host, and is in the process is evolving a longer beak (Carroll et al., 2005). We compared the efficiency with which the derived, longer beaked bugs attack the seeds of the invasive plant in comparison with the bug population still using a co-occurring native host. The derived bugs damage the seeds of the introduced host at almost twice the rate. Thus, one community-level impact of morphological evolution in response to invasion is the evolution of biological control value. Whether that control will be strong enough to select for counteradaptation in the invader is yet to be determined.

In addition to the adaptive responses of native herbivores on novel resources, the other common association is that of native aquatic species responding to novel predation risks from introduced predators. Common antipredator adaptations include morphological and behavioral changes that reduce the probability of mortality. Here, questions of how evolution interacts with prey population dynamics become especially important, because declining populations will often have reduced adaptive potential, and strong selection may lead to extinction before adaptive rescue is possible (Gomulkiewicz & Holt, 1995).

Although yet to be documented empirically, adaptations to invaders must in many cases alter the selective environments that invaders experience, resulting in reciprocal evolution. Through the (co)evolutionary responses to these interactions between invaders and natives, invasive species may gradually become "integrated" into their new biotic communities (Carroll & Watters, this volume; Vermeij, 1996), both becoming less invasive and having less impact on their new communities. The impacts of the invader, and the responses of natives and of other invaders, over both the short and long term, will determine the configuration and reconfiguration of biotic communities into the future.

FUTURE DIRECTIONS

Coevolution and the Future of Biotic Communities

> *. . . several hundred square miles are covered by one mass of these prickly plants, and are impenetrable by man or beast. . . . nothing else can now live. . . . I doubt whether any case is on record of an invasion on so grand a scale of one plant over the aborigines.*
>
> Darwin (1860, p. 120), referring to the naturalization of the Mediterranean giant thistle, "cardoon" (*Cynara cardunculus*), in Uruguay

Interactions with new enemies, mutualists, and competitors comprise the new biotic environment of an introduced species, and influence its success and impact (Mitchell et al., 2006). As exotics have expanded and altered native systems around the world, conservation biologists have focused on the ecological causes and consequences of invasions (Callaway & Maron, 2006; Hufbauer & Torchin, 2007). However, invaders and natives both evolve in response to invasion, and influence the evolution of one another (Strauss et al., 2006a; Zangerl & Berenbaum, 2005). Accordingly, ecological and evolutionary processes must be considered together. Moreover, the dynamics of such interactions will likely change over time and space, and will influence additional community members both directly and indirectly.

Placing the current state of an invasion in the context of its history and time line is basic to modeling its ecological and evolutionary dynamics. To understand the keys that release an established taxon onto an invasive trajectory, Facon and colleagues (2006) framed the problem as follows: Has the invaded environment changed in a way that might favor the alien? Has the alien evolved? What is the geographic and chronological history of introductions and any subsequent spread? When did invasiveness appear in relation to any such events? The more detailed the historical information, the better the chance of constructing a realistic model of an invasion. However, collection of the pertinent data will often be only haphazard, particularly from the time before the invasion was recognized. Nonetheless, scientific records and collections sometimes preserve historical information (see, for example, Carroll et al., 2005; Phillips & Shine, 2004). In addition, inferences from comparative, experimental, and phylogenetic methods may help to fill in the gaps.

For example, biogeographic comparisons of the performance of aliens in their invaded and indigenous ranges is a fundamental experimental design for understanding the extent to which invaders have changed plastically, genetically, or both. How do the causes of demographic variation in introduced populations compare with those in the native range (Hufbauer & Torchin, 2007)? To what extent does invasion success depend on particular qualities of an alien taxon, qualities of the invaded communities, and, especially, their interaction? As in the example of the soapberry bugs, rearing ancestral genotypes in the new environment generates the baseline phenotypes (1) to ascertain the importance of plasticity in initial adaptation and (2) to provide a basis for comparing the derived phenotypes to measure the evolutionary path they have followed. Data from the derived population in the original environment may reveal evolved loss of performance that has evolved pleiotropically as part of the response to selection in the new environment. Such data may provide insight into the phenotypic and environmental factors that promote invasiveness, and may reveal performance trade-offs and vulnerabilities that may be exploited by management practice. Lastly, hybridization of ancestral and derived populations provides information about the genetic structure of adaptive evolution, a question for which empirical information is still rare.

Analogous to the manner in which neurological damage from cerebral hemorrhages in humans has helped reveal the integrated structure and multifaceted recovery potential of the brain, ecological damage from invasions has helped illustrate the evolutionary dependence of ecological responses by providing accidental experiments that, if intentional, would be unethical. How adaptation influences the long-term effects of introduced taxa on the persistence of populations in invaded communities is just now being considered (Callaway & Maron, 2006; Hufbauer & Torchin, 2007; Kinnison & Hairston, 2007; Strayer et al., 2006). The relevant processes may be termed *eco-evolutionary* (Kinnison & Hairston, 2007). Understanding and predicting how eco-evolutionary processes will determine the structure and dynamics of invaded communities is the next big challenge for the field of invasion biology.

Biological invasions progress through phases of transport, establishment, and spread (Sakai et al., 2001). Ecological and evolutionary dynamics are likely to be relatively more or less significant at different stages of invasion, although currently we have too little understanding to generalize. Although events at any latter phase will likely be influenced by occurrences in an earlier phase, different, complementary approaches may be required to investigate each. Regarding the transport phase, for example, knowledge of the behavioral ecology of ant species—mating systems, colony size, and organizational flexibility—helped to predict emigration probability (Tsutsui & Suarez, 2003). Regarding establishment, information about the phylogeny of actual or potential invaders may yield clues about likely impacts as well as vulnerabilities that might be exploited for control. Phylogenetic proximity may reduce biotic resistance to establishment in alien plants (Ricciardi & Ward, 2006), for example, and evolution from selection during colonization may promote persistence (Carroll & Dingle, 1996; Quinn et al., 2001). Relative phylogenetic distance may promote invasiveness and spread in taxa that do establish (Strauss et al., 2006b). Unfortunately, we have a poor understanding of how to use phylogeny to predict when human intervention could prevent, or at least mediate, a pending species invasion.

The niche of an alien colonist will probably almost always differ from that of its progenitors, in part because of differences in the biotic community. If deleterious influences of natural enemies are reduced, increased performance may result in the invasive species occupying realms previously regarded as outside of the physiological tolerance of the species (see, for example, Holt et al., 2005). New evolutionary dynamics will stem from such niche shifts. For example, a leading hypothesis for the microevolutionary basis of invasion success is the evolution of increased competitive ability. This hypothesis is based on the assumption that there is an allocation trade-off between the ability to compete for resources and the ability to defend against enemies. When an alien colonizes an environment in which enemies are reduced or absent, selection should favor phenotypes that shift resources away from defense and to competitive ability (Blossey & Nötzold, 1995). This prediction has been borne out in several studies (see, for example, Siemann & Rogers, 2003; Zangerl & Berenbaum, 2005). Allocation constraints are similarly invoked in the suggestion that disturbed communities are more susceptible to subsequent invasion because they have become populated by residents with more disturbance-tolerant, but less competitive, phenotypes (Kneitel & Perrault, 2006).

Microevolutionary dynamics should be most important in the transition to invasiveness, for which adaptive changes may be key, and also as part of the longer term integration of aliens into their new communities. As invaders outcompete natives and spread, invasives may alter the ecological conditions experienced by many native taxa. In addition, they may represent an uncontested resource that some natives may be selected to exploit. Understanding how simultaneous ecological and evolutionary processes may interact, and assessing their relative demographic importance, has received recent theoretical treatment by Hairston and colleagues (2005). They propose a quantitative means of assessing concurrent rates of evolutionary and ecological change in a population, and of measuring the direct contribution of evolution to ecological change. For a particular population attribute of interest (for example, population growth rate or equilibrium population density), time–series data may be used to assess the absolute and relative importance of ecological and evolutionary factors to that attribute through time. As an example, they modeled an evolving population of Darwin's finches. Their year-to-year analyses of population growth rate showed that microevolutionary changes had twice the impact of contemporaneous substantial ecological change—namely, the amount of rain that fell.

Accordingly, models to predict the spread of invasive species will be more effective if they are sensitive to evolution of the niche. Based on studies of the spread of the marine (cane) toad (*Chaunus marinus*) into broad areas of Australia, Urban and colleagues (2007) showed how incorporating the anuran's changing niche better describes its pattern of invasion. Cane toads have expanded into regions of Australia originally regarded as unsuitable based on the climates it inhabits in its native New World range. It is possible that some of this expansion may have been permitted by reduced stress through emancipation from biotic enemies. However, the toads are evolving longer hopping limbs at the invasion front (Phillips et al., 2006), suggesting continuing evolution of invasiveness. Moreover, the pace of the invasion into challenging climatic realms is actually accelerating, leading Urban and colleagues (2007) to infer that tolerance to abiotic physiological stress is also evolving. They speculate that the huge size to which the population has grown during the invasion has provided more opportunity for beneficial mutations to arise. Increments in the rate of spread subsequent to the appearance of each beneficial mutation would create a sequence of relative lags, each transitioning into periods of greater invasiveness for the population as a whole.

From a practical standpoint, although such eco-evolutionary factors present obvious challenges to conservation biology, they may also offer opportunities to manage and craft population and community dynamics (Carroll & Watters, this volume). At one extreme, a significant proportion of economically and environmentally deleterious bioinvasions are already regarded as lost causes, because they have escaped the phase during which direct human intervention might have offered control. Although most species may remain permanently in their new realms, as they inevitably become integrated into their new communities, certain forms of control will appear. Beyond efforts at classic biological control, in which natural enemies are imported, scientists may also exploit means of *adaptive biological control*, in which the adaptation of native species to exploit aliens is enhanced through genetic or environmental manipulation (Carroll, 2007a). Because evolutionary change may typify invasions, even in cases when aliens and natives readily coexist (Lau, 2006), it is important to consider means of managing that evolution to achieve desired demographic outcomes. This may often mean the acceptance of permanently altered communities, because some species of conservation concern have already been shown to depend on the habitats now provided by invasive taxa (see, for example, Malakoff, 1999). Although reduced local biodiversity and biotic homogenization may be the outcome in many instances of biological invasion, the longer term impacts are still poorly studied and understood (McKinney & Lockwood, 2005).

Moreover, some invasions generate the evolution of additional biotic diversity (see, for example, Carroll et al., 2007; Malausa, 2005; Schwarz et al., 2005). In addition, information gleaned from the study of invasive species has direct relevance to the restoration of threatened native species. The challenge of decimated native communities is how to bring small, vulnerable populations to self-sustaining levels. If we can maintain such populations through enough generations for adaptive processes to occur, populations may recover without continued human intervention. During such a lag phase, creative genetic and environmental management, based on tools and insights gained from the study of the very invaders that, in some cases, are threatening the populations of natives, may be key to the restoration of endangered species (Carroll & Watters, this volume).

CONCLUSIONS

At all stages of biological invasions, from colonization of new environments, through population expansion to eventual integration of alien species into communities, evolutionary processes act simultaneously and interactively with ecological processes to mold responses of invaders and the invaded. However, our understanding of the role of evolution, and its interaction with ecological processes (eco-evolutionary dynamics), in species invasions is embryonic. For example, although the requirement for adaptive evolutionary change has been implicated in explaining time lags between colonization and invasion, we have little sense of why some species overcome this evolutionary hurdle when others do not. Likewise, although both loss of genetic variation (for example, in ants) and increase in genetic variation (through hybridization) that accompany or quickly follow colonization may influence invasions, we still know little about the genetics that underlie invasiveness.

An evolutionary approach to invasion biology offers insights both to predict biological invasions and to manage those invasions. Unfortunately, application of eco-evolutionary theory to those challenges is limited by the infancy of the field. New phylogenetic analyses have been effective in demonstrating how relatedness affects both invasiveness of species and invasibility of communities, for example, but research is still limited to a few examples and is not yet generalizable. We cannot yet predict invasions a priori. Moreover, even though adaptation of native species to invasive alien species, and vice versa, is well documented, the long-term ecological consequences of their coevolutionary interactions have barely been addressed.

The accidental experiments created by contemporary species introductions may offer the best context in which to study ongoing eco-evolutionary processes. This intersection of evolutionary and ecological research is beginning to draw the attention of a wide diversity of biologists. It nonetheless remains indefensible that, although providing data of unparalleled value, the individuals of a great many species will continue to be the hapless, unwitting victims of our own calamitous, unwitting behavior. A central hope is that we may use our nascent capacities to recognize our impacts and control our reproduction, and through the lens of science, perceive ecologically and evolutionarily sustainable means of sharing the earth, and from that process gain an understanding of our own heritage.

SUGGESTIONS FOR FURTHER READING

Mack and colleagues (2000) provide a well-organized and broadly ranging analysis of the causes and consequences of bioinvasions. Up-to-date papers that discuss many dimensions of the ecological, evolutionary, and conservation issues of invasions are provided by Sax and associates (2005) and Nentwig (2007). Strauss and coworkers (2006a) review the ecological lessons that can be learned from the study of invasive species, and responses of natives to those invasives. Thompson (2005) offers an intricate and stimulating perspective on coevolution as an ecological process.

Mack, R. N., D. Simberloff, W. M. Lonsdale, et al. 2000. Biotic invasions: Causes, epidemiology, global consequences, and control. Ecol Appl. 10: 689–710.

Nentwig, W. (ed.). 2007. Biological invasions. Springer Verlag, Berlin.

Sax, D. F., J. J. Stachowicz, & S. D. Gaines (eds.). 2005. Species invasions: Insights into ecology, evolution and biogeography. Sinauer Associates, Sunderland, Mass.

Strauss, S. Y., J. A. Lau, & S. P. Carroll. 2006a. Evolutionary responses of natives to introduced species: What do introductions tell us about natural communities? Ecol Lett. 9: 357–374.

Thompson, J. N. 2005. The geographic mosaic of coevolution. University of Chicago Press, Chicago, Ill.

Acknowledgments We thank Andy Suarez, Sharon Strauss, and Gary Vermeij for discussing some of the ideas presented herein; and the Institute for Contemporary Evolution and the Australian-American Fulbright Commission for financial support of S. Carroll.

16

Ecosystem Recovery: Lessons from the Past

GEERAT J. VERMEIJ

Few phenomena in the realm of ecology are more maligned and despised than extinction and invasion. The notorious human contribution to both species loss and species introduction has led to the view that these phenomena are unnatural and almost wholly destructive. The study of the fossil record of life, however, holds a more hopeful message. Not only does it show that extinction and invasion have been part of the phenomenology of the biosphere for hundreds of millions, if not billions, of years, it also chronicles the essential role that invasion plays in enabling communities, ecosystems, and the biosphere as a whole to recover from episodes of species loss. I argue in this chapter that the destructive effects of extinction would linger far longer if recovery depended solely on in situ survival and evolution. The "good" process of recovery is, in other words, inextricably linked with the "bad" phenomenon of invasion.

At the core of my argument lies a simple observation: Over time, the limits of distribution of all species change. This biogeographic fact is a special case of a rule that governs all things everywhere: "The phenomenon of change [has] been the hallmark in the origin, development, and fate of all structures, living or nonliving" (Chaisson, 2005, p. 129). This truism would seem unworthy of elaboration were it not for the common assumption by ecologists that, in the absence of human interference, species have more or less fixed distributions. Abetted by the short timescales over which the dynamics of ecosystems have been studied, ecologists have underappreciated the role that species expansion plays in the reassembly of communities after crises.

Community reassembly is a complex ecological and evolutionary process in which survivors and newcomers accommodate and adapt to each other to forge a sustainable economy, which transforms and distributes resources to all its members. A full understanding of reassembly requires knowledge of who the survivors are, how and from where newcomers arrive, how functionally different components of communities and ecosystems interact as invasion and adaptation proceed, and what factors control the rates at which production, consumption, and other ecosystemwide processes are reestablished and stabilized. These questions can be answered only when systems are followed over stretches of time long enough to accommodate evolutionary adaptation.

Although we are still far from achieving such complete understanding, we have learned enough about recovery to hazard some tentative conclusions. Here I review what we know about rates of recovery, the pattern of diversity following extinction, the re-establishment of ecological function, the role of invasion, and the factors controlling which species invade and how invaders affect evolution.

THE RECOVERY OF DIVERSITY

Most paleontologists who have investigated post-extinction recovery have relied on diversity, or taxonomic richness, as their chief indicator. By plotting the number of taxa in successive time intervals, they have shown that recovery to pre-extinction levels of diversity takes anywhere from one million to as much as 10 million years, depending on the group under study and on the severity of the precipitating crisis. A rainforest plant assemblage living 1.4 million years after the end-Cretaceous mass extinction in Colorado displays diversity levels that likely exceeded pre-extinction values (Johnson & Ellis, 2002), whereas North American mammal communities did not contain large predators and herbivores until about 10 million years after the crisis, and these large animals were dwarfed by their late Cretaceous dinosaur equivalents (Janis, 2000; Van Valkenburgh, 1999). A similarly long interval was needed for plant communities in the northern Rocky Mountains and Great Plains of North America to reach pre-crisis levels of species richness (Wilf & Johnson, 2004; Wing & Harrington, 2001; Wing et al., 2005). It is likely that end-Paleocene warming, perhaps precipitated by massive releases of the greenhouse gas methane, contributed to this recovery at 56 to 55 million years ago. It is also clear that much of the increase in animal and plant diversity in North America at this time was contributed by invaders from Europe, Asia, and the tropics (Beard, 1998; Wing et al., 2005).

One interesting pattern that plots of diversity in relation to time reveal is that recovery consists of two distinct phases. Immediately after a mass extinction, preserved communities consist almost entirely of species that have survived the crisis; this is the survival phase of recovery. This phase is followed by a renewal stage, during which new species—either the products of in situ evolution or newly established invaders—appear (Harries et al., 1996). Taxa that survive during the first phase often disappear at the beginning of or during the second stage of recovery. This pattern has been well documented for the Early Triassic recovery of communities following the end-Permian mass extinction some 250 million years ago. It applies to gastropods (Erwin, 1996; Payne, 2005), brachiopods (Chen et al., 2005a,b), and ammonoid cephalopods (McGowan, 2004a,b; Saunders et al., 2004; Villier & Korn, 2004). A similar two-phase recovery is evident for earliest Paleocene molluscs after the end-Cretaceous mass extinction (Hansen et al., 1993) and for early Oligocene molluscs after the less drastic crises of the late Eocene (Hickman, 2003; Squires, 2003). The survivors of the first phase have often been referred to as *disaster species* (Hansen et al., 1993; Harries et al., 1996; Percival & Fischer, 1977) because they undergo dramatic expansion after ecological collapse, but they are more charitably regarded as species poised to take quick advantage of ephemerally favorable circumstances much as "weeds" colonize highly disturbed habitats today.

THE ECOLOGY OF RECOVERY

Plots of diversity through time give a rough indication of recovery from extinction, but they fail to capture the ecological dimension of the process of reassembly. One can only learn so much by counting species as if they were inert particles. Species are not abstract entities; they are phylogenetic units whose individual members have adaptations, interactions, enemies, allies, and resource requirements. Community assembly involves not just the addition of species by evolution or invasion, but also the establishment and integration of interactions among species. Communities and ecosystems are economic structures that work because the various functional groups make for a sustainable system of producers and consumers that use and affect resources. In effect, recovery of a sustainable economy implies the transformation from a collection of independent species to a community in which species accommodate each other and their surroundings.

Our recent work on marine communities in Florida illustrates this point well. At the end of the Pliocene, about 1.7 million years ago, a substantial episode of extinction reduced the diversity of molluscan species in Florida by about 25%. Most of the losses were compensated by invasion of species from the adjacent Caribbean region (Vermeij, 2005c). Before the extinction, abundant bivalves in the *Chione elevata* lineage were often preyed upon by gastropods that attack their prey by drilling a hole at the junction between the two valves. This mode of predation is up to three times faster than drilling through one of the valves, and is used by living predators under conditions of intense intraspecific competition among drilling muricid gastropod predators. No edge drilling is known in any *Chione* populations in Florida after the end-Pliocene extinction. We interpret this pattern as indicating that the intensity of competition among predators has never again matched that prevailing during the Pliocene (Dietl et al., 2004).

Another indication that taxonomic diversity is a poor measure of recovery comes from the history of ammonoid cephalopods after the near extinction of this important group of marine predators at the end of the Permian. Two independent studies (McGowan, 2004a; Villier & Korn, 2004) show that the number of ammonoid genera increased much faster than the range of morphologies. This finding seems to imply that ecological roles, as indicated by morphology, were realized more slowly than species richness. Pre-extinction morphological diversity was not established among ammonoids until some seven million years after the crisis (McGowan, 2004b). The relative roles of in situ evolution and of invasion in this "filling of morphospace" have not been investigated.

The post-Cretaceous world presents several other examples that indicate that the abundance and the rate of production of photosynthesizers recover much more quickly than diversity, complex trophic interactions, and resource regulation. In the plankton communities of the open sea, recovery of phytoplankton populations (especially of rapidly reproducing Cyanobacteria) after the end-Cretaceous extinctions could have taken as little as a few months, but little of the carbon fixed by these primary producers reached the seafloor to nourish bottom-dwelling (benthic) communities, because the large consumers whose bodies and fecal pellets served as the primary vehicles for downward transport of organic matter had not yet reappeared (D'Hondt et al., 1998). It would take another three million years for this crucial link between the pelagic and benthic realms to become re-established (D'Hondt et al., 1998). Even within the plankton, complex interactions lagged far behind the recovery of phytoplanktic abundance. Planktonic foraminifera harboring carbon-fixing symbionts became extinct at the end of the Cretaceous and did not reevolve until 3.5 million years after the crisis (Norris, 1996).

Similarly, the terrestrial communities preserved in New Mexico reveal a very rapid rebound of abundant terrestrial vegetation—ferns at first, then later herbaceous angiosperms—with pre-extinction levels of productivity perhaps being achieved within a decade of the end-Cretaceous disaster (Beerling et al., 2001). Full recovery of ecological functions, however, required three million years or longer (Beerling et al., 2001). When Labandeira and colleagues (2002) investigated herbivore-induced leaf damage in fossil plants from the North American Great Plains, they found that the intensity of herbivory and the number of types of insect-induced damage dropped sharply as the result of the end-Cretaceous mass extinction. Pre-extinction levels were not achieved again until a full 10 million years after the crisis, at about the same time that large predatory and herbivorous mammals appeared in North America (Labandeira et al., 2002).

In their studies of shell-drilling predation in benthic assemblages of molluscs, Kelley and Hansen (1996) interpreted their data from the Gulf and Atlantic coastal plains of the United States as indicating an increase in the frequency of drilled bivalve shells during the post-Cretaceous recovery phase, as represented by the Brightseat Formation (early Paleocene) of Maryland (frequency of drilling, 0.327). Their data, however, actually show a sharp decrease in drilling frequency of bivalves beginning in the latest Cretaceous (Corsicana Formation in Texas, 0.055) and continuing into the early Paleocene (Kincaid Formation of Texas, 0.034). Cretaceous intensities of drilling had been matched or surpassed by mid-Paleocene time (Kelley & Hansen, 1996). To me, this pattern implies that drilling as an assemblage-wide form of predation suffered a brief reduction, beginning somewhat before the mass extinction and extending perhaps one million years after it. Three million years after the crisis, drilling had regained or surpassed its pre-crisis importance, implying that ecological recovery was well underway.

Recovery after the end-Permian extinction some 250 million years ago took even longer. Plants capable of forming peat deposits did not become abundant enough to produce preserved peat until 243 million years ago, some seven million years after the crisis (Retallack et al., 1996). Reef-like structures similarly took five to seven million years to become re-established (Erwin, 1996; Woods et al., 1999). The symbioses between dinoflagellates or other one-celled primary producers with colonial invertebrates that provide the foundation for complex reefs today and during much of the past 550 million years seem to have re-emerged only during the Carnian epoch of the early Late Triassic (Stanley & Swart, 1995). On land and in freshwater, the end-Permian extinction eliminated all vertebrates weighing more than 30 kg (Bakker, 1980; Benton et al., 2004) as well as small fish-eaters, insectivores, and medium-size herbivores (Benton et al., 2004). Even 15 million years after the crisis, small fish- and insect-eaters had not yet reappeared in the Ural Basin of Russia (Benton et al., 2004).

Why does recovery take so long? At least three contributing factors come to mind: (1) environmental instability or other inimical conditions after a crisis, (2) slow in situ evolution, and (3) lack of invaders. These factors likely all play a role during post-extinction recovery; moreover, they are complexly interconnected through feedbacks and should therefore not be viewed as independent components that can be statistically isolated.

It has long been appreciated that communities in early phases of recovery are transient and evidently unstable. Even if rates of production are high soon after a crisis, the regulatory effects on producers and on raw materials imposed by higher level consumers are severely compromised, potentially leading to intense, disruptive variation. Some of this variation, moreover, could be imposed by outside forces not directly caused by living organisms or by the absence of consumers. In the aftermath of the end-Permian extinctions, for example, the carbon-isotopic record reveals sharp fluctuations in the rate at which carbon was buried in sediments, and therefore potentially in the concentration of oxygen and of available nutrients (Payne et al., 2004). Diversity and ecosystem complexity began to increase only after these fluctuations ceased (Payne et al., 2004). The cause of these fluctuations remains contentious. Efficient means of recycling carbon, brought about by the collective action of consumers and producers in ecosystems, may have been effectively eliminated by the end-Permian convulsions, as indicated by the exceptional abundance of microbial mats in the earliest Triassic communities (Pruss et al., 2004). Similar microbial mats, which flourish in the absence of burrowers and grazers, have been documented in Early Silurian communities after episodes of mass extinction near the end of the Ordovician period about 440 million years ago (Sheehan & Harris, 2004).

In their analysis of the Early Triassic fluctuations, Payne and colleagues (2004) dismiss another obvious possibility: sudden, repeated, catastrophic releases of C^{12}-rich methane from deep, oxygen-poor sediments and waters in the ocean. The release of methane likely contributed to the end-Permian crisis (Berner, 2002; Vermeij & Dorritie, 1996). With so much methane liberated into the atmosphere and oxidized to CO_2, Payne and colleagues (2004) argue that there would be insufficient methane buildup in Triassic waters and sediments to create enough stored methane for additional releases large enough to disrupt the carbon cycle. If, however, the Early Triassic world was characterized by low oxygen concentrations in the atmosphere and oceans, as evidence from Early Triassic sediments amply indicates (Grice et al., 2005; Hallam, 1991; Kashiyama & Oji, 2004; Wignall & Twitchett, 1996), methane production by anaerobic bacteria would have been far more widespread than it is today or than it would be during other times when free oxygen was plentiful. Under such oxygen-poor conditions, methane could build up rapidly and could easily have been sufficient for changes in temperature or sea level to have brought about periodic, destabilizing releases of methane during the first few million years of the Triassic. The protracted Early Triassic recovery would then be explained by methane releases, exacerbated by the absence of effective means of distributing and recycling carbon and other essential resources.

I have argued elsewhere that extinction by itself is an insufficient trigger for evolution (Vermeij, 2004). The existence of a survival phase of recovery before in situ speciation takes place supports this hypothesis. In the aftermath of the end-Cretaceous crisis, it is not until the end-Paleocene and early-Eocene warming that pre-crisis community organization is reestablished in North America. Significant adaptive evolution, including escalation, appears to depend on an increasingly prolific, accessible, and predictable supply of resources. This supply relaxes constraints on adaptation, and therefore allows new, often initially highly imperfect adaptive syndromes to arise (Vermeij, 2004). Insofar as organisms have a hand in regulating, stabilizing, and enhancing resource supplies, these opportunities for erasing the effects of crisis depend on positive feedbacks between organisms and their environments.

RECOVERY AND REFUGES

If recovery through in situ evolution is a protracted process requiring millions of years and contingent on plentiful, accessible, and predictable resources, a potentially faster route to reassembly is by invasion of species from refuges. A refuge is a habitat or region where the agencies responsible for ecological collapse have little effect or are absent altogether. Because life persisted through all known mass extinctions, such refuges must have always existed.

Pinpointing the locations, characteristics, and taxonomic composition of refuges remains a major task for paleobiologists. Already it is clear, however, that colonization from such refuges has been important in the reassembly of ecosystems. In the low-diversity communities of the Early Silurian, after the end-Ordovician extinctions, tropical marine communities in what is now North America but what was then Laurentia were composed largely of immigrants from Baltica (Krug & Patzkowsky, 2004; Sheehan, 1975). In the generally oxygen-poor Early Triassic, a diverse marine fauna existed in well-oxygenated environments in Oman (Twitchett et al., 2004), which would well have acted as a point of origin for the founding taxa of later Triassic ecosystems throughout the world. The Paleocene marine faunas of the southeastern United States are composed almost entirely of immigrant taxa; similar conclusions apply to varying degrees to other warm-water post-Cretaceous marine communities (Jablonski, 1998). Late Eocene extinctions in the northeastern Pacific set the stage for subsequent invasion of species from the northwestern Pacific and the southern hemisphere (Hickman, 2003; Squires, 2003; Vermeij, 2001). In the Caribbean region, widespread extinctions during and near the end of the Pliocene were offset by colonizing taxa from West Africa and the Indo-Pacific region (Vermeij & Rosenberg, 1993). Contemporaneous species losses in Florida were followed by invasions from the Caribbean (Vermeij, 2005c). Middle- and late-Pliocene extinctions in the North Atlantic may have provided opportunities for immigration of hundreds of species from the North Pacific and for expansion of European species to eastern North America (Vermeij, 2005a, b).

In all these cases, refuges preserved many competitively vigorous species. Our evidence from the Pliocene indicates that the invaders typically exceeded the surviving species in areas affected by extinction in the expression of effective antipredatory characteristics (Vermeij, 2005a, c). This is the result of the selective extinction of well-defended species in the areas prone to extinction, as well as of the selective invasion of well-defended immigrants from refuges.

Detailed comparisons between the magnitude of extinction and the magnitude of invasion for each of three intervals of the Pliocene in the warm western Atlantic reveal that, although there is not a one-to-one correspondence between extinction and invasion, there is evidence that some invaders from the Caribbean were kept out of Florida as long as certain incumbent taxa were present. No tropical American cowries (Cypraeidae) established a beachhead in Florida while the highly diverse *Siphocypraea* group held sway there. After this endemic Floridian clade became extinct at the end of the Pliocene, three separate clades of cowries from the Caribbean region expanded their ranges to Florida (Vermeij, 2005c). Similarly, in a comparison of post-Cretaceous marine faunas from around the world, Jablonski (1998) was unable to detect a pattern of higher magnitude end-Cretaceous extinction being associated with more extensive post-crisis invasion, but he did emphasize the large role that invaders played in recovering marine communities of the Paleocene.

These findings highlight the role that incumbent species play in keeping out would-be immigrants. During normal times, most individuals and most species are well adapted to prevailing circumstances. Although they may not be optimally "designed," they are nonetheless likely to be better adapted than are most potential invaders. There is, in other words, a strong incumbency effect: Well-established species hold a distinct advantage over new arrivals (Rosenzweig & McCord, 1991). Only when incumbents disappear can the initially less well-adapted immigrants establish themselves in their place.

The imprecise correspondence between extinction of incumbents and the successful subsequent establishment of invaders raises important questions about the nature of incumbency. Are there incumbent species that are disproportionately effective at keeping out invaders? Are these effective incumbents more vulnerable than others to extinction during episodes of ecological collapse? For example, we might speculate that the removal of a top predator species provides opportunities for the invasion of many subordinate predators and many prey species. The extinction of a small herbivore or of a competitively subordinate plant, on the other hand, might not provide any real opening for a potential invader. These possibilities have not been investigated.

LONG-TERM TRENDS

Several authors have remarked that long-term patterns of escalation (enemy-directed evolution) are only briefly interrupted by mass extinctions (Jackson &, McKinney, 1990; Vermeij, 1987, 2004). Although highly escalated species—consumers as well as enemies—appear to be more vulnerable to extinction during crises than more subordinate taxa, species with high metabolic rates and rapid population growth have intrinsic advantages in recovering ecosystems. It is these species, which invade and evolve rapidly, that are most responsible for competition-driven escalation in the post-crisis world (Vermeij, 2004, 2005b).

The aftermath of the end-Cretaceous crisis provides several examples. One of several antipredatory defenses of sand- and mud-burrowing marine bivalves is deep burrowing depth. Lockwood (2004) showed that, although extinction at the end of the Cretaceous affected shallow- and deep-burrowing bivalves of the clade Veneroida about equally, new post-crisis veneroid genera belong disproportionately to the deep-burrowing category. In the southeastern United States, well-armored gastropods and bivalves are generally better represented in recovering Paleocene communities than in their Cretaceous counterparts (Hansen et al., 1999; Kelley & Hansen, 1996). Unfortunately, it is not clear in these cases whether the higher representation of well-defended taxa in Paleocene communities results from in situ evolution, invasion, or both. It will be necessary to analyze changes in range in a phylogenetic context to resolve this issue.

If one of the determinants of the rate of ecosystem recovery is the rate of arrival of species from elsewhere, a case can be made that recovery after ancient crises was slower than that after more recent collapses. The chief reason is that high dispersibility is a derived trait in most major clades of animals.

Among the major clades of bottom-dwelling marine invertebrates, dispersal by larval stages that feed in the plankton is always derived relative to low dispersibility by larvae that spend a short time in, and do not feed on, the plankton (Chaffee & Lindberg, 1986; Peterson, 2005; Signor & Vermeij, 1994). According to the fossil record and evidence from molecular sequence divergences, plankton feeding first appeared during the latest Neoproterozoic era, just before the Cambrian; several additional groups evolved larval plankton feeding around the Cambrian–Ordovician boundary. Dispersal over thousands of kilometers of open ocean, characteristic of so-called *teleplanic larvae*, is limited to some clades of late Mesozoic to Recent age, including tonnoidean gastropods, some echinoid and asteroid echinoderms, a few cowries, some neogastropods, and various crabs and lobsters; it also characterizes a number of advanced teleost fishes. Interestingly, these high-dispersal clades contain many major predators and herbivores with large per-capita and collective effects in the communities they inhabit as adults. Under the assumption that these teleplanic taxa can invade far-flung sites rapidly, high-level consumers have the potential to arrive early and to exercise substantial selection on many species from the earliest stages of recovery. This effect would have intensified during the course of the past 600 million years, and would have become particularly strong during the past 150 million years of earth history.

High-powered adults capable of extensive movement similarly occupy derived phylogenetic positions. This is notably the case for large Mesozoic marine reptiles; many Mesozoic and Cenozoic fishes including sharks, rays, tunas, billfishes, groupers, moray eels, and wrasses; Cenozoic marine birds and mammals; and especially Cenozoic migrating birds. As is the case for taxa with teleplanic larvae, these high-powered groups whose adults disperse far and wide contain major consumers with potentially large selective and ecological effects. I suggest that neither teleplanic larval nor high-powered adult dispersers were common or perhaps even present during the Paleozoic and early Mesozoic eras, and that consumers of large effect arrived in, and shaped the selective regime of, recovering ecosystems at later stages of reassembly during these times than in late Mesozoic and Cenozoic ecosystems.

Dispersal, then, contributes to recovery by spreading traits that confer power and competitive prowess. I suspect that it is this dispersal of biologically vigorous invaders that accounts in large part for the persistence of escalatory trends with hardly any interruption across extinction boundaries. In short, invasion is an integral and ultimately beneficial process in ecological recovery.

IMPLICATIONS FOR CONSERVATION

If the fossil record teaches one crucial lesson, it is that full recovery of communities takes a very long time, and that it requires both invasion and evolution. Recovery does not, of course, mean the re-creation of the communities and ecosystems existing before a crisis; it means the fashioning of entirely new systems from the remnants of old ones. The new systems will have many of the same characteristics as those prevailing before the crisis, but they will be different, and they will be a long time coming.

Organisms play a huge, and probably irreplaceable, role in regulating all surface processes on earth. Under normal conditions, this regulation is carried out in a fashion strikingly resembling the free market. When ecosystems collapse, so do the multiple checks and balances that have built up in sustainable ecosystems over millions of years. With our technological know-how, we humans can learn to regulate and manage ecosystems and the resources that sustain them, but our track record of top–down management and government is decidedly mixed. Given the protracted time of recovery, during which many ecosystem services will be disrupted and unpredictable, we would do well to prevent ecological collapse. Above all, we must protect the productive side of the natural economy and we must minimize our interference with the free market of trade that underlies geochemical cycling.

FUTURE DIRECTIONS

The foregoing speculative review raises important unanswered questions that I hope will guide future research. Among these questions are the following:

1. What are the relative contributions of in situ evolution and of invasion in community reassembly and recovery after major extinctions?
2. How do well-adapted incumbents in communities keep out invaders?
3. Are there types of incumbent species (such as top consumers or competitively dominant primary producers) whose losses would provide disproportionate opportunities for invading species?
4. Are the traits that enhance dispersibility and invasibility phylogenetically derived and, if so, have the speed of invasion and the role of invasion in recovery increased over time?
5. In the human-dominated biosphere, where transport of species by humans has become so important, will recovery from extinction be faster than it would have been without human interference?
6. Alternatively, are we destroying so many potential refuges for species that invasion-enhanced recovery will be less likely?

These and other questions underscore the necessity of treating such phenomena as extinction, invasion, and recovery in an ecological context. We must engage in studies knowing that species are functionally, ecologically, and phylogenetically heterogeneous; they are not interchangeable and they cannot be treated simply as characterless names. At a time when humans are placing enormous stresses on the earth's biosphere, it is more important than ever to gain a better understanding of all the phenomena that affect recovery.

SUGGESTIONS FOR FURTHER READING

Rosenzweig and McCord (1991) provide a highly readable paper that is one of the first to treat the phenomenon of evolutionary incumbency from both a theoretical and an empirical perspective. Although the authors did not consider recovery, the points they emphasize are clearly relevant to it. Jablonski (1998) notes that the rate of recovery varies geographically, and that much of the process depends on invading species. This is one of the few papers that deals with these ecological processes on a global scale. The work of Labandeira and colleagues (2002) is one of the finest examples of post-extinction recovery studied in an ecological context. It forms part of a series of papers by these three authors and their collaborators, who show that re-establishment of pre-extinction complexity of plant–insect relationships involves the arrival of important new community members from elsewhere.

Jablonski, D. 1998. Geographic variation in the molluscan recovery from the end-Cretaceous extinction. Science 279: 327–330.

Labandeira, C. C., K. R. Johnson, & P. Wilf. 2002. Impact of the terminal Cretaceous event on plant–insect associations. Proc Nat Acad Sci USA 99: 2061–2066.

Rosenzweig, M. L., & R. D. McCord. 1991. Incumbent replacement: Evidence of long-term evolutionary progress. Paleobiology 17: 202–213.

17

Host–Pathogen Evolution, Biodiversity, and Disease Risks for Natural Populations

SONIA ALTIZER
AMY B. PEDERSEN

Pathogens can play important roles in natural systems, from influencing host genetic diversity to altering the composition of ecological communities. Infectious diseases can also threaten natural populations (Harvell et al., 1999; Lafferty & Gerber, 2002). *Virgin ground* epidemics that quickly spread throughout previously unexposed plant and animal populations can cause high mortality and reductions in host abundance (see, for example, Anagnostakis, 1987; Osterhaus & Vedder, 1988; Sherald et al., 1996). Although documented cases of pathogen-driven host extinction are rare (Smith et al., 2006), several studies suggest that pathogens can cause declines in previously healthy populations and can be one of many threats to already declining species (Anderson et al., 2004; Daszak et al., 2000; Lafferty & Gerber, 2002; Pedersen et al., 2007). Recent examples include population crashes in African apes resulting from Ebola virus (Walsh et al., 2003), amphibian declines caused by chytridiomycosis (Lips et al., 2006), the near extinction of several Hawaiian forest birds as a result of avian malaria (Van Riper et al., 1986), recent declines in oaks in the western United States because of sudden oak death syndrome (Rizzo & Garbelotto, 2003), and widespread mortality among flowering dogwoods in eastern North America after the spread of anthracnose blight (Sherald et al., 1996).

The overarching theme of this chapter is that infectious diseases provide a model system to understand how evolutionary principles are relevant to biodiversity and conservation. Throughout we use the terms *parasite, pathogen,* and *infectious disease* interchangeably, and we consider both microparasites (viruses and bacteria) and macroparasites (protozoa, fungi, arthropods, and helminths) that can infect hosts and, in many cases, lower host fitness and cause outward signs of disease. Because of their potential for rapid evolution and impacts on host survival and reproduction, host–parasite interactions can generate a number of evolutionary outcomes, ranging from the maintenance of genetic variation to significant shifts in the genetic composition of both host and parasite populations. Because parasites, at times, can both contribute to and threaten biological diversity, understanding the evolutionary dynamics of host–parasite interactions is crucial for biological conservation.

In this chapter we begin by considering how host resistance and genetic diversity can help buffer wild populations against epidemics of new and existing pathogens. We discuss the genetic basis of host resistance, how parasite infection may maintain genetic diversity, and the importance of resistance variation for the conservation of threatened species. Next we demonstrate how evolutionary processes may influence pathogen emergence and host shifts. Parasites can evolve to capitalize on new transmission opportunities, alter their virulence, and adapt to novel host species or changing environments. From a conservation perspective, these issues can be important for captive breeding programs, control strategies for existing pathogens, and landscape-level management approaches. Recent and historical examples from wild systems are provided to illustrate key points, although we caution that they are not intended to represent an exhaustive list.

As a final point, parasites can influence host diversification and, more generally, are a major component of biodiversity themselves. Parasites that live uniquely on threatened host species could go extinct long before their hosts, and more accurate knowledge of parasite biodiversity will likely compound estimates of future biodiversity loss. We conclude by echoing recent assertions that interactive networks of host and parasite populations might be necessary to protect biological diversity and evolutionary processes (Crandall et al., 2000; Thompson, 2005).

HOST GENETIC DIVERSITY AND RESISTANCE TO INFECTION

Parasites represent powerful selective agents in natural populations, in part because they can spread rapidly and cause significant negative effects on host fitness. When exposure to pathogens is high in a host population, traits conferring resistance are predicted to increase in frequency. In animals, these resistance strategies include behavioral defenses to avoid exposure or physically remove parasites, physiological and innate responses to infection, and humoral and cell-mediated immune defenses (Clayton & Moore, 1997). Resistance mechanisms in plants include biochemical defenses, receptor–protein interactions, and changes in phenology that lower contact rates with infective stages (Fritz & Simms, 1992). These defenses can be *innate,* or maintained in the absence of infection (and hence a "first line of defense" after infection) or *adaptive,* or induced during the course of infection by a pathogen (for example, antibody-mediated immunity in vertebrate animals). Moreover, defenses can be highly *specific,* such that they recognize or defend against a particular type of pathogen or even a single pathogen genotype, or *general,* in that they attack a variety of infectious organisms.

Given the strong selective pressures imposed by parasites and the benefits of host resistance traits, it is important to investigate factors that maintain intra- and interpopulation variation in resistance to pathogens. In other words, why aren't all individuals resistant to infectious diseases? Models based on simple host–parasite interactions show that genetic variation can be maintained by at least three key mechanisms: frequency-dependent selection, balancing selection, or negative correlations between resistance and other fitness-conferring traits (Box 17.1) (Schmid-Hempel & Ebert, 2002). In addition, specific defenses against different parasites may act antagonistically, such that an immune response against one agent might suppress resistance to other infectious diseases (Yazdanbakhsh et al., 2002; see also Pedersen & Fenton, 2007).

The Genetics and Maintenance of Host Resistance

The genetics underlying variation in host immunity in wild populations has attracted much recent interest (Frank, 2002; Schmid-Hempel, 2004), and many studies have demonstrated differential susceptibility among host genotypes in wild populations (reviewed in Altizer et al., 2003). The frequency of resistant genotypes in a population can be affected by and can feed back to local parasite dynamics. For example, long-term studies of trematode parasites infecting the freshwater snail *Potamopyrgus antiopodarum* (Fig. 17.1) have shown that host genetic diversity can be maintained through local adaptation of parasites to their hosts and through frequency-dependent selection (Lively, 1992, 1999). Cross-infection experiments provided evidence of local adaptation by demonstrating that common host clones were significantly more susceptible to sympatric parasites than were rare host clones (Dybdahl & Lively, 1998). Further research revealed that changes in the frequencies of common and rare snail clones were driven by parasite tracking of susceptible genotypes. This evidence of frequency-dependent selection suggests that high infection rates can ultimately favor host sexual reproduction as a strategy for generating novel host genotypes that may resist infection (Dybdahl & Lively, 1998).

Plant–pathogen coevolution can similarly lead to a high diversity of host resistance and parasite virulence alleles. In fact, natural plant populations have been shown to harbor a staggering abundance of genetic polymorphisms for resistance to fungal diseases (Burdon & Thrall, 1999; Parker, 1992), and wild plant populations have been cited as a source of resistance genes for pathogens and

BOX 17.1 Evolutionary Mechanisms That Can Maintain Variation in Host Resistance

Multiple processes can maintain genetic polymorphisms in host resistance in natural populations. First, in the case of *frequency-dependent selection,* parasites can become locally adapted to common host genotypes, and thus are better able to infect them. Hence, hosts with rare genotypes may escape parasite infection, conferring a selective advantage to rare alleles and a disadvantage to common alleles. This process can cause time-lagged cycles in both host and parasite allelic frequencies, and may ultimately lead to the maintenance of genetic variation over longer timescales (Seger & Hamilton, 1988). The phenomenon of parasites tracking common host genotypes has been demonstrated in several wild systems and is important for arguments concerning the role of parasites in generating advantages to host sexual reproduction (Dybdahl & Lively, 1998).

More generally, *balancing selection* refers to processes that favor the persistence of multiple alleles, or genetic polymorphisms in a population, in contrast to directional selection, in which allelic frequencies tend to shift in a single direction. Balancing selection can be realized through frequency-dependent selection, as described earlier, or through *heterozygote advantage,* in which individuals with different alleles at any given locus tend to have greater fitness than homozygous individuals. As one example addressed in the text, individuals heterozygous for MHC alleles could experience an advantage in the face of diverse pathogen strains by virtue of recognizing a greater variety of antigens for response by the vertebrate immune system (Penn et al., 2002). Spatial and temporal changes in the risk of infection by different pathogen genotypes or species, as generated in part by *environmental heterogeneity,* could also favor the maintenance of multiple alleles over larger spatial and temporal scales.

Third, resistance-conferring host traits may be costly in terms of reductions in other fitness components (for example, fecundity, growth rates, resource competition) as a result of pleiotropy or resource-based trade-offs. Modeling studies have indicated that even small resistance costs (measured as differences in fitness between resistant and susceptible hosts in the absence of infection) should lead to genetic polymorphisms, such that both susceptible and resistant genotypes are maintained in the presence of parasites (Antonovics & Thrall, 1994). A growing number of field and experimental studies has identified measurable costs of resistance to pathogens infecting many host species (see, for example, Sheldon & Verhulst, 1996), although other studies emphasize that the presence and size of costs will depend on host and pathogen characteristics (see, for example, Carr et al., 2006; Mitchell-Olds & Bradley, 1996), and that the shape of the trade-off function will be important for the longer term dynamics.

In light of potential costs of host resistance, inducible defenses that are activated only after parasite infection may be beneficial when the risk of infection is rare or unpredictable (Harvell, 1990). Such defenses include antibody-mediated responses in vertebrates, antimicrobial proteins in invertebrates and plants, and behavioral removal of parasites in animals. As one example, recent work in sea fan corals has shown that hosts can mount inducible defenses against infection with emerging fungal pathogens, and that environmental factors that cause variation in this response may predict patterns of host susceptibility (Harvell et al., 2002; Ward et al., 2007). Understanding how inducible defenses can be mobilized rapidly against novel pathogens, their rate of evolution, and the costs that they pose on host fitness should help inform efforts to manage disease resistance for conservation.

A

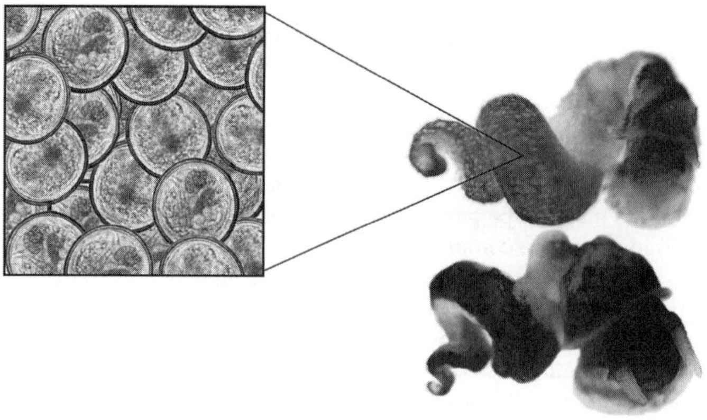

B

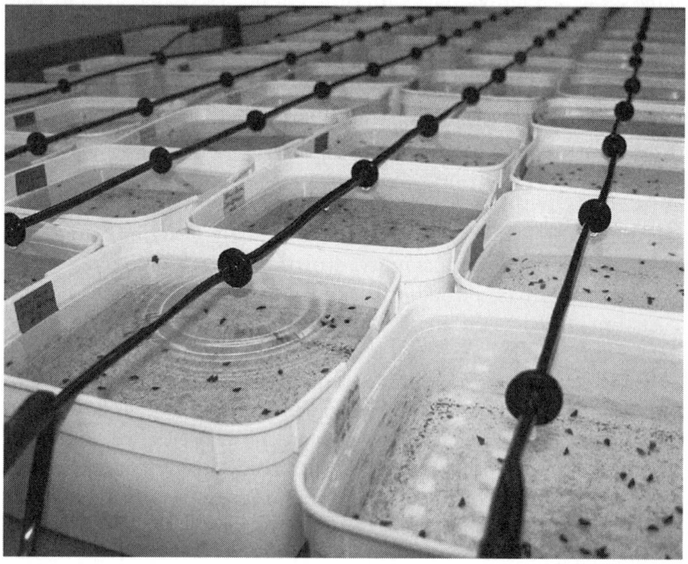

FIGURE 17.1 Field and experimental studies of the freshwater snail *Potamopyrgus antipodarum* and the trematode parasite *Microphallus,* for which the snails serve as intermediate hosts, have demonstrated local adaptation and time-lagged frequency-dependent selection in a host–parasite interaction. (A) Pathology caused by infection with *Microphallus* (top) compared with a healthy snail (pictured below). Both snails are shown without shells; parasite eggs ingested by snails develop into encysted intermediate stages and castrate the snail. (B) Experimental setup of snail clones after exposure to trematode eggs derived from specific parasite genotypes. (C, next page) Lakes on the south island of New Zealand, such as Lake Alexandrina (pictured here), serve as key sites for field studies of snails and parasites in their native environments. (Reproduced with permission from C. Lively. Photos courtesy of G. Harp and C. Lively.)

C

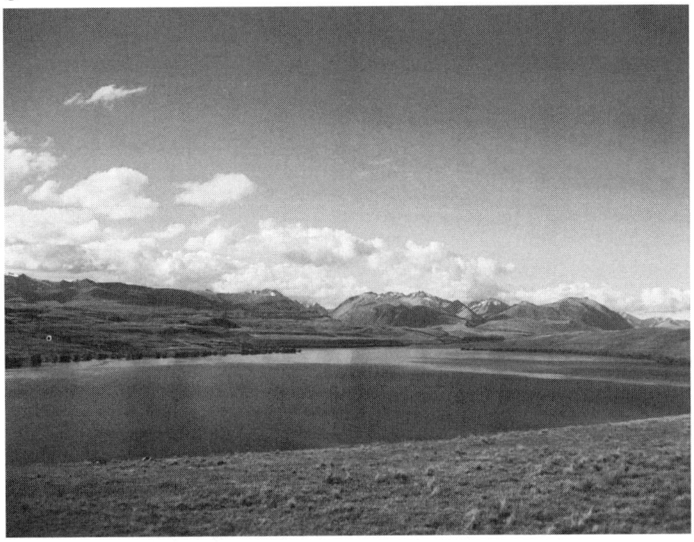

FIGURE 17.1 Continued

pests for crop plants (Jones, 2001). For example, long-term field studies of wild flax and flax rust in natural populations in Australia (Fig. 17.2) indicate that a large number of resistance alleles can persist in plant metapopulations (Burdon & Jarosz, 1991). The distribution of genotypes can shift rapidly after local epidemics, and trade-offs arising from fitness costs have been linked with variation in host resistance and pathogen virulence (Burdon & Thompson, 1995; Thrall & Burdon, 2003). Studies of anther-smut infections in their wildflower hosts also demonstrated variation in resistance among plant genotypes, impacts of host resistance on pathogen prevalence, and costs of resistance in terms of delayed flowering (see, for example, Alexander et al., 1996; Carlsson-Granér, 1997).

Among vertebrate animals, a variety of genes and gene complexes are important for mediating defenses against infectious diseases (see, for example, Acevedo-Whitehouse & Cunningham, 2007). The majority of scientific interest to date has focused on MHC as playing a key role in acquired immunity. Major histocompatibility complex molecules are immune proteins that recognize and bind to pathogen proteins (antigens) inside infected host cells and transport these antigens to cell outer membranes. Here they are presented to T cells to initiate antibody production and cell-mediated immune responses. Specific MHC molecules preferentially bind to specific pathogen peptides, and hence different MHC alleles confer resistance to different pathogens. In natural populations, MHC class I and II genes show enormous variation and are important for recognizing a wide diversity of pathogens (Hedrick & Kim, 2000; Nei & Hughes, 1991). Individual hosts that are heterozygous across multiple MHC loci recognize a greater diversity of pathogens than homozygous individuals (Doherty & Zinkernagel, 1975), and at the population-level, high levels of MHC allelic variation can increase the chance that at least some hosts' immune systems will recognize a single pathogen. Studies of humans and domesticated animals also highlight the importance of non-MHC immune genes in protecting against infectious diseases (Acevedo-Whitehouse & Cunningham, 2006), and a broader understanding of resistance evolution in wild vertebrates will ultimately require comprehensive genetic analysis of other immune regions that control aspects of pathogen defense.

Costs, Trade-offs, and Environmental Variation

Research in the developing field of *ecological immunity* (McDade, 2003; Norris & Evans, 2000) examines the ecological causes and consequences of variation in immune function. One central issue is that hosts that respond to parasite pressure through

A

B

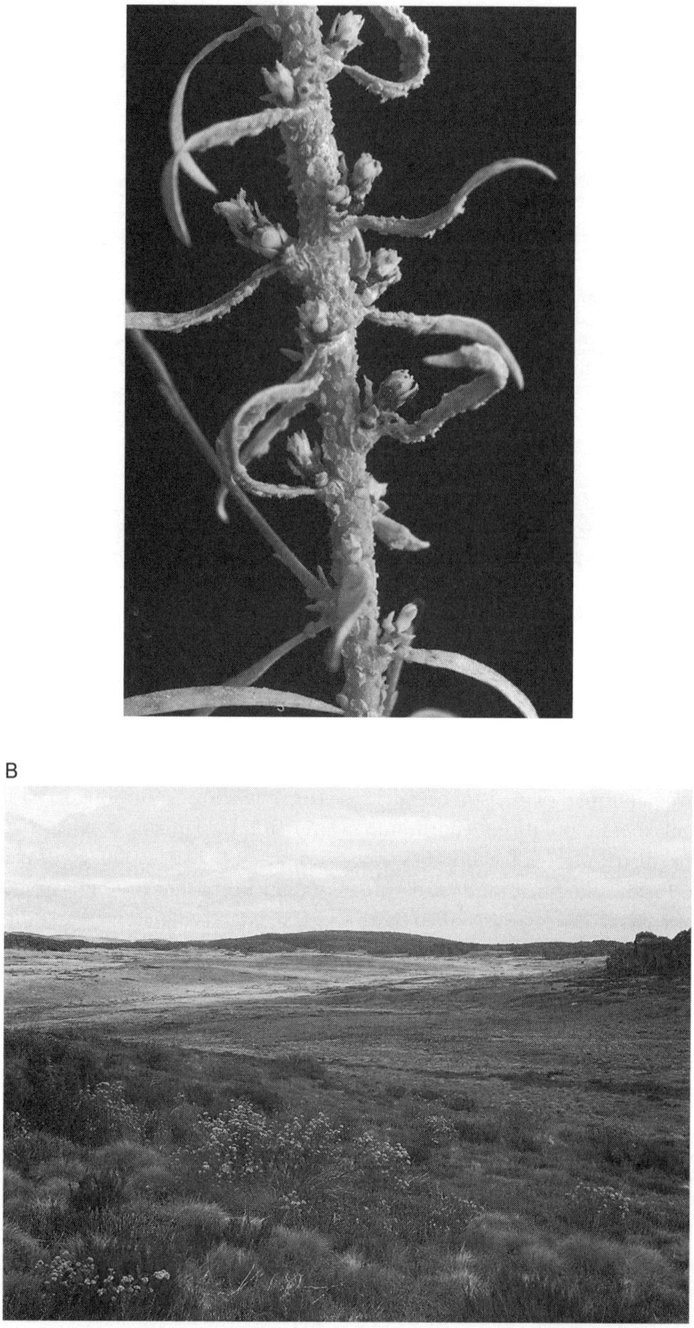

FIGURE 17.2 Long-term studies of the dynamics of wild flax (*Linum marginale*) and its fungal rust (*Melampsora lini*) illustrate coevolutionary dynamics in a naturally occurring plant–pathogen interaction. (A) The heavily infected leaves and stem of the *L. marginale* host plant are shown covered with uredial lesions (which appear bright orange when shown in color) after experimental inoculation. (B) Many populations used to investigate this host–pathogen interaction occur in southeastern Australia in environments such as this subalpine grassland surrounded by eucalypt forests (pictured here in the Kosciuzko National Park, N.S.W.). (Reproduced with permission from L. Barrett. Photos courtesy of C. Davies and L. Barrett.)

increased resistance could suffer lower competitive ability or reduced reproduction (and vice versa; Boxes 17.1 and 17.2) (reviewed in Lochmiller & Deerenberg, 2000). Thus, hosts might be expected to invest in greater defenses only when the risk of pathogen infection is high, leading to geographic variation in genetically based host resistance traits. Also, it is well established that environmental variation, including temperature, resource availability, and environmental stressors, can affect the expression of host resistance in plants and animals (Rolff & Siva-Jothy, 2003). These environmental processes could generate spatial and temporal variation in host defenses, and might also represent important constraints on host resistance evolution (Box 17.1).

Implications of Host Resistance for Species Conservation

Several studies have demonstrated that genetic variation for host resistance is common in natural populations and has significant consequences for parasite infection rates. For example, one comparative analysis showed that macroparasites were more likely to colonize fish species that had low levels of genetic variation as indicated by mean heterozygosity (Poulin et al., 2000). Field studies of intestinal nematodes affecting Soay sheep demonstrated that host allelic variation and levels of heterozygosity were associated with higher host survival and resistance to infection (Coltman et al., 1999; Paterson et al., 1998). But what are the implications of these findings for species conservation? One concern is that small or endangered host populations might suffer disproportionate impacts from infectious diseases as a result of the loss of genetic variability through population bottlenecks, genetic drift, and inbreeding (Lyles & Dobson, 1993). Hosts bred in captivity and treated to prevent and remove parasitic organisms may further experience increased susceptibility caused by relaxed selection and costs associated with resistance-conferring traits.

The relationship between the loss of genetic diversity and increased disease susceptibility has been found in many wild populations (Box 17.2) (Acevedo-Whitehouse et al., 2003). Recent studies of birds and amphibians have shown that loss of heterozygosity (based on selectively neutral microsatellite markers) is associated with reduced immunocompetence, greater risk of infection, and increased severity of infection (Hawley et al., 2005; MacDougall-Shackelton et al., 2005; Pearman & Garner, 2005). Evidence from wild plant populations suggests that inbreeding may increase susceptibility to fungal and viral disease (Carr et al., 2003; Ouborg et al., 2000).

Among vertebrate animals, allelic diversity at MHC loci can be lower than expected among endangered species that have undergone dramatic declines in population size (see, for example, Aldridge et al., 2006). On the other hand, a high diversity of MHC genotypes has been documented across small populations of red wolves, Arabian oryx, and some other endangered species (Hedrick et al., 2000, 2002). In one extreme example, Aguilar and colleagues (2004) demonstrated homozygosity across multiple selectively neutral loci in the San Nicholas Island fox (*Urocyon littoralis dickeyi*). This pattern is indicative of the loss of variation resulting from genetic drift after an extreme population bottleneck approximately 10 to 20 generations before samples were collected. However, these animals showed high levels of variation across five MHC loci, leading the authors to conclude that intense balancing selection was necessary to prevent the loss of rare alleles and ultimately maintain MHC variation in the face of past bottlenecks (circa <10 individuals).

In light of these findings, one goal for conservation management is to design captive breeding, stocking regimes, and species management programs that maintain levels of immunity or resistance variation present in wild populations. One case study shows that the Hawaiian amakihi, a native bird species whose population size and geographic range were negatively affected by the introduction of avian malaria, has persisted and begun to repopulate lowland forest habitats despite high infection rates from *Plasmodium relictum* (Woodworth et al., 2005). One possible explanation for this recovery is host evolution of genetically based resistance or tolerance of infection. Findings such as these should motivate future conservation efforts to focus not just on protecting host populations in areas less affected by pathogens, but also to identify hosts inhabiting parasitized environments where significant evolutionary responses might have occurred.

A related point is that host and parasite movement among habitat patches may be crucial for both host persistence and the spread and maintenance of resistance alleles. For example, Carlsson-Graner

BOX 17.2 Invasive Species as Model Systems for Studying Host and Parasite Evolution

Species introductions offer many opportunities to understand the strength of pathogens as agents of selection. First, most populations of introduced species are likely to originate from a few founders, thus limiting their genetic diversity and possibly increasing their susceptibility to parasites (Sakai et al., 2001). Second, invasive species that escape infection by a broad range of parasites (Mitchell & Power, 2003; Torchin et al., 2003) might reallocate resources away from enemy defense and into growth and reproduction (Wolfe et al., 2004). Indeed, greater allocation to growth and reproduction after enemy release is widely thought to underlie the increased vigor or success of introduced species in their new range (Siemann & Rogers, 2001). Together, these ideas suggest that low genetic diversity combined with evolutionary reductions in parasite defense should make introduced species vulnerable targets for future epidemics, and that detailed studies of pathogen interactions with invasive species in their native and introduced ranges are greatly needed.

Genetic variation tends to be lower among introduced populations, particularly if colonists come from a single source population or undergo an establishment phase during which population sizes remain small (Sakai et al., 2001). Founder events resulting in genetic drift and inbreeding could lower the fitness of introduced populations and limit their ability to adapt to future challenges arising from parasites and infectious diseases. For example, reduced host genetic diversity has been suggested as the reason for the unusually high susceptibility of introduced house finches in eastern North America to mycoplasmal conjunctivitis (Hawley et al., 2005, 2006). In some cases, however, repeated introductions from multiple native sites could lead to a mixing of alleles from different source populations (admixture), resulting in greater genetic variation—rather than less—in the introduced range (as has been demonstrated with brown anole lizards [Kolbe et al., 2004]). Hosts with such novel gene combinations might be highly resistant to parasite infections (Sakai et al., 2001).

Many invasive species are larger bodied, more abundant, and more vigorous in their introduced range relative to their native range (Crawley, 1987; Grosholtz & Ruiz, 2003). A specific hypothesis termed the *evolution of increased competitive ability* proposes that exotic species should adapt to the loss of natural enemies by allocating more energy to growth and reproduction, rather than investing in costly defenses (Blossey & Nötzold, 1995). Predictions of this hypothesis are that in the native range, growth and reproduction should be lower, natural enemies should be more common, and investment in defenses higher; whereas in the introduced range, natural enemies should be less common or absent, defenses lower, and growth and reproduction greater (Wolfe et al., 2004). Furthermore, these phenotypic differences should be genetically based and, given the choice, parasites and other natural enemies should better attack the invasive phenotypes—two predictions that can be tested using common-garden and reciprocal transplant experiments. Some recent studies provide support for genetic divergence in enemy defense and reproductive strategies between native and novel populations of introduced weeds and trees (Siemann & Rogers, 2001; Wolfe et al., 2004), and suggest that introduced species might be vulnerable to future pathogen introductions.

Another area for future research includes the role of evolutionary change in those pathogens that persist or become established in populations of exotic host species. For example, do traits such as high mutation rates or more rapid generation times favor pathogen adaptation to hosts in their new range? How might pathogen virulence change in response to shifts in host genetic diversity or abundance in their new habitats? Currently, comprehensive studies of host resistance and pathogen evolution among invasive species in both their native and introduced ranges are rare, despite the potential insights that can be gained from such comparisons. Future studies of exotic species and their parasites will likely provide new evidence regarding the role of host genetic diversity in immunity and resistance, how costs and trade-offs affect host investment in immune defenses, and how pathogens adapt to populations of exotic organisms.

and Thrall (2002) demonstrated that isolated populations of the wildflower *Lychnis alpina* were rarely infected with anther-smut disease, but when populations were infected, the prevalence was high. In comparison, highly connected populations show more widespread pathogen occurrence across sites, with low prevalence within each location. This pattern was consistent with model simulations that assumed higher rates of movement of both pathogen propagules and host resistance alleles among connected populations (Carlsson-Graner & Thrall, 2002). In bighorn sheep, larger population sizes and increased dispersal is associated with lower extinction risk and rapid population recovery after bronchopneumonial epidemics, indicating a potential role for host dispersal in genetic variation and resistance evolution (Singer et al., 2001). Together, these studies suggest that habitat size and ecological corridors can have significant effects on host dispersal and movement of resistance alleles in ways that affect host responses to future epidemics.

ROLE OF PATHOGEN EVOLUTION IN DISEASE EMERGENCE

The evolutionary potential of pathogens sets them apart from other major threats to biodiversity. Most pathogens have short generation times and large population sizes; hence, strong selection pressures following ecological changes can increase the probability of pathogen evolution and the likelihood of disease emergence in a novel host species. The relevance of pathogen evolution to human health is underscored by recent threats from severe acute respiratory syndrome, avian influenza, and antibiotic-resistant bacteria. However, empirical evidence of how evolutionary processes influence the likelihood of emergence and pathogen spread in natural systems is less clear and supported by only a few examples (Altizer et al., 2003; Schrag & Weiner, 1995).

Two major questions are how do parasites establish in new host populations, and does evolution affect the probability of long-term parasite success? Both epidemiology and evolutionary potential of the pathogen are important during and after pathogen entry into a novel host population. In terms of epidemiology, a newly infected individual must transmit the pathogen to more than one other host for the parasite to increase in frequency when initially rare (commonly referred to as $R_0 > 1$; Anderson & May, 1991). Novel infections are more likely to establish if chains of transmission allow new mutations to arise that increase R_0 above one, and hence evolution and mutations that influence transmission and virulence can play a crucial role in the emergence process (Antia et al., 2003).

Recent studies of host shifts among fungal pathogens infecting plants indicate that geographic proximity and opportunities for cross-species transmission, rather than genetic changes in the parasites themselves, are primarily responsible for the origin of new host–parasite combinations (Antonovics et al., 2002; Roy, 2001). Yet, after such introductions, other studies demonstrate that genetic variants of parasites from novel hosts can rapidly diverge from ancestral genotypes (Ley & Yoder, 1997; Oldroyd, 1999), suggesting that evolution has indeed occurred. Canine parvovirus (CPV), which first appeared in wolves, coyotes, and domesticated dogs during the mid 1970s, is an example of a pathogen that likely arose as a new genetic variant of feline panleukopenia parvovirus, a closely related virus from domesticated cats (Parrish, 1990). Genetic studies following CPV emergence suggest that mutation and selection associated with a major capsid gene, rather than recombination of existing feline variants, played a key role in viral emergence (Shackelton et al., 2005).

Evolutionary Potential, Disease Emergence, and Virulence

Several pathogen characteristics may predispose them to high emergence potential. First, transmission strategies that increase encounters with new host species may increase the range of hosts that a pathogen can infect (Woolhouse et al., 2005). Second, greater genetic variability, high mutation rates, and more rapid generation times might allow certain pathogens to exploit new host species more readily. Specifically, parasites with high antigenic variation should have an increased ability to recognize host proteins or evade host immune defenses (Bitter et al., 1998) and infect new host species (Cleaveland et al., 2001). Viruses dominate this group because their high mutation rates, greater antigenic diversity, and short generation times allow them to adapt rapidly to a larger number of host species. In fact, emerging infectious diseases in humans, domesticated animals, plants, and wildlife are often dominated by viruses, especially RNA viruses that are characterized by unusually high mutation rates and relatively large host ranges (Anderson et al., 2004; Holmes, 2004; Pedersen et al., 2005; Woolhouse et al., 2005).

Evolution can also cause shifts in the virulence of existing and emerging diseases. Theory on the evolution of parasite virulence suggests that parasites should evolve toward an intermediate level of virulence. This is based on an a presumed trade-off between transmission and virulence, such that the production of infectious stages required for transmission to a new host is positively correlated with the damage the parasites cause to the host. Intermediate virulence is favored because highly virulent pathogens may kill hosts before successful transmission occurs, whereas parasites with very low virulence will not produce enough infective stages to be transmitted (Levin, 1996). Spatial structure changes the evolution of virulence, however, such that if most transmission occurs at a local (rather than global) scale, then less virulent pathogens are favored (Boots & Sasaki, 1999). Boots and colleagues (2004) further demonstrated that acquired immunity can lead to the coexistence of both nonvirulent and highly virulent strains. These results help explain the sudden emergence of a virulent strain of rabbit hemorrhagic disease virus that spread throughout Europe, causing high mortality in free-living rabbit populations. Molecular evidence suggests that this highly virulent strain emerged from a recombination event involving less virulent strains. The sudden emergence of the highly virulent strain and coexistence with less virulent genotypes was likely a product of the host's social structure, as rabbits tend to have highly structured populations (Boots et al., 2004).

Anthropogenic Effects on Pathogen Evolution and Risks for Wild Populations

Large-scale environmental changes caused by humans could directly affect pathogen life cycles and transmission, leading to evolutionary shifts in other parasite traits. One example that highlights the role of human activity and selective pressure on sudden evolutionary changes is in the increased transmission and host range of *Toxoplasma gondii,* the protozoan parasite that causes toxoplasmosis in humans and other mammalian hosts (Su et al., 2003). *Toxoplasma gondii* is maintained by wild and domesticated cats that can shed infectious stages in their feces and transmit the parasite to humans and wildlife. Molecular genetic analyses of this parasite indicate that the ability of clonal lineages to transmit orally and without sexual recombination (as opposed to passing through an intermediate host to complete the sexual phase of the life cycle) was associated with a selective sweep that occurred around the time of human agricultural expansion several thousand years ago (Su et al., 2003). By creating high densities of multiple species of domesticated mammals, human activity may have selected for increased oral transmission. Thus, humans may be unknowingly selecting for new variants of wildlife pathogens through global commerce, changes in host density and habitat quality, and climate shifts over longer timescales (Harvell et al., 2002).

Spillover from domesticated plants and animals is a particular threat towild species because infected reservoir hosts can facilitate epizootics in natural populations with otherwise low population density (Anderson et al., 2004; Daszak et al., 2000). As one example, African wild dogs (*Lycaon pictus*) became extinct in the Serengeti in 1991, in part as a result of spillover by canine distemper from a domestic dog outbreak (Funk et al., 2001). Other pathogens of domesticated dogs have affected wild populations—even in the ocean, where an outbreak of morbillivirus in harbor seals was genetically similar to a strain from domestic dogs (Osterhaus & Vedder, 1988). Similarly, the anthropogenic introduction of plant pathogens from imported timber, agricultural crops, or nursery plant species has caused dramatic declines in wild plant populations (with examples reviewed in Anderson et al., 2004).

Repeated pathogen introduction events, especially from multiple sources, can influence genetic heterogeneity in the pathogen population. For example, the repeated introduction of the Dutch elm disease fungus (*Ophiostoma sps.*) via infested timber in the 20th century led to the death of billions of elm trees in North America and Europe. This was caused primarily by two different *Ophiostoma* species introduced several decades apart (Brasier, 2001). Genetic and phenotypic evidence indicates that the more aggressive causal agent (introduced later) gradually replaced the less aggressive species, and that hybridization between the two fungal species produced novel genotypes that may have further increased pathogen virulence and environmental tolerance. This example emphasizes the need for additional studies to examine the degree to which evolutionary changes in pathogens have affected their spread and affects on novel host species.

HOST–PARASITE COEVOLUTION, GEOGRAPHIC VARIATION, AND DIVERSIFICATION

Most examples in natural systems suggest that neither host nor parasite evolution operates in isolation, but that their interaction leads to coevolutionary dynamics. Over time this could result in reciprocal adaptations of interacting lineages, possibly accompanied by cospeciation events and genetic arms races (Box 17.3) (Page, 2003). Relative to other types of species interactions, the intimate associations between hosts and parasites offer many opportunities for studying coevolution on contemporary timescales. Collectively, studies of host–parasite interactions emphasize that coevolution is common, causes heterogeneity in species characteristics over space and time, and that genetic changes linked with coevolution can drive significant changes in species abundance (Thompson, 2005).

An often-cited example of coevolution of host and parasite traits involves myxomatosis in Australian and European rabbit populations (Fenner & Fantini, 1999). After intentional releases in Australia during the early 1950s, the myxoma virus rapidly shifted in virulence. Specifically, within 5 years after the introduction of a highly virulent virus, other strains with reduced severity (for example, longer times to host death and moderate mortality rates) became increasingly common (Marshall & Fenner, 1960). An evolutionary response of the host followed, characterized by greater resistance and lower viral-induced host death rates (Marshall & Fenner, 1958).

Selection operating on hosts and parasites can also lead to coevolutionary arms races (Frank, 2002; Hamilton, 1982), whereby parasites select for increased host immunity, which exerts reciprocal selection for alternative transmission strategies, manipulation of host behavior, and changes in virulence of the parasite. This cycle of parasite adaptation leads to further selection on host defenses. As one example, studies of interactions between feral pigeons and feather lice demonstrated that selection for improved defense systems (in other words, grooming for parasite removal) in the avian host is accompanied by selection for parasite mechanisms to escape these host defenses (Clayton et al., 1999, 2003).

Phylogenetic Patterns and Cospeciation

Over longer timescales, reciprocal adaptations of hosts and parasites might lead to cospeciation and phylogenetic diversification (Clayton et al., 2003; Page, 2003) (Box 17.3). Cospeciation is defined as the parallel divergence of two or more interacting lineages (Page, 2003), and this process can generate congruent host and parasite phylogenies. Several studies have demonstrated patterns consistent with host–parasite cospeciation, yet other outcomes are possible, including host shifting, parasite extinction, and a process called *missing the boat,* during which the host splits into two or more lineages but the parasite remains in only one (Page, 2003). Host shifting is likely to be common among parasites that can evolve rapidly relative to host generation times, and where transmission generates frequent opportunities for cross-species transfer.

Modern molecular tools allow researchers to examine the origins and coevolution of host–parasite associations and the relative rates of host and pathogen evolution (see, for example, Holmes, 2004). Viruses and bacteria that have fast generation times and large population sizes can evolve faster than their hosts and might commonly show evidence for rapid genetic changes after shifts to new host species or transmission among host populations. Surprisingly, one study of cospeciation among strains of simian foamy viruses (SFVs; in the family Retroviridae) isolated from Old World monkeys and apes provides evidence counter to this general expectation (Switzer et al., 2005). This study showed strong evidence for congruent primate and virus molecular phylogenies (with evidence of only a few *host jumps*), and comparable divergence times and rates of nucleotide substitution for hosts and parasites, suggesting that primates and SFVs have evolved at similar rates. Analysis of other RNA viruses (for example, morbilliviruses, influenza A

BOX 17.3 Diversification and Coextinction of Host and Parasite Lineages

Interspecific interactions (in other words, predator–prey, plant–herbivore, and host–parasite) have been proposed to drive major diversification between assemblages of coevolving organisms (Farrell, 1998; Percy et al., 2004) and could ultimately be responsible for the great species diversity found on earth. In terms of parasitism, one comparative study of primate–pathogen interactions demonstrated that parasite diversity (species richness) was positively correlated with rates of primate host diversification (Nunn et al., 2004). In other words, primate host species from more diverse lineages harbored a greater number of parasite species (including viruses, protozoa, and helminths). One mechanism that could give rise to this association is that parasites increase host evolutionary diversification, through, for example, their effects on sexual selection. Because sexually selected traits can also correlate with parasite resistance (Hamilton & Zuk, 1982), high parasite pressure might ultimately lead to greater potential for host speciation. A second explanation for the patterns observed by Nunn and colleagues (2004) is that parasites infecting hosts from more diverse lineages could experience greater opportunities for diversification. Specifically, related host species that overlap in geographic range might provide more opportunities for host sharing by generalist parasites, and host shifting by specialist parasites—leading to higher parasite species richness.

An arms race between hosts and parasites, involving an ongoing struggle to mount greater host resistance against infection, and higher virulence and transmissibility of the parasites, could also account for positive correlations between host diversification and parasite species richness. As a final possibility, coextinctions of hosts and parasites might also drive the associations reported by Nunn and colleagues (2004). Specifically, parasite lineages might be lost as their hosts decline in population size and ultimately go extinct. In other words, higher extinction rates in declining host lineages could generally reduce parasite diversity, especially if parasites go extinct before their hosts (Koh et al., 2004). Another study using the same host–parasite data supported this idea by demonstrating that more threatened primate hosts harbored fewer parasite species (Altizer et al., 2007). Future comparative research on host–parasite diversification will likely benefit from investigating the geographic patterning of host–parasite interactions (Thompson, 1994), incorporating information on parasite phylogeny into comparative tests (Hafner & Page, 1995), and examining the degree to which parasites themselves have gone extinct along with their hosts (Gompper & Williams, 1998).

Lastly, results of comparative studies, although provocative, do not directly examine mechanisms that underlie associations between host and parasite diversification. Theoretical studies have shown that frequency-dependent selection between prey and natural enemies can lead to evolutionary branching in both the host and enemy populations (Doebeli & Dieckmann, 2000). Small-scale experiments of coevolution between bacteria and virulent phages in spatially structured environments have also demonstrated that parasites can drive allopatric divergence among host populations, increasing host diversification by selecting for antiparasite defenses genetically linked to different host traits in different populations (Buckling & Rainey, 2002). Future studies aimed at developing and testing hypotheses for host and parasite evolutionary divergence will give insight to the degree to which hosts and parasites show evidence for coadaptation, concordant phylogenies, and mechanisms that drive patterns of diversification.

viruses, and flaviviruses), however, shows evidence for rapid substitution rates and more frequent cross-species transmission events (see, for example, Chen & Holmes, 2006; Twiddy et al., 2003). Although there is variation in a parasite's ability to establish within a new host species, understanding how host and parasite characteristics covary with the risk of host shifts will be crucial for predicting disease emergence events in declining or endangered species, for which the impact of a novel disease could be devastating (Lafferty & Gerber, 2002; Pedersen et al., 2007).

PATHOGENS AS INDICATORS OF HOST CONTACT AND ISOLATION

Can genetic studies of infectious diseases reveal information about host biology, behavior, and disease transmission that enhance conservation efforts? In humans, it has been demonstrated that host migration events can be revealed by pathogen genetic structure, especially for rapidly evolving viruses like human immunodeficiency virus (Holmes, 2004). A handful of studies have applied this approach more recently to wildlife populations. For example, low genetic diversity and the need to examine patterns over relatively short time frames could pose difficulties in detecting patterns of host evolutionary divergence, especially for long-lived species. In such cases, evolution of some pathogens can be rapid enough to illuminate host geographic isolation and contact patterns. One analysis investigated sequence variation of two genes (*pol* and *env*) in the feline immunodeficiency virus to assess patterns of isolation and physical contact among cougars in western North America (Fig. 17.3) (Biek et al., 2006). Low cougar population densities during the past century and restricted movements were expected to result in pronounced genetic structure, but only weak isolation-by-distance was revealed using cougar microsatellite markers. By comparison, phylogenetic analysis of viral strains isolated from cougars in the northern Rocky Mountains revealed eight viral lineages and geographic population substructure, with different lineages dominating in the central versus the periphery of the cougars' range (Biek et al., 2006). Genetic analysis further revealed that the spatial occurrence of viral lineages is expanding, most likely as a result in increases in cougar population size or movement. This work illustrates the usefulness of pathogen molecular markers for understanding contemporary population movements and geographic structuring of their hosts.

In another study, three species of whale lice (host-specific ectoparasitic crustaceans) revealed historical separation of populations of endangered right whales (Kaliszewska et al., 2005). Genetic analyses of parasite mtDNA sequences revealed that three whale populations in the North Atlantic, North Pacific, and southern oceans diverged around five to six million years ago, after the formation of the Isthmus of Panama. High genetic diversity among lice in the currently small populations of North Atlantic right whales indicates that their host population sizes probably numbered in the tens of thousands before the modern era of commercial whaling, and that these populations had not experienced prolonged historical bottlenecks prior to the past few centuries (Kaliszewska et al., 2005).

Analysis of genetic divergence among pathogen lineages can also be used to predict patterns of disease emergence and to develop control strategies for pathogens in endangered host populations. Statistical analysis of viral sequence changes over space and time has indicated dispersal patterns for the geographic expansion of fox rabies virus in North America (Real et al., 2005) and a wavelike spread of Ebola virus in Central Africa (Walsh et al., 2005). The case of Ebola in Africa is important for both human health and the future of the remaining wild ape populations, which are also susceptible to the virus. The epidemic appears to have advanced as a wavefront, and Walsh and colleagues (2005) suggest that the ladderlike phylogeny of viral isolates and concordant spatial pattern support transmission of the virus through populations of gorillas. In contrast, Leroy and associates (2005) showed that fruit bats harbor the virus, and proposed that Ebola outbreaks are driven less by direct ape-to-ape transmission. These recent analyses may help researchers pinpoint the origins of the epizootic and predict the future path of spread, with implications for concentrating control efforts in wild ape populations at the highest risk.

A

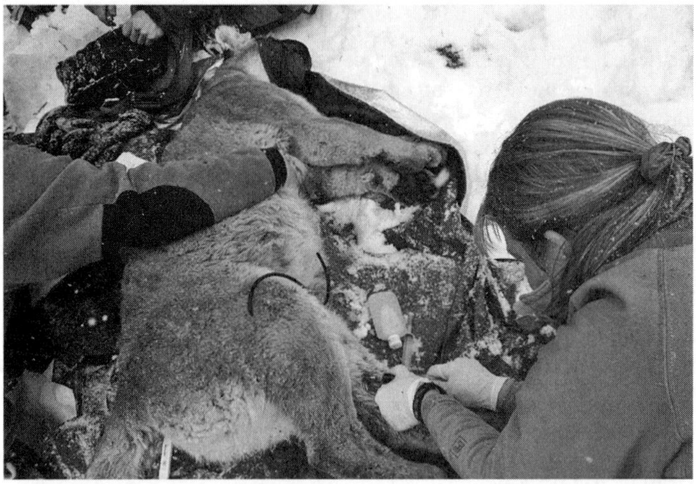

B

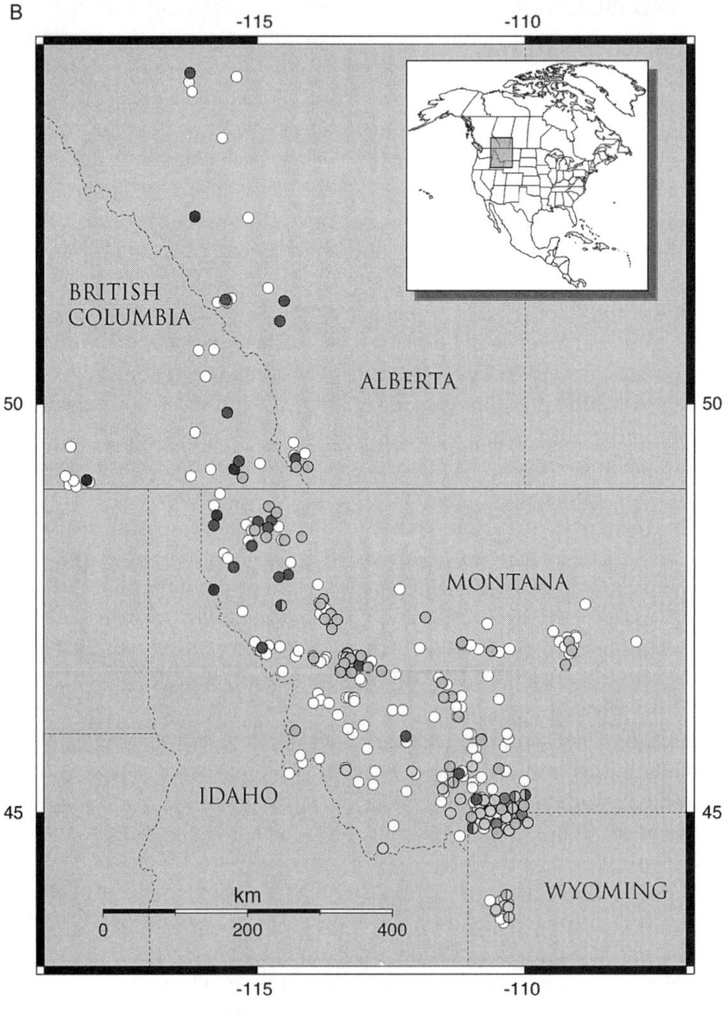

FIGURE 17.3 Continued

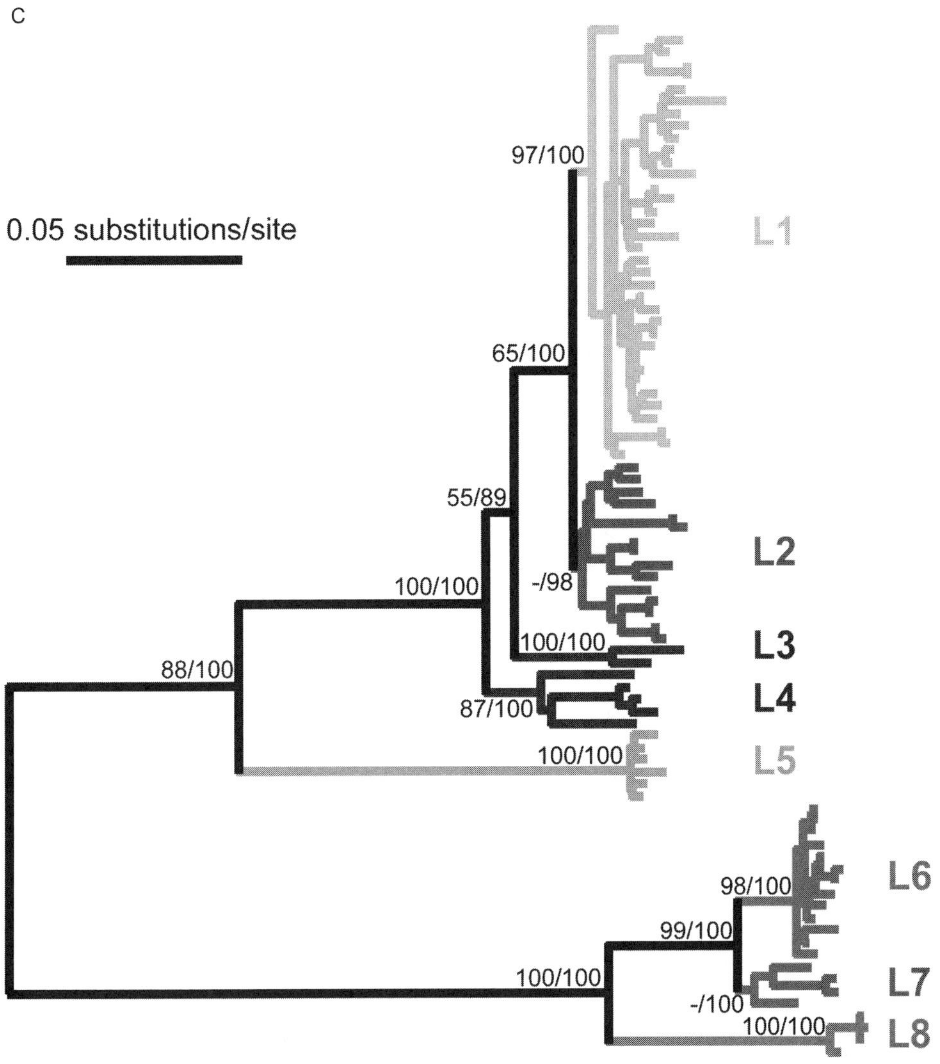

FIGURE 17.3 Evolutionary genetics of the feline immunodeficiency virus FIVPco have been used to elucidate the population structure of its cougar hosts (*Puma concolor*) in North America. (A) Researchers draw a blood sample from a cougar to be used for FIVPco analysis. (Photo courtesy of T. Ruth.) (B) Spatial distribution of FIVPco lineages in the natural habitat of cougars in the western United States and Canada. White circles represent negative samples (no FIVPco); filled circles represent samples from cougars that tested positive (with the several viral lineages indicated by various shadings). (C) Phylogeny of FIVPco constructed from the concatenated *pol* and *env* sequences. Eight viral lineages are designated based on more than 5% divergence, and their colors match those shown in (B). Labeled nodes represent bootstrap support based on 1,000 neighbor-joining trees generated from maximum-likelihood distances estimated from the original likelihood model and posterior proportions from the Bayesian analysis. (Figure provided by R. Biek; data are from Biek et al. [2006]. Reproduced with copyright permission from Biek, R., et al. 2006. A virus reveals population structure and recent demographic history of its carnivore host. (Science 311: 538–541.)

FUTURE DIRECTIONS

Parasites and Host Extinction Risk

Although managing wild host populations in light of pathogen risks may seem trivial relative to numerous other threats to biodiversity, several species have been driven to extinction or to near extinction as a result of infectious diseases (Lips et al., 2006; Smith et al., 2006; Thorne & Williams, 1988). On the one hand, models of directly transmitted infectious diseases suggest that highly virulent pathogens will disappear from small host populations before their hosts go extinct (Anderson & May, 1991). However, there are several cases in which pathogens can cause host extinction (de Castro & Bolker, 2005). First, directly transmitted parasites could drive host population sizes so low that extinction by stochastic factors becomes a concern. Second, pathogens transmitted sexually, vertically, or by biting arthropods (in other words, with frequency-dependent transmission; Getz & Pickering, 1983) should not suffer from reduced prevalence as host populations decline, and the same is generally true for parasites that cause host sterility rather than mortality (O'Keefe & Antonovics, 2002). Third, some generalist parasites maintained in reservoir populations, particularly domesticated species living at high density, can also affect threatened species (Fenton & Pedersen, 2005). In addition, spatially explicit models show that even if parasites do not cause global host extinction, they can lead to local extinction and cause large-scale declines in host density as the infection spreads to new patches (Boots & Sasaki, 2002; Sato et al., 1994).

One of the best-known examples of a host suffering near extinction resulting from disease involves the black-footed ferret (*Mustela nigripes*), one of the most endangered mammals in North America. In the mid 1980s, outbreaks of canine distemper virus (affecting the ferrets) and sylvatic plague (affecting their prairie dog prey) effectively eliminated black-footed ferrets from the wild and severely threatened a captive breeding program. At one point the entire species was reduced to fewer than 20 captive animals (Dobson & Lyles, 2000). Despite current success in breeding and release of captive animals, future threats from infectious diseases will probably play a role in ferret recovery. Dramatic losses of gorilla populations in Africa as a result of Ebola virus outbreaks further underscore the ecological relevance of infectious disease for wildlife conservation. In one instance, an outbreak in the Lossi Sanctuary eliminated an entire population of 143 gorillas (Leroy et al., 2004). Disease-mediated extinction risk is not limited to wildlife species, as several plants are also suffering dramatic declines resulting from the introduction and spread of virulent pathogens (Anderson et al., 2004). In Australia, for example, the anthropogenic introduction of Jarrah dieback disease (*Phytophthora cinnamomi*), a globally occurring fungus with a host range of more than 900 species, has caused a national threat to endemic flora, with significant risks for the health of several ecological communities (Wills, 1993).

Host–Parasite Evolution and Conservation of Biodiversity

Many studies have highlighted the importance of genetic variation in host resistance in causing disease patterns in both field and experimental settings (reviewed in Altizer et al., 2003), but these almost always involve single-host–single-pathogen interactions. In reality, host individuals, populations, and species are affected by large numbers of parasitic organisms spanning divergent phyla from viruses to nematodes to parasitic insects, and for which different types of defenses vary in their effectiveness. Future studies of the types of pathogens that pose the greatest threats to endangered host populations, together with broader genomic surveys that examine a range of *immune genes* in animals and plants are needed to shed further light on the benefits and feasibility of selective maintenance of resistance traits in captive breeding programs (Acevedo-Whitehouse & Cunningham, 2006). Another important area for future research involves developing a better understanding of costs and trade-offs of resistance and how environmental variation affects host immunity. Such investigations will require measuring ecologically relevant immune defenses in the wild and understanding their correlations with other components of host survival and reproduction (Norris & Evans, 2000). In cases in which host defenses are costly, hosts that lose their parasites during population bottlenecks or while in captive breeding programs may also lose their ability to respond to future disease threats after relaxed selection for immune defenses.

As a related point, when hosts decline toward extinction, their host-specific parasites will probably also be lost (Gompper & Williams, 1998).

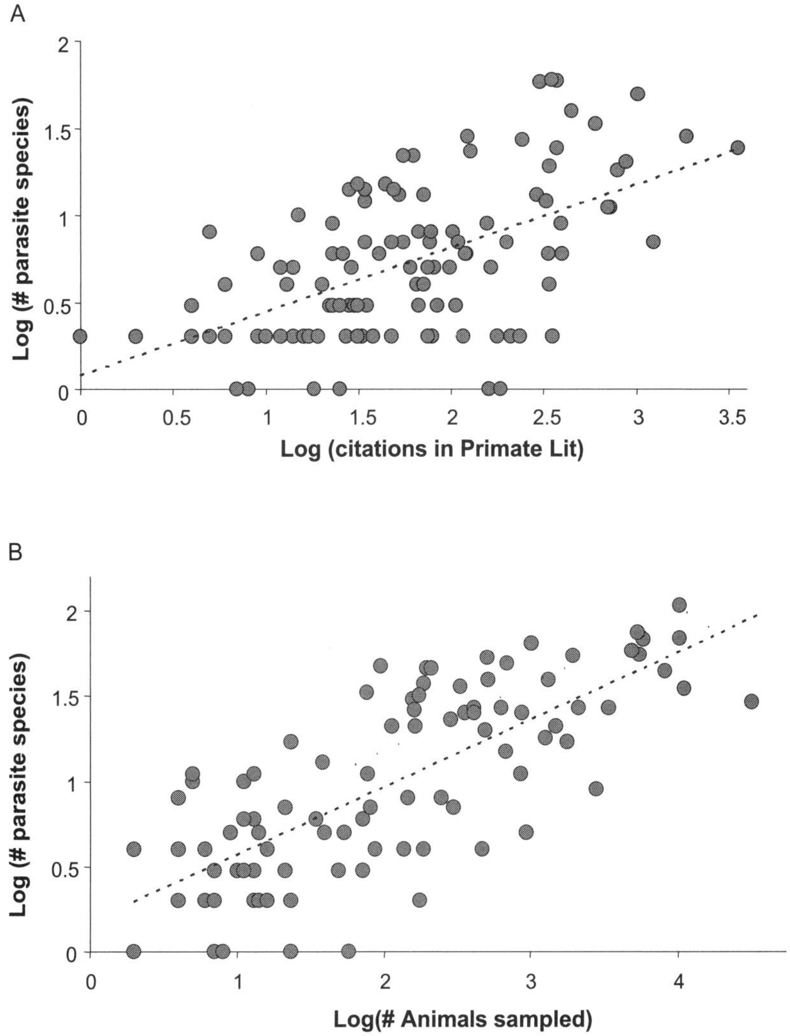

FIGURE 17.4 (A, B) Parasite diversity in free-living mammals. Each point represents parasite species richness based on counts of viruses, bacteria, fungi, protozoa, helminths, and arthropods reported from a single host species. These figures demonstrate positive correlations between study effort and the number of parasites reported from free-living host populations, suggesting that scientists have uncovered only a fraction of the total diversity of parasites and pathogens in natural populations. More generally, this association indicates that "parasites are like the stars—the more you look, the more you find" (J. Antonovics, personal communication, September 2000). Data are shown separately for nonhuman primates (n = 119) (A) and hoofed mammals (artio- and perissodactyls; n = 97) (B). In (A), the sampling effort is measured as the number of literature citations for each primate species using the online bibliographic resource PrimateLit (Wisconsin Primate Research Center and Washington National Primate Research Center). In (B), the sampling effort is based on the sum of the number of animals across all studies upon which parasite count data were derived. Lines show least-squares regression based on a linear regression model. Data are as reported based on methods described by Nunn and Altizer (2005) and are available as part of the Global Mammal Parasite Database (www.mammalparasites.org).

One question at the interface of conservation biology and disease ecology involves whether the loss of parasites is harmful or helpful to a host population. In other words, is the elimination of all naturally occurring parasites beneficial to the host? Given that most of the examples highlighted in this chapter focus on parasites as a threat to wild populations, the answer might seem to be yes. Yet, many ecologists would argue that the best approach for long-term conservation is to preserve geographically structured populations of interacting species, including parasites, in part to maintain intact evolutionary processes. Keeping host–parasite relationships intact requires landscape management strategies that might include protecting corridors and networks of multiple habitat types important to a broad range of species.

Protecting parasitic organisms has not been championed as a priority for current or future conservation efforts. However, host declines and associated coextinctions of parasites could dramatically compound estimates of future biodiversity loss (Fig. 17.4) (Koh et al., 2004). Parasites are an integral part of life on earth, with their biodiversity projected to be significantly greater than the species richness of free-living hosts (Price, 1980); yet, biologists have uncovered only a miniscule percentage of the diversity of infectious organisms from natural host communities. Many micro- and macroparasites that live uniquely on threatened host species could go extinct long before their hosts, and this poses a distinct threat for parasite extinction. As with most taxa, we do not have accurate numbers of how many species of parasitic organisms might be affected by future extinctions. From a broader perspective, interactions between hosts and parasites may be a major force promoting both genetic and species diversity in natural communities. Conservation strategies that result in loss of parasites could ultimately reduce host populations' behavioral, physical, and immune defenses needed to respond to future ecological changes.

Given the explosive growth in scientific understanding of the ecology and evolution of infectious diseases, including noninvasive techniques for collecting information on host infection status, hormones, and host and parasite genetic identity (see, for example, Krief et al., 2005), scientists are now well positioned to address many exciting questions. How do changes in host movement patterns and population structure influence the genetic structure and evolution of their host-specific parasites? Can genetic data from contemporary parasite populations provide insights into the evolutionary histories of host populations? To what degree are multiple host resistance traits genetically correlated with each other and with fitness variables, and how does this affect host infection by a diversity of parasites? Lastly, as humans disturb natural ecosystems, break transmission barriers among species, and reduce host population sizes, outbreaks of emerging infectious diseases among rare or threatened host species may become more common. Understanding the degree to which pathogen evolution and genetic variation in host resistance traits play a role in disease emergence events in natural systems will improve efforts to manage future diseases risks for human populations and in natural systems.

SUGGESTIONS FOR FURTHER READING

Several well-known texts provide a comprehensive overview of the recent advances and status of the biology of infectious diseases in natural populations, including the edited volumes of Dobson and Grenfell (1995) and Hudson and colleagues (2002). For a summary of host–pathogen evolutionary dynamics, Frank (2002) uses a multidisciplinary approach to understand immunology within an evolutionary context, and Page (2003) provides an engaging review of phylogeny, coevolution, and cospeciation. Several reviews have been published recently that aim to understand the risks posed by infectious diseases in the context of animal and plant conservation; we specifically recommend Lafferty and Gerber (2002), de Castro and Bolker (2005), and Anderson and associates (2004). Conceptual issues and specific examples of the relevance of host–parasite evolutionary interactions for biodiversity and conservation are summarized in Altizer and colleagues (2003).

Altizer, S., D. Harvell, & E. Friedle. 2003. Rapid evolutionary dynamics and disease threats to biodiversity. Trends Ecol Evol. 18: 589–596.

Anderson, P. K., A. A. Cunningham, N. G. Patel, et al. 2004. Emerging infectious diseases of plants: Pathogen pollution, climate change, and agrotechnology drivers. Trends Ecol Evol. 19: 535–544.

de Castro, F., & B. Bolker. 2005. Mechanisms of disease-induced extinction. Ecol Lett. 8: 117–126.

Dobson, A., & B. Grenfell. 1995. Ecology of infectious disease in natural populations. Cambridge University Press, Cambridge, UK.

Frank, S. A. 2002. Immunology and evolution of infectious disease. Princeton University Press, Princeton, N.J.

Hudson, P. J., A. Rizzoli, B. T. Grenfell, et al. 2002. The ecology of wildlife diseases. Oxford University Press, Oxford, UK.

Lafferty, K., & L. Gerber. 2002. Good medicine for conservation biology: The intersection of epidemiology and conservation theory. Conserv Biol. 16: 593–604.

Page, R. D. M. 2003. Tangled trees: Phylogenies, cospeciation and coevolution. University of Chicago Press, Chicago, Ill.

V

EVOLUTIONARY MANAGEMENT

Given that conservation biology is often considered the science of rare species, it seems only appropriate to mention one of the rarest and strangest: the pushmi-pullyu. Accounts of this African jungle ungulate suggest that it was never very abundant and went extinct sometime during the 19th century (see Lofting's [1920] account of Dolittle's work). This extinction represents a notable loss given the animal's outstanding anatomy, habits, and origins. Most noteworthy was the animal's alleged possession of a horned head at both ends of its body. This anatomy appears to have effectively thwarted attempts by predators and afforded exceptional manners by allowing one end to feed while the other engaged in polite conversation. However, the cost of this unique adaptation appears to have been some inevitable compromise in the needs and actions of the animal's two ends. In this respect I cannot think of any organism better suited to adorn the standard of evolutionary management.

Since the inception of modern ecological and evolutionary thinking there has been general recognition that evolution is driven by the ecological processes that underlie natural selection (and other evolutionary processes), and in turn ecological dynamics are shaped by the evolutionarily determined attributes of organisms (see, for example, Darwin & Wallace, 1858; Elton, 1930; Fisher, 1930; Lack, 1947). Evolutionary and ecological processes might thus be considered alternate heads of a single eco-evolutionary dynamic—a pushmi-pullyu of sorts. However, when it comes to the practicalities of resource management, or even conservation, we have generally directed much less effort toward addressing contemporary and future evolution (Stockwell et al., 2003). There are numerous explanations for such a bias, ranging from misperceptions that evolution is generally too slow to matter, to the practicalities of monitoring evolutionary processes and their consequences in the wild (Kinnison et al., 2008). Perceptions and practicalities aside, empirical and theoretical examinations show that evolution is not only measurable and pervasive in the face of human activities (Hendry & Kinnison, 1999; Hendry et al., 2008; Palumbi, 2001a), it can influence the dynamics of populations, communities, and ecosystems as well (Fussman et al., 2007; Hairston et al., 2005; Kinnison & Hairston, 2007).

Certainly, there are management contexts wherein evolutionary genetic principles are currently considered, such as in the designation of evolutionary units for conservation (Waples, 1995) or in attempts to minimize the evolutionary genetic problems faced by small populations (Amos & Balmford, 2001). Likewise, most managers would probably support the value of preserving evolutionary potential so species might cope with future challenges (see, for example, Smith & Grether, this volume). Still, it does not take much time with the literature to appreciate that ideal solutions to even these issues are far from resolved, and that the links between most management activities, evolution, and its consequences have barely been explored. In the meantime, we can be sure of one thing: management actions (or inactions) will proceed, even those with suspected effects on the evolution of wild populations (see, for example, Coltman et al., 2003; Tallmon et al., 2004), and even with uncertainties surrounding long-term consequences.

The following three chapters tackle very different challenges in evolutionary management. In so doing they highlight the relevance, complexities, and uncertainties of applying an evolutionary perspective to issues surrounding wild, bioengineered, and harvested organisms. In the first chapter (chapter 18), Maile Neel considers a classic problem in conservation: the number and nature of sites needed to conserve a species adequately. This problem spans from ecological aspects of a minimal viable population, to the complex eco-evolutionary problem of balancing local diversity and genetic connectivity in a reserve network. Perhaps most penetrating, Neel provides an assessment of the effectiveness of three conservation programs in meeting various standards for securing the genetic variation and evolutionary potential of rare and threatened plants.

Michelle Marvier's chapter (chapter 19) addresses an evolutionary management issue unique to our time: the evolutionary risks presented by the escape of transgenes. Although interspecific exchange of genetic variation has been important in the origins of biodiversity, particularly in plants, the evolutionary gulfs crossed during gene transfers seem almost boundless. Indeed, the transfer of fluorescence genes from marine invertebrates into organisms from all kingdoms of life (Stewart, 2006) seems almost as fanciful as the origins of the pushmi-pullyu (by its own account related to the Abyssinian gazelles and Asiatic chamois on its mother's side and the last of the unicorns on its father's). These genetic transfers may allow us to step far beyond the constraints of artificial selection. However, evidence discussed by Marvier shows that transgenes do not always remain in their intended genomes, but are dispersed into the genomes of wild or feral counterparts. How do the genes we transfer tilt the scales of fitness to foster or impede the threats of transgene flow and introgression? What are the consequences for evolution of more difficult pests or pollution of wild genetic resources?

The last chapter of Part V (chapter 20) presents the most classically Darwinian example of the set. Mikko Heino and Ulf Dieckmann delve into the selective nature of population harvest and its consequences for evolution in contemporary time. Although evolutionary responses to strongly selective harvest may seem a foregone conclusion to some readers, many may not appreciate the considerable discourse that recent studies have prompted among fisheries biologists and managers. The management of harvested species is already widely recognized as one of the most challenging, high-stakes applications of ecological and social science principles, so it should be no surprise that the addition of an evolutionary dimension has been met with, let's just say, mixed enthusiasm. Heino and Dieckmann not only lay out some of the classic case studies and theory in support of such Darwinian fisheries effects, but also explore the seeming paradox of how apparently adaptive evolution may thwart the sustainable use and viability of harvested populations.

According to Lofting (1920), no one ever succeeded in capturing a pushmi-pullyu by tackling it from just one direction, although many tried. However a singular specimen did come into a workable arrangement with its pursuers when approached respectfully and from multiple sides. The allegorical value of that approach will hopefully be apparent to the readers of the following chapters.

Michael T. Kinnison
Orono, Maine

18

Conservation Planning and Genetic Diversity

MAILE C. NEEL

Because it is not possible to conserve all natural areas or species occurrences, conservation practitioners must regularly decide both how many and which sites are necessary to have the highest probability of conserving the greatest amount of biological diversity. Beyond initially capturing diversity in networks of sites that are protected from habitat destruction and degradation, long-term conservation requires maintaining the ecological and evolutionary processes that facilitate persistence into the future. Accomplishing these tasks is straightforward in concept but becomes overwhelmingly complex in practice. This complexity arises as a result of varying scales of analysis and planning, varying concepts of which aspects of biological diversity are most important to conserve, differing approaches to quantifying diversity, uncertainty regarding how diversity is distributed, and even more extreme uncertainty regarding probability of future persistence. The purpose of this chapter is to explore the potential for state-of-the-art conservation planning methods and real-world conservation efforts to conserve genetic diversity. In general, three primary aspects of within-species genetic diversity are relevant in conservation: (1) maintaining the range of diversity within taxa to provide variation in ecological functioning and the raw material for future evolutionary potential, (2) preventing excessive levels of inbreeding within populations to ensure that inbreeding depression does not increase species extinction probabilities, and (3) maintaining movement among populations. The first aspect is critical to capturing diversity initially in reserve networks, and the latter two aspects affect maintaining diversity over time.

The spatial scales across which conservation planning occurs range from selecting specific sites in local planning areas (see, for example, Grand et al., 2004), to selecting broad priority areas within ecoregions (Groves et al., 2002), to allocating resources among global diversity hotspots (see, for example, Myers et al., 2000a). Specific criteria used to prioritize sites for protection also differ across these scales and among conservation organizations. Some commonly used criteria include selecting viable populations of particular rare species (Endangered Species Act, United States Congress (USC) 1973), selecting areas that support exceptional numbers of endemic and imperiled species (Myers et al., 2000a) or rare ecological communities, and selecting portfolios of sites that represent all viable native species or community types (see, for example, Anderson et al., 1999; Poiani et al., 2000). Through time, planning has shifted from individual species and local plans to large-scale efforts that aim to meet a range of conservation objectives simultaneously. This shift has been motivated by the continuing rapid loss of biodiversity in the face of high economic and social costs of acquiring and managing land for conservation purposes. It has been facilitated by increasingly sophisticated computer-based systematic conservation planning methods that optimize site selection based on representation of multiple conservation features through the process of complementarity (see, for example, Ball & Possingham 2000; Bedward et al., 1992; Cowling et al., 1999; Groves et al., 2002; Margules & Pressey, 2000; Pressey & Nicholls, 1989). These approaches are appealing because they include more diversity in less area than do species richness hotspot or ad hoc approaches (see, for example, Pressey et al., 1993) by sequentially prioritizing sites that contribute most to meeting specified quantitative objectives (for example, adding the largest remaining amounts of specified features not yet included in a network of sites). Features to be represented in reserve networks can be any mappable aspect of biological or environmental diversity, such as actual or predicted distributions of particular taxa (Csuti et al., 1997), vegetation communities (Scott et al., 1993), or landscape features (Faith, 2003; Faith et al., 2004a; Hortal & Lobo, 2005).

Typically, genetic diversity is not considered explicitly in systematic conservation planning efforts, but an implicit part of *representing* species is representing the diversity within them. If appropriate data are available, genetic diversity can easily be incorporated into systematic conservation planning by mapping genetic management units or evolutionarily significant units rather than simply species-level distributions (see, for example, Moritz, 2002). Alternatively, all alleles or haplotypes known from mapped populations can be represented in selected reserves (see, for example, Neel & Cummings, 2003a). Exactly what aspects of genetic diversity are important for conservation remains the subject of extensive debate. Historically, genetic diversity patterns have been measured using neutral or nearly neutral genetic markers (for example, Aldrich et al., 1998; Hamrick & Godt, 1990). This easily measured type of diversity has been used as an index of overall genetic variation within species, as a means to identify high-priority populations, and to set genetic conservation goals (see, for example, Petit et al., 1998). However, patterns based on neutral markers are often not correlated with phenotypic patterns (for example, Navarro et al., 2005; Storfer, 1996). Because they are more likely to be the result of natural selection and to contribute directly to fitness, many researchers and practitioners advocate measuring phenotypic characteristics.

Although adaptive phenotypic diversity is what we ultimately are seeking to conserve, often it is not practical to quantify it in a way that is relevant to conservation decision making. Estimating the heritability of these traits through the confounding haze of genotype × environment interactions requires laborious experimental and specialized analytical techniques, some of which (for example, common-garden and reciprocal transplant studies) are not feasible to conduct on most rare species. Determining which physical traits are currently adaptive is also challenging, and determining which may be important under future conditions is downright speculative. Even if we knew what traits to measure, quantifying the spatial distribution of variance in these characteristics in populations across the range of a species for use in conservation planning is logistically challenging. Additionally, fitness patterns for multiple adaptive characteristics that respond to multiple environmental factors may not be spatially correlated. Thus, reserves established based on variation in one trait would not necessarily reflect the distribution of variation in other traits of complex phenotypes. Recently it has been suggested that rather than conserving patterns based on current adaptations, we should target allelic variation at loci affecting physical traits of interest to provide potential for future change (McKay & Latta, 2002). Detailed information on genetic control of particular adaptive traits is unlikely to be obtainable for more than a handful of extremely well-studied species and their close relatives. In the end, all these arguments over which type of variation is more important to use in planning efforts are relevant only if data are available.

As conservation efforts have shifted from individual species to multispecies, community, ecosystem, and ecoregional approaches, fewer and fewer data are available for individual species. Thus, although we are often interested in maintaining many types of diversity and processes simultaneously, lack of sufficient data typically limits what can be specifically targeted in conservation planning efforts. In this way, the strength of systematic conservation planning approaches (for example, their reliance on substantial spatially explicit distribution data) is also a primary limitation (Cabeza & Moilanen, 2001; Grand et al., 2007). Lack of information on genetic diversity within species is particularly severe, and in most cases it is neither explicitly measured nor directly used in conservation planning. Even when conservation planning is carried out for individual species, data limitations result in decisions being made using only general ecological information. In the absence of data, it is suggested that genetic variation and local adaptation within species can be captured by including populations from throughout the geographic range and across environmental gradients that potentially represent a variety of selective regimes (Faith, 2003; Faith et al., 2004a; Ferrier, 2002). A key factor in whether this approach will work is how many populations will be conserved.

At the same time conservation efforts have shifted from single species to larger spatial ecological scales, focus has shifted to conserving processes that affect persistence of populations and species directly rather than focusing solely on diversity patterns. Conservation planning goals can incorporate such processes by including features related to ecological integrity of systems (Cowling et al., 1999; Margules & Pressey, 2000) if spatially explicit data on their distribution is available. Within-population processes are likely to be maintained as long as populations remain large relative to historical sizes so that rates of inbreeding and genetic drift remain within the range of natural variation. In all but extreme cases, this can be accomplished without explicit knowledge or manipulation of genetic diversity. Population size is best estimated using long-term demographic data, but such detailed information is typically lacking for plants. Easily mapped surrogates for population size include area of habitat or, better yet, area of high-quality habitat as assessed by probability of occurrence (Araujo et al., 2002; Cabeza et al., 2004). Priority for site selection can be placed on larger populations. Although these ecological surrogates for population size are superior to simple species representation, they need to be applied with caution because reserves selected based on habitat area or effective habitat area can have extinction risks that are an order of magnitude higher than reserves selected based on species-specific demographic parameters (Nicholson et al., 2006).

Incorporating among-population processes into conservation planning requires understanding historical spatial patterns of connectivity among populations and patches of suitable habitat. These connections facilitate range shifts, gene flow, population rescue, and recolonization after local extinction events (Cabeza & Moilanen, 2001; Moilanen & Cabeza, 2002). Preserving among-population processes for many species simultaneously is difficult because habitat requirements and dispersal characteristics are highly species specific. The focal species approach (Lambeck, 1997) is intended to plan for the most area and dispersal-sensitive species, assuming that if their needs are met, other species will be secure. This approach has been criticized extensively (Andelman & Fagan, 2000; Lindenmayer & Fischer, 2003; Lindenmayer et al., 2002) but it is still used in practice because there are few superior alternatives. Connections, or lack thereof, resulting from processes such as dispersal and vicariance can be detected based on patterns of genetic variation among populations (Coates & Atkins, 2001; Coates et al., 2003; Diniz-Filho & De Campos Telles, 2002; Dyer & Nason, 2004). Population genetic analysis methods that estimate the nature and degree of differentiation (for example, STRUCTURE [Pritchard et al., 2000]) or conversely migration (for example, MIGRATE [Beerli & Felsenstein, 2001]) among all pairs of populations provide a more useful means of understanding connectivity patterns than do statistics that summarize across all populations (for example, F_{ST}). Spatial patterns detected from such analyses can be incorporated into reserve selection algorithms by specifying separation distances for populations either to minimize disruptions of movement among connected populations or to select populations that are likely to be genetically independent depending on the conservation goals. A limitation of such genetic analyses is that the results integrate over a range of evolutionary time, so it is not possible to separate current from historical processes based on pattern alone, and pattern does not necessarily predict future behavior. The most serious limitation of implementing these approaches, however, is the lack of genetic data for most species.

In the absence of genetic data, connectivity among populations is most often quantified based on landscape or habitat characteristics (see, for example, Cabeza, 2003; Neel et al., 2004; Neel, 2008). When it is measured in this way, connectivity can be incorporated into reserve selection by imposing penalties on the boundary length between reserve and nonreserve lands to ensure selected sites form larger contiguous blocks of habitat and by specifying acceptable distances for population separation. Rouget and colleagues (2003, 2006) incorporated a range of processes beyond landscape connectivity into systematic conservation planning by identifying and mapping environmental features that were thought to underpin ecological and evolutionary processes in the subtropical thicket biome of the Cape Floristic Region of South Africa (for example, edaphic interfaces, upland–lowland interfaces, riverine corridors, and macroclimatic gradients). Explicitly representing such features in reserve networks substantially increased the area required to meet all targets beyond what was necessary to represent current diversity patterns alone (Rouget et al., 2006). Simultaneously representing local variation within species and maintaining historical connections among populations will also require more populations than simple representation, especially if capturing the range of ecological variation results in selection of spatially isolated remnant populations that require additional habitat to create connections.

The focus on ecological characteristics in conservation planning is often justified because ecological and anthropogenic factors typically pose more immediate extinction threats to species than do genetic factors, and populations that are ecologically secure are typically also genetically secure (Lande, 1988; Schemske et al., 1994; Soulé & Simberloff, 1986). Because maintaining genetic diversity is often an implicit conservation goal, how well we can meet the goal without explicit knowledge and actions is a major question. The answer to this question is a function of both how much genetic diversity we think is sufficient and how many populations are required to capture that amount of diversity. Clearly, more than a single population of a particular species is necessary to represent within-species genetic diversity and to provide an acceptable probability that diversity will persist. In general, quantifying the number of populations and the amount of habitat necessary is exceedingly difficult and is a fundamental recurring issue in conservation (Neel & Cummings, 2003a; Soulé & Sanjayan, 1998; Tear et al., 2005). Thus, although computerized reserve selection algorithms have vastly improved our ability to represent multiple diversity elements in minimum sets of sites, we have not made similar advances in providing scientific guidance regarding how many replicate occurrences of a species or how many hectares of particular habitat are needed to ensure their persistence. Unfortunately, all we can say about the genetic consequences of loss of whole populations is that the proportion of total diversity that will be lost when any particular population is destroyed will increase as the level of differentiation among populations increases. More important, the ultimate ecological consequences of such genetic losses associated with whole-population loss are not well understood beyond knowing in general that within-species genetic diversity is important for species persistence (see, for example, Hughes & Stachowicz, 2004), for ecosystem functioning (see, for example, Luck et al., 2003), and for future evolution (Ellstrand & Elam, 1993; Frankham, 2005a). In addition to direct loss of diversity, loss of whole populations can affect among-population processes if the distributional gaps that are created by such losses exceed pollen movement and seed dispersal distances, thus isolating remaining populations.

One of the only quantitative standards for the amount of genetic diversity to conserve comes from The Center for Plant Conservation (CPC), which recommends having a 90% to 95% probability of conserving all allozyme alleles that occur at a frequency greater than 0.05 (Center for Plant Conservation, 1991). This standard is based on theoretical work by Marshall and Brown (1975). In contrast, Petit and colleagues (1998) recommend conserving examples of all sampled alleles. No guidance is available for an acceptable proportion of variance in quantitative traits within a species. Although allozyme electrophoresis is a dated technique, no standards have been set for AFLPS or microsatellites, nor is it clear that amounts of diversity at these noncoding loci are as relevant conservation targets as are alleles at protein coding loci. A range of conservation intensities (in other words, numbers of populations or amount of area) has been suggested, only one of which is specifically for genetic diversity. Based on theoretical predictions developed by Marshall and Brown (1975) and Brown and Briggs (1991), the CPC (1991) has suggested that sampling from five plant populations is sufficient to meet the genetic diversity standard described earlier. The World Conservation Union recommends protection of 10% to 12% of the land area or 10% to 12% of the area of each ecosystem in a nation or region to maintain general species diversity (cited in Noss, 1996; Soulé & Sanjayan, 1998). Duffy and colleagues (1999) considered rare plant species to be adequately represented if 10% to 12% of the populations of each species were in protective status. Kiester and associates (1996) considered three populations sufficient to represent midsize mammalian predators. The Nature Conservancy suggests conserving at least 10 viable populations of each species distributed throughout as much of each regional planning area in which the species occurs. Some Nature Conservancy regions advocate conserving larger proportions of populations of species with more limited distributions.

Empirical data indicate that it is necessary to conserve far higher proportions of populations than are suggested by the guidelines described here to represent genetic diversity adequately within species (Neel & Cummings, 2003a). Specifically, it was shown that 53% to 100% of the populations of four plant species were required to represent allozyme allele richness and to meet CPC genetic diversity guidelines when populations were chosen with no prior knowledge of genetic diversity. It was necessary to conserve 20% to 64% of populations to be within ±10% of species-level heterozygosity estimates. Proportions of populations needed were similar regardless of whether the populations were chosen randomly or based on ecological criteria (Neel & Cummings, 2003b). In contrast, all alleles could be represented in 10% to 40% of populations if allele distribution data were used to select populations. These alleles are, of course, just a small subset of the genome-level diversity within each of these species and they are not likely to be adaptive. Thus, the indicated proportions of populations are likely at the lower end of what is necessary to conserve evolutionarily meaningful genetic diversity and, furthermore, to conserve microevolutionary processes.

CASE STUDIES

Given that the proportion of populations within a species that are protected is an overriding factor in the potential for conserving raw genetic material on which evolutionary processes operate, I used three sets of data on rare plants to quantify the conservation intensities (in other words, proportions of populations protected) for real species to proportions commonly advocated by conservation organizations, and to proportions proposed by Neel and Cummings (2003a). This assessment provides insight into how well existing and proposed protection efforts are likely to conserve genetic diversity. To bracket the range of suggested conservation intensities, I assessed the percentage of species with five populations and with 20%, 50%, and 100% of populations protected. Populations of these species have been selected by different agencies using a variety of methods, but information on which methods were used was not reliably available. Because these are rare species, it is likely that a large proportion of the remaining populations were selected for conservation based on species-specific information and criteria; however, genetic diversity data are not available for most of these species, so most of the criteria were ecological. Thus, the resulting conservation intensities are likely to be relatively high when compared with multiobjective planning efforts for both common and rare species. As such, these are likely to be best-case scenarios for genetic conservation. Detailed understanding of the potential for conserving among-population processes requires a degree of spatially explicit information that is not currently available and thus is beyond the scope of this chapter. Consequently, this is a modest attempt to assess the potential that existing and proposed levels of protection will initially capture within-species genetic diversity.

The first data set includes all 264 plant species on the state-endangered species list in Massachusetts. Conservation intensity is evaluated as the proportion of the current number of populations in protective status and the proportion of current plus extirpated (historical) populations in protective status based on element occurrence data from the Massachusetts Natural Heritage and Endangered Species Program. These separate comparisons allow for assessment of the degree to which considering only the current species status in setting conservation targets would underestimate genetic losses. The conservation intensities specified in an approved statewide conservation plan (The BioMap [Finton & Corcoran, 2001]) were also assessed.

The second data set comprises 949 globally rare plant species and subspecific taxa that are of immediate conservation concern in California out of 1,495 plant species tracked by The California Nature Conservancy. These taxa meet the criteria for NatureServe categories G1 and G2—at high to very high risk of extinction due to extreme rarity (corresponding categories for subspecific taxa are those labeled T1 or T2). Fuller definitions of G and T rankings are available from NatureServe (2007). Levels of protection are in accordance with Gap Analysis Program (GAP) status definitions (Crist, 2000) as follows. Status 1 protected areas have permanent protection from conversion from a natural state and a mandate for management to maintain that state, including permitting natural disturbance processes. Status 1 lands include designated wilderness areas, some national parks, and many reserves managed by conservation organizations. Status 2 areas have permanent protection from conversion and a mandate to maintain a natural state primarily, but some management practices that degrade the environment are permitted and natural disturbances may be suppressed. Status 3 areas have permanent protection from conversion of natural land cover for the majority of the area but are subject to extractive uses of low intensity over a broad area or of high intensity over localized areas; protection to federally listed endangered and threatened species is typically conferred. Typical status 3 lands include multiple-use areas managed by agencies such as the United States Department of Agriculture Forest Service and the Bureau of Land Management. Information on historical versus extant occurrences was not available for this data set, so it was only possible to assess conservation intensity as a function of the proportion of extant occurrences in different levels of protective status.

The third set of species comprises the 638 plant taxa on the U.S. Endangered Species list covered in 1 of the 267 recovery plans available as of March 2007 (available from www.fws.gov/endangered/recovery/). The number of current (at the time the plan was written) and current plus historical populations of each species, and number of populations determined by the U.S. Fish and Wildlife Service to be required for recovery was recorded from each recovery plan. When quantitative data were available, the number of populations required for recovery (and thus "delisting") of each species was compared with both the current number of populations and with the historical number of populations to assess the degree to which considering only extant populations would underestimate potential genetic losses.

Together these three data sets provide insight into conservation intensities used by a state agency, a nongovernmental conservation organization, and a U.S. federal agency. They also provide information on plant species with different degrees of extinction risk. The taxa in Massachusetts are state rare but are, for the most part, not globally rare and have distributions beyond the state boundary. In contrast, most of the plant taxa of concern in California are globally rare and threatened, and many are endemic to California. The federally endangered species have a range of distributions and they are united only by their legal status.

State-Rare Species in Massachusetts

The 264 plant taxa managed by the State of Massachusetts Natural Heritage and Endangered Species Program are, for the most part, not globally rare (for example, 95% are ranked G3–G5), but are rare within the state (Table 18.1). Only three of these species are federally listed as endangered (E) or threatened (T) (*Agalins acuta* [E], *Isotria medeoloides* [T], and *Scirpus ancistrochaetus* [E]). The high frequency of G3 to G5 rankings indicates that these taxa have a sufficiently large number of populations over a broad enough geographic distribution to be considered secure at a global scale. Many of these taxa are at their distributional limits in Massachusetts and have few occurrences within the state. Because extinction risk increases as range size decreases (Gaston, 1994) and peripheral populations are important for evolutionary processes (Leppig & White, 2006), many states take seriously their responsibility for maintaining distributions of widespread but locally rare taxa.

Most of the 264 taxa have declined within the state. Only 42 taxa remain at all sites from which they were historically recorded. The remaining taxa have been lost from between 7.7% and 92.3% of the occurrences known from the state, with a mean loss of 60% of occurrences. The combined number of known extant plus historical occurrences averaged 15.1 per taxon; a mean of 8.1 occurrences per taxon remain extant. The number of extant occurrences per taxon ranged from 1 to 138, with a median of four occurrences, a value just below the CPC recommendations.

On average, 58.2% of the extant occurrences per taxon are protected (median value, 54.5%), but the degree of protection varies widely across taxa (Fig. 18.1), ranging from no protected occurrences for 8% of taxa to all extant occurrences being protected for 20% of taxa. Approximately 30% of taxa have five populations protected and thus would meet the CPC standard. At the same time, 50.7% of taxa do not even have five extant population in the state. Almost 92% of taxa have at least 20% of their extant occurrences in protective status, and 67.8% of taxa have at least 50% of their extant occurrences protected. The percentages of populations that are protected indicate a generally optimistic outlook for representing within-species diversity with current levels of protection. Full and immediate implementation of the state's conservation plan (the BioMap [Finton & Corcoran, 2001]) would protect 93% of all extant occurrences of these rare species within the state, including all populations for 67.8% of species, indicating high likelihood of representing within-species

TABLE 18.1 Distribution of Global Rarity Ranks for Plants Tracked by the Massachusetts Natural Heritage and Endangered Species Program and The California Nature Conservancy

Global Rarity Rank	No. of Massachusetts Plants	No. of California Plants	Federally Listed Species
G1	4	276	355
G1TH	0	0	1
G2	6	381	127
G2TH	0	0	1
G3	20	126	34
G3T1	0	24	19
G3T2	0	30	2
G3TH	0	0	1
G4	70	150	0
G4T1	0	50	12
G4T2	0	51	6
G4T3	0	0	3
G5	162	242	0
G5T1	1	51	13
G5T2	0	58	5
Extinct or presumed extinct in the wild	0	0	8
Not ranked	2	1	51
Federal threatened	2	37	111
Federal endangered	1	128	527

Analyses for Massachusetts included all plants because they are all considered state rare. Analyses for California included only G1/G2 or T1/T2 taxa to limit the analysis to the highest priority species. Species classified with multiple ranks were assigned to the highest priority rank (for example, a G1G2 species was considered a G1), and species with a questionable rank (for example, G1) were simply assigned to that rank without question.

genetic diversity. The few species currently receiving no state-level protection are obvious priorities for conservation efforts, but such efforts should be placed into the context of activities throughout the entire species range.

Unfortunately, when conservation intensity is calculated as a function of the number of populations originally known from the state, the potential for representing genetic diversity does not look as promising. On average, 34.2% of the original occurrences per taxon are protected. There are only 11 taxa for which all occurrences that have ever been known in the state are protected. Approximately 71% of taxa have at least 20% of their total known occurrences protected and 40.4% have at least 50% of all occurrences protected. Thirty-six taxa would be fully protected if BioMap was fully implemented and, on average, 59.6% of all occurrences ever known per species would be protected. Comparing the proportion of current versus total occurrences that would be protected highlights the danger of using present-day conditions as the baseline for setting conservation targets. Ignoring substantial existing losses is likely to overestimate grossly the proportion of within-species diversity that can be represented. Although one cannot conserve populations that no longer exist, understanding the extent of preexisting losses argues for higher conservation intensities for the remaining populations of a taxon to represent original range of genetic diversity and for restoration of populations in suitable but currently unoccupied habitat within the historical range of the species.

Globally Rare Species in California

In contrast to Massachusetts, most of the plant taxa of concern in California are globally rare and threatened: 63.4% (n = 949) of tracked species are globally rare at the species (G1 or G2) or subspecific (T1 or T2) levels (Table 18.1), and many are endemic to California. In total there were 15,620

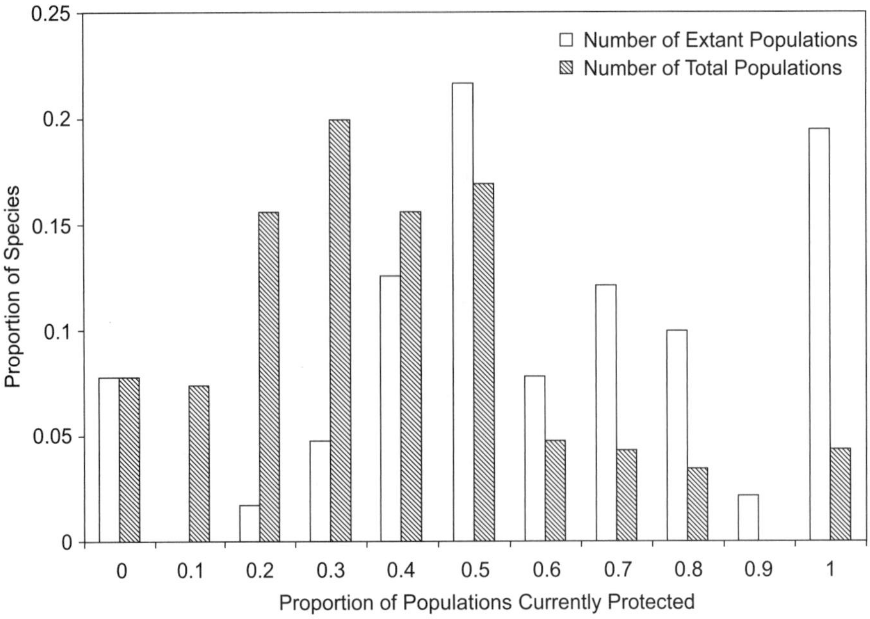

FIGURE 18.1 Proportion of current and historical populations of 264 rare plant species tracked by the Massachusetts Natural Heritage and Endangered Species Program that are protected.

occurrences of these 949 species in the state (mean, 16.5; median, 11; range, 1–139 occurrences per taxon). Just more than 51% of all occurrences were on lands receiving at least some level of protection (in other words, GAP status 1, 2, or 3). Status 3 lands supported the largest percentage of these occurrences of species (31.1% of all occurrences and 60.1% of the 8,080 occurrences on status 1–3 lands). Because GAP status 3 lands are open to multiple uses, conservation is not typically the highest priority, and protection of rare species occurrences is not guaranteed. Almost 80% of all occurrences of the 949 taxa combined are either unprotected or occur on multiple-use lands. The remaining 3,223 occurrences on GAP status 1 and 2 lands receive more reliable protection. The proportion of populations that occur on different status conservation lands varies dramatically across taxa (Fig. 18.2). On average, 19.8% of the occurrences per taxon are on status 1 lands, 27.1.9% are on a combination of status 1 and status 2 lands, and 57.7% of occurrences per taxon are on a combination of status 1 through 3 lands.

A staggering 52.4% of the 949 taxa have no occurrences in the highest protective status, and 35.6% of taxa have no occurrences in either of the two highest protection categories (Fig. 18.2). Just more than 14% of taxa have at least five populations on status 1 lands and 24.4% of taxa have at least five populations on a combination of status 1 and 2 lands. The standard of having 20% of occurrences protected is met on status 1 lands for 27.8% of taxa, on status 1 or 2 lands for 41.2% of taxa, and on status 1 through 3 lands for 80.6% of species. Only 18.1% of species have more than 50% of their occurrences on status 1 lands, 24.0% of taxa have more than 50% of their occurrences on either status 1 or 2 lands, and 59.7% of taxa are protected at this level on status 1 through 3 lands. Less than 10% of taxa have all populations protected on status 1 (8.3%) or a combination of status 1 and 2 (9.1%) lands, whereas 24.3% of taxa have all populations protected on a combination of status 1 through 3 lands.

These protection intensities are unlikely to represent genetic diversity for most taxa and they are even more unlikely to protect a sufficient number of populations to maintain interpopulation processes. However, the potential for maintaining this diversity depends on spatial distribution of the protected

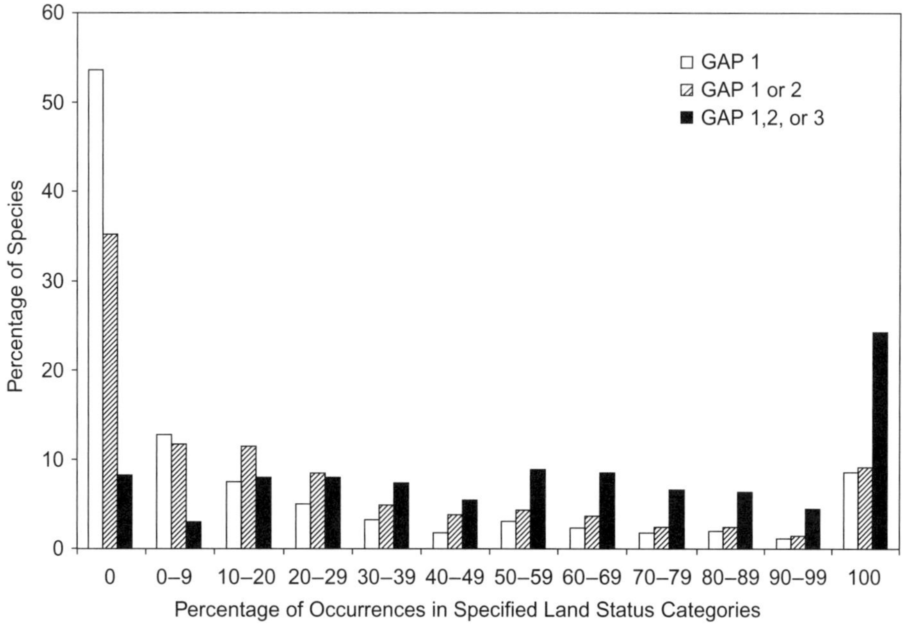

FIGURE 18.2 Proportion of populations of 949 globally rare plant species (G1 or G2) or subspecific plant taxa (T1 or T2) from California that are protected to varying degrees. Definitions of G and T categories are available from NatureServe (www.natureserve.org). Levels of protection follow the GAP Analysis Program definitions as follows: Status 1 areas have permanent protection from conversion and a mandate for management to maintain a natural state, including permitting natural disturbance processes. Status 2 areas have permanent protection from conversion and a mandate to maintain a natural state primarily, but some management practices that degrade the environment are permitted and natural disturbances may be suppressed. Status 3 areas have permanent protection from conversion of natural land cover for the majority of the area but are subject to extractive uses of low intensity over a broad area or of high intensity over a localized area; protection to federally listed species is conferred.

populations and such spatial information is not available. The large proportion of populations on status 3 lands highlights the need to develop understanding of species responses to management activities. It is interesting that the species in California that are a much higher conservation priority are proportionally less well protected than the lower priority species in Massachusetts. This discrepancy is potentially due in part to the fact that in California there are nearly four times as many species over which conservation effort needs to be allocated. Additionally there may be less overlap in the occurrences of species in locations and habitats in California as a result of the larger area and greater variation in geological substrates and environmental conditions, so that reserves protecting particular species do not coincidentally include many other species.

Federally Listed Endangered and Threatened Species

This data set allows examination of the magnitude of current population losses for threatened and endangered taxa, and assessment of conservation intensities considered necessary to remove them from protection of the U.S. Endangered Species Act. As with the Massachusetts data set, conservation intensity is measured as a function of current populations and current plus historical populations. The 267 approved recovery plans examined provided quantitative data on the numbers of historical populations for 404 taxa and ranged from 1 to 475 populations (mean, 17.0 populations; median, 8 populations). Numbers of extant populations were available for 596 taxa and ranged from 0 to 175 populations (mean, 10.9 populations; median, 4

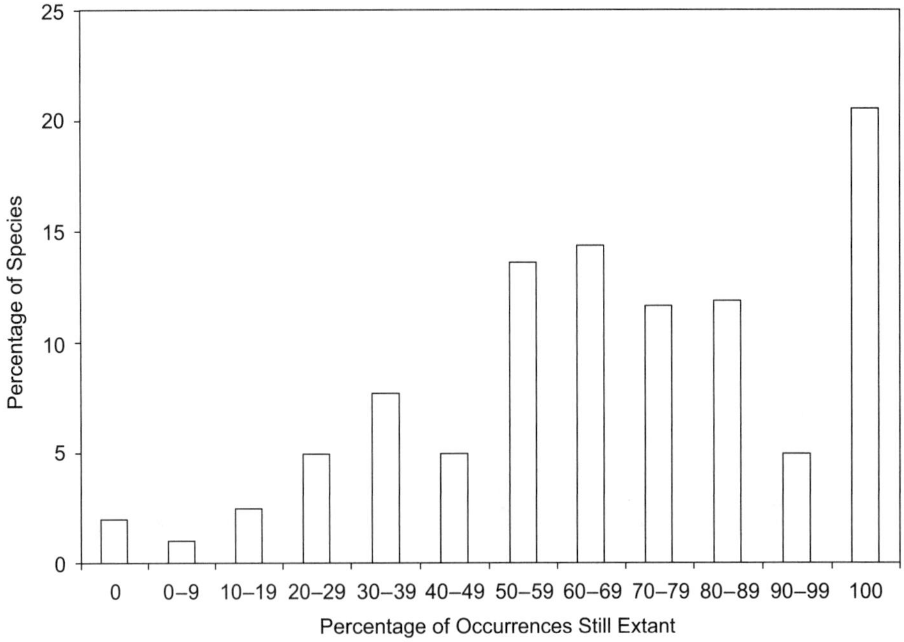

FIGURE 18.3 Percentage of populations of federally listed plant species that remain extant based on data from recovery plans for 404 taxa for which quantitative data on current and historical population numbers were provided.

populations). Two hundred eight species have fewer than five extant populations. The percentage of populations remaining extant (n = 404) ranged from 0% to 100% (average, 66.5%; Fig. 18.3). Twenty percent of these species still have all populations in existence. Just more than 5% of taxa had less than 20% of the original number of populations remaining, and 23.0% of the taxa had less than 50% of the original number of populations remaining. Qualitative data indicate that 127 additional taxa that lack quantitative data on population numbers have also suffered population losses, but it is not possible to assess the magnitude of those losses from available documentation. Overall, 446 of the 638 taxa examined are known to have suffered loss of whole populations, and 378 are known to have suffered reductions in population size; 285 species for which data were available had suffered both types of losses.

Recovery criteria for 409 of the examined taxa specified the number of populations necessary to remove species from the endangered species list. Target numbers ranged from 1 to 115 (average, 10.7 populations). Approximately 94% of taxa have recovery targets of at least five populations. The percentages of populations required for recovery represented from 11.2% to 800% of historical population numbers (n = 284 species) and from 12.0% to 1,500% of current population numbers (n = 383 species). Recovery targets included at least 20% of all historical plus current populations for 96.8% of taxa, and 20% of all current populations for 98.7% of taxa. At least 50% of current occurrences were required for 91.9% of taxa and 50% of historical plus current occurrences were required for 82.4% of taxa (Fig. 18.4). At least all current populations were specified for 72.1% of taxa, and numbers equivalent to at least all current and historical populations were required for 53.5% of taxa. On average, recovery objectives required 247.5% of current populations per taxon and 154.5% of all current and historical occurrences combined. Median per-taxon objectives were 100% of all current and historical populations and 150% of current population numbers. Targets for 62.0% of the examined species were greater than the current number of populations, indicating the need for massive restoration efforts or creation of populations.

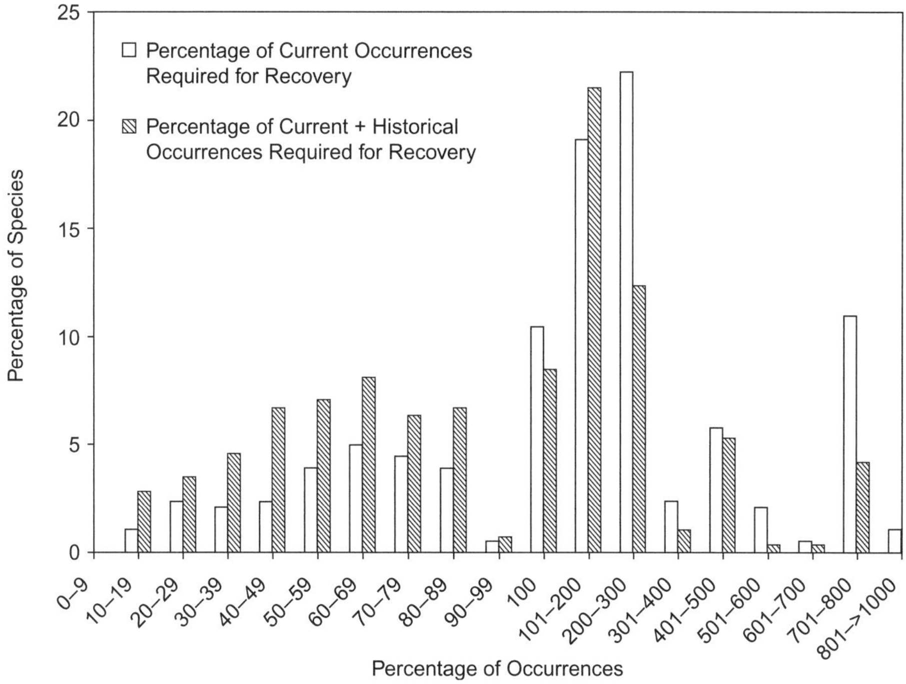

FIGURE 18.4 Percentage of current (n = 383) and historical populations (n = 284) that are considered necessary for recovery of federally listed plant species based on recovery plans for species that provided quantitative recovery objectives and population numbers.

Many of the taxa for which targets exceeded the current number of populations are found in a very small number of occurrences. For example, 53 of these taxa are currently known from a single location; 190 taxa occur in less than five populations, 267 taxa occur in less than 10 populations, and thus the level of effort may be warranted to reduce extinction risks for these taxa.

Recovery targets for 45.0% of species (n = 128) exceeded the number of populations ever known to exist. Thirty-three of the taxa for which recovery targets exceed historical population numbers currently exist in one population, 107 taxa occur in less than five populations, and 124 occur in less than 10 populations. The accuracy of historical distributions varies by taxon; it is problematic because thorough surveys were not conducted for many taxa until after they had likely incurred large losses. Confirming the necessity of requiring more populations than were known to ever exist is essential and will involve a case-by-case assessment.

GENERALIZATIONS FROM THE CASE STUDIES

Because habitat loss and fragmentation typically pose more immediate and direct threats to species than do genetic threats, and because ecological data are more commonly available, ecological criteria will likely remain the primary biological justification for prioritizing sites for conservation. Still, representing levels and patterns of within-species genetic diversity in viable populations is important because of the relationship between genetic diversity and fitness and maintaining future evolutionary potential (Ellstrand & Elam, 1993). To conserve genetic diversity and evolutionary processes in species, it is necessary to reduce threats that decrease population size as well as those that decrease numbers of populations. Because this analysis does not include the condition of populations, the potential for conserving diversity may be overestimated.

Threats that decrease population size have received most attention in the conservation genetics literature; however, losses resulting from destruction of whole populations are often more significant, especially in the context of large-scale conservation planning efforts. On average, species in Massachusetts and federally endangered species have lost approximately 35% to 40% of the number of populations that once existed (Figs. 18.1 and 18.3, respectively). The genetic consequences resulting from such losses differ from the consequences of reduction in population size. Within-population loss of alleles resulting from genetic drift and reductions in heterozygosity will proceed at a predictable rate as a function of effective population size. Loss of heterozygosity can have immediate detrimental effects on fitness, whereas loss of alleles or overall genetic variance has a longer term effect on the potential for future adaptation. When a whole population is lost, the diversity and potential local adaptation within that population are also lost, but there is no immediate change in heterozygosity in the remaining populations, and thus short-term genetic consequences may be minimal. Although the proportion of species-level genetic diversity that will be lost as different numbers of populations are lost is a general function of the degree of differentiation among populations, large proportions of populations can be necessary to represent genetic diversity even if there is little differentiation among populations, because diversity of low-frequency private or otherwise spatially restricted alleles is difficult to capture (Neel & Cummings, 2003a).

Unquestionably, the best way to represent genetic diversity in a subset of populations is to base conservation decisions on known levels of diversity within, and distribution of diversity among, populations. This assumes that we have determined which aspects of genetic diversity are crucial to maintain, and that we have described that diversity in a spatially explicit and historically accurate manner that can be incorporated into a systematic conservation planning framework. Without explicit data, the challenge is to determine how many populations are sufficient to have a reasonable probability both of capturing within-species diversity, and maintaining processes that facilitate persistence of that diversity and ultimately of the species themselves. Predicting that populations of particular sizes or particular numbers of populations of a species will persist into the future is fraught with uncertainty because of the stochastic nature of population processes, natural disturbance events, and anthropogenic factors. The probability of maintaining processes in a particular species is a function of complex interactions among life history characteristics, population sizes, mating systems, dispersal ability, and threats. It is clear that if species are not included in reserve networks or are only included at a few sites, full representation of within-species genetic diversity cannot be guaranteed and processes will be substantially altered as populations in unprotected habitats are lost. It is also clear that we cannot protect all populations of all species. Between these two extremes lies the balance between how many populations are biologically sufficient and how many are politically palatable (Tear et al., 2005).

Empirical results (Neel & Cummings, 2003a,b) have indicated that reserve designs based on the most often suggested conservation intensities of 15% to 30% will likely result in substantial loss of genetic diversity. Thirty percent of state-rare taxa in Massachusetts currently fail to meet even this standard, as do 70% of globally threatened plant taxa in California. Even fewer of these species meet suggested levels of protection that are reasonably likely to represent genetic diversity (for example, >50% of populations) (Neel & Cummings, 2003a). Although species in Massachusetts have more acceptable levels of protection, they still highlight the risk of using extant populations alone as the baseline for establishing representation targets in that a substantial fraction of the original populations has already been lost. Without considering the original distribution and abundance of species, preexisting losses would be severely underestimated—an effect referred to as a *shifting baseline* (Pauly, 1995). At the same time, this data set highlights the need to take the whole species range into account when planning for conservation. Most of the rare plant species in Massachusetts are globally secure and yet even current conservation intensities within the state for these species exceed those for globally rare and threatened plant species in California (Figs. 18.1 and 18.2), and proposed conservation intensities include almost all the populations of each species within the state. For conservation to be both effective and efficient, actions must be put into a rangewide context. Most of the rare species in California have their worldwide distribution completely within that state, and yet the comparatively small proportion of protected populations (Fig. 18.2) is not sufficient to ensure representation of genetic diversity. In contrast, recovery objectives for most federally endangered and threatened species and proposed conservation intensities for species in Massachusetts are likely to conserve genetic diversity even if that diversity is not explicitly targeted. However, suggested conservation intensities for the listed species varied over nearly three orders of magnitude (approximately 12%–1,500% of current numbers of populations), so some species are likely to be underrepresented. Additionally, rationale for how such a large range of conservation intensities can be considered both adequate and necessary for recovery is warranted.

After we have captured genetic diversity, the most important processes that will allow that diversity to persist are mating among individuals within populations and dispersal among populations. Mating patterns are particularly important because of the role that outcrossing, selfing, and other nonrandom mating plays in structuring genetic diversity within and among populations, transmitting diversity from one generation to the next, and determining rates of loss of that diversity across generations (Dudash & Murren, this volume). Dispersal among populations is important in facilitating historical levels of seed dispersal and pollen movement processes that are essential to maintaining diversity within populations and the populations themselves. Simultaneously maintaining within- and among-population processes requires managing ecological conditions that support relatively large populations in spatial arrangements that maintain historical among-population connections. Appropriate management of ecological conditions is especially critical on lands that are not specifically protected and managed for conservation (for example, GAP status 3 or equivalent lands). The importance of such lands is highlighted by the fact that they support the majority of occurrences of globally rare plant taxa in California (Fig. 18.2).

In addition to providing insight into conservation, quantifying existing and proposed levels of protection provides guidance for future data collection. For example, within- and among-population diversity patterns are generally considered to be helpful in prioritizing population selection such that the most diverse and divergent populations are conserved. However, proposed conservation intensities for federally endangered and threatened species and rare species in Massachusetts are so high, collecting data on genetic diversity patterns is unlikely to contribute substantially to conservation planning. Because conservation goals for more than 75% of species include all extant populations, few, if any, decisions regarding trade-offs for protecting some populations but not others are required. Furthermore, there is likely to be little change in distances among populations that would alter gene flow processes. In these cases, understanding risks of diversity loss in small populations may be a higher priority, especially when population sizes have been drastically reduced compared with historical sizes. Because recovery objectives for more than 62% of federally endangered plant species require more populations than currently exist, the relevant conservation genetic issues for these species relate to identifying appropriate sources of genetic material that are appropriate for restoring or creating populations.

FUTURE DIRECTIONS

Conservation science has yielded great advances in the way we select sites for inclusion in reserve networks. The array of increasingly sophisticated yet practical computer-based methods allows practitioners to optimize application of conservation strategies or management techniques to meet multiple, explicitly defined objectives. We need to make commensurate advances in providing practical scientific guidance for what specific features should be included in such planning or, more important, for how much of each feature is sufficient to meet conservation goals. Perhaps the question, "How much is enough?" is in fact unanswerable because of the degree of uncertainty regarding persistence (Tear et al., 2005). Yet, practitioners need to have some way to know which management actions will be sufficient to meet the intended goals or at least to reduce uncertainty surrounding meeting those goals to an acceptable level. The following research areas will contribute to providing such knowledge.

Linking Pattern and Process

Critics of systematic conservation planning often suggest that maintaining diversity patterns conflict with maintaining processes, and they argue for prioritizing conservation of processes. For example, protecting conditions that foster speciation processes may be preferable to representing existing species richness patterns. It is likely true that simply representing static patterns is insufficient as a complete conservation strategy; however, conserving the elements of diversity is a necessary prerequisite to maintaining processes, because they provide the raw materials on which processes operate. Furthermore, it is important to ask whether pattern and process are really disconnected. We know that patterns we observe today represent integration of ecological and evolutionary processes over time. One could argue that sites supporting high species diversity, especially of endemic species, historically supported processes that generated lots of species or that fostered species persistence. Likewise, sites with high genetic diversity allowed persistence of many alleles or a large amount of phenotypic variance. Selecting genetically divergent sites may protect conditions that fostered diversification and local adaptation. However, past performance is no guarantee of future results, and the conditions that have generated current spatial patterns may no longer exist or may not continue to exist under many scenarios of anthropogenic environmental change. Increased understanding of relationships between pattern and process is essential to resolving conflicts over which is more important to conserve and, more important, to inform conservation practice in general.

Techniques to integrate genetic diversity patterns with descriptions of spatial connectivity, such as population graphs (Dyer & Nason, 2004), provide great promise for increasing such understanding by quantifying the partial contribution of each population to the total genetic variance of a species in a spatial context. To date, these methods have been used with allelic data, but they could be adapted for use with quantitative genetic traits if data on the proportions of species-level variance contributed by individual populations were available. Improved understanding of relationships between spatial patterns of population loss and loss of genetic diversity could also allow us to predict more precisely and accurately how much diversity has been or would be lost as a function of the proportion of whole populations that are lost. The next critical step is to improve understanding of the consequences of such losses in terms of current ecosystem functioning and potential for future adaptation.

We also need to improve our understanding of links between landscape pattern and the ecological and evolutionary processes that affect the probability of persistence. We have a broad array of effective ways to quantify landscape patterns, including traditional landscape pattern metrics (Neel et al., 2004), graph theoretic metrics (Urban & Keitt, 2001; Neel, 2008), and methods based on circuit theory (McRae, 2006). Nonetheless, we have little understanding of the relationships between these landscape patterns and actual connectivity from the perspective of most species (see, for example, Fagan & Stephens, 2006). Beyond understanding current landscape pattern–process relationships, we need to improve our understanding of the nature of historic connectivity among natural populations and the degree to which anthropogenic habitat loss and fragmentation has altered that connectivity. Molecular markers provide outstanding tools for quantifying current dispersal distances and frequencies that otherwise cannot be easily quantified (see, for example, Smouse et al., 2001; Sork et al., 1999). Linking changes in gene flow to changes in landscape pattern promises to improve our understanding greatly of evolutionarily significant dispersal distances and probabilities for an array of species. This knowledge will in turn inform the spatial design of reserves to maintain connectivity. Furthermore, comparing current dispersal patterns with genetic diversity patterns that integrate across both recent and deep historical time frames can provide insight into the degree to which processes have been altered and can guide management efforts to restore connectivity. Using comparative methods to study pattern–process linkages across taxa with a range of life history characteristics, phylogenetic histories, and ecological characteristics would provide a powerful way to determine whether we can better predict conditions under which particular processes are most important, yet free us from having to study every species in depth.

Evaluating the Effectiveness of Surrogates in Representing Pattern and Process

Given that we will never have detailed information on genetic patterns or evolutionary processes for most species, we need to understand better how well we can achieve conservation goals without detailed information. In essence, if we cannot characterize the relevant aspects of diversity in a timely and cost-effective manner, we must identify reliable surrogates for that diversity that are feasible to measure. Molecular markers have long served as proxies for overall genetic variation, but links between the relatively easily measured molecular variation, such as numbers of alleles or haplotypes, and adaptive variation, remain tenuous. Moreover, even marker data are not available for most species. Existing meta-analyses of patterns of genetic diversity as a function of rarity or life history (see, for example, Gitzendanner & Soltis, 2000; Hamrick & Godt, 1990) are useful, but are too general to guide specific actions. The value of ecological surrogates, such as numbers of populations, general expanse of habitat, or environmental features, hinges on our ability to improve further our understanding of their relationships to different aspects of genetic diversity and ecological and evolutionary processes (Cowling et al., 1999; Rouget et al., 2003). Evaluating the suitability of surrogates will require a combination of empirical studies (Garnier-Géré & Ades, 2001; Neel & Cummings, 2003a, b) and modeling approaches (Rouget et al., 2006), because the time frames over which many of the processes of interest operate prohibit empirical assessment of persistence.

Improving Methods of Mapping Processes

Because conservation is inherently a place-based endeavor, all inputs into systematic conservation planning must be mappable. The lack of spatially explicit data on evolutionary and ecological processes has been identified as a major barrier to integrating processes into conservation planning (Pressey et al., 2003; Rouget et al., 2003, 2006). Given the criticisms of conserving pattern alone, it is ironic that integrating processes into systematic conservation planning requires mapping the spatial pattern of relevant processes after they are identified, and then basing conservation decisions on patterns of process. The challenge is to integrate the dynamic nature of processes into planning efforts and especially into implementation of the resulting plans. As mentioned earlier, approaches that include understanding patterns of spatial autocorrelation offer a potentially promising approach to identifying genetically independent populations when genetic data are available (Diniz-Filho & De Campos Telles, 2002; Dyer & Nason, 2004). Identifying environmental features that are associated with key evolutionary and ecological processes is also an appealing approach (Rouget et al., 2003, 2006), but it requires detailed understanding of relationships that will likely only be available in a relatively few well-studied species. A major challenge will be finding ways to apply this approach more broadly to species for which we have little or no information.

Conservation Outside of Traditional Reserves

Although reserves can be an effective way to maintain native biodiversity, our ability to establish new reserves is frequently limited, and reserves are but one conservation option. Therefore, it is also important to develop land-use strategies in the context of multiple-use management that will allow native species to persist or, better, flourish coincident with a variety of human activities. Systematic conservation planning software such as MARXAN (Ball & Possingham, 2000) is increasingly being adapted to allow planners to optimize expenditures on different types of management (for example, full protection, restoration, or limited uses compatible with conservation). Understanding ecological and evolutionary responses to habitat management and manipulation is essential to providing scientific input into such planning tools, and is likely to be a more fruitful line than simply documenting genetic diversity patterns for many additional species.

SUMMARY

Systematic conservation planning approaches are incredibly valuable in providing an objective means of optimizing allocation of conservation resources in defined areas. They are, however, no better than the information used as the basis for setting objectives and making decisions, and often the sophistication of the algorithms far outstrips the information content of the data. As conservation scientists, we have challenging opportunities to improve the quantity and quality of information generated in conservation planning processes. The most important contributions we can make are in providing guidance for how much we need to conserve to ensure persistence of both the patterns of diversity and the processes that create and maintain it. In all cases, calls for increasing sophistication in scientific understanding need to be accompanied by practical suggestions for how such understanding can be gained in ways that contribute to conservation decision making. This means combining key attributes that we are seeking to conserve with a mechanism for measuring the relevant characteristics or effective surrogates in cost- and time-effective ways.

SUGGESTIONS FOR FURTHER READING

Two papers that provide an introduction to the science and art of systematic conservation planning are those by Possingham and colleagues (2000) and Leslie and associates (2003). In addition, studies by Pressey coworkers (2003) and Rouget and colleagues (2003) are excellent examples of integrating ecological and evolutionary processes into systematic conservation planning. One of the major challenges in this field is to integrate the dynamic nature of landscapes and populations into prioritization processes that rely on mapped entities. These authors demonstrate that it is possible to move beyond static planning frameworks.

Leslie, H., M. Ruckelshaus, I. R. Ball, et al. 2003. Using siting algorithms in the design of marine reserve networks. Ecol Appl. 13: S185–S198.

Possingham, H. P., I. R. Ball, & S. Andelman. 2000. Mathematical methods for identifying representative reserve networks (pp. 291–305). In S. Ferson & M. Burgman (eds.). Quantitative methods for conservation biology. Springer-Verlag, New York.

Pressey, R. L., R. M. Cowling, & M. Rouget. 2003. Formulating conservation targets for biodiversity pattern and process in the Cape Floristic Region, South Africa. Biol Conserv. 112: 99–127.

Rouget, M., R. M. Cowling, R. L. Pressey, & D. M. Richardson. 2003. Identifying spatial components of ecological and evolutionary processes for regional conservation planning in the Cape Floristic Region, South Africa. Divers Distrib. 9: 191–210.

Acknowledgments I thank Paul Somers and Jessica Patalano from the Massachusetts Natural Heritage Program, and Rebecca Shaw from the California Nature Conservancy for generously providing data on rare plant occurrences. L. Campbell, L. Templeton, and S. Zeigler assisted with data collection for the endangered species data set. L. Campbell and two anonymous reviewers provided helpful suggestions that vastly improved earlier versions of this manuscript. Development of the database for endangered species was funded in part by the Strategic Environmental Research and Development Program of the Department of Defense.

19

Implications of Transgene Escape for Conservation

MICHELLE MARVIER

Humans are now the dominant evolutionary force acting on the earth's biota (Palumbi, 2001a). Through land transformation, climate impacts, profligate use of antibiotics and pesticides, and massive changes to global nutrient cycles, we have become the major selective agent directing the evolutionary future of most species. Recently, however, we have moved beyond exerting selective pressures as a by-product of our activities to acting as *intelligent designers* of new life forms. As genetic engineers we now use modern molecular techniques to move genes across phyla or even kingdoms; insert synthetic genes into plant, animal, and bacterial genomes; and alter the expression of "native" genes, for example, by inserting promoter sequences. These sorts of genetic manipulations are commonplace for crop plants and domesticated animals, and are also being implemented in nondomesticated species such as forest trees, fish, and some insect species.

Of course, people have been altering the genetic makeup of domesticated species for millennia through artificial selection, hybridization, and, more recently, mutagenesis. However, the technology of genetic engineering unleashes new potential to create organisms that possess unprecedented combinations of traits. As one of many possible examples of these novel trait combinations, consider the creation of transgenic zebrafish: Researchers isolated a gene that codes for a green fluorescent protein from a jellyfish and inserted this gene along with appropriate promoter and termination sequences into the nuclear DNA of zebrafish embryos (Amsterdam et al., 1995). The resulting adult fish fluoresce green under black light. Such mixing of genes across very distantly related organisms was once the stuff of science fiction, but today has become relatively routine.

Thus, within just a few decades, genetic engineering has become a major route by which humans are influencing the future ecology and evolution of species. Genetic engineering has the potential for either conservation benefits or harms, but mostly there is uncertainty. One thing that is certain is that little can be said about all transgenic organisms as a comprehensive group. Instead, assessment of potential risks and benefits must focus on the specifics—the transgenic traits, the particular transgenic event, the population genetics of both transgenic and related nontransgenic populations, the interactions of these populations with other species, and so on. Another certainty is that it is no simple matter to contain transgenes physically. In genetically engineered plants, which are the focus of this chapter, transgenes move with pollen and seed, but they also get around a good deal thanks to human error and the failure of people to follow government-mandated safeguards consistently (Marvier & Van Acker, 2005). Once transgenes enter wild or feral populations, little can be done to ensure their eradication. The ultimate fate of transgenes in natural populations will depend upon population genetic processes, over which humans can exert little control. Conservationists are interested in maintaining natural biodiversity, and massive releases of new genes into species with the potential for escape into the wild could have consequences for biodiversity. As we will see in this chapter, quantitative estimates of actual impacts, as opposed to theoretical discussions of *potential impacts,* are scarce. This paucity of hard data, combined with the principle of "it depends" (it depends on the transgene, the genetic background, the evolutionary processes at play, the organism of concern and its population structure, and so on) leaves us with no blanket statements or conclusions.

CONCEPTS

Recent Developments in Biotechnology

To begin to understand the potential conservation implications of transgenic organisms, one needs a sense of the types of transgenic manipulations that have occurred to date and what sorts of transgenic organisms we might expect in the future. Although the focus here will be on gene flow from transgenic plants, readers should keep in mind that molecular biologists have also created varieties of transgenic animals (including livestock, fish, and insects), fungi, and bacteria, and that these organisms may also raise conservation concerns. Some of the animal varieties, such as transgenic cattle, pigs, and chickens engineered to produce pharmaceutical compounds, will be maintained in highly secure laboratory facilities, from which there is little risk of gene flow. However, for other varieties, there is a reasonable expectation that transgenes might escape (for example, from fish species genetically engineered for faster growth in aquaculture), and yet other varieties are designed to be released directly into the environment (for example, transgenic insects, such as mosquitoes engineered to resist malarial parasites). Lastly, the genetic engineering of microbes, with occasional horizontal gene transfer and unknown ecologies, has implications that have received very little attention.

Genetic transformation of plants is often accomplished via bacterial transformation. Using this approach, a transgene, which may be derived from another species or synthetically constructed, is inserted into the Ti (tumor-inducing) plasmid of a bacterial plant pathogen, such as *Agrobactierum tumefascians*. When attacking a plant host, *A. tumefascians* transfers the Ti plasmid DNA to the host cells, where the transgene then becomes incorporated into the plant's genome (Gelvin, 2003). Although *Agrobacterium*-mediated transformation has proved useful for many applications, some plants, especially monocots, are more difficult to transform with this approach. Although these limitations are being overcome through continued research, biolistic transformation has also proved to be a successful alternative (Altpeter et al., 2005). Biolistic transformation involves bombarding host cells with microparticles of metal coated with transgenic DNA. Genetic transformation continues to become increasingly sophisticated, and the list of plant species that has been successfully transformed is steadily expanding.

Among plant species, major grains and field crops have received the most attention from biotechnology labs. These crops have been genetically modified to resist insects, tolerate herbicide spraying, and resist plant pathogens. Genetic modifications are also being performed to enhance the nutritional quality or reduce the allergenicity of various food species, but these varieties are not yet planted in significant quantities. Environmental tolerances are another target of genetic engineering; crops are being engineered to tolerate drought, saline soils, and heavy metals, but these also have not yet become widespread.

In 2005, transgenic crops were planted on some 90 million ha worldwide, with the United States accounting for 55% of this area (James, 2005). Transgenic soybean (60% of global transgenic area), maize (24%), cotton (11%), and canola (5%) are the principal transgenic crops. Herbicide tolerance is the most popular transgenic trait (71% of the global transgenic area), with *Bt* crops (containing a bacterial gene that codes for an insect toxin) running a distant second (18%), and crops with both herbicide resistance and *Bt* genes (such transgenic varieties with multiple transgenes are often referred to as *stacked*) in third place (11%) (James, 2005).

More recently, crops have been genetically engineered to produce highly valuable pharmaceutical and industrial proteins. One example is transgenic maize designed to produce lipase, a pancreatic enzyme that is deficient in people with cystic fibrosis (Ma et al., 2005). Crop varieties engineered to produce these pharmaceutical and industrial proteins are grown under more stringent protocols designed to prevent the escape of plant material, pollen, and seeds that might pollute the general food supply with a drug or a human hormone. In the United States, for example, fields of pharmaceutical-producing maize must be physically separated from other maize plants by a minimum isolation distance of 1 km, or 0.5 km combined with a 28-day difference in the timing of planting (Rose et al., 2006). These and other such rules are intended to minimize the probability of cross-pollination between transgenic and nontransgenic varieties. Plants engineered to produce pharmaceutical and industrial proteins cover relatively little acreage (fewer than 200 acres per year in the United States in 2004 through 2005; APHIS 2008), but, depending on the toxicity and other effects of the compounds produced, these plants may have the potential to harm wildlife on a local scale.

Trees and turfgrasses are also being genetically engineered. As with crop species, transgenic traits for trees include herbicide resistance and insect resistance, but labs are also developing tree varieties with altered lignin content. Of the at least 33 different tree species that have been successfully genetically engineered thus far (van Frankenhuyzen & Beardmore, 2004), most of the effort has been invested in species that are well suited for plantations, such as poplar, pine, and eucalyptus. Reports have surfaced that China has already planted a large quantity of transgenic poplar trees (see, for example, Clayton, 2005), but these rumors are yet to be substantiated. In the United States, the number of applications for field trials of transgenic trees has been steadily growing, but no transgenic tree varieties (except virus-resistant papaya) have yet been approved for commercial release (van Frankenhuyzen & Beardmore, 2004). Grasses are being engineered primarily to tolerate herbicide application, with golf courses being the primary anticipated market. Gene flow from transgenic to conspecific but uncultivated populations is a major concern for both transgenic trees and grasses because very long-distance pollen dispersal is typical of many grass and tree species. For example, Watrud and colleagues (2004) documented hybridization of transgenic and wild bentgrass (*Agrostis stolonifera*) as far as 21 km beyond the edge of field trial plots. Moreover, Reichman and associates (2006) detected establishment of these same transgenes coding for herbicide tolerance via both pollen dispersal and seed dispersal in natural populations of *A. stolonifera*. Thus, the potential for transgenes to become integrated into natural populations appears to be especially high for tree and grass species.

How Might Transgenic Plants Affect Conservation Efforts?

There are many possible avenues through which transgenic plants may either help or hinder conservation efforts. For example, improving crop yields through genetic engineering might mean that less land is needed to produce food for an anticipated human population of 9 to 12 billion people, thereby sparing valuable habitat for native plant and animal species. Genetic enhancements that reduce the need for water or nutrients, or that increase the ability of plants to tolerate toxins such as heavy metals, could both reduce pressure for further land conversion and shift agronomic pressures to different types of lands. For example, transgenic plants that tolerate saline soils might spur further conversion of salt marshes for use in agriculture, to the detriment of the biological diversity and valuable ecosystem services that salt marshes provide. The ambition of genetic engineering for enhanced food production is vividly demonstrated by the International Rice Research Institute's plans to engineer a rice variety that uses C4 photosynthesis as opposed to C3 photosynthesis—a change that could increase yields by 50%. This degree of complexity—genetically engineering a C3 species to a C4 species—would mark an unprecedented makeover of a plant. The International Rice Research Institute's director views this as a 10-year project. Although there is no evidence that this engineering goal will be accomplished, it is a good example of where many biotechnologists hope to take genetic engineering.

The potential risks of transgenic plants will depend on the particular species and traits under consideration, but they may include harm to nontarget animal species. For example, there was a great deal of concern several years ago that pollen from *Bt* maize might harm monarch butterfly populations. This concern grew out of the observation that pollen from *Bt* maize often coats the leaves of milkweed plants that grow in and around maize fields, and experimental demonstration indicates that monarch butterfly caterpillars feeding on these pollen-coated milkweed plants experience greatly increased mortality (Losey et al., 1999). These specific fears have mostly been laid to rest, but unintentional harm to nontarget species, including humans, remains a concern with some transgenic plants. For example, a recent study showed that there is potential for temporal and spatial overlap between pollen-shedding *Bt* maize and the larval stages of the endangered Karner blue butterfly (Peterson et al., 2006).

The widespread adoption of transgenic plants also provides an opportunity either to improve or to worsen our dependence on synthetic pesticides and fertilizers. It is hoped that reliance on *Bt* crops may reduce the use of synthetic insecticides, many of which have well-documented toxic, carcinogenic, and/or mutagenic effects. However, farmers may spray either more or less herbicide on herbicide-tolerant crops compared with nontransgenic fields, and as of now it is not clear that pesticide use is substantially reduced by widespread adoption of transgenic crops. A second hope is that crops and trees can be engineered to maintain high yields while using less nitrogen and phosphorous, which in theory should reduce the use of synthetic fertilizers and thereby reduce the aquatic pollution associated with fertilizers. Again, we have no compelling data to know whether this hope can be realized. For either good or bad, transgenic crops will almost certainly affect farming practices, which in turn will affect the suitability of lands surrounding farms for biodiversity. Because agriculture is the predominant land use on a global scale, anything that alters agronomic practices will also alter the selective regime on a wide variety of wild species.

It is hard to imagine all the routes by which transgenic plants might directly or indirectly affect conservation success. However, the focus of this book is the evolutionary aspects of conservation and, within this context, the major concern is the flow of transgenes from cultivated species to uncultivated (or wild) relatives. The next section examines the processes that affect transgene flow to natural populations.

Transgenes Do Not Stay Put

Gene flow from transgenic plants (or *transgene flow*) can occur either within species (from transgenic plants to nontransgenic populations of the same species) or between species (from transgenic plants to other compatible, usually uncultivated, species). For transgenic grasses and trees and some transgenic crop species (for example, carrot, celery, and oilseed rape), the transgenic species itself exists as an uncultivated plant. In some countries, a substantial portion of crop species also exist as uncultivated wild plants—either because they are native to the region or because they have become widely naturalized. In these cases, there are few barriers preventing the flow of transgenes into wild populations.

For the majority of transgenic crop varieties, however, transgene flow to wild populations would first require hybridization with a distinct species. Transgene flow across species requires the movement of transgene-containing pollen or seed, successful cross-fertilization, and subsequent seedling establishment (Table 19.1). Hybridization is fairly common among plant species. For example, a recent review by Armstrong and colleagues (2005) found that 54% of the major crops of New Zealand will hybridize with wild relatives if cross-pollination is carried out experimentally under laboratory conditions. However, beyond the barriers to hybridization that might occur in a laboratory setting, there are additional barriers to hybridization in natural settings that must be overcome in order for transgene flow to occur. For example, the transgenic plant and the wild relative must have overlapping flowering periods and be physically close enough so that viable seed or pollen from the transgenic plants can reach the other species (Table 19.1). Thus, the road to hybrid formation is rife with pitfalls and problems. Despite this, many crop species can and do hybridize with wild relatives (Table 19.2) (Ellstrand et al., 1999, Kwon & Kim, 2001, Stewart et al., 2003). Indeed, some crop species, such as oilseed rape (*Brassica napus*) readily outcross with uncultivated plant species, and in the United Kingdom alone, tens of thousands of such hybrids are formed each year (Wilkinson et al., 2003). For other crops, the rate of hybrid formation is low, but if a crop is planted over large areas for long periods, even low probability events can add up to significant numbers of hybrids (Arnold, 1997, 2006).

For transgenes to affect the evolution of a wild population, hybridization is only a necessary first step. Hybrid offspring are not always fertile, especially when the two parent species have differing numbers of chromosomes. In addition, if the transgenic trait is selectively disadvantageous, the transgene will likely soon be eliminated from the wild population. On the other hand, transgenes can persist in a wild population either through continual transgenic × wild hybridization or through introgression. Introgression is the stable incorporation of genes from one gene pool to another, such that

TABLE 19.1 Barriers to Hybridization and Introgression of Transgenes in Natural Populations of Plants

Barriers to Hybridization	Example or Explanation
Pollen incompatibility	Pollen–stigma interactions (pollen does not germinate), pollen–style interactions (pollen tube does not complete development)
Spatial isolation	Depends on characteristics of pollen, mode of pollination, characteristics of seeds, mode of seed dispersal
Temporal isolation	Species do not flower simultaneously
Lack of "bridge species"	In the absence of direct fertility, it may be possible for transgenes to move via an intermediate species
Barriers to Introgression	**Example or Explanation**
Hybrids infertile	Incompatible number of chromosomes could interfere with gametogenesis
Backcrossed offspring infertile	Backcrossed individuals are either genetically incompatible with parental strains or suffer low pollen fertility
Natural selection against backcrossed individuals	Poor fitness; either low survival or reduced fertility of backcrossed individuals

the transgene can be passed from generation to generation within the wild population even in the absence of further hybrid formation. Just as there are many barriers to hybridization, introgression can also be a difficult process, limited primarily by the ability of hybrids to backcross with the wild population (Table 19.1). Specifically, the hybrids and wild individuals may be genetically incompatible, or the pollen of the hybrid individuals may have low fertility. In some cases, poor relative fitness of hybrid and backcross genotypes will quickly eliminate transgenes from the wild plant population. But in other cases, when the transgene confers a substantial fitness advantage or the recipient population is small and therefore subject to intense genetic drift, the transgene may become stably incorporated into the gene pool of the wild plants.

Population genetics theory tells us that the velocity of a gene's spatial spread in a population is proportional to the selective advantage of the gene (Fisher, 1937). If a gene has no selective advantage, then it will not spread. However, selective advantage can vary spatially, and in those cases the velocity of spread is proportional to the arithmetic mean of the selective advantage, averaged over space (Shigesada & Kawasaki, 1997). This theory suggests that even though a novel gene may in most places be neutral, it could spread widely if there were

TABLE 19.2 Weediness of Selected Crop Species and Their Tendency to Hybridize with Wild Species

Examples of Genetically Modified Crops Grown Commercially in the United States	Related Wild Species in the United States with which Genetically Modified Crop Can Hybridize
Beta vulgaris (beet)	*Beta vulgaris* var. *maritima* (hybrid is a weed)
Brassica napus (oilseed rape, canola)	*Brassica rapa* (field mustard) and *B. juncea* (Indian mustard)
Cucurbita pepo (squash)	*C. texana* (wild squash)
Gossypium hirsutum (cotton)	Hybridizes with wild congeners that are not weedy but that may be threatened by hybridization
Oryza sativa (rice)	*Oryza sativa* f. *spontanea* (red rice)

A summary of a few examples are provided in Keeler and colleagues (1996), Snow and Palma (1997), and Ellstrand and coworkers (1999).

ample pollen movement combined with the presence of a few hotspots of selective advantage.

Managing Transgene Flow to Wild Plants

Where transgene flow is of greatest concern, the cultivation of certain transgenic plants has been banned outright. For example, transgenic cotton has been approved for unregulated commercial production in all of the United States except in the states of Florida and Hawaii, where wild relatives of cotton are found. Laws prohibiting the creation, testing, or use of transgenic organisms have also been put into effect at the national level. For example, Mexico banned the planting of transgenic maize in 1998. There is some doubt, however, about the effectiveness of such bans, because it appears that Mexican farmers probably planted transgenic maize seed that was imported to Mexico as food aid (CEC, 2004).

Short of outright bans, a number of strategies have been proposed to reduce, and hopefully prevent, transgene flow. In fact, a combination of these strategies is required for all U.S. field trails of transgenic varieties prior to their approval for commercial release, and, as mentioned earlier, transgenic plants designed to produce pharmaceutical or industrial proteins are grown under especially stringent protocols for containment. Containment strategies include geographic isolation, temporal isolation, the use of border rows or bare areas around fields, and either removing reproductive parts or harvesting plants before they become reproductive. Again, the effectiveness of these strategies is questionable, both because pollen and seed will occasionally escape despite our best efforts, and because people frequently fail to comply with all regulations and/or make mistakes that allow transgene escape (Marvier & van Acker, 2005). The propensity for either sloppiness or honest mistakes is likely to continue unabated unless something is done about what is now rather spotty regulatory oversight of field trials (Pollack, 2006).

Many researchers hold out hope that biotechnology can, itself, solve the problem of transgene flow and reduce the potential for human error. A number of potential genetic barriers to transgene flow have been suggested, and some of these are currently under development and field testing. Proposed genetic barriers to transgene flow include linking transgenes to other plant genes that reduce fitness, inserting additional transgenes to reduce the fertility of either seeds (terminator technology) or pollen (gene use–restriction technologies, or GURTs), and transformation of chloroplast DNA because chloroplasts are, for the most part, maternally inherited (Stewart et al., 2003). However, research has demonstrated that even tightly linked genes can occasionally become separated during recombination. Furthermore, recovery of fertility occasionally does occur, and in some species the inheritance of chloroplasts is not exclusively maternal. Redundancy (using multiple confinement strategies) can help to reduce the probability of transgene flow, but no strategy for biological confinement of transgenes is completely fail proof (National Research Council, 2004).

Does Transgene Flow Really Matter?

Even in cases when transgene flow and introgression are unlikely, if a transgenic plant is cultivated on large areas for sufficiently long periods, some amount of transgene flow is expected. Therefore, from a precautionary perspective, transgene flow might be considered all but a foregone conclusion. The key question, then, is whether transgene flow represents a major conservation concern given everything else we need to worry about in a world of 6.5 billion people (likely to be nine billion by 2050), massive habitat alterations, overexploitation, species introductions, and global climate disruption. What might be the implications of transgene flow for biodiversity and for conservation efforts? And what are the prospects for recovering an uncontaminated pretransgene state should it become necessary?

There are five major routes by which transgene flow might threaten the persistence of species and further exacerbate the current conservation crisis.

Transgenes May Exacerbate Weed Problems

In the context of conservation biology, weeds are plants (often, but not always, nonnative) that aggressively outcompete other plant species. Examples in the United States include kudzu, gorse, German ivy, and purple loosestrife. Plant invasions, and weedy plants in general, present a major threat to biodiversity and an enormous challenge for land managers (Hobbs & Humphries, 1995). Therefore one major conservation concern regarding transgene flow is that a transgene may confer a fitness advantage to a wild plant, and in turn either create or exacerbate weed problems (Ellstrand, 2003b; Pilson & Prendeville, 2004). Examples of transgenic traits that improve plant fitness include resistance to insect herbivores and diseases, and altered environmental tolerances that allow a plant to invade new habitats. Worries about enhanced weediness are often brushed aside because cultivated plants tend to be far less fit than their wild relatives, and it is therefore assumed that crop–wild hybrids will be similarly poor competitors. However, concerns that hybridization between transgenic plants and wild plants might lead to *superweeds* are not unfounded. Some of the world's most pernicious weeds originated from hybridization events (Ellstrand & Schierenbeck, 2000; Ellstrand et al., 1999). Moreover, transgenic modification is no longer restricted to domesticated crops. Transgenic modification of grasses and forest trees create especially worrisome potentials for weed problems because these species have not been subjected to extensive selection for domestication, and hence reduced hardiness in the wild.

A second, distinct concern related to weed problems is the movement of transgenes that confer herbicide resistance. Although herbicide resistance will only increase plant fitness in the presence of herbicide spraying, the spread of these transgenes to wild plants can frustrate efforts to manage weeds in natural areas, and can force land managers to switch from relatively benign formulations to more toxic or more persistent herbicide compounds. For example, introgression of transgenic herbicide resistance into volunteer oilseed rape populations has already resulted in individual plants resistant to multiple herbicides (Hall et al., 2000). Feral oilseed rape resistant to multiple herbicides will present weed management challenges to farmers, and could undermine the commercial value of herbicide-resistant seed, because the idea is to have the crops resistant but weeds vulnerable to the herbicide spray.

Maladaptive Transgenes May Reduce the Fitness of Small Populations

Outbreeding depression is a reduction in the mean fitness of a population resulting from hybridization. The potential for maladaptive transgenes to reduce the average fitness of related wild plant populations should be considered, especially if the potential recipient population is small (Ellstrand, 1992). This concern extends to all small populations—both those that have been reduced in number as a result of human activities and those that naturally consist of few individuals. In large populations, natural selection will likely eliminate the maladaptive gene with no serious consequences to the population. However, in small populations, repeated immigration of a maladaptive gene can swamp the effects of natural selection. Furthermore, if a transgene does become introgressed into a small wild population, random genetic drift can overwhelm selection and lead to the fixation of detrimental alleles.

Genetic Swamping May Cause Extinction

Extinction via genetic swamping may be a common, albeit difficult-to-document, mechanism by which biodiversity is lost (Rhymer, this volume). Genetic introgression or repeated high-frequency hybridization between a transgenic population and a wild relative can cause the smaller wild population to become genetically assimilated into the larger transgenic population (Levin et al., 1996). Unlike outbreeding depression, which is the result of maladaptive genes, this process is primarily the result of selectively neutral or even adaptive traits. Although this particular threat applies to both transgenic and nontransgenic plants (Ellstrand [2003b] reviews 10 cases of genetic assimilation via hybridization with a nontransgenic crop), the fact that many transgenes are designed specifically to improve plant fitness may make extinction by genetic swamping more likely when a transgenic plant variety is involved (Gepts & Papa, 2003).

Even in the case of perfectly neutral transgenes, if large expanses of land are devoted to homogenous plantings of single transgenic varieties, the steady flow of transgenes into related species could forever alter the genetic background of these wild populations. What would be lost is the local variety and differentiation of populations, which would instead accumulate the same transgenic backgrounds. Because gene function and phenotype are modified by genetic background, any tendency to homogenize the genetic background in wild plants will reduce opportunities for unique and contrasting evolutionary innovations. This is a risk with any large-scale agriculture, but because transgenic crops tend to be more expensive, they are often associated with industrial-scale agriculture. Large farms engaged in transgenic crop production will, year after year, spew out massive amounts of transgenic pollen.

Transgene Products May Harm Populations of Interacting Species

Certain transgenic plants are engineered to produce insecticidal toxins (for example, the *Bt* plants) or pharmaceutical substances that may be harmful to wildlife. If the transgenes coding for these proteins were to enter wild plant populations, a broader array of nontarget organisms might be affected. Transgenes for *Bt* toxin or other forms of insect resistance are more likely to become introgressed into wild populations than, say, pharmaceutical-producing transgenes, because they are more likely to improve the fitness of recipient individuals.

Transgenes May Alter Ecological Function

Transgenic modifications that significantly alter the ecological function of a species (or of its wild relatives via hybridization) could have major and possibly unpredictable effects on ecosystem function and composition. For example, increased biomass accumulation of a transgenic tree species might lead to more frequent and intense fires that favor one suite of species over another. Substantial changes in ecological function have the potential to push certain species out of ecosystems.

Given all these maybe's and caveats, the answer to the question of whether transgene flow is a serious conservation concern is "sometimes no and sometimes yes." The devil is in the details. The following two case studies illustrate some of the issues that transgene flow may entail.

CASE STUDIES

Transgenic Contamination of Mexican Maize

In 2001, David Quist and Ignacio Chapela ignited a firestorm of controversy by providing genetic documentation of transgenic contamination in Mexican landraces of maize. Their work aroused controversy for several reasons. First, Mexico is the center of maize diversity—maize was first domesticated in Mexico, and farmers have traditionally saved seed and generated many locally adapted landraces. The many distinct genetic variations of maize that exist in Mexico are considered an invaluable resource for the future of global food security. "Contamination" of these diverse landraces with transgenes was considered by many to be a serious threat to this natural genetic resource. Adding to the concerns was the fact that planting of transgenic maize had been banned in Mexico since 1998. It was unclear, therefore, whether transgenes had introgressed into the local maize landraces or if farmers had merely flouted the ban and planted transgenic maize seed during the years (1999 and 2000) when Quist and Chapela's maize samples were collected.

The other major controversy revolved around a second claim made by Quist and Chapela (2001). Using inverse PCR, they examined DNA sequences flanking the transgenes. They reported finding diverse sequence segments in these flanking regions, which they interpreted as evidence of transgene movement, possibly occurring during recombination. Other researchers were suspicious of the data supporting this second claim because inverse PCR is notoriously subject to error (Kaplinsky et al., 2002; Metz & Futterer, 2002).

A recent study by Ortiz-Garcia and colleagues (2005) examined maize from areas near those sampled by Quist and Chapela (2001) and found no evidence of transgenic contamination. This newer study examined more than 150,000 seeds collected in 2003 and 2004 from 870 plants in 125 fields and found no sign of the transgenes. Although it is possible that the transgenes are present in the population at a low frequency, the extent of the original contamination seems to have diminished substantially in just a few short years. This finding lends support to the idea that farmers had illegally planted transgenic maize during 1999 and 2000, which was detected as hybrids, but that the transgenes did not subsequently become introgressed into the Mexican maize landraces at a detectable frequency. However, the areas sampled by Ortiz-Garcia and colleagues were not very representative of the Oaxacan maize-growing areas (Soleri & Cleveland, 2006) and their approach has been criticized on the statistical grounds that they should have estimated the genetically effective number of individuals sampled, rather than the absolute number (Cleveland et al., 2005).

What are the implications of potential transgenic contamination of Mexican landraces of maize for conservation? Frankly, even if transgenes were to become introgressed into these landraces, it is hard to imagine exactly how such contamination would differ in terms of conservation risks from possible introgression of other genes from modern maize varieties. There are, however, important cultural considerations. The relationship between Mexican farmers and maize reaches far beyond food production. Maize is an important cultural and spiritual symbol in the Mexican countryside, and contamination of local landraces with genes manufactured in high-tech laboratories is widely considered to be an unacceptable violation of this resource (CEC, 2004). If the American bald eagle were somehow to become genetically contaminated, one can only imagine the public uproar that would result. One of the hardest lessons for high-tech science to learn is that not everything can be reduced to objective scientific "facts"; there are cultural values that also matter and that deserve consideration and respect.

Although it is unlikely that the incidental transgene flow into Mexican landraces of maize would entail any of the conservation threats listed earlier, this example illustrates the important distinction between hybridization and introgression, with introgression in this case representing a substantially more pernicious outcome that, based on the most recent evaluation of transgenes in Mexican maize, was perhaps not realized. Also the furor and resentment from Mexico after the initial report by Quist and Chapela (2001) reminds us that biotechnology is often a symbol of technological imperialism and has dimensions of risk that extend well beyond the biological arena.

Transgenic Trees: Conservation Friend or Foe?

Transgenic trees may offer viable solutions to some of our most pressing conservation problems. For example, trees genetically engineered to produce biomass quickly might constitute an important carbon sink that could help alleviate global warming (Herrera, 2005). Moreover, if transgenic trees could be used to increase production on plantations substantially, some of the logging pressure on natural forests might be alleviated, and trees engineered to sequester heavy metals and other environmental toxins could be used for phytoremediation of polluted sites. Trees are also being genetically engineered to produce less lignin, which would allow for more efficient extraction of paper pulp and thereby require less energy and a reduced volume of toxic chemicals (Chiang, 2002). Genetic engineering has even been suggested for noncommercial trees. For example, Adams and colleagues (2002) advocated genetic engineering of native tree species as a way to improve the trees' resistance to introduced pests and diseases. They argued that the tide of species introductions is unstoppable and that, given the constant onslaught of introduced pests, genetic engineering for resistance could prevent extinction or endangerment of native tree species.

On the downside, transgene flow to wild populations is a likely occurrence for most transgenic tree species. Transgenic trees are frequently grown in close proximity to nontransgenic, natural populations of the same species. In such cases, hybridization across species is obviously not a barrier to transgene flow. Moreover, trees are long-lived, can produce copious pollen, and typically release their pollen high in the air, allowing for long-distance dispersal for wind-pollinated species. Recent field work and modeling efforts suggest that pollen from transgenic conifers can move and remain viable on a scale of kilometers, making physical isolation an unworkable strategy for transgene containment (Linacre & Ades, 2004). Furthermore, there are relatively few barriers to transgene introgression because, unlike many field crops, trees have not undergone extensive selection for domestication and there is little reason to expect transgenic × wild hybrids to be substantially less fit than parental populations. Indeed, even with technological restrictions on gene flow, such as linking transgenes to genes that inhibit floral development, there will likely be occasional reversions of individuals to a fertile state. Thus, many researchers and regulators are beginning to come to grips with the idea that, for trees at least, transgene escape is likely to be both inevitable and irreversible (Bonfils, 2006). Acceptance of this notion would shift the focus of risk assessment away from gene containment to understanding the ecological and evolutionary consequences of transgene flow (Williams, 2006b). Unfortunately, little empirical research has yet addressed the consequences of transgene flow for natural tree populations. Trees are noteworthy because, unlike field crops, they do not require special cultivation practices, and are not weakened by centuries or millennia of domestication. We cannot extrapolate our experiences with maize and soybeans to trees.

James and associates (1998) suggested that a tiered approach to risk assessment that explicitly considers the selective fitness of transgenic phenotypes might be appropriate for the regulation of transgenic trees. Phenotypes that confer large fitness advantages are far more likely to spread than those that are neutral or harmful. On the other hand, gene-by-environment interactions can change the selective advantage of certain traits over time. If transgene flow is indeed inevitable and irreversible, then some genetic transformations that are particularly risky may need to be shelved. Although a similar tiered approach might make sense for all taxa, it is most likely to receive especially strong support when it comes to trees.

FUTURE DIRECTIONS

Obviously there is a great deal of uncertainty regarding the likelihood and implications of transgene flow to wild populations. Because so many of the concerns depend on the specifics of each case, every new transgenic organism warrants fresh inquiry, and there will therefore continue to be a great deal of research focused on this topic. Fortunately, the questions addressed by the research do not have to be reinvented for each new transgenic organism. In particular, four basic issues must be addressed in each case: (1) determining the potential for hybridization (for example, assessing reproductive compatibility and, in the absence of direct gene flow, assessing whether bridge species exist), (2) quantifying the rate of hybridization, (3) assessing the opportunity and persistence for backcrossing and introgression of transgenes into wild populations, and (4) examining the potential ecological impacts of transgenes in wild populations, including the potential to exacerbate weediness, alter habitat requirements of species, or increase extinction risks (Ellstrand, 2003b).

Beyond these specific questions, there are several overarching issues that also warrant some concerted exploration. First, genetic instability continues to be a big concern. The question of whether transgenes behave differently from other genes, especially after introgression into wild populations, merits further research. At this point, our regulatory policy assumes that transgenes behave the same as other genes. Because this is a fundamental assumption of policy, it warrants critical scrutiny.

Lastly, we need a better understanding of the prospects for recovery from transgenic contamination. Chemical pollutants can be cleaned up. Even invasive species can be completely eradicated, if one acts quickly and with a sustained effort (Myers et al., 2000b). Although transgenes can disappear from a population even after introgression has occurred as a result of either selection or random drift, these are processes over which we have little control. Can we devise ways to eliminate transgenes should they be found to be harmful? We can recall faulty automobiles. Is it feasible to recall a faulty transgene after it has been released? If not, then any decision to release a transgenic organism into the environment should be made in full mindfulness that the outcomes may be irreversible (Marvier, 2004).

SUGGESTIONS FOR FURTHER READING

The Pew Trust (2001) provides a comprehensive and readable overview of the types of transgenic plants and animals currently under development. There is perhaps no better overview of the array of genetic engineering efforts underway, as well as the reasons why different modifications are being pursued. Without sacrificing rigor, this 100+–page document, available free from the Web, can be understood by anyone with a modest background in biology.

For balanced discussions of the pros and cons of transgenic trees, see the excellent volumes edited by Strauss and Bradshaw (2004) and by Williams (2006a). These two books give both sides of the debate about transgenic trees, without hyperbole or exaggerated claims. Strauss and Bradshaw's book (2004) also delves into the ethical and social dimensions of transgenic trees.

The mechanisms and possible consequences of transgene flow from crops to wild species are recently reviewed by Armstrong and colleagues (2005), Ellstrand (2003b), Hails and Morley (2005), and Stewart and associates (2003). The book by Ellstrand (2003b) is especially noteworthy because it is wonderfully written, thorough, and original. Ellstrand has, throughout the years, been one of the world's premier researchers on gene flow in plants; he has synthesized the data of others as well as conducted his own field experiments. Ellstrand's chapter on the reproductive biology of the world's most important crops is, by itself, worth the price of the book.

Armstrong, T. T., R. G. Fitzjohn, L. E. Newstrom, et al. 2005. Transgene escape: What potential for crop–wild hybridization? Mol Ecol. 14: 2111–2132.

Ellstrand, N. C. 2003b. Dangerous liaisons? When cultivated plants mate with their wild relatives. John Hopkins University Press, Baltimore, Md.

Hails, R. S., & K. Morley. 2005. Genes invading new populations: A risk assessment perspective. Trends Ecol Evol. 20: 245–252.

Pew Trust. 2001. Harvest on the horizon: Future uses of agricultural biotechnology. The Pew Initiative on Food and Biotechnology. The Pew Trusts, Washington, D.C. pewagbiotech.org/research/harvest/harvest.pdf.

Stewart, C. N., Jr., M. D. Halfhill, & S. I. Warwick. 2003. Transgene introgression from genetically modified crops to their wild relatives. Nat Rev Genet. 4: 806–817.

Strauss, S. H., & H. D. Bradshaw. 2004. The bioengineered forest: Challenges for science and society. Resources for the Future, Washington, D.C.

Williams, C. G. 2006a. Landscapes, genomics and transgenic conifers. Springer Press, Dordrecht, the Netherlands.

20

Evolution and Sustainability of Harvested Populations

MIKKO HEINO
ULF DIECKMANN

Sustainably harvested populations are characterized by a balance of births and deaths. If harvesting is too intensive, deaths exceed births and the harvested population declines. When this continues for too long, extinction becomes inevitable. For harvesting to be sustainable, harvesting mortality must thus be offset either by decreased natural mortality or by increased fecundity. Mechanisms underlying such compensation in nature are often not well known. Yet it is clear that the growth rate of most natural populations is reduced by density-dependent processes. Typically, when population densities become large, survival of newborn and juvenile individuals declines. Other common manifestations of density dependence are slower somatic growth and reduced fecundity in dense populations. When harvesting reduces population densities, pressures originating from density-dependent natural processes are thus relaxed. Accordingly, the key to ecologically sustainable harvesting is not to exceed the capacity of relaxed density dependence to compensate for the deaths caused by harvesting.

Even though achieving ecologically sustainable harvesting is by no means easy, it is important to realize that such short-term sustainability does not even suffice to guarantee sustainability in the long term. This is because harvesting may have evolutionary implications that gradually undermine the viability of the exploited population and/or the quality and quantity of the harvest. This occurs through selection-driven changes in demographically relevant adaptive traits. For example, large individuals often provide the most valuable targets to harvesters and thus experience the highest harvest-induced mortalities. In this way, harvesting may qualitatively change the mortality regime to which a population had adapted in the past (Fig. 20.1), and favor evolution of smaller adult body size. At the same time, large individuals, in addition to having the lowest natural mortality, often mate most successfully and have access to the widest range of resources. The loss of such individuals directly through harvesting, and indirectly through harvest-induced evolution, is thus likely to compromise a population's productivity and resilience.

In general, harvest-induced selection occurs whenever harvesting causes trait-specific differences in survival or fecundity. Evolution will then ensue, provided that selection is sufficiently consistent and persistent through time, and that the trait-specific differences possess a heritable basis. The history of successful animal and plant domestication and breeding is testimony to the heritable basis of a very large range of traits that might become exposed to harvest-induced selection. These include body size, growth rate, size and style of sexual ornaments, age and size at maturation, reproductive effort, and many aspects of behavior.

Indeed, mechanisms of harvest-induced evolution are in no way different from those that have been harnessed for millennia for the purpose

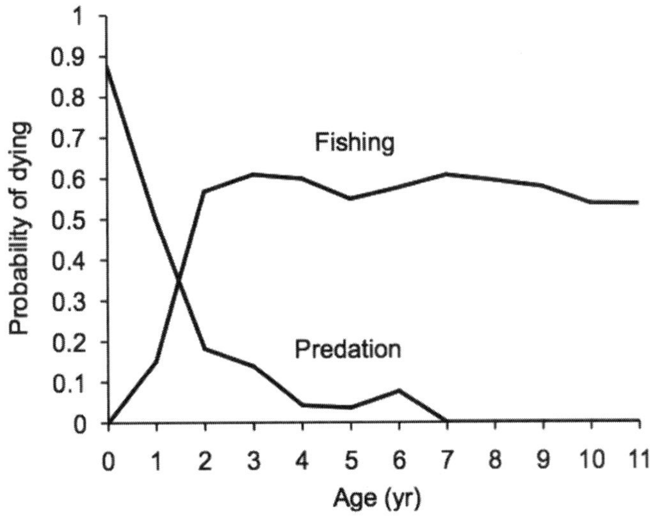

FIGURE 20.1 Estimated age-dependent profiles of annual mortality for North Sea cod (*Gadus morhua*) from predation and from fishing. The natural annual mortality originating from other causes is assumed to be around 5%. The probability that an individual of age 3 years survives until age 12 years is 0.02%. Without fishing mortality, that survival probability would be 47%. (Data from ICES [1997].)

of plant and animal breeding. The main difference is that, although harvest-induced evolution is usually unintentional and disadvantageous for the harvester, plant and animal breeders have actively promoted the breeding of individuals with desired characteristics to maintain or improve a stock's long-term quality. It is therefore not unexpected that a limited, and often merely intuitive, awareness of the evolutionary dimensions of harvesting has already existed for a long while. For example, foresters sometimes protect trees with straight trunks, based on the understanding that the subsequent inheritance of this characteristic will benefit future tree generations. Similarly, game managers may encourage the culling of individuals with only modest antlers, such that individuals with more rewarding antlers continue to arise in decent numbers.

Such awareness, however, has largely been confined to terrestrial systems. An early exception was Californian fish biologist Cloudsley Rutter, who had the foresight to note already in 1902 that regulations encouraging the selective harvest of the largest salmon returning to spawn would inevitably lead to a deterioration in the salmons' body size, because only smaller salmon were thus allowed to breed. Despite this early warning, the management of capture fisheries has been remarkably unaffected by evolutionary thinking. This lack of attention is difficult to justify, especially when considering the socioeconomic importance of capture fisheries. Around the globe, harvesting of wild fish continues at an industrial scale, resulting in important sources of animal proteins for a significant proportion of humankind. By contrast, at least in industrialized countries, the capture of terrestrial animals is mostly of local importance, often providing recreational opportunities, rather than serving as a crucial source of nutrition.

For decades, the large-scale and economic importance of marine fisheries has motivated the continuous and detailed collection of data. This explains why our current understanding of the evolutionary dimensions of harvesting, based on quantitative observations in the field, has gained so much from the monitoring of marine fisheries. Although the resulting emphasis on marine populations is accurately reflected in this chapter, it must be understood that harvest-induced evolution concerns taxa irrespective of their biome.

It is this broader perspective that underlies the following overview of the evolutionary dimensions of harvesting.

CONCEPTS

Selection Pressures Caused by Harvesting

That genetic selection occurs when harvesting is selective is evident—and it should be understood that harvesting is virtually always selective. In contrast, it is less obvious, and thus often insufficiently appreciated, that even changes in overall mortality that are entirely unselective, affecting all individuals of a population uniformly, are powerful drivers of genetic selection. This is because increased overall mortality reduces longevity, so that the risks, and thus the costs, of all strategies involving waiting or saving are elevated. Prominent examples of such strategies are waiting to mature and saving acquired energy for the next season. Here mortality simply acts as a discounting factor of future benefits. Because harvesting may drastically increase this discounting factor (Fig. 20.2), it generally favors live-fast-and-die-young strategies.

For example, individuals may mature late, resulting in more time to achieve some characteristic such as large body size that increases their reproductive value *at the time of maturation*. Alternatively, they may mature early, resulting in a suboptimal reproductive value at maturation, but also in a shorter waiting time, and thus a higher probability of surviving to maturation and realizing that reproductive value. If there were no mortality, reproductive value at the time of maturation would alone determine the evolutionarily favorable option. However, mortality risk adds a penalty to delayed reproduction, and if mortality risk is very high, delayed reproduction is close to suicidal in evolutionary terms. Similarly, saving energy by reducing current reproductive effort in favor of current growth or future reproductive effort may pay if there is a future—but increased mortality quickly erodes these expected future benefits.

Although harvesting can drive evolution even when it is unselective, the evolutionary consequences of harvesting are often exacerbated by a harvest's selective nature. Such selectivity can be intentional or unintentional. Intentional selection is

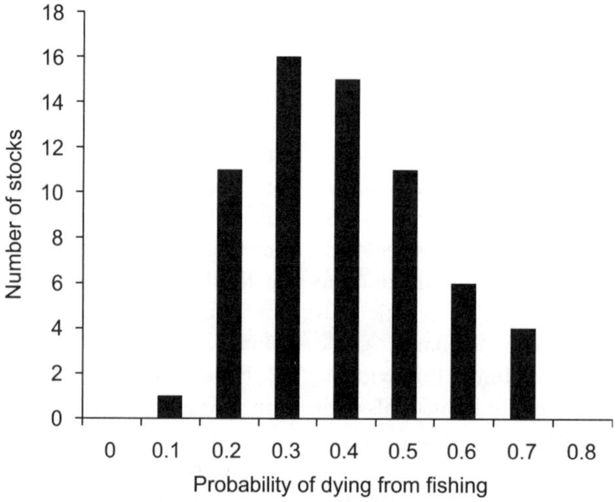

FIGURE 20.2 Estimated annual fishing mortality for 64 fish populations in the northeast Atlantic. On average, approximately 40% of individuals in the targeted age classes are removed each year. Natural mortality is typically believed to be around 20% on an annual basis. (Data from the International Council for the Exploration of the Sea [www.ices.dk].)

most obvious with respect to visible characteristics such as size; managers often impose such selectivity on populations to improve long-term harvest potential. For example, size limits are regularly used to protect a population's youngest individuals. Sometimes also maximum size limits are enforced to ensure that the largest and most fecund individuals are retained in a population, thus securing the successful recruitment of future generations. It must be appreciated, however, that any trait that lowers the likelihood of being harvested will be favored by selection, even when such selectivity is coincidental from the harvesters' perspective. In particular, such unintentional selection might influence a multitude of behavioral, morphological, and physiological traits such as escape behavior, burst swimming or running speed, risk aversion, and body morphology (Heino & Godø, 2002). Because of this multitude of target traits, some selectivity is largely inevitable. Even if unselective harvest were attempted, this would prove difficult to realize, because individuals are rarely randomly distributed with respect to their body size and many other characteristics.

Inheritance of Traits Affected by Harvesting

The exact genetic basis of traits affected by harvest-induced selection is usually unknown. In general, however, such traits will usually be influenced by many loci, each with small effect, so that genotypic variation among individuals is continuous. Inheritance of such quantitative traits can be described through the quantitative genetics framework (see, for example, Falconer & Mackay, 1996).

Because nongenetic factors affect phenotypic variability, the transmission of phenotypic traits from parents to offspring is almost never complete. The degree to which offspring phenotypes are correlated with, and thus can be predicted by, the corresponding midparental phenotypes is called *heritability*. Heritability is determined both by past selection history (which shapes genetic variability in the traits in question) and by current environmental conditions (which impinge on trait expression through developmental noise and phenotypic plasticity). In the first approximation, the speed of evolution is proportional to the strength of selection and to heritability. Although both of these factors are difficult to quantify accurately in the wild, all available evidence suggests that selection pressures associated with harvesting are often strong, and that the relevant traits have at least moderate heritability. With these ingredients, harvest-induced evolution can readily proceed.

Evidence for Harvest-Induced Evolution

The recent decade has seen increased acceptance of the fact that rapid contemporary evolution is commonplace (Hendry & Kinnison, 1999; Stockwell et al., 2003), with a growing number of such examples resulting from studies of harvest-induced evolution.

Most terrestrial examples concern hunting that is selective for sexual ornaments (Coltman et al., 2003; Harris et al., 2002; Jachmann et al., 1995). The idea is simple: If hunters preferentially shoot animals with the largest ornaments (like male horns or antlers), then an individual heavily investing in such ornaments is less likely ever to reap a payoff from its costly investments into these structures. Males with less developed ornaments will then gain a selective advantage, even if they would not gain many mating opportunities under more natural conditions. There are several models that confirm that this scenario could work, but much of the evidence remains suggestive (Box 20.1). An exception is provided by a study of bighorn sheep by Coltman and colleagues (2003) that we discuss in more detail later.

Another scenario occurs when managers use antler size as a surrogate of age when enforcing a harvesting strategy that targets a certain-age interval. Unfortunately, however, in many species, such as white-tailed deer (*Odocoileus virginianus*), genetic variation in antler size at age implies that genotypes delaying the development of complex antlers can avoid the hunting pressure for longer (Strickland et al., 2001), resulting in an undesired hunting-induced selection pressure. In other cases, managers take a step toward animal breeding and encourage hunting of deer with simple antlers in an attempt to improve the quality of trophies in the long run. Either way, harvest-induced selection for antler characteristics is clearly present, whereas explicit empirical evidence remains suggestive.

There are also examples suggesting harvest-induced evolution in wild plants (Law & Salick, 2005; McGraw, 2001). Harvesting plants can be much like harvesting animals by being destructive and positively size selective. Selectivity may arise

BOX 20.1 Detecting Harvest-Induced Evolution

Demonstrating harvest-induced evolution with phenotypic field data is inherently difficult. Typically, one has to work with a time series showing a trend in a characteristic assumed to be under harvest-induced selection. There are a number of pitfalls that must be avoided before one can credibly attribute such change to harvest-induced evolution.

The first group of pitfalls arises from the need to prove that the observed phenotypic change really is evolutionary, and thus possesses a genetic basis. Changes in phenotypic distributions can simply result from the direct demographic effects of sustained selection. For example, if males with large antlers are always culled, then mean antler size must obviously end up being lower compared with a situation without culling, even if variation in antler size is not genetic. Similarly, whenever age at maturation is variable, increasing mortality results in a population with lower mean age, and thus also with lower age at maturation. Such demographic changes illustrate selection in action and are necessary for evolution to take place, but they alone are not sufficient evidence for evolution. The second source of nongenetic variability in phenotypic data is plasticity. Plastic changes result from the effect of environmental conditions on the translation from genotype to phenotype and may thus occur in response to just about any change in the environment. Also, harvesting may trigger plastic changes by resulting in lower population abundance and improved resource availability. After environmental conditions are restored, the corresponding plastic changes are expected to disappear within a generation or less.

There are only two ways to prove genetic change. The seemingly most appealing method is to use molecular genetic data. Two practical obstacles are immediately evident: lack of historic tissue samples and lack of identified genes determining the trait in question. The other method is to conduct common-garden experiments, which compare populations that supposedly differ genetically by exposing them to exactly the same environment; phenotypic differences remaining under such circumstances must have a genetic basis. This method is most useful for comparing extant populations of recent common ancestry. Species introductions have sometimes resulted in such seminatural experiments. By contrast, this method cannot be applied to comparing populations in time to corroborate the genetic basis of phenotypic trends, unless live samples of the ancestral population have been faithfully preserved.

If strict proof of genetic change is not possible, one has to try to make the best use of phenotypic data. One option is to capture plastic effects and genetic effects through multiple regression analysis. If plastic effects are not sufficient (and genetic effects are thus required) to explain observed phenotypic patterns, then the case for evolution is strengthened. Swain and colleagues (2007) provide a recent example of this approach. The second option is to use reaction norms. By definition, the estimation of reaction norms requires environmental variability to be observable. Maturation reaction norms typically include the age and size of individuals, and thus their growth rate as well, as explanatory variables and account for demographic effects, in addition to growth-related plasticity effects (Box 20.2). For both of these approaches, a fundamental limitation always remains: One can never exclude the possibility that some unaccounted environmental factor is triggering the phenotypic changes through a plastic response that is unknown or not considered. This possibility can be minimized through the careful analysis of potentially relevant environmental factors.

A second group of pitfalls arises from the need to prove that the observed phenotypic changes were caused by harvesting, and not by some other selective force. Unfortunately, a study based on historical data—without replication and controls—is the weakest

(continued)

possible setting for showing causal relationships. However, credibility of harvest-induced selection as the most likely causal factor may be increased in a number of ways. First, we can independently evaluate alternative hypotheses and determine whether harvest-induced evolution arises as the most credible hypothesis. Second, although replication in the strict sense is typically infeasible for important resource populations, populations subject to the same "treatment" (namely, increased harvest mortality) are plentiful. A large number of fish stocks from different species and geographic areas are showing similar changes in their maturation reaction norms in response to sustained and elevated harvest mortality (Table 20.1). This ubiquity of analogous trends is suggestive of a common explanation. Third, one can carefully construct a model, incorporating harvesting as well as other potential selective forces, to determine to what extent observed patterns are reproduced in the model. Such models are useful for assessing whether documented phenotypic change could result from selection-induced genetic change, for evaluating whether the speed of phenotypic changes is compatible with such an explanation, and for examining whether harvest-induced selection is among the main driving forces.

either because large plants are easier to spot or because they are more valuable (Mooney & McGraw, 2007). Also, nondestructive utilization of plants can elicit selection pressures, but these will be more subtle.

The majority of examples of harvest-induced evolution deal with fish, and the bulk of them have focused on commercial fisheries. The fisheries-induced selection elicited by modern exploitation has thus been likened to a large-scale, uncontrolled experiment in life history theory (Rijnsdorp, 1993). As shown in Figure 20.2, the additional mortality imposed by industrial-scale fisheries can be very high. In typical fisheries, both immature and mature individuals above a certain size limit are harvested. Theoretical predictions on maturation evolution are then rather clear-cut: Evolution is expected to cause earlier maturation at smaller size (Ernande et al., 2004; Heino, 1998; Law & Grey, 1989). Experiments with fish agree with theoretical predictions (Reznick & Ghalambor, 2005). In addition, field studies corroborate these expectations: Trends toward earlier maturation are ubiquitous in commercially exploited fish stocks (Trippel, 1995). Furthermore, analyses utilizing maturation reaction norms (Box 20.2) have helped to conclude that these trends cannot be explained by mere demographic changes or by growth-related phenotypic plasticity (Table 20.1). Any single study based on phenotypic field observations will of course always be subject to alternative interpretations, so that the suggestion that observations are most parsimoniously interpreted in terms of fisheries-induced evolution may be challenged. However, when a large number of independent studies suggest the same pattern, the case for fisheries-induced evolution is significantly strengthened, especially when those studies are taxonomically and geographically diverse (Table 20.1).

In semelparous fish, Rutter's (1902) prediction that positively size-selective fishing favors slower growth seems theoretically robust and has been verified in experiments (Conover & Munch, 2002). Field evidence comes from Pacific salmon. Ricker (1981, 1995) concluded that evolution of slower growth was likely contributing to declining trends in size at maturation of pink salmon (*Oncorhynchus gorbuscha*) and coho salmon (*O. kisutch*). For iteroparous fish, the story is more complicated. Slower adult growth will result from earlier maturation, but evolution of juvenile growth is more complicated (D. Boukal et al., work in progress; E. Dunlop et al., work in progress). Empirical evidence is limited to one population of Atlantic cod (*Gadus morhua*), for which Swain and colleagues (2007) concluded that a genetic decline in juvenile growth had likely occurred.

Consequences of Harvest-Induced Evolution

Is harvest-induced evolution beneficial? As we shall see next, there is no simple answer to this general question. There certainly exist conditions under which harvest-induced evolution makes resource populations more resilient to harvesting. This is

BOX 20.2 Fisheries-Induced Evolution and Maturation Reaction Norms

There is a ubiquitous trend toward earlier maturation in exploited fish stocks (Trippel, 1995). At first glance, this would seem to support unambiguously the hypothesis that fishing selects for earlier maturation (Beacham, 1987; Borisov, 1978; Law & Grey, 1989). However, because maturation is a very plastic trait, readily influenced by resource availability and other factors (Bernardo, 1993), it was believed for many years that mere plastic responses to the increased resources availability in fished-down stocks were sufficient to explain the observed maturation trends. This problem of disentangling plastic and genetic changes was essentially considered unsolvable in the field. However, as already pointed out by Rijnsdorp (1993), it is possible, through careful statistical analysis, to isolate certain plastic effects in maturation trends. In particular, probabilistic maturation reaction norms offer an elegant approach to identifying growth-related phenotypic plasticity in maturation based on commonly available data (reviewed in Dieckmann & Heino, 2007).

In general, a reaction norm is the profile of phenotypes that a genotype produces across a given range of environmental conditions. A reaction norm for age and size at maturation describes how variability in growth conditions, reflected by variations in size at age, influences maturation (Stearns & Koella, 1986). A probabilistic maturation reaction norm (Box Fig. 20.2) measures the *probability* with which an immature individual that has reached a certain age and size matures during a given time interval (Heino et al., 2002a). More important, this probability is conditional on having reached the considered combination of age and size—in other words, on surviving until, and growing to, this age and size. Through this definition, probabilistic maturation reaction norms allow considering the maturation process separately from survival and growth effects.

The introduction of probabilistic maturation reaction norms has opened the way for a large range of case studies (Table 20.1). Although only certain confounding effects are accounted for (such as those related to survival and to growth-related plasticity), the consistency of findings throughout these case studies strongly supports the hypothesis that fisheries-induced evolution toward earlier maturation is commonplace.

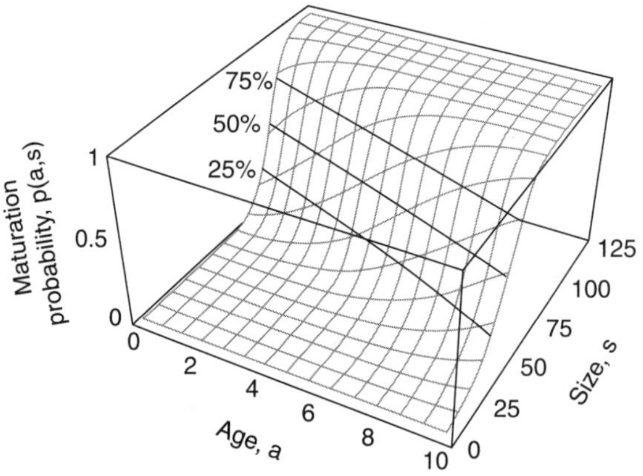

Box Fig 20.2 Schematic illustration of a probabilistic reaction norm for age and size at maturation. The reaction norm describes the probability for a juvenile individual to mature depending on its age and size. Often, only some probability contour lines projected on the age–size plane are shown, instead of the whole three-dimensional probability surface (compare with Fig. 20.4).

TABLE 20.1 Overview of Case Studies in which Probabilistic Maturation Reaction Norms Have Been Used to Help Disentangle Demographic, Plastic, and Evolutionary Effects in Maturation

Species	Population or Stock	Period with Data	Evolutionary Trend in Maturation Reaction Norm?	Reference
Atlantic cod *Gadus morhua*	Northeast Arctic	1932–1998	Yes	Heino et al., 2002b
	Georges Bank	1970–1998	Yes	Barot et al., 2004
	Gulf of Maine	1970–1998	Yes	
	Northern (2J3KL)	1977/1981–2002	Yes	Olsen et al., 2004
	Southern Grand Bank (3NO)	1971–2002	Yes	Olsen et al., 2005
	St. Pierre Bank (3Ps)	1972–2002	Yes	
Haddock *Melanogrammus aeglefinus*	Georges Bank	1968–2002	Yes	L. O'Brien et al. (in preparation)
Plaice *Pleuronectes platessa*	North Sea	1957–2001	Yes	Grift et al., 2003, 2007
American plaice *Hippoglossoides platessoides*	Labrador–NE Newfoundland (2J3K)	1973–1999	Yes	Barot et al., 2005
	Grand Bank (3LNO)	1969–2000	Yes	
	St. Pierre Bank (3Ps)	1972–1999	Yes	
Sole *Solea solea*	Southern North Sea	1958–2000	Yes	Mollet et al., 2007
Atlantic herring *Clupea harengus*	Norwegian spring spawning	1935–2000	Yes, but weak	Engelhard & Heino, 2004
Grayling *Thymallus thymallus*	Lake Lesjaskogsvatnet, Norway	1903–2000 (ca. 15 years)	Yes	T. Haugen et al. (in preparation)
Smallmouth bass *Micropterus dolomieu*	Opeongo Lake, Ontario, Canada	1937–1990	No	Dunlop et al., 2005

With the exception of grayling and smallmouth bass, these fish stocks all are, or have been, subject to intensive commercial exploitation. All but one case study show a significant trend in maturation reaction norms, which means that changes in age at maturation cannot be explained by demographic changes or growth-related phenotypic plasticity alone. This suggests the existence of an evolutionary trend, even though this can never be shown conclusively with phenotypic field data.

positive both for the resource population and for those harvesting it. However, harvest-induced evolution also affects the utility of resource populations in other ways. In particular, our current understanding suggests that both the quality and the quantity of harvest are often likely to decline as a result of harvest-induced evolution.

Harvest-induced evolution is adaptation to harvesting, which sounds like a good thing to happen—at least from the perspective of the harvested populations. In concrete terms, adaptation to harvesting can often be envisaged as a resource population's evolving to avoid harvesting pressure. Individuals may become more difficult to find and catch, they may avoid expressing or developing characteristics that make them prime targets, or they may minimize the duration of those parts of their life cycle spent in stages particularly vulnerable to harvesting. For example, individuals could become more wary about human contraptions such as traps, nets, and hooks, or, if hunters preferentially kill animals with large ornaments, delay developing such ornaments. Thus, for a given population abundance and harvesting effort, the catch after harvest-induced evolution is expected to be lower than it was before.

Based on these considerations, one might expect that harvest-induced evolutionary changes always render resource populations more resilient against harvesting. However, there are several caveats to this simple conclusion. First, adaptation to harvesting usually implies that individuals become less well adapted to aspects of their "natural" environment. Evolution is a balancing act, so that sacrificing adaptedness to natural selection will pay off in evolutionary terms when harvesting pressures are high. However, environmental conditions will change over time. During a period of favorable climate, for example, natural selection may be relaxed, so that a population can evolve mostly in response to harvest-induced selection. After some harvest-induced evolution under favorable environmental conditions, such a population may become increasingly vulnerable to periods of unfavorable climate. This scenario is not as far-fetched as it may initially sound. Slow life histories with long reproductive life spans are often understood as adaptations to variable recruitment success. Fast life histories favored by harvesting may then do well most of the time, but are occasionally bound to receive severe "punishment" during periods of environmental adversity. If harvesting had already pushed such populations to the limit of their demographically sustainable exploitation, harvest-induced evolution in conjunction with adverse environmental periods may thus induce population declines, or even collapses.

A second caveat is that humans are very cunning predators. They will not simply sit and wait while their resource species are gradually escaping harvesting through harvest-induced evolution, but instead will adjust their harvesting practices and preferences. Human preferences and aspiration levels tend to be more relative rather than absolute. For example, everybody knows that fish were bigger and trophy antlers more common in the past, but beating decadal records (or at least one's neighbor) will already bring full satisfaction. Technological development has a similar effect. Given the same harvesting effort and resource abundance, more and more catch will be obtained as harvest technology progresses. Thus, the adaptive system of resources and their harvesters may not converge to equilibrium, but harvesters will continue to drive evolution of the resources further and further away from their "natural" state.

A third caveat results from an insidious aspect of evolution. Evolution, in general, is not driven by what is the best for a population as a whole, but by what best serves the selfish interests of individuals. It is therefore not guaranteed that evolution results in a population's abundance being maximized, and it is quite possible that this abundance declines as a result of the population's evolution (Mylius & Dieckmann, 1995). In the extreme, a population may undergo what is known as *evolutionary suicide*—meaning, through gradual adaptation, the population may evolve to a combination of adaptive traits for which it no longer is viable, suddenly crashing to extinction (for reviews see Dieckmann & Ferrière, 2004; Parvinen, 2005). Although no empirical examples of evolutionary suicide have been documented to date, it should be kept in mind that collapses or extinctions driven by selection pressures will often appear indistinguishable from ecologically driven extinctions, unless special care and attention is exercised in collecting and analyzing the relevant data. In theory at least, the potential for evolutionary suicide has been demonstrated for populations harvested based on fixed quotas (B. Ernande et al., work in progress). Adaptation to harvesting leads to the reduction of harvestable biomass, which, under fixed quota regimes, translates into elevated harvesting mortality. This triggers a further evolutionary decline in the harvestable biomass and thus a further increase in harvesting mortality, and so forth. One could expect that this ecoevolutionary feedback process leads to steadily declining population biomasses, but this is not always the case. Instead, discontinuous transitions to extinction may occur suddenly and from rather large population sizes, without obvious prior warning signals.

Although the three previous caveats explain why evolution cannot be relied on when trying to ensure an exploited population's persistence, it has to be borne in mind that, in addition, harvested populations are not only managed for their continual existence, but also for sustained harvest of good quantity and quality. Alas, theoretical studies suggest that the effects of harvest-induced evolution on harvest quantity and quality are largely negative (Heino, 1998; Law & Grey, 1989). Some of these predictions have already been confirmed empirically (Conover & Munch, 2002; Edley & Law, 1988): Yields may decrease, and the average size of harvested individuals may decline.

Populations adapted to harvesting may also have less capacity to rebound after harvesting pressures are relaxed. After a period of intense harvesting, the exploited population will have become increasingly adapted to harvesting, and thus less adapted to its natural environment. Genotypes best adapted to the natural environment will often have a higher potential rate of population growth than genotypes adapted to harvesting, so that the reduction in the frequency of the former at the expense of the latter will slow down recovery after harvesting is relaxed (K. Enberg et al., work in progress). For example, Hutchings (2005) has shown that maturation changes in northwest Atlantic cod stocks have likely led to a reduction in their potential rate of increase. This may be one factor contributing to the very modest rates of recovery these cod stocks are showing, despite a prolonged period of no or little fishing.

Independent from how harvest-induced evolution may change the resilience or value of a resource population, one can argue that maintaining the natural genotypic integrity of a resource population has an intrinsic value. For example, hunting elephants for ivory favors an increased frequency of tuskless female elephants (Jachmann et al., 1995). Most people would argue that elephants are best off as they are—with tusks—irrespective of whether tusks increase the commercial value or the natural viability of elephants.

We can conclude that harvest-induced evolution can improve the resilience of a harvested population (in the sense of being compared with the, usually only hypothetical, state of the same population subject to the same harvest regime but unchanged by harvest-induced evolution). From a strictly conservationist standpoint, this positive effect may dominate the overall picture, at least when considered to trade off favorably against concerns about preserving a population's ancestral genotypic composition. However, harvested populations are managed for utilitarian benefits. These benefits—sustained harvest and quality of harvest—are likely to deteriorate under harvest-induced evolution.

CASE STUDIES OF HARVEST-INDUCED EVOLUTION

Northeast Arctic Cod

Northeast Arctic cod is a stock of Atlantic cod (*G. morhua*) that uses the Barents Sea as its main feeding area. During the spawning season, mature cod migrate against the ocean currents to their spawning grounds off the northwest coast of Norway. Eggs and larvae are then taken by the currents back to the nursery and feeding areas in the Barents Sea, where juveniles remain until they mature. This separation of feeding and spawning grounds enables two fisheries, with evolutionary consequences of harvest that are strikingly different (Law & Grey, 1989). The feeder fishery is not selective with respect to maturity status of fish and will favor earlier maturation through the mechanisms explained earlier. The spawner fishery significantly affects only mature fish, and implies a selection pressure for delayed maturation. The latter serves as an example of how a population may evolve to become less exposed to harvesting. The centuries-long history of the spawner fishery may even have been responsible for the late maturation historically documented for this stock. The feeder fishery, in contrast, was only developed from the 1920s onward and became dominant after World War II. One must expect this reversal of the selective landscape to have strong evolutionary consequences—and this is exactly what data on age and size at first spawning are showing (Fig. 20.3). Estimated probabilistic maturation reaction norms (Box 20.2)

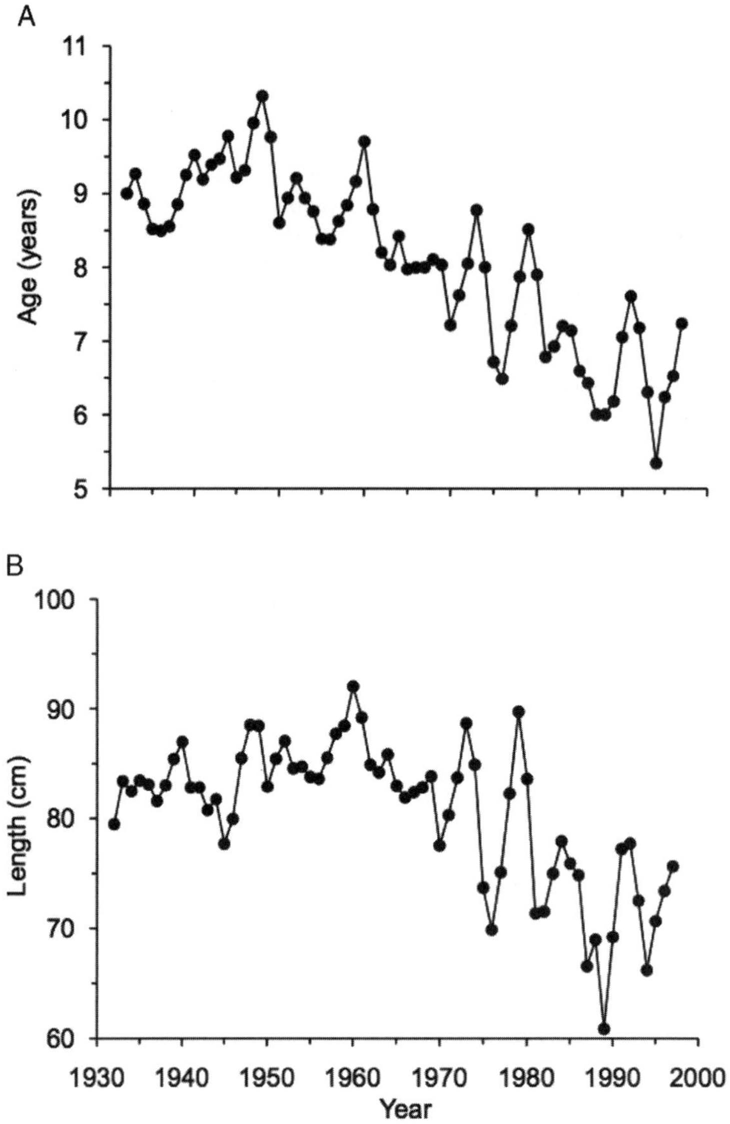

FIGURE 20.3 (A, B) Changes in mean age at maturation (A) and size at maturation (B) in Northeast Arctic cod (*Gadus morhua*). (From Heino et al. [2002b].)

suggest that the trend in maturation indeed has a large evolutionary component (Heino et al., 2002b).

Models indicate that this stock could not sustain current fishing pressures if it had retained the historical pattern of delayed maturation. However, there are also undesirable consequences of harvest-induced evolution. The average size of cod has dropped (small cod are less valuable per kilogram than large ones), and maximum sustainable yield may have decreased significantly (Heino, 1998; Law & Grey, 1989). Recruitment has probably been negatively affected because small, young female cod produce eggs of lower quality than large, old females (Ottersen et al., 2006). Furthermore, delayed maturation allowing for large adult body sizes is particularly important for a stock that undertakes a long, energetically demanding spawning migration (as relative migration cost declines with size) and inhabits a climatically extreme and variable environment. A period of poor feeding conditions hits the smallest cod hardest, and therefore is especially dangerous for a stock in which the mean size of spawning adults has declined. Thus, although fisheries-induced evolution may have saved this stock from a harvest-induced collapse, this rescue comes at high costs. It might therefore have been better to avoid, or at least significantly redress, the fishing regime that has led to this evolutionary response in the first place. Harvest-induced evolution can thus be viewed as having obscured and delayed this realization.

What if managers were to attempt restoring the maturation schedule of Northeast Arctic cod? The current selection pressures would have to be reversed by switching back to the historical harvesting pattern. Unfortunately, there is a pronounced asymmetry in these selection pressures. Although the current harvesting pattern creates strong selection for early maturation, the historical harvesting pattern results in no more than mild selection for delayed maturation (Law & Grey, 1989). Our own analyses suggest that the evolutionary recovery of Northeast Arctic cod would thus take centuries.

This sobering estimate may even be deemed "optimistic," because resuming the historical harvesting pattern is hardly feasible. At the national level, within Norway, a challenge results from the fact that such drastic regulation would benefit only a certain segment of the fishing fleet (mostly small vessels operating in the spawning grounds), whereas another segment (big trawlers operating in the feeding grounds) would suffer. Cod fishing would also become increasingly seasonal, against the interests of consumers and the fishing industry. At the international level, the challenge is that the spawner fishery takes place deep in the Norwegian fishing zone, partly within the country's territorial waters. In such a setting, it is not obvious how access rights could be granted to other fishing nations such as Russia that would suffer most from much reduced fishing in the Barents Sea.

Northern Cod

The populations of Atlantic cod in the northwest Atlantic, off the coast of Canada and the northeastern United States, supported major fisheries for hundreds of years, but largely collapsed in the late 1980s and early 1990s. Many of these stocks have not yet recovered. Perhaps the most famous of these collapses was that of so-called *northern cod*, a stock complex off southern Labrador and eastern Newfoundland. Closure of the fisheries in the early 1990s has not brought the stock back, and its current abundance is estimated to be about 1% of that in the 1960s. Having once been the mainstay of Newfoundland's economy, the collapse and closure of the cod fisheries caused much economic and social hardship. Considerable effort has thus been invested in trying to understand why the stocks collapsed (and later on, why the recovery has remained pending for so long). It is now evident that excessive fishing pressure was the main cause of the collapse, whereas a period of unfavorable ocean climate made things even worse by contributing to triggering the collapse.

In contrast to Northeast Arctic cod, northern cod exhibits no clear separation of spawning and feeding grounds. Correspondingly, the northern cod fisheries have always targeted a mixture of immature and mature cod above some size threshold. The theoretical prediction under such a harvesting pressure is that the affected population will evolve toward earlier maturation. This prediction is supported by data on northern cod collected through research surveys: The age at which 50% of the females were mature dropped from about 6.5 years in the 1960s to about 6 years in the mid 1980s and to about 5 years in the mid 1990s (Morgan, 2000). Much of the drop occurred during a period of poor growth and body condition, which would instead have been expected to elicit the opposite phenotypic response had it been based on growth-related phenotypic plasticity. Probabilistic maturation reaction norms (Fig. 20.4) indeed suggest that that the observed changes in maturation were not merely phenotypically plastic, but also reflected genetic changes in maturation schedules (Olsen et al., 2004).

Did the change in maturation schedules affect the collapse? This is a question that is only now being investigated. Earlier maturation was favored by high mortality and might have increased the stock's capacity to sustain the harvesting. On the other hand, it is quite possible that earlier maturation at small size has been a costly strategy, in terms of lost resilience to poor feeding conditions. What appears to be clear is that the changed maturation schedule is not benefiting the stock's recovery potential. Hutchings (2005) estimated that an early-maturing population may suffer from a 25% to 30% reduction in its maximum annual population growth rate.

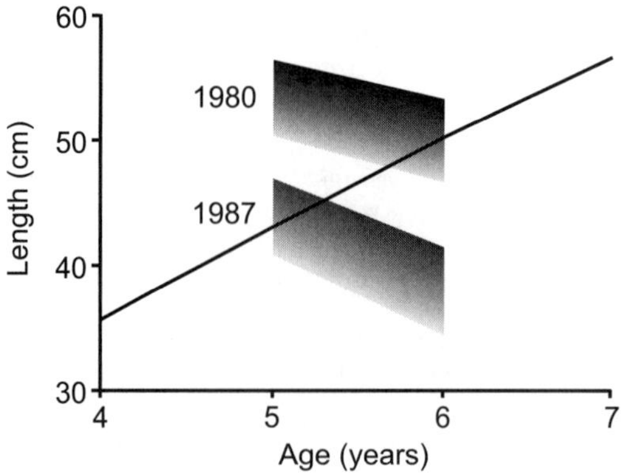

FIGURE 20.4 Probabilistic maturation reaction norms for female northern cod off southern Labrador born in 1980 and in 1987. Shaded rectangles show how the estimated maturation probability increases from 25% to 75% for the two considered ages. The solid line depicts the average growth trajectory. For the same growth trajectory, cod of the 1987 cohort reached a similar probability of maturing already 1 year earlier than cod of the 1980 cohort. (Data from Olsen and colleagues [2004].)

Irrespective on the possible effect of harvest-induced changes in maturation schedules on the collapse, these changes could have been used as an early warning signal. Drastic changes in maturation do not occur without strong selection pressures, and can thus serve as indicators that fishing mortality may affect a resource population more strongly than is advisable. In particular, Olsen and colleagues (2004) have shown that the changes in the maturation schedule of northern cod could have been detected up to a decade before the collapse of this stock became reality.

Mountain Sheep

Mountain sheep (*Ovis* spp.) in the Rocky Mountains are valuable targets of strictly regulated trophy hunting. Rams with big horns are the most sought-after targets for sport hunters, and harvesting is also legally limited to individuals fulfilling specific requirements for horn size. Consequently, rams with big horns are more likely to be shot at an early age. On the other hand, the mating success of rams increases with their dominance rank, age, and horn length. Thus, trophy hunting selects against those rams that naturally achieve the highest expected mating success.

Coltman and colleagues (2003) analyzed more than 30 years of data from a bighorn sheep (*O. canadensis*) population in Alberta, Canada. Not unexpectedly, horn length at a certain age shows a clear decline in this study population (Fig. 20.5). What is unique to this study is that Coltman and colleagues (2003) also had access to genetic data allowing pedigree reconstruction, which enabled them to show that horn size was highly heritable and that the observed decline in horn size was genetic.

Does declining horn size matter? One might be tempted to think that the absolute size of sexual ornaments is not that important: In male–male contests, it is the relative size differences that matter. As long as some variability in horn size persists, males should be able to establish their hierarchy just as before. Therefore, although undesirable from the trophy hunters' perspective, the documented decline in horn size might be rather inconsequential for the viability of the population.

Nonetheless, sexually selected traits do not exist in isolation from other traits. Indeed, Coltman and colleagues (2003) showed that there is a strong and positive genetic correlation between horn size and body weight. Consequently, selection for smaller horns will also select for smaller body size. Smaller body size could have direct negative effects

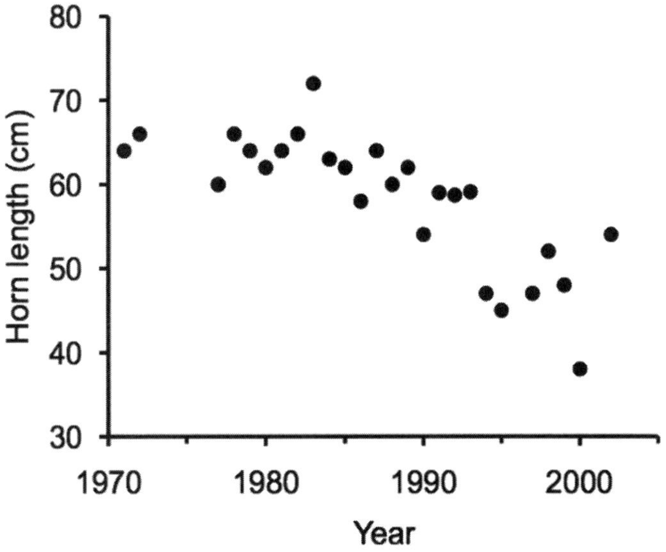

FIGURE 20.5 Average horn length in 4-year old rams of bighorn sheep. (Data from Coltman and colleagues [2003].)

on, for example, overwintering survival and parasite resistance (positive genetic correlation between parasite resistance and body size has been demonstrated in another sheep species). Thus, evolution driven by trophy hunting may indeed reduce the viability of the bighorn sheep population.

However, in thinhorn sheep (*O. dalli*) populations in Yukon Territory, Canada, horn growth seems to have remained essentially constant for the past 40 years (Loehr et al., 2007), despite selective harvest similar to that in bighorn sheep. Loehr and associates (2007) ascribe this to the positive correlation between natural mortality and horn growth: Size-selective harvest leads to higher mortality for rams with fast horn growth, but these rams would also face higher mortality without hunting. The correlation of horn size and horn growth with other traits related to fitness seems to vary significantly among different species. Such diversity impedes general conclusions, highlighting the need for wildlife managers to develop population-specific insights into the evolutionary implications of harvesting.

CONCLUDING REMARKS

Harvest-induced evolution has many facets. The hunting of mountain sheep is intentionally selective, whereas the historic fishing selectivity for Northeast Arctic cod was a by-product of spatial population structure. Harvest of northern cod was primarily size selective, although the evolutionary selection pressure mostly originated from the overall increase of mortality.

In all cases, significant evolutionary changes have taken place, although it is only for bighorn sheep that we can conclude with high certainty that changes were genetic; because of the lack of genetic sampling, evidence for fisheries-induced evolution is less direct. Despite this evidence of harvest-induced evolution, management in all three cases has been devoid of evolutionary awareness. This has come at a cost. The quality and quantity of Northeast Arctic cod catch has likely declined substantially, and without harvest-induced evolutionary responses, the stock might in fact have collapsed. The quality of bighorn sheep harvest has declined, and a decrease in the viability of the population is suspected. Also for northern cod, the quantity of potential harvest has likely declined, although it seems moot to agonize about such losses when a stock has collapsed and has yet to recover. And the role harvest-induced evolutionary changes have played in this tragedy still remains to be understood.

What could be done better? Lowering harvest pressure will almost certainly help to slow the pace of harvest-induced evolution. Further options depend on details that are specific to a population and its harvesting regime. Shifting the bulk of harvesting of Northeast Arctic cod back to its spawning grounds would bring long-term benefits, but is hardly a realistic option because of social and political constraints. An analogous option is not available at all for northern cod, for which feeding and spawning grounds overlap; the remaining option here is to assess whether size limits could be set to minimize unwanted evolutionary impacts. The case of bighorn sheep would benefit from prioritizing management objectives. The size limit for hunting was increased in 1996 to a level that reduced harvest-induced selection as well as the harvest itself. If a supply of prime trophies were the first priority, managers could allow culling rams with small horns to give a selective advantage to rams with the potential to become top-class trophies. On the other hand, if the quantity of harvest is also deemed important, a less selective harvest might offer a good compromise.

FUTURE DIRECTIONS

Harvest-induced evolution poses an additional challenge to achieving sustainable harvesting: What appears to be ecologically sustainable may prove evolutionarily detrimental, which, in turn, may exert a negative feedback on ecological sustainability. A gradual awakening to this challenge has occurred only during the past two decades. We now have a fair understanding of harvest-induced selection pressures, and evidence is accumulating that certain harvest-induced changes are most parsimoniously interpreted as being of evolutionary nature. However, there still are many gaps in our knowledge of harvest-induced evolution that need to be addressed by future research:

- Empirical evidence of harvest-induced evolution is restricted to just a few types of traits (mostly maturation traits and sexually selected traits such as antlers and horns). Presumably this reflects more what types of trait are amenable to observation and analysis, rather than which traits are prone to evolve rapidly in response to harvesting.
- Similarly, the reported empirical evidence stems from just a few taxonomic groups (primarily fish and ungulates). Again, this is likely to reflect the availability of data, rather than the fact that these particular taxonomic groups are more vulnerable to harvest-induced evolution than others.
- The repercussions of selective harvesting based on sexually selected traits remains poorly understood. How do natural, harvest-induced, and sexual selection interact?
- Our understanding of the demographic consequences of harvest-induced evolution is still scant. Is evolutionary suicide a likely outcome? How do harvest-induced evolutionary changes affect the likelihood of population collapses and the potential for subsequent recoveries?
- Evolution always implies genetic change, but we still know little about the genetics of fitness-related traits in the wild. Observing evolution is mostly based on indirect evidence rather than on the direct identification of genetic change. We expect that the rapid development of molecular genetics will soon facilitate the direct detection of harvest-induced evolution. On the other hand, classic approaches to studying evolution based on phenotypic observations are far from being fully explored. For example, developing the quantitative genetics of complex traits such as maturation reaction norms will facilitate detecting and managing harvest-induced evolution in such traits.
- Although harvest-induced evolution may have beneficial effects on certain aspects of population resilience, the overall effect will often be deemed negative (1) because the beneficial effects on resilience may be weak or uncertain, (2) because a population's capacity for dealing with adverse environmental conditions may be compromised, and (3) because the quantity and quality of harvest are typically diminished. The obvious conclusion is that harvest-induced evolution should be managed. But are there better ways of managing harvest beyond the obvious recommendation of lowering overall harvest pressure? Are there ways to accelerate evolutionary recovery by carefully crafting the selectivity of harvesting?
- Learning to cope with harvest-induced evolution by trial and error is foolish. However, experimentation is seldom feasible, because resource populations are usually large and too valuable for risky trials, and evolution does not take place overnight. We therefore believe that modeling is an essential tool for developing the scientific basis of the evolutionarily sustainable management of harvested populations. Properly devised models—incorporating sufficient ecological and genetic

detail—will allow virtual experimentation, which will help to understand past evolutionary changes, predict the timescales on which harvest-induced evolution unfolds, and evaluate the expected effects of envisaged management measures.

SUGGESTIONS FOR FURTHER READING

Nelson and Soulé (1987) provide an early account of the evolutionary dimension of harvesting in aquatic systems; many of the points raised there remain topical to date. Stokes and colleagues (1993) and Smith (1994) summarize the state of the art at the beginning of the 1990s, whereas Dieckmann and Heino (2007) present an updated overview of fisheries-induced maturation evolution and Jørgensen and colleagues (2007) raise the issue of taking evolutionary perspectives aboard in fisheries management. Harris and associates (2002) offer a terrestrial perspective on the evolutionary consequences of hunting, and Dieckmann and coworkers (2008) give a comprehensive modern account of fisheries-induced evolution.

Dieckmann, U., O. R. Godø, M. Heino, & J. Mork. 2009. Fisheries-induced adaptive change. Cambridge University Press, Cambridge.

Dieckmann, U., & M. Heino. 2007. Probabilistic maturation reaction norms: Their history, strengths, and limitations. Mar Ecol Prog Ser. 335: 253–269.

Harris, R. B., W. A. Wall, & F. W. Allendorf. 2002.Genetic consequences of hunting: What do we know and what should we do?Wildlife Soc Bull. 30: 634–643.

Nelson, K., & M. Soulé. 1987. Genetical conservation of exploited fishes (pp. 345–368). In N. Ryman & F. Utter (eds.). Population genetics and fishery management. Washington Sea Grant Program, Seattle, Wash.

Smith, P. J. 1994. Genetic diversity of marine fisheries resources: Possible impacts of fishing. FAO fish technical paper no. 344, Food and Agriculture Organization, Rome.

Stokes, T.K., J. M. McGlade, & R. Law (eds.). 1993. The exploitation of evolving resources. Lecture notes in biomathematics 99. Springer-Verlag, Berlin.

Acknowledgments We thank our colleagues who have helped to develop our understanding of harvest-induced evolution. Our work has been supported by the European Community's Sixth Framework Programme (contract MRTN-CT-2004-005578) and the Norwegian Research Council (project 173417/S40).

References

Acevedo-Whitehouse, K., & A. A. Cunningham. 2006. Is MHC enough for understanding wildlife immunogenetics? Trends Ecol Evol. 21: 433–438.

Acevedo-Whitehouse, K., F. Gulland, D. Greig, & W. Amos. 2003. Disease susceptibility in California sea lions. Nature 422: 35.

Achard, F., H. D. Eva, H.- J. Stibig, P. Mayaux, J. Gallego, T. Richards & J.-P. Malingreau. 2002. Determination of deforestation rates of the world's humid tropical forests. Science 297: 999–1002.

Adams, J. R., C. Lucash, L. Schute, & L. P. Waits. 2007. Locating hybrid individuals in the red wolf *(Canis rufus)* experimental population area using a spatially targeted sampling strategy and faecal DNA genotyping. Mol Ecol. 16: 1823–1834.

Adams, J. M., G. Piovesan, S. Strauss, & S. Brown. 2002. The case for genetic engineering of native and landscape trees against introduced pests and diseases. Conserv Biol. 16: 874–879.

ADBSS [Arizona Desert Bighorn Sheep Society] 2004. Wildlife water developments and desert bighorn sheep in the southwestern United States (p. 17). Arizona Desert Bighorn Sheep Society, Mesa Ariz.

Adger, N., P. Aggarwal, S. Agrawala, J. Alcamo, A. Allali, O. Anisimov, N. Arnell, M. Boko, O. Canziani, T. Carter, G. Casassa, U. Confalonieri, R. V. Cruz, E. de Alcaraz, W. Easterling, C. Field, A. Fischlin, B. B. Fitzharris, C. G. García, Clair Hanson, H. Harasawa, K. Hennessy, S. Huq, R. Jones, L. K. Bogataj, D. Karoly, R. Klein, Z. Kundzewicz, M. Lal, R. Lasco, G. Love, X. Lu, G. Magrín, L. J. Mata, R. McLean, B. Menne, G. Midgley, N. Mimura, M. Q. Mirza, J. Moreno, L. Mortsch, I. Niang-Diop, R. Nicholls, B. Nováky, L. Nurse, A. Nyong, M. Oppenheimer, J. Palutikof, M. Parry, A. Patwardhan, P. R. Lankao, C. Rosenzweig, S. Schneider, S. Semenov, J. Smith, J. Stone, J.-P. v Ypersele, D. Vaughan, C. Vogel, T. Wilbanks, P. P. Wong, S. Wu & G. Yohe. 2007. Climate change 2007: Climate change impacts, adaptation and vulnerability. Working Group II contribution to the intergovernmental panel on climate change fourth assessment report. Intergovernmental Panel on Climate Change Secretariat, Geneva.

Agrawal, A. A. 2001. Phenotypic plasticity in the interactions and evolution of species. Science 294: 321–326.

Agrawal, A. A., & P. M. Kotanen. 2003. Herbivores and the success of exotic plants: A phylogenetically controlled experiment. Ecol Lett. 6: 712–715.

Aguilar, A., G. Roemer, S. Debenham, M Binns, D. Garcelon & RK Wayne 2004. High MHC diversity maintained by balancing selection in an otherwise genetically monomorphic

mammal. Proc Natl Acad Sci USA 101: 3490–3494.
Albers, M. A., & T. J. Bradley. 2006. Fecundity in *Drosophila* following desiccation is dependent on nutrition and selection regime. Phys Biochem Zool. 79: 857–865.
Albert, A. Y. K., & D. Schluter. 2005. Selection and the origin of species. Curr Biol. 15: R283–R288.
Aldrich, P. R., J. L. Hamrick, P. Chavarriaga, & G. Kochert. 1998. Microsatellite analysis of demographic genetic structure in fragmented populations of the tropical tree *Symphonia globulifera*. Mol Ecol. 7: 933–944.
Aldridge, B., L. Bowen, B. Smith, G. Antonelis, F. Gulland & J. Stott. 2006. Paucity of class I MHC gene hetergeneity between individuals in the endangered Hawaiian monk seal population. Immunogenetics 58: 203–215.
Alekseev, A. N., H. V. Dubinina, A. V. Semenov, & C. V. Bolshakov. 2001. Evidence of ehrlichiosis agents found in ticks (Acari: Ixodidae) collected from migratory birds. J Medical Entomol. 38: 471–474.
Alexander, H. M., P. H. Thrall, J. Antonovics, A. M. Jarosz & P. V. Oudemans. 1996. Population dynamics and genetics of plant diseases: A case study of anther-smut disease of *Silene alba* caused by the fungus *Ustilago violacea*. Ecology 77: 990–996.
Allendorf, F. W. (ed.). 1998. Conservation and genetics of marine organisms. J Heredity 89: 377–464.
Allendorf, F. W., R. F. Leary, P. Spruell, & J. K. Wenburg. 2001. The problems with hybrids: Setting conservation guidelines. Trends Ecol Evol. 16: 613–622.
Allendorf, F. W., & N. Ryman. 2002. The role of genetics in population viability analysis (pp. 50–85). In S. R. Beissinger & D. R. McCullough (eds.). Population viability analysis. University of Chicago Press, Chicago, Ill.
Alley, R., T. Berntsen, N. L. Bindoff, Z. Chen, A. Chidthaisong, P. Friedlingstein, J. Gregory, G. Hegerl, M. Heimann, B. Hewitson, B. Hoskins, F. Joos, J. Jouzel, V. Kattsov, U. Lohmann, M. Manning, T. Matsuno, M. Molina, N. Nicholls, J. Overpeck, D. Qin, G. Raga, V. Ramaswamy, J. Ren, M. Rusticucci, S. Solomon, R. Somerville, T. F. Stocker, P. Stott, R. J. Stouffer, P. Whetton, R. A. Wood & D. Wratt. 2007a. Climate change 2007: The physical science basis. Summary for policy makers. Contribution of Working Group I to the fourth assessment report of the intergovernmental panel on climate change. Intergovernmental Panel on Climate Change Secretariat, Geneva.
Alley, R., T. Berntsen, N. L. Bindoff, Z. Chen, A. Chidthaisong, P. Friedlingstein, J. Gregory, G. Hegerl, M. Heimann, B. Hewitson, B. Hoskins, F. Joos, J. Jouzel, V. Kattsov, U. Lohmann, M. Manning, T. Matsuno, M. Molina, N. Nicholls, J. Overpeck, D. Qin, G. Raga, V. Ramaswamy, J. Ren, M. Rusticucci, S. Solomon, R. Somerville, T. F. Stocker, P. Stott, R. J. Stouffer, P. Whetton, R. A. Wood & D. Wratt. 2007b. Climate change 2007: The physical science basis. Summary for policy makers. IPCC fourth assessment report. Intergovernmental Panel on Climate Change Secretariat, Geneva.
Altizer, S., D., Harvell, & E. Friedle. 2003. Rapid evolutionary dynamics and disease threats to biodiversity. Trends Ecol Evol. 18: 589–596.
Altizer, S., C. L. Nunn, and P. Lindenfors. 2007. Do threatened hosts have fewer parasites? A comparative study in primates. Journal of Animal Ecology 76: 304–314.
Altpeter, F., N. Baisakh, R. Beachy, R. Bock, T. Capell, P. Christou, H. Daniell, K. Datta, S. Datta, P. J. Dix, C. Fauquet, N. Huang, A. Kohli, H. Mooibroek, L. Nicholson, T. T. Nguyen, G. Nugent, K. Raemakers, A. Romano, D. A. Somers, E. Stoger, N. Taylor & R. Visser. 2005. Particle bombardment and the genetic enhancement of crops: Myth and realities. Mol Breed. 15: 305–327.
Amos, W., & A. Balmford. 2001. When does conservation genetics matter? Heredity 87: 257–265.
Amsterdam, A., S. Lin, & N. Hopkins. 1995. The *Aequorea victoria* green fluorescent protein can be used as a reporter in live zebrafish embryos. Dev Biol. 171: 123–129.
Anagnostakis, S. L. 1987. Chestnut blight: The classical problem of an introduced pathogen. Mycologia 79: 23–37.
Andelman, S., & W. F. Fagan. 2000. Umbrellas and flagships: Efficient conservation surrogates, or expensive mistakes? Proc Natl Academy Sci USA 97: 5954–5959.
Anderson, E. C. 2005. An efficient Monte Carlo method for estimating N_e from temporally spaced samples using a coalescent-based likelihood. Genetics 170: 955–967.
Anderson, M. P., P. Comer, D. Grossman, C Groves, K Poiani, M Reid, R Schneider, B Vickery & A Weakley 1999. Guidelines for representing ecological communities in ecoregional conservation plans. The Nature Conservancy, Arlington, Virginia.

Anderson, A. R., J. E. Collinge, A. A. Hoffmann, M Kellett & SW McKechnie 2003. Thermal tolerance trade-offs associated with the right arm of chromosome 3 and marked by the *hsr-omega* gene in *Drosophila melanogaster*. Heredity 90: 194–202.

Anderson, P. K., A. A. Cunningham, N. G. Patel, F. J. Morales, P. R. Epstein & P. Daszak. 2004. Emerging infectious diseases of plants: Pathogen pollution, climate change, and agrotechnology drivers. Trends Ecol Evol. 19: 535–544.

Anderson, A. R., A. A. Hoffmann, S. W. McKechnie, P. A. Umina & A. R. Weeks. 2005. The latitudinal cline in the *In(3R)Payne* inversion polymorphism has shifted in the last 20 years in Australian *Drosophila melanogaster* populations. Mol Ecol. 14: 851–858.

Anderson, R. M., & R. M. May. 1991. Infectious diseases of humans: Dynamics and control. Oxford University Press, Oxford, UK.

Anderson, E. C., & E. A. Thompson. 2002. A model-based method for identifying species hybrids using multilocus genetic data. Genetics 160: 1217–1229.

Andersson, M. 1994a. Sexual selection. Princeton University Press, Princeton, N.J.

Andersson, S. 1994b. Unequal morph frequencies in populations of tristylous *Lythrum salicaria* (Lythraceae) from southern Sweden. Heredity 72: 81–85.

Andolfatto, P., F. Depaulis, & A. Navarro. 2001. Inversion polymorphisms and nucleotide variability in *Drosophila*. Genet Res. 77: 1–8.

Andreasen, K. 2005. Implications of molecular systematic analyses on the conservation of rare and threatened taxa: Contrasting examples from Malvaceae. Conserv Genet. 6: 399–412.

Angilletta, M. J., Jr., P. H. Niewiarowski, & C. A. Navas. 2002. The evolution of thermal physiology in ectotherms. J Therm Biol. 27: 249–268.

Angilletta, M. J., Jr., R. S. Wilson, C. A. Navas, & R. S. James. 2003. Tradeoffs and the evolution of thermal reaction norms. Trends Ecol Evol. 18: 234–240.

Anonymous, 2007. Bunnies build up resistance to Calicivirus. Australian Broadcasting Corporation News Online, (www.abc.net.au /news/australia/qld/summer/200702/ s1853413.htm), Accessed June 1, 2007.

Antia, R., R. R. Regoes, J. C. Koella, & C. T. Bergstrom. 2003. The role of evolution in the emergence of infectious diseases. Nature 426: 658–661.

Antonovics, J., A. D. Bradshaw, & R. G. Turner. 1971. Heavy metal tolerance in plants. Adv Ecol Res. 7: 1–85.

Antonovics, J., M. Hood, & J. Partain. 2002. The ecology and genetics of a host shift: *Microbotryum* as a model system. Am Nat. 160: S40–S53.

Antonovics, J., & P. H. Thrall. 1994. The cost of resistance and the maintenance of genetic polymorphism in host–pathogen systems. Proc R Soc Lond B. 257: 105–110.

Aparicio, A., R. G. Albaladejo, M. Porras, & G. Ceballos. 2000. Isozyme evidence for natural hybridization in *Phlomis* (Lamiaceae): Hybrid origin of the rare *P.* × *margaritae*. Ann Bot. 85: 7–12.

APHIS [Animal and Plant Health Inspection Service] 2008. Release permits for pharmaceuticals, industrials, value added proteins for human consumption, or for phytoremediation granted or pending by APHIS. United States Department of Agriculture, Animal and Plant Health Inspection Service (APHIS). http://www.aphis.usda.gov/brs/ph_permits.html. Accessed March 4, 2008.

Araujo, M. B. 2002. Biodiversity hotspots and zones of ecological transition. Conserv Biol. 16: 1662–1663.

Araujo, M. B., P. H. Williams, & R. J. Fuller. 2002. Dynamics of extinction and the selection of nature reserves. Proc R Soc Lond B-Biol Sci. 269: 1971–1980.

Arctander, P., & J. Fjeldså. 1994. Andean tapaculos of the genus *Scytalopus* (Aves, Rhinocryptidae): A study of speciation using DNA sequence data. EXS 68: 205–225.

Armbruster, P., & D. H. Reed. 2005. Inbreeding depression in benign and stressful environments. Heredity 95: 235–242.

Armstrong, T. T., R. G. Fitzjohn, L. E. Newstrom, A. D. Wilton & W. G. Lee. 2005. Transgene escape: What potential for crop–wild hybridization? Mol Ecol. 14: 2111–2132.

Arnold, M. L. 1997. Natural hybridization and evolution. Oxford University Press, New York.

Arnold, M. L. 2004. Natural hybridization and the evolution of domesticated, pest and disease organisms. Mol Ecol. 13: 997–1007.

Arnold, M. L. 2006. Evolution through genetic exchange. Oxford University Press, New York.

Arnqvist, G., M. Edvardsson, U. Friberg, & T. Nilsson 2000. Sexual conflict promotes speciation in insects. Proc Natl Acad Sci USA 97: 10460–10464.

Ashman, T.- L., T. M. Knight, J. Steets, P. Amarasekare, M. Burd, D. R. Campbell, M. R. Willson. 2004. Pollen limitation of plant reproduction: Ecological and

evolutionary causes and consequences. Ecology 85: 2408–2421.
Avise, J. C. 1992. Molecular population structure and the biogeographic history of a regional fauna: A case history with lessons for conservation biology. Oikos 63: 62–76.
Avise, J. C. 1994. Molecular markers, natural history and evolution. Chapman & Hall, New York.
Avise, J. C. 2000. Phylogeography: The history and formation of species. Harvard University Press, Cambridge, Mass.
Avise, J. C. 2004a. Molecular markers, natural history and evolution. 2nd ed. Sinauer Associates, Sunderland, Mass.
Avise, J. C. 2004b. The hope, hype, and reality of genetic engineering. Oxford University Press, New York.
Avise, J. C. 2005. Phylogenetic units and currencies above and below the species level (pp. 76–100). In A. Purvis, T. Brooks. & J. Gittleman (eds.). Phylogeny and conservation. Cambridge University Press, Cambridge, UK.
Avise, J. C., J. Arnold, R. M. Ball, Jr., E. Bermingham, T. Lamb, J. E. Neigel, C. A. Reeb & N. C. Saunders. 1987. Intraspecific phylogeography: The mitochondrial DNA bridge between population genetics and systematics. Annu Rev Ecol Syst. 18: 489–522.
Avise, J. C., & J. L. Hamrick (eds.). 1996. Conservation genetics: Case histories from nature. Chapman & Hall, New York.
Avise, J. C., R. A. Lansman, & R. O. Shade. 1979. The use of restriction endonucleases to measure mitochondrial DNA sequence relatedness in natural populations. I. Population structure and evolution in the genus *Peromyscus*. Genetics 92: 279–295.
Badyaev, A. V. 2005. Stress-induced variation in evolution: From behavioural plasticity to genetic assimilation. Proc R Soc London B. 272: 877–886.
Badyaev, A. V., K. R. Foresman, & R. L. Young. 2005. Evolution of morphological integration: Developmental accommodation of stress-induced variation. Am Nat 166: 382–395.
Baer, B., & P. Schmid-Hempel. 1999. Experimental variation in polyandry affects parasite loads and fitness in a bumble-bee. Nature 397: 151–154.
Baillie, J. E. M., C. Hilton-Taylor, & S. N. Stuart (eds.). 2004. IUPN red list of threatened species. A global species assessment. International Union for the Protection of Nature (IUPN), Gland, Switzerland.
Baker, A. M., J. M. Hughes, J. C. Dean, & S. E. Bunn. 2004. Mitochondrial DNA reveals phylogenetic structuring and cryptic diversity in Australian freshwater macroinvertebrate assemblages. Mar Freshw Res. 55: 629–640.
Baker, C. S., G. M. Lento, F. Cirpiano, & S. R. Palumbi. 2000. Predicted decline of protected whales based on molecular genetic monitoring of Japanese and Korean markets. Proc R Soc Lond B. 267: 1191–1199.
Bakker, R. T. 1980. Dinosaur heresy—dinosaur renaissance: Why we need endothermic archosaurs and a comprehensive theory of bioenergetic evolution (pp. 351–462). In R. D. K. Thomas & E. C. Olson (eds.). A cold look at the warm-blooded dinosaurs. Westview Press, Boulder, Colo.
Balanyà, J., J. M. Oller, R. B. Huey, G. W. Gilchrist & L Serra 2006. Global genetic change tracks global climate warming in *Drosophila subobscura*. Science 313: 1773–1775.
Balanyà, J., L. Serra, G. W. Gilchrist, R. B. Huey, M. Pascual, F. Mestres & E. Sole. 2003. Evolutionary pace of chromosomal polymorphism in colonizing populations of *Drosophila subobscura*: An evolutionary time series. Evolution 57: 1837–1845.
Ball, I. R., & H. Possingham. 2000. MARXAN (V1.8.2): Marine reserve design using spatially explicit annealing: A manual. Great Barrier Reef Marine Park Authority, Townsville, Australia. Available online as of April 2004 at http://www.ecology.uq.edu.au/marxan.htm.
Ballou, J. D., M. Gilpin. & T. J. Foose (eds.). 1995. Population management for survival and recovery. Columbia University Press, New York.
Balmford, A. & W. Bond. 2005. Trends in the state of nature and their implications for human well-being. Ecol Lett. 8: 1218–1234.
Barot, S., M. Heino, M. J. Morgan, & U. Dieckmann. 2005. Maturation of Newfoundland American plaice (*Hippoglossoides platessoides*): Long-term trends in maturation reaction norms despite low fishing mortality? ICES J Mar Sci. 62: 56–64.
Barot, S., M. Heino, L. O'Brien, & U. Dieckmann. 2004. Long-term trend in the maturation reaction norm of two cod stocks. Ecol Appl. 14: 1257–1271.
Barrett, S. C. H. 1988. Evolution of breeding systems in *Eichornia* (Pontederiaceae): A review. Ann Miss Bot Gard. 75: 741–760.
Barrett, S. C. H. 2002. The evolution of plant sexual diversity. Nat Rev Genet. 3: 274–284.

Barrett, S. C. H. 2003. Mating strategies in flowering plants: The outcrossing-selfing paradigm and beyond. Phil Trans R Soc Lond B. 358: 991–1004.

Barrett, S. C. H., & C. G. Eckert. 1990. Variation and evolution of mating systems in seed plants (pp. 229–254). In S. Kawano (ed.). Biological approaches and evolutionary trends in plants. Academic Press, Tokyo.

Barrett, S. C. H., & J. R. Kohn, Jr. 1991. Genetic and evolutionary consequences of small population size in plants: Implications for conservation (pp. 3–30). In D. A. Falk & K. H. Holsinger (eds.). Genetics and conservation of rare plants. Oxford University Press, New York.

Bartlein, P. J., & I. C. Prentice. 1989. Orbital variations, climate and paleoecology. Trends Ecol Evol. 4: 195–199.

Barton, N. H., & M. Turelli. 1989. Evolutionary quantitative genetics: How little do we know? Annu Rev Genet. 23: 337–370.

Bascompte, J., P. Jordano, C. J. Melián, & J. M. Olesen. 2003. The nested assembly of plant–animal mutualistic networks. Proc Natl Acad Sci USA 100: 9383–9387.

Bascompte, J., P. Jordano, & J. M. Olesen. 2006. Asymmetric coevolutionary networks facilitate biodiversity maintenance. Science 312: 431–433.

Bascompte, J., H. Possingham, & J. Roughgarden. 2002. Patchy populations in stochastic environments: Critical number of patches for persistence. Am Nat 159: 128–137.

Beacham, T.D. 1987. Variation in length and age at sexual maturity of Atlantic groundfish: a reply. Environmental Biology of Fishes 19:149–153.

Beard, K. C. 1998. East of Eden: Asia as an important center of taxonomic origination in mammalian evolution. Bull Carnegie Mus Nat Hist. 34: 5–39.

Beckmann, J. P., & J. Berger. 2003. Rapid ecological and behavioural changes in carnivores: The responses of black bears (*Ursus americanus*) to altered food. J Zool. 261: 207–212.

Bedward, M., R. L. Pressey, & D. A. Keith. 1992. A new approach for selecting fully representative reserve networks: Addressing efficiency, reserve design and land suitability with an iterative analysis. Biol Conserv, 62: 115–125.

Beebee, T., & G. Rowe. 2001. Application of genetic bottleneck testing to the investigation of amphibian declines: A case study with natterjack toads. Conserv Biol. 15: 266–270.

Beerli, P., & J. Felsenstein. 2001. Maximum likelihood estimation of a migration matrix and effective population sizes in n subpopulations by using a coalescent approach. Proc Natl Academy Sci USA 98: 4563–4568.

Beerling, D. J., B. H. Lomax, G. R. Upchurch, Jr., D. J. Nicholas, C. L. Pillmore, L. L. Handley & C. M. Scrimgeour. 2001. Evidence for the recovery of terrestrial ecosystems ahead of marine primary production following a biotic crisis at the Cretaceous–Tertiary boundary. J Geol Soc Lond. 158: 737–740.

Bell, A. M., & J. A. Stamps. 2004. Development of behavioural differences between individuals and populations of sticklebacks, *Gasterosteus aculeatus*. Anim Behav. 68: 1339–1348.

Belovsky, G. E., C. Mellison, C. Larson, & P. A. Van Zandt. 1999. Experimental studies of extinction dynamics. Science 286: 1175–1177.

Benkman, C. W. 1993 Adaptation to single resources and the evolution of crossbill (*Loxia*) diversity. Ecol Monogr. 63: 305–325.

Benkman, C. W. 1999. The selection mosaic and diversifying coevolution between crossbills and lodgepole pine. Am Nat. 154: S75–S91.

Benkman, C. W. 2003. Divergent selection causes the adaptive radiation of crossbills. Evolution 57: 1176–1181.

Benkman, C. W., W. C. Holimon, & J. W. Smith. 2001. The influence of a competitor on the geographic mosaic of coevolution between crossbills and lodgepole pine. Evolution 55: 282–294.

Benkman, C. W., & A. M. Siepielski. 2004. A keystone selective agent? Pine squirrels and the frequency of serotiny in lodgepole pine. Ecology 85: 2082–2087.

Benkman C. W., A. M. Siepielski, & T. L. Parchman. 2007. The local introduction of strongly interacting species and the loss of geographic variation in species and species interactions. Mol Ecol. 17: 395–404.

Benton, M. J., V. P. Tvedokhlebov, & M. V. Surkov. 2004. Ecosystem remodeling among vertebrates at the Permian–Triassic boundary in Russia. Nature 432: 97–100.

Berenbaum, M. R., & A. R. Zangerl. 2006. Parsnip webworms and host plants at home and abroad: Trophic complexity in a geographic mosaic. Ecology 87: 3070–3081.

Berlocher, S. H., & J. L. Feder. 2002. Sympatric speciation in phytophagous insects: Moving beyond controversy? Annu Rev Entomol. 47: 773–815.

Bernardo, J. 1993. Determinants of maturation in animals. Trends Ecol Evol. 8: 166–173.

Berner, R. A. 2002. Examination of hypotheses for the Permo-Triassic boundary extinction by

carbon cycle modeling. Proc Natl Acad Sci USA 99: 7172–7177.
Berry, A., & M. Kreitman. 1993. Molecular analysis of an allozyme cline: Alcohol dehydrogenase in *Drosophila melanogaster* on the east coast of North America. Genetics 134: 869–893.
Bettencourt, B. R., I. Kim, A. A. Hoffman, & M. E. Feder. 2002. Response to laboratory and natural selection at the *Drosophila* hsp70 genes. Evolution 56: 1796–1801.
Bickham, J. W., S. Sandhu, P. D. N. Hebert, L Chikhi & R Athwal 2000. Effects of chemical contaminants on genetic diversity in natural populations: Implications for biomonitoring and ecotoxicology. Mutation Res. 463: 33–51.
Biek, R., A. Drummond, & M. Poss. 2006. A virus reveals population structure and recent demographic history of its carnivore host. Science 311: 538–541.
Bijlsma, R., & V. Loeschcke. 2005. Environmental stress, adaptation and evolution: An overview. J Evol Biol. 18: 744–749.
Bitter, W. H. G., R. Kieft, & P. Borst. 1998. The role of transferin–receptor variation in the host range of *Tyrpanosoma brucei*. Nature 391: 499–502.
Bjørnstad, O. N., & B. T. Grenfell. 2001. Noisy clockwork: Time series analysis of population fluctuations in animals. Science 293: 638–644.
Blackburn, T. M., P. Cassey, R. P. Duncan, K. L. Evans & K. J. Evans. 2004. Avain extinctions and mammalian introductions on oceanic islands. Science 305: 1955–1958.
Blossey, B., & R. Nötzold. 1995. Evolution of increased competitive ability in invasive nonindigenous plants: A hypothesis. J Ecology 83: 887–889.
Blows, M. W., & A. A. Hoffmann. 1993. The genetics of central and marginal populations of *Drosophila serrata*. I. Genetic variation for stress resistance and species borders. Evolution 47: 1255–1270.
Blows, M. W., & A. A. Hoffmann. 2005. A reassessment of genetic limits to evolutionary change. Ecology 86: 1371–1384.
Bolnick, D. I., & T. J. Near. 2005. Tempo of hybrid inviability in centrarchid fishes (Teleostei: Centrarchidae). Evolution 59: 1754–1767.
Bolnick, D. I., R. Svanback, J. A. Fordyce, L. H. Yang, J. M. Davis, C. D. Hulsey & M. L. Forister. 2003. The ecology of individuals: Incidence and implications of individual specialization. Am Nat. 161: 1–28.
Bond, W. J., & J. E. Keeley. 2005. Fire as a global 'herbivore': The ecology and evolution of flammable ecosystems. Trends Ecol Evol. 20: 387–394.
Bonfils, A.- C. 2006. Canada's regulatory approach (pp. 229–243). In C. G. Williams (ed.). Landscapes, genomics and transgenic conifers. Springer, Dordrecht, the Netherlands
Boots, M., P. J. Hudson, & A. Sasaki. 2004. Large shifts in pathogen virulence relate to host population structure. Science 303: 842–844.
Boots, M., & A. Sasaki. 1999. 'Small worlds' and the evolution of virulence: Infection occurs locally and at a distance. Proc R Soc Lond B. 266: 1933–1938.
Boots, M., & A. Sasaki. 2002. Parasite-driven extinction in spatially explicit host–parasite systems. Am Nat. 159: 706–713.
Booy, G., R. J. J. Hendriks, M. J. M. Smulders, JM van Groenendael & B Vosman 2000. Genetic diversity and the survival of populations. Plant Biol 2: 379–395.
Borisov, V. M. 1978. The selective effect of fishing on the population structure of species with a long life cycle. J Ichthyol. 18: 896–904.
Both, C., S. Bouwhuis, C. M. Lessells, & M. E. Visser. 2006. Climate change and population declines in a long-distance migratory bird. Nature 441: 81–83.
Botkin, D. B., H. Saxe, M. B. Araújo, R. Betts, R. H. W. Bradshaw, T. Cedhagen, P. Chesson, T. Dawson, J. R. Etterson, D. P. Faith, S. Ferrier, A. Guisan, A. Skjoldborg Hansen, D. W. Hilbert, C. Loehle, C. Margules, M. New, M. J. Sobel & D. R. B. Stockwell. 2007. Forecasting the effects of global warming on biodiversity. BioScience 57: 227–236.
Boughman, J. W. 2001. Divergent sexual selection enhances reproductive isolation in sticklebacks. Nature 411: 900–901.
Boughman, J. W. 2002. How sensory drive can promote speciation. Trends Ecol Evol. 17: 571–577.
Boughman, J. W., H. D. Rundle & D. Schluter. 2005. Parallel evolution of sexual isolation in sticklebacks. *Evolution* 59: 361–373.
Boulding, E. G. 1990. Are the opposing selection pressures on exposed and sheltered shores sufficient to maintain genetic differentiation between gastropod populations with high intermigration rates? Hydrobiologia 193: 41–52.
Boulding, E. G., & T. K. Hay. 1993. Quantitative genetics of shell form of an intertidal snail: Constraints on short-term response to selection. Evolution 47: 576–592.
Boulding, E. G., & T. K. Hay. 2001. Genetic and demographic parameters determining

population persistence after a discrete change in the environment. Heredity 86: 313–324.

Boulding, E. G., T. K. Hay, M. Holst, S. Kamel, D. Pakes & A. D. Tie. 2007. Modelling the genetics and demography of step cline formation: Gastropod populations preyed on by experimentally introduced crabs. J. Evol. Biol. 20: 1976–1987.

Bowen, B. W., W. S. Nelson, & J. C. Avise. 1993. A molecular phylogeny for marine turtles: Trait mapping, rate assessment, and conservation relevance. Proc Natl Acad Sci USA 90: 5574–5577.

Bradshaw, W. E., S. Fujiyama, & C. M. Holzapfel. 2000. Adaptation to the thermal climate of North America by the pitcher-plant mosquito, *Wyeomyia smithii*. Ecology 81: 1262–1272.

Bradshaw, W. E., & C. M. Holzapfel. 2001. Genetic shift in photoperiodic response correlated with global warming. Proc Natl Acad Sci USA 98: 14509–14511.

Bradshaw, W. E., & C. M. Holzapfel. 2006. Evolutionary response to rapid climate change. Science 312: 1477–1478.

Bradshaw, W. E., P. A. Zani, & C. M. Holzapfel. 2004. Adaptation to temperate climates. Evolution 58: 1748–1762.

Brasier, C. M. 2000. The rise of the hybrid fungi. Nature 405: 134–135.

Brasier, C. M. 2001. Rapid evolution of introduced plant pathogens via interspecific hybridization. Bioscience 51: 123–1233.

Bridle, J. R., and T. H. Vines. 2007. Limits to evolution at range margins: when and where does adaptation fail? Trends Ecol. Evol. 22: 140–147.

Briggs, W. H., & I. L. Goldman. 2004. Genetic variation and selection response in model breeding populations of *Brassica rapa* following a diversity bottleneck. Genetics 172: 457–465.

Briskie, J. V., & M. Mackintosh. 2004. Hatching failure increases with severity of population bottlenecks in birds. Proc Natl Acad Sci USA 101: 558–561.

Brodie, E. D., III, A. J. Moore, & F. J. Janzen. 1995. Visualizing and quantifying natural selection. Trends Ecol Evol. 10: 313–318.

Brodie, E. D., Jr., B. J. Ridenhour, & E. D. Brodie, III. 2002. The evolutionary response of predators to dangerous prey: Hotspots and coldspots in the geographic mosaic of coevolution between garter snakes and newts. Evolution 56: 2067–2082.

Brook, B. W., D. W. Tonkyn, J. J. O'Grady, & R. Frankham. 2002. Contribution of inbreeding to extinction risk in threatened species. Conserv Ecol. 6: 16.

Brook, B. W., L. W. Traill, & C. J. A. Bradshaw. 2006. Minimum viable populations and global extinction risk are unrelated. Ecol Lett 9: 375–382.

Brookes, M. 1997. A clean break: Conservation and genetics make bad bed fellows. New Sci. 156: 64.

Brooks, R. J., G. P. Brown, & D. A. Galbraith. 1991. Effects of a sudden increase in natural mortality of adults on a population of the common snapping turtle (*Chelydra serpentina*). Can J Zool. 69: 1314–1320.

Brown A.H.D & J. D.Briggs 1991. Sampling strategies for genetic variation in ex situ collections of endangered plant species. In: D. A. Falk & K. E. Holsinger (eds.) Genetics and conservation of rare plants. Oxford University Press, New York, pp 99–122.

Brown, J. H. 1995. Macroecology. University of Chicago Press, Chicago, Ill.

Brown, J. H., & A. Kodric-Brown. 1977. Turnover rates in insular biogeography: Effect of immigration on extinction. Ecology 58: 445–449.

Brown, W. M., & J. Wright. 1979. Mitochondrial DNA analyses of the origin and relative age of parthenogenetic lizards (genus *Cnemidophorus*). Science 203: 1247–1249.

Bryant, E. H., S. A. McCommas, & L. M. Combs. 1986. The effect of an experimental bottleneck upon quantitative genetic variation in the housefly. Genetics 114: 1191–1211.

Bryant, E. H., & L. Meffert. 1995. An analysis of selection response in relation to a population bottleneck. Evolution 49: 626–634.

Buckling, A., & P. B. Rainey. 2002. The role of parasites in sympatric and allopatric host diversification. Nature 420: 496–499.

Bulmer, M. J. 1985. The mathematical theory of quantitative genetics. Clarendon Press, Oxford, UK.

Burdon, J. J., & A. M. Jarosz. 1991. Host–pathogen interactions in natural populations of *Linum marginale* and *Melampsora lini:* I. Patterns of resistance and racial variation in a large host population. Evolution 45: 205–217.

Burdon, J. J. & J. N. Thompson. 1995. Changed patterns of resistance in a population of *Linum marginale* attacked by the rust pathogen *Melampsora lini.* J Ecology 83: 199–206.

Burdon, J. J., & P. H. Thrall. 1999. Spatial and temporal patterns in coevolving plant and pathogen associations. Am Nat 153: S15–S33.

Burdon, J. J., & P. H. Thrall. 2000. Coevolution at multiple spatial scales: *Linum marginale–Melampsora lini:* From the individual to the species. Evol Ecol 14: 261–281.

Burdon, J. J., P. H. Thrall, & G. J. Lawrence. 2002. Coevolutionary patterns in the *Linum marginale–Melampsora lini* association at a continental scale. Can J Bot. 80: 288–296.

Bürger, R., & M. Lynch. 1995. Evolution and extinction in a changing environment: A quantitative-genetic analysis. Evolution 49: 151–163.

Burgess, K. S., J. R. Etterson, & L. F. Galloway. Artificial selection shifts flowering phenology and other correlated traits in an autotetraploid herb. Heredity 99:641–648.

Burke, T. 1994. Special issue on conservation genetics: Introduction. Mol Ecol. 3: 277–435.

Burness, G. P., J. Diamond, & T. Flannery. 2001. Dinosaurs, dragons, and dwarfs: The evolution of maximal body size. Proc Natl Acad Sci USA 98: 14518–14523.

Bush, M. B., & H. Hooghiemstra. 2005. Tropical biotic responses to climate change (pp. 125–137). In T. E. Lovejoy & L. Hannah (eds.). Climate change and biodiversity. Yale University Press, New Haven, Conn.

Cabeza, M. 2003. Habitat loss and connectivity of reserve networks in probability approaches to reserve design. Ecol Lett. 6: 665–672.

Cabeza, M., M. B. Araujo, R. J. Wilson, C. D. Thomas, M. J. R. Cowley & A. Moilanen 2004. Combining probabilities of occurrence with spatial reserve design. J Appl Ecol. 41: 252–262.

Cabeza, M., & A. Moilanen. 2001. Design of reserve networks and the persistence of biodiversity. Trends Ecol Evol. 16: 242–248.

Calboli, F. C. F., W. J. Kennington, & L. Partridge. 2003. QTL mapping reveals a striking coincidence in the positions of genomic regions associated with adaptive variation in body size in parallel clines of *Drosophila melanogaster* on different continents. Evolution 57: 2653–2658.

Callaway, R. M., J. L. Hierro, & A. S. Thorpe. 2005a. Evolutionary trajectories in plant and soil microbial communities: *Centaurea* invasions and the geographic mosaic of coevolution (pp. 341–363). In D. F. Sax, J. J. Stachowicz, & S. D. Gaines (eds.). Species invasions: Insights into ecology, evolution, and biogeography. Sinauer Associates, Sunderland, Mass.

Callaway, R. M., & J. L. Maron. 2006. What have exotic plant invasions taught us over the past 20 years? Trends Ecol Evol. 21: 369–374.

Callaway, R. M., & W. M. Ridenour. 2004. Novel weapons: Invasive species and the evolution of increased competitive ability. Front Ecol Env. 2: 436–443.

Callaway, R. M., W. M. Ridenour, T. Laboski, T. Weir, & J. M. Vivanco. 2005b. Natural selection for resistance to the allelopathic effects of invasive plants. J. Ecol. 93: 576–583.

Calsbeek, R., & T. B. Smith. 2003. Ocean currents mediate evolution in island lizards. Nature 426: 552–555.

Campbell, D., & L. Bernatchez. 2004. Generic scan using AFLP markers as a means to assess the role of directional selection in the divergence of sympatric whitefish ecotypes. Mol Biol Evol. 21: 945–956.

Campbell, L. G., A. A. Snow, & C. E. Ridley. 2006. Weed evolution after crop gene introgression: Greater survival and fecundity of hybrids in a new environment. Ecol Lett. 9: 1198–1209.

Canadian Climate Center. 1999. Model CGCM1. Available at http://www.cccma.bc.ec.gc.ca/models/cgcm1. Shtml.

Captive Breeding Specialist Group. 1989. Florida panther viability analysis and species survival plans. Captive Breeding Specialist Group, Apple Valley. Minn.

Carlsson-Granér, U. 1997. Anther-smut disease in *Silene dioica*: Variation in susceptibility among genotypes and populations, and patterns of disease within populations. Evolution 51: 1416–1426.

Carlsson-Graner, U., & P. H. Thrall. 2002. The spatial distribution of plant populations, disease dynamics and evolution of resistance. Oikos 97: 97–110.

Caro, T. M., & M. K. Laurenson. 1994. Ecological and genetic factors in conservation: A cautionary tale. Science 263: 485–486.

Carr, D. E., & M. R. Dudash. 1997. The effects of five generations of enforced selfing on potential male and female function in *Mimulus guttatus*. Evolution 51: 1795–1805.

Carr, D. E., & M. R. Dudash. 2005. Recent approaches to the genetic basis of inbreeding depression in plants. Phil Trans R Soc Lond B. 358: 1071–1084.

Carr, D. E., & M. D. Eubanks. 2002. Inbreeding alters resistance to insect herbivory and host plant quality in *Mimulus guttatus* (Scrophulariaceae). Evolution 56: 22–30.

Carr, D. E., J. F. Murphy, & M. D. Eubanks. 2003. The susceptibility and response of inbred and outbred *Mumulus guttatus* to infection by Cucumber mosaic virus. Evol Ecol. 17: 85–103.

Carr, D. E., J. F. Murphy, & M. D. Eubanks. 2006. Genetic variation and covariation for resistance and tolerance to Cucumber mosaic virus in *Mumulus guttatus* (Phrymaceae): A

test for costs and constraints. Heredity 96: 29–38.
Carroll, S. P. 2007a. Natives adapting to invasives: Ecology, genes, and the sustainability of conservation. Ecol Res. 22: 892–901.
Carroll, S. P. 2007b. Brave New World: The epistatic foundations of natives adapting to invaders. Genetica 129: 193–204.
Carroll, S. P. 2008. Facing change: forms and foundations of contemporary adaptation in native populations. Mol Ecol. 17: 361–372.
Carroll, S. P., & C. Boyd. 1992. Host race radiation in the soapberry bug: Natural history with the history. Evolution 46: 1052–1069.
Carroll, S. P., & P. S. Corneli. 1999. The evolution of behavioral reaction norms as a problem in ecological genetics: Theory, methods and data. In S. Foster & J. Endler (eds.). The evolution of behavioral phenotypes: Perspectives from the study of geographic variation. Oxford University Press, Oxford, U.K.
Carroll, S. P., & H. Dingle. 1996. The biology of post-invasion events. Biol Conserv. 78: 207–214.
Carroll, S. P., H. Dingle, & T. R. Famula. 2003a. Rapid appearance of epistasis during adaptive divergence following colonization. Proc R Soc Lond B. 270: S80–S83.
Carroll, S. P., H. Dingle, T. R. Famula, & C. W. Fox. 2001. Genetic architecture of adaptive differentiation in evolving host races of the soapberry bug, *Jadera haematoloma*. Genetica 112: 257–272.
Carroll, S. P., H. Dingle, & S. P. Klassen. 1997. Genetic differentiation of fitness-associated traits among rapidly evolving populations of the soapberry bug. Evolution 51: 1182–1188.
Carroll, S. P., A. P. Hendry, D. Reznick, & C. Fox. 2007. Evolution on ecological time scales. Funct Ecol. 21: 387–393.
Carroll, S. P., S. P. Klassen, & H. Dingle. 1998. Rapidly evolving adaptations to host ecology and nutrition in the soapberry bug. Evol Ecol. 12: 955–968.
Carroll, S. P., J. E. Loye, H. Dingle, M. Mathieson, T. R. Famula & M. P. Zalucki 2005. And the beak shall inherit: Evolution in response to invasion. Ecol Lett. 8: 944–951.
Carroll, S. P., M. Marler, R. Winchell, & H. Dingle. 2003b. Evolution of cryptic flight morph and life history differences during host race radiation in the soapberry bug, *Jadera haematoloma* Herrich-Schaeffer (Hemiptera: Rhopalidae). Ann Entomol Soc Am. 96: 135–143.
Carson, H. L. 1965. Chromosomal morphism in geographically widespread species of *Drosophila* (pp. 503–531). In H. G. Baker & G. L. Stebbins (eds.). The genetics of colonizing species. Academic Press, New York.
Cassel, A., J. Windig, S. Nylin, & C. Wiklund. 2001. Effects of population size and stress on fitness-related characters in the scarce heath, a rare butterfly in western Europe. Conserv Biol. 15: 1667–1673.
Caughley, G. 1994. Directions in conservation biology. J Anim Ecol 63: 215–244.
CEC. 2004. Maize and biodiversity: The effects of transgenic maize in Mexico. Commission for Environmental Cooperation Secretariat Report: Quebec.
Center for Plant Conservation. 1991. Genetic sampling guidelines for conservation collections of endangered plants (pp. 225–238). In D. A. Falk & K. E. Holsinger (eds.). Genetics and conservation of rare plants. Oxford University Press, New York.
Chaffee, C., & D. R. Lindberg. 1986. Larval biology of Early Cambrian molluscs: The implications of small body size. Bull Marine Sci. 39: 536–549.
Chaisson, E. 2005. Epic of evolution: Seven ages of the cosmos. Columbia University Press, New York.
Chapin, J. P. 1954. The birds of the Belgium Congo. Am Mus Nat Hist. 75: 1–846.
Chapman, M. A., & J. M. Burke. 2006. Letting the gene pool out of the bottle: The population genetics of genetically modified crops. New Phytol. 170: 429–443.
Charlesworth, D., & B. Charlesworth. 1987. Inbreeding depression and its evolutionary consequences. Annu Rev Ecol Syst. 18: 237–268.
Charmantier, A., & D. Garant. 2005. Environmental quality and evolutionary potential: Lessons from wild populations. Proc R Soc B. 272: 1415–1425.
Chen, R., & E. Holmes. 2006. Avian influenza virus exhibits rapid evolution. Mol Biol Evol. 23: 2336–2341.
Chen, Z.- Q., K. Kaiho, & A. D. George. 2005a. Early Triassic recovery of the brachiopod faunas from the end-Permian mass extinction: A global review. Palaeogeography, Palaeoclimatology, Palaeoecology 224: 270–290.
Chen, Z.- Q., K. Kaiho, & A. D. George. 2005b. Survival strategies of brachiopod faunas from the end-Permian mass extinction. Palaeogeography, Palaeoclimatology, Palaeoecology 224: 232–269.
Cheverud, J. M., & E. J. Routman. 1996. Epistasis as a source of increased additive genetic

variance at population bottlenecks. Evolution 50: 1042–1051.
Chiang, V. L. 2002. From rags to riches. Nat Biotechnol. 20: 557–558.
Cincotta, R. P., & R. Engelman. 2000. Nature's place: Human population and the future of biological diversity. Population Action International, Washington, D.C.
Clayton, M. 2005. Now, bioengineered trees are taking root. Christian Science Monitor March 10: 14.
Clayton, D. H., S. E. Bush, B. M. Goates & K. P. Johnson. 2003. Host defense reinforces host–parasite cospeciation. Proc Natl Acad Sci USA 100: 15694–15699.
Clayton, D. H., P. L. M. Lee, D. M. Tompkins, & E. D. Brodie III. 1999. Reciprocal natural selection on host–parasite phenotypes. Am Nat. 154: 261–270.
Clayton, D. H., & J. A. Moore. 1997. Host–parasite evolution: General principles and avian models. Oxford University Press, Oxford, U.K.
Cleaveland, S., M. K. Laurenson, & L. H. Taylor. 2001. Diseases of humans and their domestic mammals: Pathogen characteristics, host range, and the risk of emergence. Phil Trans R Soc Lond B. 356: 991–999.
Cleveland, D. A., D. Soleri, F. Aragon Cuevas, J. Crossa & P. Gepts 2005. Detecting (trans)gene flow to landraces in centers of crop origin: Lessons from the case of maize in Mexico. Environ Biosafety Res. 4: 197–208.
Clewell, A. F. 2000. Restoring for natural authenticity. Ecol Rest. 18: 216–217.
Coates, D. J., & K. A. Atkins. 2001. Priority setting and the conservation of western Australia's diverse and highly endemic flora. Biol Conserv. 97: 251–263.
Coates, D. J., S. Carstairs, & V. L. Hamley. 2003. Evolutionary patterns and genetic structure in localized and widespread species in the *Stylidium caricifolium* complex (Stylidiaceae). Am J Bot. 90: 997–1008.
Cohen, J. 1997. Can cloning help save beleaguered species? Science 276: 1329–1330.
Colautti, R. I., A. Ricciardi, I. A. Grigorovich, & H. J. MacIsaac. 2004. Is invasion success explained by the enemy release hypothesis? Ecol Lett. 7: 721–733.
Coltman, D. W., P. O'Donoghue, J. T. Jorgenson, J T Hogg, C Strobeck & M Festa-Bianchet 2003. Undesirable evolutionary consequences of trophy hunting. Nature 426: 655–658.
Coltman, D. W., J. G. Pilkington, J. A. Smith, & J. M. Pemberton. 1999. Parasite-mediated selection against inbred Soay sheep in a free-living island population. Evolution 53: 1259–1267.
Coltman, D. W. & J. Slate. 2003. Microsatellite measures of inbreeding: a meta-analysis. Evolution 57: 971–983.
Comstock, R. E., & H. F. Robinson. 1952. Estimation of average dominance of genes (pp. 494–516). In J. W. Gowen (ed.). Heterosis. Iowa State College Press, Ames, Iowa.
Conover, D. O., & S. B. Munch. 2002. Sustaining fisheries yields over evolutionary time scales. Science 297: 94–96.
Cooke, B. D., S. McPhee, A. J. Robinson, & L. Capucci. 2002. Rabbit haemorrhagic disease: Does a pre-existing RHDV-like virus reduce the effectiveness of RHD as a biological control in Australia? Wildl Res. 29: 673–682.
Cornuet, J. M., & G. Luikart. 1997. Description and power analysis of two tests for detecting recent population bottlenecks from allele frequency data. Genetics 144: 2001–2014.
Cossins, A. R., & K. Bowler. 1987. Temperature biology of animals. Chapman & Hall, London, UK.
Cotton, S., K. Fowler, & A. Pomiankowski. 2004. Do sexual ornaments demonstrate heightened condition-dependent expression as predicted by the handicap hypothesis? Proc R Soc Lond B, 271: 771–783.
Courchamp, F., T. Clutton-Brock, & B. Grenfell. 1999. Inverse density dependence and the Allee effect. Trends Ecol Evol. 14: 405–410.
Cowling, R. M., & R. L. Pressey. 2001. Rapid plant diversification: Planning for an evolutionary future. Proc Natl Acad Sci USA 98: 5452–5457.
Cowling, R. M., R. L. Pressey, A. T. Lombard, PG Desmet & AG Ellis 1999. From representation to persistence: Requirements for a sustainable system of conservation areas in the species-rich Mediterranean-climate desert of southern Africa. Divers Distrib. 5: 51–71.
Coyne, J. A., & H. A. Orr. 2004. Speciation. Sinauer Associates, Sunderland, Mass.
Craig, J. K., C. J. Foote, & C. C. Wood. 2005. Countergradient variation in carotenoid use between sympatric morphs of sockeye salmon (*Onchorhynchus nerka*) exposes nonanadromous hybrids in the wild by their mismatched spawning colour. Biol J Linn Soc. 84: 287–305.
Crandall, K. A., O. R. P. Bininda-Emonds, G. M. Mace, & R. K. Wayne. 2000. Considering evolutionary processes in conservation biology. Trends Ecol Evol. 15: 290–295.
Crawley, M. J. 1987. What makes a community invasible? (pp. 429–453). In A. J. Gray, M. J. Crawley, & P. J. Edwards (eds.).

Colonization, succession and stability. Blackwell Scientific Publications, Oxford, UK.

Creel, S. 2006. Recovery of the Florida panther: Genetic rescue, demographic rescue, or both? [Response to Pimm et al. (2006).] Anim Conserv 9: 125–126.

Crist, P. J. 2000. Mapping and categorizing land stewardship. Version 2.1.0. A handbook for conducting GAP analysis. Online. http://www.gap.uidaho.edu/handbook/Stewardship/default.htm. Accessed July 19, 2006.

Crnokrak, P., & D. A. Roff. 1999. Inbreeding depression in the wild. Heredity 83: 260–270.

Cronin, T. W., N. J. Marshall, & R. L. Caldwell. 2000. Spectral tuning and the visual ecology of mantis shrimps. Phil Trans R Soc London Series B. 355: 1263–1267.

Crow, J., & M. Kimura. 1970. An introduction to population genetics theory. (Harper & Row, New York); reprinted (1977) by Burgess, Minneapolis, Minn.

Csuti, B., S. Polasky, P. H. Williams, R. L. Pressey, J. D. Camm, M. Kershaw, A. R. Kiester, B. Downs, R. Hamilton, M. Huso & K. Sahr 1997. A comparison of reserve selection algorithms using data on terrestrial vertebrates in Oregon. Biol Conserv. 80: 83–97.

Culver, M., W. E. Johnson, J. Pecon-Slattery, & S. J. O'Brien. 2000. Genomic ancestry of the American puma (*Puma concolor*). J Heredity 91: 186–197.

Cwynar, L. C., & G. M. MacDonald. 1987. Geographic variation of lodgepole pine in relation to population history. Am Nat. 129: 463–469.

Daehler, C. C. 1999. Inbreeding depression in smooth cordgrass (*Spartina alterniflora*, Poaceae) invading San Francisco Bay. Am J Bot. 86: 131–139.

Dahlgaard, J., V. Loeschcke, P. Michalak, & J. Justesen. 1998. Induced thermotolerance and associated expression of the heat-shock protein Hsp70 in adult *Drosophila melanogaster*. Funct Ecol. 12: 786–793.

Da Lage, J. L., P. Capy, & J. R. David. 1990. Starvation and desiccation tolerance in *Drosophila melanogaster*: Differences between European, North African, and Afrotropical populations. Genet Sel Evol. 22: 381–391.

Darwin, C. 1859. On the origin of species by means of natural selection, or the preservation of favoured races in the struggle for life. J Murray, London, UK.

Darwin, C. 1860. The voyage of the *Beagle*. [1962 Natural History library edition.]. Doubleday, Garden City, N.Y.

Darwin, C. 1868. The variation of animals and plants under domestication. Murray, London, UK.

Darwin, C. 1876. The effects of cross and self-fertilisation in the vegetable kingdom. John Murray, London, UK.

Darwin, C., & A. Wallace. 1858. On the tendency of species to form varieties: And on the perpetuation of varieties and species by natural means of selection. As communicated by C. Lyell & J. D. Hooker. J Proc Linn Soc Zool. 3: 45–62.

Daszak, P., A. A. Cunningham, & A. D. Hyatt. 2000. Emerging infectious diseases of wildlife-threats to biodiversity and human health. Science 287: 443–449.

David, J. R., & P. Capy. 1988. Genetic variation of *Drosophila melanogaster* natural populations. Trends Genet. 4: 106–111.

Davis, M. B., R. G. Shaw, & J. R. Etterson. 2005. Evolutionary responses to climate change. Ecology 86: 1704–1714.

de Castro, F., & B. Bolker. 2005. Mechanisms of disease-induced extinction. Ecol Lett, 8: 117–126.

della Torre, A. D., C. Constantini, N. J. Besansky, A Caccone, V Petrarca, JR Powell & M Coluzzi 2002. Speciation within *Anopheles gambiae:* The glass is half full. Science 298: 115–117.

Demarais, B. D., T. W. Dowling, M. E. Douglas, WL Minckley & PC Marsh 1992. Origin of *Gila seminuda* (Teleostei, Cyprinidae) through introgressive hybridization: Implications for evolution and conservation. Proc Natl Acad Sci USA 89: 2747–2751.

Dewoody, Y. D., & J. A. Dewoody. 2005. On the estimation of genome-wide heterozygosity using molecular markers. J. Hered. 96: 85–88.

D'Hondt, S., P. Donaghay, J. C. Zachos, D. Luttenberg & M. Lindinger 1998. Organic carbon fluxes and ecological recovery from the Cretaceous–Tertiary mass extinction. Science 282: 276–279.

Dieckmann, U., & R. Ferrière. 2004. Adaptive dynamics and evolving biodiversity (pp. 188–224). In R. Ferrière, U. Dieckmann, & D. Couvet (eds.). Evolutionary conservation biology. Cambridge University Press, Cambridge, UK.

Dieckmann, U., O. R. Godø, M. Heino, & J. Mork. 2008. Fisheries-induced adaptive change. Cambridge University Press, Cambridge, UK.

Dieckmann, U., & M. Heino. 2007. Probabilistic maturation reaction norms: Their history, strengths, and limitations. Mar Ecol Prog Ser. 335: 253–269.

Dieringer, D., & C. Schlötterer. 2002. Microsatellite analyser (MSA): A platform independent analysis tool for large microsatellite data sets. Mol Ecol Notes 3: 167–169.

Dietl, G. P., G. S. Herbert, & G. J. Vermeij. 2004. Reduced competition and altered feeding behavior among marine snails after a mass extinction. Science 306: 2229–2231.

Dingemanse, N. J., C. Both, P. J. Drent, & J. M. Tinbergen. 2004. Fitness consequences of avian personalities in a fluctuating environment. Proc R Soc London B. 271: 847–852.

Diniz-Filho, J. A., & M. P. De Campos Telles. 2002. Spatial autocorrelation analysis and the identification of operational units for conservation in continuous populations. Conserv Biol. 16: 924–935.

Dirzo, R., & P. H. Raven. 2003. Global state of biodiversity and loss. Ann Rev Envir Resour. 28: 137–167.

Dobson, A., & B. Grenfell. 1995. Ecology of infectious disease in natural populations. Cambridge University Press, Cambridge, UK.

Dobson, A. P., & A. Lyles. 2000. Enhanced: Black-footed ferret recovery. Science 5468: 985–988.

Dobson, A. P., G. M. Mace, J. Poole, & R. A. Brett. 1992. Conservation biology: The ecology and genetics of endangered species (pp. 405–430). In R. J. Berry, T. J. Crawford, & G. M. Hewitt (eds.). Genes in ecology. Blackwell Science, Oxford, UK.

Dobzhansky, T. 1947. Adaptive changes induced by natural selection in wild populations of *Drosophila*. Evolution 1: 1–16.

Dobzhansky, T. 1948. Genetics of natural populations. XVI. Altitudinal and seasonal changes produced by natural selection in certain populations of *Drosophila pseudoobscura*. Genetics 33: 158–176.

Dobzhansky, T. 1965. "Wild" and "domestic" species of *Drosophila* (pp. 533–546). In H. G. Baker & G. L. Stebbins (eds.). The genetics of colonizing species. Academic Press, New York.

Dobzhansky, T. 1970. Genetics of the evolutionary process. Columbia University Press, New York.

Doebeli, M., & U. Dieckmann. 2000. Evolutionary branching and sympatric speciation caused by different types of ecological interactions. Am Nat. 156: S77–S101.

Doebeli, M., & U. Dieckmann. 2003. Speciation along environmental gradients. Nature 421: 259–264.

Doherty, P. C., & R. M. Zinkernagel. 1975. A biological role for the major histocompatibility antigens. Lancet 1: 1406–1409.

Dole, J. A. 1992. Reproductive assurance mechanisms in three taxa of the *Mimulus guttatus* complex (Scrophulariaceae). Am J Bot. 79: 650–659.

Donnelley, P., & S. Tavaré. 1995. Coalescents and genealogical structure under neutrality. Annu Rev Genet. 29: 401–421.

Drake, J. M., & D. M. Lodge. 2004. Effects of environmental variation on extinction and establishment. Ecol Lett. 7: 26–30.

Dudash, M. R. 1987. The reproductive biology of *Sabatia angularis* L. (Gentianaceae). PhD diss. University Illinois, Chicago, Ill.

Dudash, M. R. 1990. Relative fitness of selfed and outcrossed progeny in a self-compatible, protandrous species, *Sabatia angularis* L. (Gentianaceae): A comparison in three environments. Evolution 44: 1129–1139.

Dudash, M. R. 1991. Effects of plant size on female and male function in hermaphroditic *Sabatia angularis* L. (Gentianaceae). Ecology 72: 1004–1012.

Dudash, M. R., & D. E. Carr. 1998. Genetics underlying inbreeding depression in *Mimulus* with contrasting mating systems. Nature 393: 682–684.

Dudash, M. R., D. E. Carr, & C. B. Fenster. 1997. Five generations of enforced selfing and outcrossing in *Mimulus guttatus*: Inbreeding depression variation at the population and family level. Evolution 51: 54–65.

Dudash, M. R., & C. B. Fenster. 1997. Multiyear study of pollen limitation and cost of reproduction in the iteroparous, *Silene virginica*. Ecology 78: 484–493.

Dudash, M. R., & C. B. Fenster. 2000. Inbreeding and outbreeding depression in fragmented populations (pp. 55–74). In A. Young & G. Clarke (eds.). Genetics, demography, and viability of fragmented populations. Cambridge University Press, Cambridge, UK.

Dudash, M. R. & C. B. Fenster. 2001. The role of breeding system and inbreeding depression in the maintenance of an outcrossing mating strategy in *Silene virginica* (Caryophyllaceae). Am J Bot. 88: 1953–1959.

Dudash, M. R., C. J. Murren, & D. E. Carr. 2005. Using *Mimulus* as a model system to understand the role of inbreeding in

conservation and ecological approaches. Ann Miss Bot Gard. 92: 36–51.
Duffy, D.C., K. Boggs, R.H. Hagenstein, R.Y. Lipkin, & J. A. Michaelson. 1999. Landscape assessment of the degree of protection of Alaska's terrestrial biodiversity. *Conservation Biology* 13: 1332–1343.
Dugatkin, L. A., & H. K. Reeve. 1998. Game theory and animal behavior. Oxford University Press, Oxford, UK.
Dukes. J. S., & H. A. Mooney. 1999. Does global change increase the success of biological invaders? Trends Ecology Evol. 14: 135–139.
Duncan, R. P., & P. A. Williams. 2002. Darwin's naturalization hypothesis challenged. Nature 417: 608–609.
Dunlop, E. S., B. J. Shuter, & M. S. Ridgway. 2005. Isolating the influence of growth rate on maturation patterns in the smallmouth bass (*Micropterus dolomieu*). Can J Fish Aquat Sci. 62: 844–853.
Dupont, Y. L. 2002. Evolution of apomixis as a strategy of colonization in the dioecious species *Lindera glauca* (Laruaceae). Pop Ecol. 44: 293–297.
Dybdahl, M. F., & C. M. Lively. 1998. Host parasite coevolution: Evidence for rare advantage and time-lagged selection in a natural population. Evolution 52: 1057–1066.
Dyer, R. J., & J. D. Nason. 2004. Population graphs: The graph theoretic shape of genetic structure. Mol Ecol. 13: 1713–1727.
Ebert, D., C. Haag, M. Kirkpatrick, M. Riek, J. W. Hottinger &V. I. Pajunen. 2002. A selective advantage to immigrant genes in a *Daphnia* metapopulation. Science 295: 485–488.
Eckert, C. G., & S. C. H. Barrett. 1992. Stochastic loss of style morphs from populations of tristylous *Lythrum salicaria* and *Decodon verticcilatus* (Lythraceae). Evolution 46: 1014–1029.
Eckert, C. G., D. Manicacci, & S. C. H. Barrett. 1996. Genetic drift and founder effect in native versus introduced populations of an invading plant, *Lythrum salicaria* (Lythraceae). Evolution 50: 1512–1519.
Eckstrand, I. A., & R. H. Richardson. 1980. Comparison of some water balance characters in several *Drosophila* species which differ in habitat. Envir Entomol. 9: 716–720.
Edley, T., & R. Law. 1988. Evolution of life histories and yields in experimental populations of *Daphnia magna*. Biol J Linn Soc. 34: 309–326.
Edmands, S. 1999. Heterosis and outbreeding depression in interpopulation crosses spanning a wide range of divergence. Evolution 53: 1757–1768.
Edmands, S. 2007. Between a rock and a hard place: Evaluating the relative risks of inbreeding and outbreeding for conservation and management. Mol Ecol. 16: 463–475.
Edmands, S., & C. C. Timmerman. 2003. Modeling factors affecting the severity of outbreeding depression. Conserv Biol. 17: 883–892.
Ehrlich, P. R. 2001. Intervening in evolution: Ethics and actions. Proc Natl Acad Sci USA 98: 5477–5480.
Eisen, E. J. 1975. Population size and selection intensity effects on long-term selection response in mice. Genetics 79: 305–323.
Elgar, M. A., & D. Clode. 2001. Inbreeding and extinction in island populations: A cautionary note. Conserv Biol. 15: 284–286.
Elle, E., & J. D. Hare. 2002. Environmentally induced variation in floral traits affects the mating system in *Datura wrightii*. Funct Ecol. 16: 79–88.
Ellner, S. P., J. Fieberg, D. Ludwig, & C. Wilcox. 2002. Precision of population viability analysis. Conserv Biol. 16: 258–261.
Ellstrand, N. C. 1992. Gene flow by pollen: Implications for plant conservation genetics. Oikos 63: 77–86.
Ellstrand, N. C. 2003a. Current knowledge of gene flow in plants: Implications for transgene flow. Phil Trans R Soc Lond B. 358: 1163–1170.
Ellstrand, N. C. 2003b. Dangerous liaisons? When cultivated plants mate with their wild relatives. Johns Hopkins University Press, Baltimore, Md.
Ellstrand, N. C., & D. R. Elam. 1993. Population genetic consequences of small population size: Implications for plant conservation. Annu Rev Ecol Syst. 24: 217–242.
Ellstrand, N. C., H. C. Prentice, & J. F. Hancock. 1999. Gene flow and introgression from the domesticated plants into their wild relatives. Annu Rev Ecol Syst. 30: 539–563.
Ellstrand, N. C., & K. C. Schierenbeck. 2000. Hybridization as a stimulus for the evolution of invasiveness in plants. Proc Natl Acad Sci USA 97: 7043–7050.
Ellstrand, N. C., & K. A. Schierenbeck. 2006. Hybridization as a stimulus for the evolution of invasiveness in plants? Euphytica 148: 35–46.
Elton, C. S. 1930. Animal ecology and evolution. Clarendon Press, Oxford, UK.
Elton, C. S. 1958. The ecology of invasions by animals and plants. Methuen, UK.

Emlen, D. J., J. Marangelo, B. Ball, & C. W. Cunningham. 2005. Diversity in the weapons of sexual selection: Horn evolution in the beetle genus *Onthophagus* (Coleoptera: Scarabaeidae). Evolution 59: 1060–1084.

Endler, J. A. 1982. Pleistocene forest refuges: Fact or fancy (pp. 641–657). In G. T. Prance (ed.). Biological diversification in the tropics. Columbia University Press, New York.

Endler, J. 1992. Signals, signal conditions, and the direction of evolution. Am Nat. 139: 125–153.

Engelhard, G. H., & M. Heino. 2004. Maturity changes in Norwegian spring-spawning herring *Clupea harengus*: Compensatory or evolutionary responses? Mar Ecol Prog Ser. 272: 245–256.

Epifanio, J., & D. Philipp. 2001. Stimulating the extinction of parental lineages from introgressive hybridization: The effects of fitness, initial proportions of parental taxa, and mate choice. Rev Fish Biol Fisheries 10: 339–354.

Erickson, D. L., & C. B. Fenster. 2006. Intraspecific hybridization and the recovery of fitness in the native legume *Chamaecrista fasciculata*. Evolution 60: 225–233.

Ernande, B., U. Dieckmann, & M. Heino. 2004. Adaptive changes in harvested populations: Plasticity and evolution of age and size at maturation. Proc R Soc Lond B. 271: 415–423.

Erwin, D. H. 1996. Understanding biotic recoveries: Extinction, survival, and preservation during the end-Permian mass extinction (pp. 398–418). In D. Jablonski, D. H. Erwin, & J. H. Lipps (eds.). Evolutionary paleobiology: In honor of James W Valentine. University of Chicago Press, Chicago, Ill.

Estes, S., P. C. Phillips, D. R. Denver, W. K. Thomas & M. Lynch 2004. Mutation accumulation in populations of varying size: The distribution of mutational effects for fitness correlates in *Caenorhabditis elegans*. Genetics 166: 1269–1279.

Etges, W. J., K. L. Arbuckle, & M. Levitan. 2006. Long-term frequency shifts in the chromosomal polymorphisms of *Drosophila robusta* in the Great Smoky Mountains. Biol J Linn Soc. 88: 131–141.

Etterson, J. R. 2001. Evolutionary potential of the annual legume, *Chamaecrista fasciculata* in relation to global warming. PhD diss. University of Minnesota, St. Paul, Minn.

Etterson, J. R. 2004a. Evolutionary potential of *Chamaecrista fasciculata* in relation to climate change: I. Clinal patterns of selection along an environmental gradient in the Great Plains. Evolution 58: 1446–1458.

Etterson, J. R. 2004b. Evolutionary potential of *Chamaecrista fasciculata* in relation to climate change: II. Genetic architecture of three populations reciprocally planted along an environmental gradient in the Great Plains. Evolution 58: 1459–1471.

Etterson, J. R., & R. G. Shaw. 2001. Constraint to adaptive evolution in response to global warming. Science 294: 151–154.

Ewald, P. W. 1994. Evolution of infectious diseases. Oxford University Press, Oxford, UK.

Excoffier, L., P. Smouse, & J. Quattro. 1992. Analysis of molecular variance inferred from metric distances among DNA haplotypes: Application to human mitochondrial-DNA restriction data. Genetics 131: 479–491.

Facon, B., B. J. Genton, J. Shykoff, P Jarne, A Estoup & P David 2006. A general eco-evolutionary framework for understanding bioinvasions. Trends Ecol Evol. 21: 130–135.

Fagan, W. F., & E. E. Holmes. 2006. Quantifying the extinction vortex. Ecol Lett. 9: 51–60.

Fagan, W. F., & A. J. Stephens. 2006. How local extinction changes rarity: An example with Sonoran Desert fishes. Ecography 29: 845–852.

Fahrig, L. 2003. Effects of habitat fragmentation on biodiversity. Annu Rev Ecol Syst. 34: 487–515.

Faith, D. P. 1992a. Conservation evaluation and phylogenetic diversity. Biol Conserv. 61: 1–10.

Faith, D. P. 1992b. Systematics and conservation: On predicting the feature diversity of subsets of taxa. Cladistics 8: 361–373.

Faith, D. P. 1994. Phylogenetic pattern and the quantification of organismal biodiversity. Phil Trans R Soc Lond B. 345: 45–58.

Faith, D. P. 1996. Conservation priorities and phylogenetic pattern. Conserv Biol. 10: 1286–1289.

Faith, D. P. 2002. Quantifying biodiversity: A phylogenetic perspective. Conserv Biol. 16: 248–252.

Faith, D. P. 2003. Environmental diversity (ED) as surrogate information for species-level biodiversity. Ecography 26: 374–379.

Faith, D. P. 2005. Global biodiversity assessment: integrating global and local values and human dimensions. Global Environmental Change: Human and Policy Dimensions 15: 5–8.

Faith, D. P. 2006a. The role of the phylogenetic diversity measure, PD, in bio-informatics:

Getting the definition right. Evol Bioinform Online. 2: 301–307.
Faith, D. P. 2006b. Taxonomic research and 2010. In: Actions for the 2010 biodiversity target in Europe how—does research contribute to halting biodiversity loss? Report of an e-conference. (eds. Young, J., Ahlbeg, M., Niemela, N., Parr, T., Pauleit, S. and Watt, A. D.), pp. 70–71. Available at: http://www.edinburgh.ceh.ac.uk/biota/Archive 2010target/8418.htm
Faith, D. P. 2007. Phylogeny and conservation. Syst Biol. 56: 690–694.
Faith, D. P., & A. Baker. 2006. Phylogenetic diversity (PD) and biodiversity conservation: Some bioinformatics challenges. Evol Bioinform Online 2: 70–77.
Faith, D. P., S. Ferrier, & P. A. Walker. 2004a. The ED strategy: Jow species-level surrogates indicate general biodiversity patterns through an 'environmental diversity' perspective. J Biogeogr. 31: 1207–1217.
Faith, D. P., C. A. M. Reid, & J. Hunter. 2004b. Integrating phylogenetic diversity, complementarity, and endemism. Conserv Biol. 18: 255–261.
Faith, D. P., & P. A. Walker. 1996a. DIVERSITY - TD (pp. 63–74). In D. P. Faith & A. O. Nicholls (eds.). BioRap: Rapid assessment of biodiversity. Vol. 3. Tools for assessing biodiversity priority areas. Centre for Resource and Environmental Studies, Australian National University, Canberra, Australia.
Faith, D. P. & P. A. Walker. 1996b. *DIVERSITY: a software package for sampling phylogenetic and environmental diversity. Reference and user's guide. v. 2.1.* CSIRO Division of Wildlife and Ecology, Canberra.
Faith, D. P., & K. J. Williams. 2006. Phylogenetic diversity and biodiversity conservation (pp. 233–235). In McGraw-Hill yearbook of science and technology. McGraw-Hill, New York, N.Y.
Falconer, D. S. 1981. Introduction to quantitative genetics. 2nd ed. Longman, London, UK.
Falconer, D. S., & T. F. C. Mackay. 1996. Introduction to quantitative genetics. 4th ed. Longman, Harlow, UK.
Falk, D. A., & K. E. Holsinger (eds.). 1991. Genetics and conservation of rare plants. Oxford University Press, New York.
Falk, D. A., E. E. Knapp, & E. O. Guerrant. 2001. An introduction to restoration genetics. Society for ecological restoration November 2001. National Park Service USA. Online. www.nps.gov/plants/restore/pubs/restgene/index.htm. Accessed March 4, 2008.
FAO. 1997. The state of the world's forests. Food and Agriculture Organization of the United Nations, Rome, Italy.
Farrell, B. 1998. Inordinate fondness explained: Why are there so many beetles? Science 281: 555–559.
Feder, M. E. 1996. Ecological and evolutionary physiology of stress proteins and the stress response: The *Drosophila melanogaster* model (pp. 79–102). In I. A. Johnson & A. F. Bennett (eds.). Phenotypic and evolutionary adaptation to temperature. Cambridge University Press, Cambridge UK.
Feder, M. E., & G. E. Hofmann. 1999. Heat-shock proteins, molecular chaperones, and the stress response: Evolutionary and ecological physiology. Annu Rev Physiol. 61: 243–282.
Fenner, F., & B. Fantini. 1999. Biological control of vertebrate pests: The history of myxomatosis: An experiment in evolution. CABI Publishing, New York.
Fenster, C. B., W. S. Armbruster, M. R. Dudash, J Thomson & P Wilson 2004. Pollination syndromes and the evolution of floral diversity. Annu Rev Ecol Evol Syst. 35: 375–403.
Fenster, C. B., & M. R. Dudash. 1994. Genetic considerations in plant population conservation and restoration (pp. 34–62). In M. L. Bowles & C. Whelan (eds.). Restoration of endangered species: Conceptual issues, planning and implementation. Cambridge University Press, Cambridge, UK.
Fenster, C. B., & M. R. Dudash. 2001. Spatiotemporal variation in the role of hummingbirds as pollinators of *Silene virginica*. Ecology 82: 844–851.
Fenster, C. B., & L. F. Galloway. 2000. Inbreeding and outbreeding depression in natural populations of *Chamaecrista fasciculate* (Fabaceae). Conserv Biol. 14: 1406–1412.
Fenster, C. B., L. F. Galloway, & L. Chao. 1997. Epistasis and its consequences for the evolution of natural populations. Trends Ecol Evol. 12: 282–286.
Fenton, A., & A. B. Pedersen. 2005. Community epidemiology in theory and practice: A conceptual framework for describing transmission dynamics in multiple hosts. Em Infect Dis. 11: 1815–1821.
Ferrier, S. 2002. Mapping spatial pattern in biodiversity for regional conservation planning: Where to from here? Syst Biol. 51: 331–363.
Ferrier, S., R. L. Pressey, & T. W. Barrett. 2000. A new predictor of the irreplaceability of areas for achieving a conservation goal, its

application to real-world planning, and a research agenda for further refinement. Biol Conserv. 93: 303–325.

Ferriére, R., U. Dieckmann, & D. Couvet. 2004. Introduction (pp. 1–14). In R. Ferriere, U. Dieckmann, & D. Couvet (eds.). Evolutionary conservation biology. Cambridge University Press, Cambridge, UK.

Festa-Bianchet, M., T. Coulson, J.- M. Gaillard, JT Hogg & F Pelletier 2006 Stochastic predation events and population persistence in bighorn sheep. Proc R Soc Lond B. 273: 1537–1543.

Fields, P. A. 2001. Review: Protein function at thermal extremes: Balancing stability and flexibility. Comp Biochem Physiol A-Mol Integr Physiol. 129: 417–431.

Finton, A., & C. Corcoran. 2001. BioMap: Guiding land conservation for biodiversity in Massachusetts. Natural Heritage and Endangered Species Program, Boston, Massachusetts Division of Fisheries and Wildlife.

Fischer, M., & D. Matthies. 1998. Effects of population size on performance in the rare plant *Gentianella germanica*. J Ecology 89: 195–204.

Fisher, R. A. 1930. The genetical theory of natural selection. Clarendon Press, Oxford, UK.

Fisher, R. A. 1937. The wave of advance of advantageous genes. Ann Eugenics (Lond) 7: 355–369.

Fitzpatrick, B. M. 2004. Rates of evolution of hybrid inviability in birds and mammals. Evolution 58: 1865–1870.

Fiumera, A. C., B. A. Porter, G. Looney, M. A. Asmussen & J. C. Avise. 2004. Maximizing offspring production while maintaining genetic diversity in supplemental breeding programs of highly fecund managed species. Conserv Biol. 18: 94–101.

Fleming, I. A., K. Hindar, I. Mjolnerod, B. Jonsson, T. Balstad & A. Lamberg 2000. Lifetime success and interactions of farm salmon invading a natural population. Proc R Soc Lond B. 267: 1517–1523.

Folk, D. G., P. Zwollo, D. M. Rand, & G. W. Gilchrist. 2006. Selection on knockdown performance in *Drosophila melanogaster* impacts thermotolerance and heat-shock response differently in females and males. J Exp Biol. 209: 3964–3973.

Forde, S. E., J. N. Thompson, & B. J. M. Bohannan. 2004. Adaptation varies through space and time in a coevolving host–parasitoid interaction. Nature 431: 841–844.

Foreman, D. 2004. Rewilding North America. Island Press, Washington, D.C.

Forest, F., R. Grenyer, M. Rouget, TJ Davies, RM Cowling, DP Faith, A Balmford, JC Manning, S Proches, M van der Bank, G Reeves, TAJ Hedderson & V Savolainen 2007. Preserving the evolutionary potential of floras in biodiversity hotspots. Nature 445: 757–760.

Fox, G. A., & B. E. Kendall. 2002. Demographic stochasticity and the variance reduction effect. Ecology 83: 1928–1934.

Frank, S. A. 2002. Immunology and evolution of infectious disease. Princeton University Press, Princeton, N.J.

Frank, K. 2005. Metapopulation persistence in heterogeneous landscapes: Lessons about the effect of stochasticity. Am Nat. 165: 374–388.

Frankel, O. H. 1974. Genetic conservation: Our evolutionary responsibility. Genetics 78: 53–65.

Frankel, O. H., & J. G. Hawkes (eds.). 1975. Crop genetic resources for today and tomorrow. Cambridge University Press, Cambridge, UK.

Frankel, O. H., & M. E. Soulé. 1981. Conservation and evolution. Cambridge University Press, New York.

Frankham, R. 1995a. Effective population size/adult population size ratios in wildlife: A review. Genet Res. 66: 95–107.

Frankham, R. 1995b. Inbreeding and extinction: A threshold effect. Conserv Biol. 9: 792–799.

Frankham, R. 1997. Do island populations have less genetic variation than mainland populations? Heredity 78: 311–327.

Frankham, R. 2005a. Conservation biology: Ecosystem recovery enhanced by genotypic diversity. Heredity 95: 183.

Frankham, R. 2005b. Genetics and extinction. Biol Conserv. 126: 131–140.

Frankham, R. 2005c. Stress and adaptation in conservation genetics. J Evol Biol. 18: 750–755.

Frankham, R., J. D. Ballou, & D. A. Briscoe. 2002. Introduction to conservation genetics. Cambridge University Press, Cambridge, UK.

Frankham, R., H. Manning, S. H. Margan, & D. A. Briscoe. 2000. Does equalization of family sizes reduce genetic adaptation to captivity? Anim Conserv. 3: 357–363.

Franklin, I. R. 1980. Evolutionary change in small populations (pp. 135–149). In M. E. Soulé & B. A. Wilcox (eds.). Conservation biology: An evolutionary–ecological perspective. Sinauer Associates, Sunderland, Mass.

Freeman, S., & J. C. Herron. 2006. Evolutionary analysis. 4th ed. Pearson Prentice Hall, Upper Saddle River, N. J.

Friesen, V. L., J. F. Piatt, & A. J. Baker. 1996. Evidence from cytochrome *b* sequences and allozymes for a 'new' species of alcid: The

long-billed Murrelet (*Brachyramphus perdix*). Condor 98: 681–690.
Fritts, T. H., & G. H. Rodda. 1998. The role of introduced species in the degradation of island ecosystems: A case history of Guam. Annu Rev Ecol Syst. 39: 113–140.
Fritz, R. S., & E. L. Simms. 1992. Plant resistance to herbivores and pathogens: Ecology, evolution and genetics. University of Chicago Press, Chicago, Ill.
Fry, J. D. 2001. Rapid mutational declines of viability in *Drosophila*. Genet Res. 77: 53–60.
Funk, S. M., C. V. Fiorella, S. Cleaveland, & M. E. Gompper. 2001. The role of disease in carnivore ecology and conservation (pp. 241–281). In: J. Gittleman, S. Funk, D. MacDonald, & R. Wayne (eds.). Carnivore conservation. Cambridge University Press, Cambridge, UK.
Funk, D. J., P. Nosil, & W. J. Etges. 2006. Ecological divergence exhibits consistently positive associations with reproductive isolation across disparate taxa. Proc Natl Acad Sci USA 103: 3209–3213.
Fussman, G. F., M. Loreau, & P. A. Abrams. 2007. Eco-evolutionary dynamics of communities and ecosystems. Funct Ecol. 21: 465–477.
Gaggiotti, O. E., & P. E. Smouse. 1996. Stochastic migration and maintenance if genetic variation in sink populations. Am Nat. 147: 919–945.
Garant, D., & L. E. B. Kruuk. 2005. How to use molecular marker data to measure evolutionary parameters in wild populations. Mol Ecol. 14: 1843–1859.
García-Ramos, G., & M. Kirkpatrick. 1997. Genetic models of adaptation and gene flow in peripheral populations. Evolution 51: 21–28.
García-Ramos, G., & D. Rodríguez. 2002. Evolutionary speed of species invasions. Evolution 56: 661–668.
Garland, T., Jr., R. B. Huey, & A. F. Bennett. 1991. Phylogeny and coadaptation of thermal physiology in lizards: A reanalysis. Evolution 45: 1969–1976.
Garner, A., J. L. Rachlow, & J. F. Hicks. 2005. Patterns of genetic diversity and its loss in mammalian populations. Conserv Biol. 19: 1215–1221.
Garnier-Géré, P. H., & P. K. Ades. 2001. Environmental surrogates for predicting and conserving adaptive genetic variability in tree species. Conserv Biol. 15: 1632–1644.
Garrido, J. L., P. J. Rey, X. Cerda, & C. M. Herrera. 2002. Geographical variation in diaspore traits of an ant-dispersed plant (*Helleborus foetidus*): Are ant community composition and diaspore traits correlated? J Ecol. 90: 446–455.
Gaston, K. J. 1994. Geographic range sizes and trajectories to extinction. Biodivers Lett. 2: 163–170.
Gates, D. M. 1993. Climate change and its biological consequences. Sinauer Associates, Sunderland, Mass.
Gaubert, P., P. J. Taylor, C. A. Fernandes, M. W. Bruford & G. Veron 2005 Patterns of cryptic hybridization revealed using an integrative approach: A case study on genets (Carnivora, Viverridae, *Genetta* spp.) from the southern African subregion. Biol J Linn Soc. 86: 11–33.
Gavrilets, S. 2000a. Rapid evolution of reproductive barriers driven by sexual conflict. Nature 403: 886–889.
Gavrilets, S. 2000b. Waiting time to parapatric speciation. Proc R Soc Lond B. 267: 2483–2492.
Gavrilets, S. 2003. Models of speciation: What have we learned in 40 years? Evolution 57: 2197–2215.
Gavrilets, S., H. Li, & M. D. Vose. 2000. Patterns of parapatric speciation. Evolution 54: 1126–1134.
Gelvin, S. B. 2003. *Agrobacterium*-mediated plant transformation: The biology behind the "gene jockeying" tool. Microbiol Mol Biol Rev. 67: 16–37.
Gephard, S., & J. McMenemy. 2004. An overview of the program to restore Atlantic salmon and other diadromous fishes to the Connecticut River with notes on the current status of these species in the river. Am Fish Soc Mon. 9: 287–317.
Gepts, P., & R. Papa. 2003. Possible effects of (trans)gene flow from crops on the genetic diversity from landraces and wild relatives. Environ Biosafety Res. 2: 89–103.
Getz, W. M., & J. Pickering. 1983. Epidemic models: Thresholds and population regulation. Am Nat. 121: 892–898.
Ghalambor, C. K., J. K. McKay, S. P. Carroll, & D. Reznick. 2007. Adaptive versus non-adaptive phenotypic plasticity and the potential for adaptation to new environments. Funct Ecol. 21: 394–407.
Gharrett, A. J., & W. W. Smoker. 1991. Two generations of hybrids between even- and odd-year pink salmon (*Oncorhynchus gorbuscha*): A test for outbreeding depression? Can J Fish Aq Sci. 48: 1744–1749.
Ghazoul, J. 2004. Alien abduction: Disruption of native plant–pollinator interactions by invasive species. Biotropica 36: 156–164.
Gibbs, A. G., & L. M. Matzkin. 2001. Evolution of water balance in the genus *Drosophila*. J Exp Biol. 204: 2331–2338.

Gibert, P., B. Moreteau, G. Petavy, D. Karan & J. R. David 2001 Chill-coma tolerance, a major climatic adaptation among *Drosophila* species. Evolution 55: 1063–1068.

Gibson, R. M., & G. C. Bachman. 1992. The costs of female choice in a lekking bird. Behav Ecol. 3: 300–309.

Gilchrist, G. W. 1995. Specialists and generalists in changing environments. 1. Fitness landscapes of thermal sensitivity. Am Nat 146: 252–270.

Gilchrist, G. W. 1996. A quantitative genetic analysis of thermal sensitivity in the locomotor performance curve of *Aphidius ervi*. Evolution 50: 1560–1572.

Gilchrist, G. W. 2000. The evolution of thermal sensitivity in changing environments (pp. 55–70). In K. B. Storey & J. M. Storey (eds.). Cell and molecular responses to stress. Vol 1. Environmental stressors and gene responses. Elsevier Science, Amsterdam.

Gilchrist, G. W., & R. B. Huey. 1999. The direct response of *Drosophila melanogaster* to selection on knockdown temperature. Heredity 83: 15–29.

Gilchrist, G. W., R. B. Huey, J. Balanyá, M Pascual & L Serra 2004. A time series of evolution in action: A latitudinal cline in wing size in South American *Drosophila subobscura*. Evolution 58: 768–780.

Gilchrist, G. W., & C. E. Lee. 2007. All stressed out and nowhere to go: Does evolvability limit adaptation in invasive species? Genetica 129: 127–132.

Giles, B. E., & J. Goudet. 1997. Genetic differentiation in Silene dioica metapopulations: Estimation of spatiotemporal effects in a successional plant species. Am Nat 149: 507–526.

Gilligan, D. M., D. A. Briscoe, & R. Frankham. 2005. Comparative losses of quantitative and molecular genetic variation in finite populations of *Drosophila melanogaster*. Genet Res Camb. 85: 47–55.

Gilpin, M. E., & M. E. Soulé. 1986. Minimum viable populations: Processes of species extinction (pp. 19–34). In M. E. Soulé (ed.). Conservation biology: The science of scarcity and diversity. Sinauer Associates, Sunderland, Mass.

Giraud, A., I. Matic, O. Tenaillon, A. Clara, M. Radman, M. Fons & F. Taddei. 2001. Costs and benefits of high mutation rates: Adaptive evolution of bacteria in the mouse gut. Science 291: 2606–2608.

Gitzendanner, M. A., & P. S. Soltis. 2000. Patterns of genetic variation in rare and widespread congenors. Am J Bot. 87: 783–792.

Gompper, M. E., & E. S. Williams. 1998. Parasite conservation and the black-footed ferret recovery program. Conserv Biol. 12: 730–732.

Gomulkiewicz, R., & R. D. Holt. 1995. When does evolution by natural selection prevent extinction? Evolution 49: 201–207.

Gomulkiewicz, R., J. N. Thompson, R. D. Holt, SL Nuismer & ME Hochberg 2000. Hot spots, cold spots, and the geographic mosaic theory of coevolution. Am Nat 156: 156–174.

Goodnight, C. J. 1988. Epistasis and the effect of founder events on the additive genetic variance. Evolution 42: 441–454.

Goodnight, C. J. 2000. Modeling gene interaction in structured populations (pp. 129–145). In J. B. Wolf, E. D. Brodie, & M. J. Wade (eds.). Epistasis and the evolutionary process. Oxford University Press, Oxford, UK.

Goodwillie, C., S. Kalisz, & C. Eckert. 2005. The evolutionary enigma of mixed mating systems in plants: Occurrence, theoretical explanations, and empirical evidence. Annu Rev Ecol Evol Syst. 36: 47–79.

Grand, J., J. Buonaccorsi, S. A. Cushman, C. R. Griffin & M. C. Neel. 2004. A multiscale landscape approach to predicting bird and moth rarity hotspots, in a threatened pitch pine–scrub oak community. Conserv Biol. 18: 1063–1077.

Grand, J., M. P. Cummings, T. G. Rebelo, T. H. Ricketts & M. C. Neel. 2007. Biased data reduce efficiency and effectiveness of conservation reserve networks. Ecol Lett. 10: 364–374.

Grant, P. R., & B. R. Grant. 1995. Predicting microevolutionary responses to directional selection on heritable variation. Evolution 49: 241–251.

Grant, B. R., & P. R. Grant. 1996. Cultural inheritance of song and its role in the evolution of Darwin's finches. Evolution 50: 2471–2487.

Grant, P. R., B. R. Grant, J. A. Markert, L. F. Keller, & K. Petren. 2004. Convergent evolution of Darwin's finches caused by introgressive hybridization and selection. Evolution 58: 1588–1599.

Grether, G. F. 2000. Carotenoid limitation and mate preference evolution: A test of the indicator hypothesis in guppies (*Poecilia reticulata*). Evolution 54: 1712–1724.

Grether, G. F. 2005. Environmental change, phenotypic plasticity and genetic compensation. Am Nat 166: E115–E123.

Grice, K., C. Cao, D. Love, M. E. Böttcher, R. J. Twitchett, E. Grosjean, R. E. Summons, S. C. Turgeon, W. Dunning & Y. Jin. 2005. Photic

zone euxinia during the Permian–Triassic superanoxic event. Science 307: 706–709.
Griffin, A. S., & C. S. Evans. 2003. Social learning of antipredator behaviour in a marsupial. Anim Behav. 66: 485–492.
Grift, R. E., M. Heino, A. D. Rijnsdorp, S. B. M. Kraak & U. Dieckmann. 2007. Three-dimensional maturation reaction norms for North Sea plaice. Mar Ecol Prog Ser. 334: 213–224.
Grift, R. E., A. D. Rijnsdorp, S. Barot, M. Heino & U. Dieckmann. 2003. Fisheries-induced trends in reaction norms for maturation in North Sea plaice. Mar Ecol Prog Ser. 257: 247–257.
Grinnell, J. 1919. The English sparrow has arrived in Death Valley: An experiment in nature. Am Nat. 43: 468–473.
Groom, M. J., G. K. Meffe, & C. R. Carroll (eds.). 2005. Principles of conservation biology. 3rd ed. Sinauer Associates, Sunderland, Mass.
Groombridge, B. (ed.). 1992. Global biodiversity: Status of the earth's living resources. Chapman & Hall, New York.
Grosholz, E. D., & G. M. Ruiz. 2003. Biological invasions drive size increases in marine and estuarine invertebrates. Ecol Lett. 6: 700–705.
Groves, C. R., D. B. Jensen, L. L. Valutis, K. H. Redford, M. L. Shaffer, J. M. Scott, J. V. Baumgartner, J. V. Higgins, M. W. Beck & M. G. Anderson. 2002. Planning for biodiversity conservation: Putting conservation science into practice. Bioscience 52: 499–512.
Gurevitch, J., S. M. Scheiner, & G. A. Fox. 2002 The ecology of plants. Sinauer Associates, Sunderland, Mass.
Gwynne, D. T., & D. C. F. Rentz. 1983. Beetles on the bottle: Male buprestids mistake stubbies for females (Coleoptera). J Aust Entomol Soc. 22: 79–80.
Haag, C. R., M. Saastamoinen, J. H. Marden, & I. Hanski. 2005. A candidate locus for variation in dispersal rate in a butterfly metapopulation. Proc R Soc Lond B. 272: 2449–2456.
Hafner, M. S., & R. D. M. Page. 1995. Molecular phylogenies and host–parasite cospeciation: Gophers and lice as a model system. Phil Trans R Soc Lond Ser B. 349: 77–83.
Haig, S. M., & F. W. Allendorf. 2006. Hybrids and policy (pp. 150–163). In J. M. Scott, D. D. Goble, & F. W. Davis (eds.). The Endangered Species Act at thirty: Conserving biodiversity in human-dominated landscapes. Vol 2. Island Press, Washington, D.C.
Haig, S. M., T. D. Mullins, E. D. Forsman, P. W. Trail & L. Wennerberg 2004. Genetic identification of spotted owls, barred owls, and their hybrids: Legal implications of hybrid identity. Conserv Biol. 18: 1347–1357.
Hails, R. S., & K. Morley. 2005. Genes invading new populations: A risk assessment perspective. Trends Ecol Evol. 20: 245–252.
Hairston, N. G. Jr., S. P. Ellner, M. A. Geber, T. Yoshida & J. A. Fox. 2005. Rapid evolution and the convergence of ecological and evolutionary time. Ecol Lett. 8: 1114–1127.
Hale, K. A., & J. V. Briskie. 2007. Decreased immunocompetence in a severely bottlenecked population of an endemic New Zealand bird. Anim Conserv. 10: 2–10.
Hall, R. J., A. Hastings, & D. R. Ayres. 2006. Explaining the explosion: Modeling hybrid invasions. Proc R Soc Lond B. 273: 1385–1389.
Hall, L., K. Topinka, J. Huffman, L. Davis & A. Good. 2000. Pollen flow between herbicide-resistant *Brassica napus* is the cause of multiple-resistant *B. napus* volunteers. Weed Sci. 48: 688–694.
Hallam, A. 1991. Why was there a delayed radiation after the end-Palaeozoic extinctions? Hist Biol. 5: 257–262.
Hallas, R., M. Schiffer, and A. A. Hoffmann. 2002. Clinal variation in *Drosophila serrata* for stress resistance and body size. Genet Res. 79: 141–148.
Hamilton, W. D. 1982. Pathogens as causes of genetic diversity in their host populations (pp. 269–296). In R. M. Anderson & R. M. May (eds.). Population biology of infectious diseases. Springer-Verlag, Berlin.
Hamilton, W.D. and M. Zuk. 1982. Heritable true fitness and bright birds: a role for parasites? Science 218: 384–387.
Hamrick, J. L., & M. J. W. Godt. 1989. Allozyme diversity in plant species (pp. 43–63). In A. H. D. Brown, M. T. Clegg, A. L. Kahler, & B. S. Weir (eds.). Plant population genetics, breeding, and genetic resources. Sinauer Associates, Sunderland, Mass.
Hamrick, J. L., & M. J. W. Godt. 1990. Plant Population Genetics, Breeding and Genetic Resources. Sinauer, Sunderland Mass.
Hamrick, J. L., & J. W. Godt. 1996. Conservation genetics of endemic plant species (pp. 43–63). In J. C. Avise & J. L. Hamrick (eds.). Conservation genetics: Case histories from nature. Chapman and Hall, New York.
Hamrick, J. L., & A. Schnabel. 1985. Understanding the genetic structure of plant populations: Some old problems and a new approach (pp. 50–70). In H. R. Gregorious (ed.). Population genetics in forestry (lecture

notes in biomathematics). Springer Verlag, Berlin.

Hanley, M. E., & D. Goulson. 2003. Introduced weeds pollinated by introduced bees: Cause or effect? Weed Biol Manag. 3: 240–212.

Hansen, T. A., B. R. Farrell, & B. Upshaw, III. 1993. The first two million years after the Cretaceous–Tertiary boundary in East Texas: Rate and paleoecology of the molluscan recovery. Paleobiology 19: 251–265.

Hansen, T. A., P. H. Kelley, V. D. Melland, & S. E. Graham. 1999. Effect of climate-related mass extinctions on escalation in molluscs. Geology 27: 1139–1142.

Hanski, I. 1998. Metapopulation dynamics. Nature 396: 41–49.

Hanski, I. 2004. Metapopulation theory, its use and misuse. Basic Appl Ecol. 5: 225–229.

Hanski, I. A., & M. E. Gilpin. 1997. Metapopulation biology: Ecology, genetics, and evolution. Academic Press, New York.

Hanski, I., A. Moilanen, & M. Gyllenberg. 1996a. Minimum viable metapopulation size. Am Nat 147: 527–541.

Hanski, I., A. Moilanen, T. Pakkala, & M. Kuussaari. 1996b. The quantitative incidence function model and persistence of an endangered butterfly metapopulation. Conserv Biol. 10: 578–590.

Hanski, I., & O. Ovaskainen. 2000. The metapopulation capacity of a fragmented landscape. Nature 404: 755–758.

Hanski, I., & O. Ovaskainen. 2003. Metapopulation theory for fragmented landscapes. Theor Popul Biol. 64: 119–127.

Harbison, S. T., A. H. Yamamoto, J. J. Fanara, KK Norga & TFC Mackay 2004. Quantitative trait loci affecting starvation resistance in *Drosophila melanogaster*. Genetics 166: 1807–1823.

Harper, M. P., & B. L. Peckarsky. 2006. Emergence cues of a mayfly in a high-altitude stream ecosystem: Potential response to climate change. Ecol Appl. 16: 612–621.

Harries, P. J., E. G. Kauffman, & T. A. Hansen. 1996. Models for biotic survival following mass extinction (pp. 41–60). In M. B. Hart (ed.). Biotic recovery from mass extinction events. Geological Society special publication 102. Geological Society, London, UK.

Harrington, R., I. Woiwod, & T. Sparks. 1999. Climate change and trophic interactions. Trends Ecol Evol. 14: 146–150.

Harris, R. B., W. A. Wall, & F. W. Allendorf. 2002. Genetic consequences of hunting: What do we know and what should we do? Wildlife Soc Bull. 30: 634–643.

Harrison, S., & A. Hastings. 1996. Genetic and evolutionary consequences of metapopulation structure. Trends Ecol Evol. 11: 180–183.

Hartl, D. L., & A. G. Clark. 1997. Principles of population genetics. Sinauer Associates, Sunderland, Mass.

Hartl, D. L., & A. G. Clark. 2007. Principles of population genetics. 4th ed. Sinauer Associates, Sunderland, Mass.

Harvell, C. D. 1990. The ecology and evolution of inducible defenses. Q. Rev Biol. 65: 323–340.

Harvell, C. D., K. Kim, J. M. Burkholder, R. R. Colwell, P. R. Epstein, D. J. Grimes, E. E. Hofmann, E. K. Lipp, A. D. Osterhaus, R. M. Overstreet, J. W. Porter, G. W. Smith, G. R. Vasta. 1999 Emerging marine diseases: Climate links and anthropogenic factors. Science 285: 1505–1510.

Harvell, C. D., C. E. Mitchell, J. R. Ward, S. Altizer, A. Dobson, R. S. Ostfeld & M. D. Samuels. 2002. Climate warming and disease risks for terrestrial and marine biota. Science 296: 2158–2162.

Hasson, E., & W. F. Eanes. 1996. Contrasting histories of three gene regions associated with *In(3L)Payne* of *Drosophila melanogaster*. Genetics 144: 1565–1575.

Hawley, D., D. Hanley, A. Dhondt, & I. Lovette. 2006. Molecular evidence for a founder effect in invasive house finch (*Carpodacus mexicanus*) populations experiencing and emergent disease epidemic. Mol Ecol. 15: 263–275.

Hawley, D., K. Sydenstricker, G. Kollias, & A. Dhondt. 2005. Genetic diversity predicts pathogen resistance and cell-mediated immunocompetence in house finches. Biol Lett. 1: 326–329.

Hebert, P. D. N., A. Cywinska, S. L. Ball, & J. R. deWaard. 2003a. Biological identifications through DNA barcodes. Proc R Soc Lond B. 270: 313–321.

Hebert, P. D. N., E. H. Penton, J. M. Burns, DH Janzen & W Hallwachs 2004. Ten species in one: DNA barcoding reveals cryptic species in the neotropical skipper butterfly *Astraptes fulgerator*. Proc Natl Acad Sci USA 101: 14812–14817.

Hebert, P. D. N., S. Ratnasingham, & J. R. DeWaard. 2003b. Barcoding animal life: Cytochrome c oxidase subunit 1 divergences among closely related species. Proc R Soc Lond B Biol Sci. 270: S596–S599.

Hedgecock, D., & F. Sly. 1990. Genetic drift and effective population sizes in hatchery propagated stocks of the Pacific oyster, *Crassostrea gigas*. Aquaculture 88: 21–38.

Hedrick, P. W. 1995. Gene flow and genetic restoration: The Florida panther as a case study. Conserv Biol. 9: 996–1007.
Hedrick, P. W. 2001. Conservation genetics: Where are we now? Trends Ecol Evol. 16: 629–636.
Hedrick, P. W. 2006. 'Genetic restoration': A more comprehensive perspective than 'genetic rescue.' Trends Ecol Evol. 20: 109.
Hedrick, P. W., & S. T. Kalinowski. 2000. Inbreeding depression in conservation biology. Annu Rev Ecol Syst. 31: 139–162.
Hedrick, P. W., & T. J. Kim. 2000. Genetics of complex polymorphisms: Parasites and maintenance of the major histocompatibility complex variation (pp. 204–234). In R. S. Singh & C. B. Krimbas (eds.). Evolutionary genetics: From molecules to morphology. Cambridge University Press, Cambridge, UK.
Hedrick, P. W., & P. S. Miller. 1992. Conservation genetics: Techniques and fundamentals. Ecol Appl. 2: 30–46.
Hedrick, P. W., R. N. Lee, & D. Garrigan. 2002. Major histocompatibility complex variation in red wolves: Evidence for common ancestry with coyotes and balancing selection. Mol Ecol. 11: 1905–1913.
Hedrick, P. W., K. M. Parker, G. Gutierrez-Espeleta, A Rattink & K Lievers 2000. Major histocompatibility complex variation in the Arabian oryx. Evolution 54: 2145–2151.
Heering, T. E., Jr., & D. H. Reed. 2005. Modeling extinction: Density-dependent changes in the variance of population growth rates. J MS Acad Sci. 50: 183–194.
Hegde, S., J. D. Nason, J. M. Clegg, & N. C. Ellstrand. 2006. The evolution of California's wild radish has resulted in the extinction of its progenitors. Evolution 60: 1187–1197.
Heino, M. 1998. Management of evolving fish stocks. Can J Fish Aquat Sci. 55: 1971–1982.
Heino, M., U. Dieckmann, & O. R. Godø. 2002a. Measuring probabilistic reaction norms for age and size at maturation. Evolution 56: 669–678.
Heino, M., U. Dieckmann, & O. R. Godø. 2002b. Reaction norm analysis of fisheries-induced adaptive change and the case of the Northeast Arctic cod. ICES CM 2002/Y:14. ICES, Copenhagen.
Heino, M., & O. R. Godø. 2002 Fisheries-induced selection pressures in the context of sustainable fisheries. Bull Mar Sci. 70: 639–656.
Hendry, A. P. 2004. Selection against migrants contributes to the rapid evolution of ecologically dependent reproductive isolation. Evol Ecol Res. 6: 1219–1236.
Hendry, A. P., T. Farrugia, & M. T. Kinnison. 2008. Human influences on rates of phenotypic change in wild animal populations. Mol Ecol. 17: 20–29.
Hendry, A. P., P. R. Grant, B. R. Grant, H. A. Ford, M. J. Brewer & J. Podos 2006. Possible human impacts on adaptive radiation: Beak size bimodality in Darwin's finches. Proc R Soc Lond B. 273: 1887–1894.
Hendry, A. P., & M. T. Kinnison. 1999. The pace of modern life: Measuring rates of contemporary microevolution. Evolution 53: 1637–1653.
Hendry, A. P., B. H. Letcher, & G. Gries. 2003. Estimating natural selection acting on stream-dwelling Atlantic salmon: Implications for the restoration of extirpated populations. Conserv Biol. 17: 795–805.
Hendry, A. P., P. Nosil, & L. H. Rieseberg. 2007. The speed of ecological speciation. Funct Ecol. 21:455–464.
Hendry, A. P., E. B. Taylor, & J. D. McPhail. 2002. Adaptive divergence and the balance between selection and gene flow: Lake and stream stickleback in the misty system. Evolution 56: 1199–1216.
Herrera, S. 2005. Struggling to see the forest through the tress. Nat Biotechnol. 23: 165–167.
Heschel, M. S., & K. N. Paige. 1995. Inbreeding depression, environmental stress, and population size variation in scarlet gilia (*Ipomopsis aggregata*). Conserv Biol. 9: 126–133.
Hickman, C. S. 2003. Evidence for abrupt Eocene–Oligocene molluscan faunal change in the Pacific Northwest (pp. 71–87). In D. R. Prothero, L. C. Ivany, & E. A. Nesbitt (eds.). From greenhouse to icehouse: The marine Eocene–Oligocene transition. Columbia University Press, New York.
Hickson, R. E., D. Penny, & D. B. Scott. 1992. Molecular systematics and evolution in New Zealand: Applications to cryptic skink species. New Zealand J Zool, 19: 33–44.
Higgins, K., & M. Lynch. 2001. Metapopulation extinction caused by mutation accumulation. Proc Natl Acad Sci USA 98: 2928–2933.
Hill, W. G. 2005. A century of corn selection. Science 307: 683–684.
Hill, J. K., C. D. Thomas, R. Fox, M. G. Telfer, S. G. Willis, J. Asher, B. Huntley 2002. Responses of butterflies to twentieth century climate warming: Implications for future ranges. Proc R Soc Lond B. 269: 2163–2171.

Hillis, D. M., & C. Moritz (eds.). 1990. Molecular systematics. Sinauer Associates, Sunderland, Mass.

Hillis, D. M., C. Moritz, & B. K. Mable (eds.). 1996. Molecular systematics. 2nd ed. Sinauer Associates, Sunderland, Mass.

Hobbs, R. J., & S. E. Humphries. 1995. An integrated approach to the ecology and management of plant invasions. Conserv Biol. 9: 761–770.

Hoffmann, A. A., A. Anderson, & R. Hallas. 2002. Opposing clines for high and low temperature resistance in *Drosophila melanogaster*. Ecol Lett. 5: 614–618.

Hoffmann, A., & M. W. Blows. 1994. Species borders: Ecological and evolutionary perspectives. Trends Ecol Evol. 9: 223–227.

Hoffmann, A. A., R. J. Hallas, J. A. Dean, & M. Schiffer. 2003a. Low potential for climatic stress adaptation in a rainforest *Drosophila* species. Science 301: 100–102.

Hoffmann, A. A., R. Hallas, C. Sinclair, & P. Mitrovski. 2001. Levels of variation in stress resistance in *Drosophila* among strains, local populations, and geographic regions: Patterns for desiccation, starvation, cold resistance, and associated traits. Evolution 55: 1621–1630.

Hoffmann, A. A., & L. G. Harshman. 1999. Desiccation and starvation resistance in *Drosophila*: Patterns of variation at the species, population and intrapopulation levels. Heredity 83: 637–643.

Hoffmann, A. A., & M. J. Hercus. 2000. Environmental stress as an evolutionary force. BioScience 50: 217–226.

Hoffmann, A. A., & J. Merila. 1999. Heritable variation and evolution under favourable and unfavourable conditions. Trends Ecol Evol. 14: 96–101.

Hoffmann, A. A., & P. A. Parsons. 1991. Evolutionary genetics and environmental stress. Oxford University Press, Oxford, UK.

Hoffmann, A. A., & P. A. Parsons. 1997. Extreme environmental change and evolution. Cambridge University Press, Cambridge, UK.

Hoffmann, A. A., M. Scott, L. Partridge, & R. Hallas. 2003b. Overwintering in *Drosophila melanogaster*, outdoor field cage experiments on clinal and laboratory selected populations help to elucidate traits under selection. J Evol Biol. 16: 614–623.

Hoffmann, A. A., J. G. Sørensen, & V. Loeschcke. 2003c. Adaptation of *Drosophila* to temperature extremes: Bringing together quantitative and molecular approaches. J Therm Biol. 28: 175–216.

Hoffmann, A. A., & A. R. Weeks. 2007. Climatic selection on genes and traits after a 100 year-old invasion: A critical look at the temperate–tropical clines in *Drosophila melanogaster* from eastern Australia. Genetica 129: 133–147.

Hogg, I. D., & P. D. N. Hebert. 2004. Biological identification of springtails (Hexapoda: Collembola) from the Canadan arctic using mitochondrial DNA barcodes. Can J Zool. 82: 749–754.

Holmes, E. C. 2004. The phylogeography of human viruses. Mol Ecol. 13: 745–756.

Holsinger, K. E. 1993. Ecological models of plant mating systems and the evolutionary stability of mixed mating systems (pp. 169–191). In R. W. Wyatt (ed.). Ecology and evolution of plant reproductive systems. Chapman and Hall, New York.

Holt, R. D., M. Barfield, & R. Gomulkiewicz. 2004. Temporal variation can facilitate niche evolution in harsh sink environments. Am Nat. 164: 187–200.

Holt, R. D., M. Barfield, & R. Gomulkiewicz. 2005. Theories of niche conservatism and evolution (pp. 259–290). In D. F. Sax, J. J. Stachowicz, & S. D. Gaines (eds.). Species invasions: Insights into ecology, evolution and biogeography. Sinauer Associates, Sunderland, Mass.

Holt, R. D., & R. Gomulkiewicz. 1997. How does immigration influence local adaptation? A reexamination of a familiar paradigm. Am Nat. 149: 563–572.

Holt, R. D., & R. Gomulkiewicz. 2004. Conservation implications of niche conservatism and evolution in heterogeneous environments (pp. 244–264). In R. Ferrière, U. Dieckmann, & D. Couvet (eds.). Evolutionary conservation biology. Cambridge University Press, Cambridge, UK.

Honnay, O., K. Verheyen, J. Butaye, H Jacquemyn, B Bossuyt & M Hermy 2002. Possible effects of habitat fragmentation and climate change on the range of forest plant species. Ecol Lett. 5: 525–530.

Hortal, J., & J. M. Lobo. 2005. An ED-based protocol for optimal sampling of biodiversity. Biodivers Conserv. 14: 2913–2947.

Houghton, J. T., Y. Ding, D. J. Griggs, M Noguer, PJ v d Linden, X Dai, K Maskell, & CA Johnson 2001. Climate change 2001: The scientific basis. Contribution of Working Group I to the third assessment report of the intergovernmental panel on climate change. Cambridge University Press, Cambridge, UK.

Houle, D., & A. Kondrashov. 2006. Mutation (pp. 32–48). In C. W. Fox & J. B. Wolf (eds.). Evolutionary genetics: Concepts and case studies. Oxford University Press, New York.

Hudson, R. R. 1983. Properties of a neutral allele model with intragenic recombination. Theor Popul Biol. 23: 183–201.

Hudson, R. R. 1990. Gene genealogies and the coalescent process. Oxford Surv Evol Biol. 7: 1–44.

Hudson, P. J., A. Rizzoli, B. T. Grenfell, H. Heesterbeek & A. P. Dobson. 2002. The ecology of wildlife diseases. Oxford University Press, New York, N.Y.

Huenneke, L. F. 1991. Ecological implications of genetic variation in plant populations (pp. 31–44). In D. A. Falk & K. H. Holsinger (eds.). Genetics and conservation of rare plants. Oxford University Press, New York.

Huey, R. B., & A. F. Bennett. 1987. Phylogenetic studies of coadaptation: Preferred temperatures versus optimal performance temperatures of lizards. Evolution 41: 1098–1115.

Huey R. B. & J. G. Kingsolver. 1993 Evolution of resistance to high temperature in ectotherms. Am Nat 142:S21-S46

Huey, R. B., G. W. Gilchrist, M. L. Carlson, D. Berrigan, & L. Serra. 2000. Rapid evolution of a geographic cline in size in an introduced fly. Science 287: 308–309.

Huey, R. B., P. E. Hertz, & B. Sinervo. 2003. Behavioral drive versus behavioral inertia in evolution: A null model approach. Am Nat. 161: 357–366.

Huey, R. B., & J. G. Kingsolver. 1993. Evolution of resistance to high temperature in ectotherms. Am Nat. 142: S21–S46.

Hufbauer, R. A., & M. E. Torchin. 2007. Integrating ecological and evolutionary theory of biological invasions (pp. 79–96). In W. Nentwig (ed.). Biological invasions. Springer Verlag, Berlin.

Hufford, K. M., & S. J. Mazer. 2003. Plant ecotypes: Genetic differentiation in the age of ecological restoration. Trends Ecol Evol. 18: 147–155.

Hughes, J. B., G. C. Daily, & P. R. Ehrlich. 1997. Population diversity: Its extent and extinction. Science 278: 689–692.

Hughes, J., K. Goudkamp, D. Hurwood, M. Hancock & S. Bunn. 2003. Translocation causes extinction of a local population of the freshwater shrimp *Paratya australiensis*. Conserv Biol. 17: 1007–1012.

Hughes, A. L., & M. Nei. 1988. Pattern of nucleotide substitution at major histocompatibility complex class-I loci reveals overdominant selection. Nature 335: 167–170.

Hughes, A. L., & M. Nei. 1989. Nucleotide substitution at major histocompatibility complex class-II loci: Evidence for overdominant selection. Proc Natl Acad Sci USA 86: 958–962.

Hughes, A. L., T. Ota, & M. Nei. 1990. Positive Darwinian selection promotes charge profile diversity in the antigen-binding cleft of class-I major-histocompatibility-complex molecules. Mol Biol Evol. 7: 515–524.

Hughes, A. R., & J. J. Stachowicz. 2004. Genetic diversity enhances the resistance of a seagrass ecosystem to disturbance. Proc Natl Academy Sci USA 101: 8998–9002.

Huntley, B. 2005. North temperate responses (pp. 109–124). In T. E. Lovejoy & L. Hannah (eds.). Climate change and biodiversity. Yale University Press, New Haven, Conn.

Husband, B. C., & S. C. H. Barrett. 1993. Multiple origins of self-fertilization in tristylous *Eichhornia paniculata* (Pontederiaceaea): Inferences from style morph and isozyme variation. J Evol Biol. 6: 591–608.

Husband, B. C., & D. W. Schemske. 1996. Evolution of the magnitude and timing of inbreeding depression in plants. Evolution 50: 54–70.

Hutchings, J. A. 2005. Life history consequences of overexploitation to population recovery in Northwest Atlantic cod (*Gadus morhua*). Can J Fish Aquat Sci. 62: 824–832.

Huxel, G. R. 1999. Rapid displacement of native species by invasive species: Effects of hybridization. Biol Conserv. 89: 143–152.

ICES. 1997. Report of the Multispecies Assessment Working Group. ICES C.M.1997/Assess. 16. ICES, Copenhagen.

Imasheva, A. G., V. Loeschcke, L. A. Zhivotovsky, & O. E. Lazebny. 1998. Stress temperatures and quantitative variation in *Drosophila melanogaster*. Heredity 81: 246–253.

Imhof, M., & C. Schlötterer. 2001. Fitness effects of advantageous mutations in evolving *Escherichia coli* populations. Proc Natl Acad Sci USA 98: 1113–1117.

Inchausti, P., & J. Halley. 2001. Investigating long-term ecological variability using the global population dynamics database. Science 293: 655–657.

Ingvarsson, P. K. 2001. Restoration of genetic variation lost: The genetic rescue hypothesis. Trends Ecol Evol. 16: 62–63.

Intergovernmental Panel on Climate Change (IPCC). 2007. Climate change 2007: The physical science basis. Summary for policy makers. Contribution of Working Group I to the fourth assessment report of the

intergovernmental panel on climate change. IPCC Secretariat, Geneva, Switzerland.
Isaac, N. J. B., S. T. Turvey, B. Collen, C. Waterman & J. E. M. Baillie 2007. Mammals on the EDGE: Conservation priorities based on threat and phylogeny. PLoS ONE 2: e296.doi:10.1371/journal.pone.0000296 [Ch 7].
IUCN. 1980. World conservation strategy: Living resource conservation for sustainable development. International Union for Conservation of Nature, Gland, Switzerland.
Ives, A. R., & M. C. Whitlock. 2002. Inbreeding and metapopulations. Science 295: 454–455.
Ivey, C. T., D. E. Carr, & M. D. Eubanks. 2003. Inbreeding alters *Mimulus guttatus* tolerance to herbivory and in natural environments. Ecology 85: 567–579.
Izem, R., & J. G. Kingsolver. 2005. Variation in continuous reaction norms: Quantifying directions of biological interest. Am Nat. 166: 277–289.
Jablonski, D. 1998. Geographic variation in the molluscan recovery from the end-Cretaceous extinction. Science 279: 1327–1330.
Jablonski, D., K. W. Flessa, & J. W. Valentine. 1985. Biogeography and paleobiology. Paleobiology 11: 75–90.
Jachmann, H., P. S. M. Berry, & H. Imae. 1995. Tusklessness in African elephants. Afr J Ecol. 33: 230–235.
Jackson, J. B. C., & F. K. McKinney. 1990. Ecological processes and progressive macroevolution of marine clonal benthos (pp. 173–209). In R. M. Ross & W. D. Allmon (eds.). Causes of evolution: A paleontological perspective. University of Chicago Press, Chicago, Ill.
Jackson, S. T., & C. Weng. 1999. Late Quaternary extinction of a tree species in eastern North America. Proc Natl Acad Sci USA 96: 135–142.
James, C. 2005. Executive summary of global status of commercialized biotech/GM crops: 2005. ISAAA briefs no. 34. ISAAA: Ithaca, N.Y.
James, A. C., R. B. R. Azevedo, & L. Partridge. 1995. Cellular basis and developmental timing in a size cline of *Drosophila melanogaster*. Genetics 140: 659–666.
James, R. R., S. P. Difazio, A. M. Brunner, & S. H. Strauss. 1998. Environmental effects of genetically engineered woody biomass crops. Biomass Bioenergy 14: 403–413.
Janis, C. M. 2000. Patterns in the evolution of herbivory in large terrestrial mammals: The Paleogene of North America (pp. 168–222). In H.- D. Sues (ed.). Evolution of herbivory in terrestrial vertebrates: Perspectives from the fossil record. Cambridge University Press, Cambridge, UK.
Jarne, P., & J. R. Auld. 2006. Animals mix it up too: The distribution of self-fertilization among hermaphroditic animals. Evolution 60: 1816–1824.
Jeffreys, A. J., V. Wilson, & S. L. Thein. 1985. Hypervariable "minisatellite" regions in human DNA. Nature 314: 67–73.
Johannesson, K., & A. Tatarenkov. 1997. Allozyme variation in a snail (*Littorina saxatilis*): Deconfounding the effects of microhabitat and gene flow. Evolution 51: 402–409.
Johansson, M., C. R. Primmer, & J. Merilä. 2007. Does habitat fragmentation reduce fitness and adaptability? A case study of the common frog (*Rana temporaria*). Mol Ecol. 16: 2693–2700.
Johns, J. S., & S. N. Handel. 2002. The effects of habitat fragmentation on pollinator visitation rates and fruit and seed production on a tropical tree, *Tabebuia rosea* (Bertol.) DC (Bignoniaceae) (p. 169). In Abstracts of the Ecological Society of America and the Society for Ecological Restoration annual meeting. Ecological Society of America, Washington, D.C.
Johnson, K. R., & B. Ellis. 2002. A tropical rainforest in Colorado 1.4 milliion years after the Cretaceous–Tertiary boundary. Science 296: 2379–2383.
Johnson, D. H., & L. D. Igl. 2001. Area requirements of grassland birds: A regional perspective. Auk 118: 24–34.
Johnston, R. F., & R. K. Selander. 1964. House sparrows: Rapid evolution of races in North America. Science 144: 548–550.
Jokela, J., J. Wiehn, & K. Kopp. 2006. Among- and within-population variation in outcrossing rate of a mixed mating freshwater snail. Heredity 97: 275–282.
Jonas, C. S., & M. A. Geber. 1999. Variation among populations of *Clarkia unguiculata* (Onagraceae) along altitudinal and latitudinal gradients. Am J Bot. 86: 333–343.
Jones, J. D. G. 2001. Putting knowledge of plant disease resistance genes to work. Curr Opin Plant Biol. 4: 281–287.
Jones, L. P., R. Frankham, & J. S. F. Barker. 1968. The effects of population size and selection intensity in selection for a quantitative character in *Drosophila*. II. Long-term response to selection. Genet Res. 12: 249–266.
Jones, M. E., G. C. Smith, & S. M. Jones. 2004. Is anti-predator behaviour in Tasmanian eastern quolls (*Dasyurus viverrinus*) effective against introducted predators? Anim Conserv. 7: 155–160.

Jordan, M. A., H. L. Snell, H. M. Snell, & W. C. Jordan. 2005. Phenotypic divergence despite high levels of gene flow in Galapagos lava lizards (*Microlophus albemarlensis*). Mol Ecol. 14: 859–867.

Jorgensen, C., K., , E. S. Enberg, R. Dunlop, D. S. Arlinghaus, K. Boukal, B. Brander, A. Ernande, F. Gårdmark, S. Johnston, H. Matsumura, K. Pardoe, A. Raab, A. Silva, U. Vainikka, M. Dieckmann, M. Heino, & A. D. Rijnsdorp. 2007. Managing evolving fish stocks. Science. 318: 1247–1248.

Junge-Berberovic, R. 1996. Effect of thermal environment on life histories of free living *Drosophila melanogaster* and *D subobscura*. Oecologia (Berlin) 108: 262–272.

Kalisz, S., & D. W. Vogler. 2003. Benefits of autonomous selfing under unpredictable pollinator environments. Ecology 84: 2928–2942.

Kalisz, S., D. W. Vogler, & K. M. Hanley. 2004. Context-dependent autonomous self-fertilization yields reproductive assurance and mixed mating. Nature 430: 884–887.

Kaliszewska, Z. A., J. Seger, V. J. Rowntree, S. G. Barco, R. Benegas, P. B. Best, M. W. Brown, R. L. Brownell Jr, A. Carribero, R. Harcourt, A. R. Knowlton, K. Marshalltilas, N. J. Patenaude, M. Rivarola, C. M. Schaeff, M. Sironi, W. A. Smith & T. K. Yamada 2. 2005. Population histories of right whales (Cetacea: Eubalaena) inferred from mitochondrial sequence diversities and divergences of their whale lice (Amphipoda: Cyamus). Mol Ecol. 14: 3439–3456.

Kaplinsky, N., D. Braun, D. Lisch, et al. 2002. Maize transgene results in Mexico are artifacts. Nature 416: 601.

Karan, D., N. Dahiya, A. K. Munjal, P. Gibert, B. Moreteau, R. Parkash & J. R. David. 1998. Desiccation and starvation tolerance of adult *Drosophila*: Opposite latitudinal clines in natural populations of three different species. Evolution 52: 825–831.

Karan, D., & R. Parkash. 1998. Desiccation tolerance and starvation resistance exhibit opposite latitudinal clines in *Drosophila kikkawai*. Ecol Entomol. 23: 391–396.

Kashiyama, Y., & T. Oji. 2004. Low-diversity shallow marine benthic fauna from the Smithian of northeast Japan: Paleoecologic and paleobiogeographic implications. Paleontol Res. 8: 199–218.

Kawecki, J., & D. Ebert. 2004. Conceptual issues in local adaptation. Ecol Lett. 7: 1225–1241.

Keeler, K. H., C. E. Turner, & M. R. Bolick. 1996. Movement of crop transgenes into wild plants (pp. 303–330). In S. O. Duke (ed.). Herbicide-resistant crops: Agricultural, environmental, economic, regulatory, and technical aspects. Lewis Publishers, Boca Raton Fla.

Keller, L. F., P. Arcese, J. N. M. Smith, W. M. Hochachka & S. C. Stearns. 1994. Selection against inbred song sparrows during a natural-population bottleneck. Nature 372: 356–357.

Keller, M., J. Kollmann, & P. J. Edwards. 2000. Genetic introgression from distant provenances reduces fitness in local weed populations. J Appl Ecol. 37: 647–659.

Keller, L. F., & D. M. Waller. 2002. Inbreeding effects in wild populations. Trends Ecol Evol 17: 230–241.

Kelley, P. H., & T. A. Hansen. 1996. Recovery of the naticid gastropod predator–prey system from the Cretaceous–Tertiary and Eocene–Oligocene extinctions (pp. 373–386). In M. B. Hart (ed.). Biotic recovery from mass extinction events. Geological Society special publication 102. Geological Society, London, UK.

Kennington, W. J., L. Partridge, & A. A. Hoffmann. 2006. Patterns of diversity and linkage disequilibrium within the cosmopolitan inversion *In(3R)Payne* in *Drosophila melanogaster* are indicative of coadaptation. Genetics 172: 1655–1663.

Kettlewell, H. B. D. 1956. Further selection experiments on industrial melanism in the Lepidoptera. Heredity 10: 287–301.

Kiesecker, J. M., & A. R. Blaustein. 1997. Population differences in responses of red-legged frogs (*Rana aurora*) to introduced bullfrogs. Ecology 78: 1752–1760.

Kiester, A.R., J.M. Scott, B. Csuti, R. F. Noss, B. Butterfield, K. Sahr & D. White. 1996. Conservation prioritization using gap data. Conservation Biology 10:1332–1342.

Kilpatrick, A. M. 2006. Facilitating the evolution of resistance to avian malaria in Hawaiian birds. Biol Conserv. 128: 475–485.

Kim, T. J., K. M. Parker, & P. W. Hedrick. 1999. Major histocompatibility complex differentiation in Sacramento River chinook salmon. Genetics 151: 1115–1122.

Kimura, M. 1968. Evolutionary rate at the molecular level. Nature 217: 624–626.

Kimura, M. 1969. The number of heterozygous nucleotide sites maintained in a finite population due to steady flux of mutations. Genetics 61: 893–903.

Kimura, M. 1983. The neutral theory of molecular evolution. Cambridge University Press, Cambridge, UK.

Kimura, M., & T. Ohta. 1971. Theoretical aspects of population genetics. Princeton University Press, Princeton, N.J.

Kinnison, M. P., & N. J. Hairston, Jr. 2007. Eco-evolutionary conservation biology: Contemporary evolution and the dynamics of persistence. Funct Ecol. 21: 444–454.

Kinnison, M. T., A. P. Hendry, & C. S. Stockwell. 2007. Contemporary evolution meets conservation biology II: Impediments to integration and application. Ecol Res. 22: 947–954.

Kirkpatrick, M., & N. H. Barton. 1997. Evolution of a species' range. Am Nat. 150: 1–23.

Knapp, E. E., & K. J. Rice. 1994. Starting from seeds: Genetic issues in using native grasses for restoration. Rest Manag Not. 12: 40–45.

Kneitel, J. M., & D. Perrault. 2006. Disturbance-induced changes in community composition increase species invasion success. Comm Ecol. 7: 245–252.

Koh, L. P., R. R. Dunn, N. S. Sodhi, R. K. Colwell, H. C. Proctor & V. S. Smith 2004. Species coextinctions and the biodiversity crisis. Science 305: 1632–1634.

Kolbe, J. J., R. E. Glor, L. Rodriguez Schettino, L. A. Chamizo, A. Larson & J. B. Losos. 2004. Genetic variation increases during biological invasion by a Cuban lizard. Nature 431: 177–181.

Koolhaas, J, M., S. M. Korte, S. F. De Boer, BJ Van Der Vegt, CG Van Reenen, H Hopster, IC De Jong, MA Ruis & HJ Blokhuis 1999. Coping styles in animals: Current status in behavior and stress-physiology. Neurosci Biobehav Rev. 23: 925–935.

Kovaliski, J. 1998. Monitoring the spread of rabbit hemorrhagic disease virus as a new biological agent for control of wild European rabbits in Australia. J Wildl Dis. 34: 421–428.

Kowarik, I. 1995. Time lags in biological invasions with regard to the success and failure of alien species (pp. 15–38). In P. Pysek, K. Prach, M. Rejmanek, & M. Wade (eds.). Plant invasions: General apsects and special problems. SPB Academic Publishing, Amsterdam, the Netherlands.

Krebs, R. A. 1999. A comparison of Hsp70 expression and thermotolerance in adults and larvae of three *Drosophila* species. Cell Stress Chaperones 4: 243–249.

Krebs, R. A., & M. E. Feder. 1997. Natural variation in the expression of the heat-shock protein Hsp70 in a population of *Drosophila melanogaster* and its correlation with tolerance of ecologically relevant thermal stress. Evolution 51: 173–179.

Krebs, R. A., & M. E. Feder. 1998. Hsp70 and larval thermotolerance in *Drosophila melanogaster*: How much is enough and when is more too much? J Ins Physiol. 44: 1091–1101.

Krebs, R. A., M. E. Feder, & J. Lee. 1998. Heritability of expression of the 70KD heat-shock protein in *Drosophila melanogaster* and its relevance to the evolution of thermotolerance. Evolution 52: 841–847.

Krief, S., M. A. Huffman, T. Sevenet, J. Guillot, C. Bories, C. M. Hladik & R. W. Wrangham. 2005 Noninvasive monitoring of the health of *Pan troglodytes schweinfurthii* in the Kibale National Park, Uganda. Int J Primat 26: 467–490.

Krimbas, C. B., & J. R. Powell. 1992. Drosophila inversion polymorphism. CRC Press, Boca Raton, Fla.

Kristensen, T. N., P. Sørensen, K. S. Pedersen, M Kruhøffer, & V Loeschcke 2006. Inbreeding by environmental interactions affect gene expression in *Drosophila melanogaster*. Genetics 173: 1329–1336.

Krug, A. Z., & M. E. Patzkowsky. 2004. Rapid recovery from the Late Ordovician mass extinction. Proc Natl Acad Sci USA 101: 17605–17610.

Kruuk, L. E. B. 2004. Estimating genetic parameters in natural populations using the 'animal model'. Phil Trans R Soc Lond B. 359: 873–890.

Kruuk, L. E. B., T. H. Clutton-Brock, J. Slate, J. M. Pemberton, S. Brotherstone & F. E. Guinness. 2000 Heritability of fitness in a wild mammal population. Proc Natl Acad Sci USA 97: 698–703.

Kwon, Y. W., & D.- S. Kim. 2001. Herbicide-resistant genetically-modified crop: Its risks with an emphasis on gene flow. Weed Biol Manag 1: 42–52.

Labandeira, C. C., K. R. Johnson, & P. Wilf. 2002. Impact of the terminal Cretaceous event on plant–insect associations. Proc Natl Acad Sci USA 99: 2061–2066.

Lack, D. 1947. Darwin's finches. Cambridge University Press, Cambridge, UK.

Lacy, R. C. 1997. Importance of genetic variation to the viability of mammalian populations. J Mammal. 78: 320–335.

Lacy, R. C. 2000. Vortex: A computer simulation model for population viability analysis. Ecol Bull. 48: 191–203.

Laerm, J., J. C. Avise, J. C. Patton, & R. A. Lansman. 1982. Genetic determination of the status of an endangered species of pocket gopher in Georgia. J Wildlife Mgt. 46: 513–518.

Lafferty, K., & L. Gerber. 2002. Good medicine for conservation biology: The intersection of epidemiology and conservation theory. Conserv Biol. 16: 593–604.

Laikre, L. 1999. Conservation genetics of Nordic carnivores: Lessons from zoos. Hereditas 130: 203–216.

Laine, A.- L. 2005. Spatial scale of local adaptation in a plant–pathogen metapopulation. J Evol Biol. 18: 930–938.

Lambeck, R. J. 1997. Focal species: A multi-species umbrella for nature conservation. Conserv Biol. 11: 849–856.

Lambrinos, J.G. 2004. How interactions between ecology and evolution influence contemporary invasion dynamics. *Ecology* 85: 2061–2070.

Land, E. D., & R. C. Lacy. 2000. Introgression level achieved through Florida panther genetic restoration. Endangered Species Update 17: 100–105.

Lande, R. 1979. Quantitative genetic analysis of multivariate evolution applied to brain:body size allometry. Evolution 33: 402–416.

Lande R 1981 Models of speciation by sexual selection of polygenic traits. Proc Natl Acad Sci USA 78: 3721–3725.

Lande, R. 1987. Extinctions thresholds in demographic models of territorial populations. Am Nat. 143: 624–635.

Lande, R. 1988. Genetics and demography in biological conservation. Science 241: 1455–1460.

Lande, R. 1993. Risks of population extinction from demographic and environmental stochasticity and random catastrophes. Am Nat. 142: 911–927.

Lande, R. 1994. Risk of population extinction from fixation of new deleterious mutations. Evolution 48: 1460–1469.

Lande, R. 1995. Mutation and conservation. Conserv Biol. 9: 782–791.

Lande, R. 1998. Anthropogenic, ecological and genetic factors in extinction and conservation. Res Popul Ecol. 40: 259–269.

Lande, R., & S. J. Arnold. 1983. The measurement of selection on correlated characters. Evolution 37: 1210–1226.

Lande, R., S. Engen, & B.- E. Saether. 2003. Stochastic population dynamics in ecology and conservation. Oxford University Press, Oxford, UK.

Lande, R., & S. Shannon. 1996. The role of genetic variation in adaptation and population persistence in a changing environment. Evolution 50: 434–437.

Landweber, L. F., & A. P. Dobson (eds.). 1999. Genetics and the extinction of species. Princeton University Press, Princeton, N.J.

Lau, J. A. 2006. Evolutionary responses of native plants to novel community members. Evolution 60: 56–63.

Laurance, W. F., & R. O. Bierregaard, Jr. 1997. Tropical forest remnants: Ecology, management and conservation of fragmented communities. University of Chicago Press, Chicago, Ill.

Law, R., & D. R. Grey. 1989. Evolution of yields from populations with age-specific cropping. Evol Ecol 3: 343–359.

Law, W., & J. Salick. 2005. Human-induced dwarfing of Himalayan snow lotus, *Saussurea laniceps* (Asteraceae). Proc Natl Acad Sci USA 102: 10218–10220.

Lawton, J. H., & R. M. May. 1995. Extinction rates. Oxford University Press, Oxford, UK.

Lee, C. E. 2002. Evolutionary genetics of invasive species. Trends Ecol Evol. 17: 386–391.

Lee, C. E., J. L. Remfert, & G. W. Gelembiuk. 2003. Evolution of physiological tolerance and performance during freshwater invasions. Int Comp Biol. 43: 439–449.

Leimu, R., P. Mutikainen, J. Koricheva, & M. Fischer. 2006. How general are positive relationships between plant population size, fitness and genetic variation? J Ecol. 94: 942–952.

Lenski, R. E. 2001. Testing Antonovics' five tenets of ecological genetics: Experiments with bacteria at the interface of ecology and genetics (pp. 25–45). In M. C. Press, N. J. Huntly, & S. Levin (eds.). Ecology: Achievement and challenge. Blackwell Scientific, Oxford, UK.

Leppig, G., & J. W. White. 2006. Conservation of peripheral plant populations in California. Madrono 53: 264–274.

Lerner, I. M. 1954. Genetic homeostasis. Oliver and Boyd, Edinburgh.

Leroy, E. M., B. Kumulungui, X. Pourrut, P. Rouquet, A. Hassanin, P. Yaba, A. Delicat, J. T. Paweska, J.-P. Gonzalez & R. Swanepoel. 2005. Fruit bats as reservoirs of Ebola virus. Nature 438: 575–576.

Leroy, E. M., P. Rouquet, P. Formenty, S. Souquiere, A. Kilbourne, J. M. Froment, M. Bermejo, S. Smit, W. Karesh, R. Swanepoel, S. R. Zaki, P. E. Rollin. 2004. Multiple Ebola virus transmission events and rapid decline of central African wildlife. Science 303: 387–390.

Leslie, H., M. Ruckelshaus, I. R. Ball, S. Andelman & H. P. Possingham. 2003. Using siting algorithms in the design of marine reserve networks. Ecol Appl. 13: S185–S198.

Letcher, B. H., & T. L. King. 2001. Parentage and grandparentage assignment with known and

unknown matings: Application to Connecticut River Atlantic salmon restoration. Can J Fish Aq Sci. 58: 1812–1821.
Levin, B. R. 1996. The evolution and maintenance of virulence in microparasites. Em Infect Dis. 2: 93–102.
Levin, D. A. 2002. Hybridization and extinction. Am Sci. 90: 254–261.
Levin, D. A., J. Francisco-Ortega, & R. K. Jansen. 1996. Hybridization and the extinction of rare plant species. Conserv Biol. 10: 10–16.
Levin, P. S., R. W. Zabel, & J. G. Williams. 2001. The road to extinction is paved with good intentions: Negative association of fish hatcheries with threatened salmon. Proc R Soc Lond B. 268: 1153–1158.
Levine, J. M., & C. M. D'Antonio. 2003. Forecasting biological invasions with increasing international trade. Conserv Biol. 17: 322–326.
Levine, J. M., M. Vila, C. M. D'Antonio, J. S. Dukes, K. Grigulis & S. Lavorel. 2003. Proc R Soc Lond B. 270: 775–781.
Levins, R. 1968. Evolution in changing environments. Princeton University Press, Princeton, N.J.
Levins, R. 1970. Extinction (pp. 77–107). In M. Gesternhaber (ed.). Some mathematical problems in biology. American Mathematical Society, Providence, R.I.
Lewis, L. A., & P. O. Lewis. 2005. Unearthing the molecular phylodiversity of desert soil green algae (chlorophyta). Syst Biol. 54: 936–947.
Lewontin, R. C., & L. C. Birch. 1966. Hybridization as a source of variation for adaptation to new environments. Evolution 20: 315–336.
Lewontin, R. C., & J. L. Hubby. 1966. A molecular approach to the study of genic heterozygosity in natural populations. II. Amount of variation and degree of heterozygosity in natural populations of *Drosophila pseudoobscura*. Genetics 54: 595–609.
Ley, D. H., & H. W. Yoder, Jr. 1997. *Mycoplasma gallisepticum* infection (pp. 194–207). In B. W. Calnek, H. J. Barnes, C. W. Beard, LR McDougald & YM Saif (eds.). Diseases of poultry. 10th ed. Iowa State University Press, Ames, Iowa.
Linacre, N. A., & P. K. Ades. 2004. Estimating isolation distances for genetically modified trees in plantation forestry. Ecol Model. 179: 247–257.
Lindenmayer, D. B., & J. Fischer. 2003. Sound science or social hook: A response to Brooker's application of the focal species approach. Landsc Urban Plann. 62: 149–158.
Lindenmayer, D. B., A. D. Manning, P. L. Smith, H. P. Possingham, J. Fischer, I. Oliver, & M. A. McCarthy. 2002. The focal-species approach and landscape restoration: A critique. Conserv Biol. 16: 338–345.
Linhart, Y. B., & M. C. Grant. 1996. Evolutionary significance of differentiation in plants. Annu Rev Ecol Syst. 27: 237–277.
Lips, K. R., F. Brem, R. Brenes, J. D. Reeve, R. A. Alford, J. Voyles, C. Carey, L. Livo, A. P. Pessier, & J. P. Collins. 2006. Emerging infectious disease and the loss of biodiversity in a neotropical amphibian community. Proc Natl Acad Sci USA 103: 3165–3170.
Lively, C. M. 1992. Parethenogenesis in a freshwater snail: Reproductive assurance versus parasitic release. Evolution 46: 907–913.
Lively, C. M. 1999. Migration, virulence, and the geographic mosaic of adaptation by parasites. Am Nat. 153: S34–S47.
Lively, C. M., & M. F. Dybdahl. 2000. Parasite adaptation to locally common host genotypes. Nature 405: 679–681.
Lloyd, D. G., & C. J. Webb. 1986. The avoidance of interference between the presentation of pollen and stigmas in angiosperms I. Dichogamy. New Zealand J Bot. 24: 136–162.
Lochmiller, R. L., & C. Deerenberg. 2000. Trade-offs in evolutionary immunology: Just what is the cost of immunity? Oikos 88: 87–98.
Lockwood, R. 2004. The K/T event and infaunality: Morphological and ecological patterns of extinction and recovery in veneroid bivalves. Paleobiology 30: 507–521.
Loehr, J., J. Carey, M. Hoefs, J. Suhonen & H. Ylönen 2007. Horn growth rate and longevity: Implications for natural and artificial selection in thinhorn sheep (*Ovis dalli*). J Evol Biol. 20: 818–828.
Loeschcke, V., & J. G. Sorensen. 2005. Acclimation, heat shock and hardening: A response from evolutionary biology. J Therm Biol. 30: 255–257.
Loeschcke, V., J. Tomiuk, & S. K. Jain (eds.). 1994. Conservation genetics. Birkhäuser Verlag, Basel, Switzerland.
Loeuille, N., & M. Loreau. 2004. Nutrient enrichment and food chains: Can evolution buffer top-down control? Theor Pop Biol. 65: 285–298.
Lofting, H. 1920. The story of Doctor Dolittle, being the history of his peculiar life at home and adventures in foreign parts never before printed. Frederick A. Stokes, New York.

Lombard, A. T., C. Hilton-Taylor, A. G. Rebelo, R. L. Pressey & R. M. Cowling. 1999. Reserve selection in the Succulent Karoo, South Africa: Coping with high compositional turnover. Plant Ecol. 142: 35–55.

Longshore, K. M., J. R. Jaeger, & J. M. Sappington. 2003. Desert tortoise (*Gopherus agassizii*) survival at two eastern Mojave Desert sites: Death by short-term drought? J Herp. 37: 169–177.

Lopez, J. E., & C. A. Pfister. 2001. Local population dynamics in metapopulation models: Implications for conservation. Conserv Biol. 15: 1700–1709.

Losey, J. E., L. S. Rayor, & M. E. Carter. 1999. Transgenic pollen harms monarch larvae. Nature 399: 214.

Losos, J. B., T. W. Schoener, & D. A. Spiller. 2004. Predator-induced behaviour shifts and natural selection in field-experimental lizard populations. Nature 432: 505–508.

Lu, G., & L. Bernatchez. 1999. Correlated trophic specialization and genetic divergence in sympatric lake whitefish ecotypes (*Coregonus clupeaformis*): Support for the ecological speciation hypothesis. Evolution 53: 1491–1505.

Luck, G. W., G. C. Daily, & P. R. Ehrlich. 2003. Population diversity and ecosystem services. Trends Ecol Evol. 18: 331–336.

Luikart, G., & J. M. Cornuet. 1998. Empirical evaluation of a test for identifying recently bottlenecked populations from allele frequency data. Conserv Biol. 12: 228–237.

Luong, L. T., & M. Polak. 2007. Costs of resistance in the *Drosphila–Macrocheles* system: A negative genetic correlation between ectoparasite resistance and reproduction. Evolution 61: 1391–1402.

Lushai, G., D. A. S. Smith, I. J. Gordon, D. Goulson, J. A. Allen & N. Maclean. 2003. Incomplete sexual isolation in sympatry between subspecies of the butterfly *Danaus chrysippus* (L) and the creation of a hybrid zone. Heredity 90: 236–246.

Lyell, C. 1830. *Principles of Geology, v. 1.* John Murray, London.

Lyles, A. M., & A. P. Dobson. 1993. Infectious disease and intensive management: Population dynamics, threatened hosts, and their parasites. J Zoo Wildl Med. 24: 315–326.

Lynch, M. 1996. A quantitative genetic perspective on conservation issues (pp. 471–501). In J. C. Avise & J. L. Hamrick (eds.). Conservation genetics: Case histories from nature. Chapman & Hall, New York.

Lynch, M., J. Conery, & R. Burger. 1995a. Mutation accumulation and the extinction of small populations. Am Nat. 146: 489–518.

Lynch, M., J. Conery, & R. Burger. 1995b. Mutational meltdowns in sexual populations. Evolution 49: 1095–1107.

Lynch, M., & W. Gabriel. 1987. Environmental tolerance. Am Nat. 129: 283–303.

Lynch, M., & R. Lande. 1993. Evolution and extinction in response to environmental change (pp. 234–250). In P. M. Kareiva, J. G. Kingsolver, & R. B. Huey (eds.). Biotic interactions and global change. Sinauer Assocoates, Sunderland, Mass.

Lynch, M., & R. Lande. 1998. The critical effective size for a genetically secure population. Anim Conserv. 1: 70–72.

Lynch, M., & B. Walsh. 1998. Genetics and analysis of quantitative traits. Sinauer Associates, Sunderland, Mass.

Ma, J. K.- C., E. Barros, R. Bock, P. Christou, P. J. Dale, P. J. Dix, R. Fischer, J. Irwin, R. Mahoney, M. Pezzotti, S. Schillberg, P. Sparrow, E. Stoger & R. M. Twyman. 2005. Molecular farming for new drugs and vaccines. EMBO Rep. 6: 593–599.

Maan, M. E., K. D. Hofker, J. J. M. van Alphen, & O. Seehausen. 2006. Sensory drive in cichlid speciation. Am Nat. 167: 947–954.

MacDougall-Shackelton, E. A., E. P. Derryberry, J. Foufopoulos, A. P. Dobson, & T. P. Hahn. 2005. Parasite-mediated heterozygote advantage in an outbred songbird population. Biol Lett 1: 105–107.

Mace, G. M., J. L. Gittleman, & A. Purvis. 2003. Preserving the tree of life. Science 300: 1707–1709.

Mack, R. N., D. Simberloff, W. M. Lonsdale, H Evans, M Clout & FA Bazzaz 2000. Biotic invasions: Causes, epidemiology, global consequences, and control. Ecol Appl. 10: 689–710.

Madsen, T., R. Shine, M. Olsson, & H. Wittzell. 1999. Restoration of an inbred adder population. Nature 402: 34–35.

Mahaney, M. C., J. Blangero, D. L. Rainwater, G. E. Mott, A. G. Comuzzie, J. W. MacCluer & J. L. VandeBerg. 1999. A multivariate quantitative genetic analysis of HDL subfractions in two dietary environments. Arterioscler Thromb Vasc Biol. 19: 1134–1141.

Maherali, H., B. L. Williams, K. N. Paige, & E. H. Delucia. 2002. Hydraulic differentiation of Ponderosa pine populations along a climate gradient is not associated with ecotypic divergence. Funct Ecol. 16: 510–521.

Mal, T. K., & J. Lovett-Doust. 2005. Phenotypic plasticity in vegetative and reproductive traits in an invasive weed, *Lythrum salicaria* (Lythraceae) in response to soil moisture. Am J Bot. 92:819–825.

Malakoff, D. 1999. Plan to import exotic beetle drives some scientists wild. Science 284: 1255.

Malausa, T., M. T. Bethenod, A. Bontemps, D. Bourguet, J. M. Cornuet & S. Ponsard. 2005. Assortative mating in sympatric host races of the European corn borer. Science 308: 258–260.

Mallet, J. 2005. Hybridization as an invasion of the genome. Trends Ecol Evol. 20: 229–237.

Manel, S., M. K. Schwartz, G. Luikart, & P. Taberlet. 2003. Landscape genetics: Combining landscape ecology and population genetics. Trends Ecol Evol. 18: 189–197.

Margules, C. R., & R. L. Pressey. 2000. Systematic conservation planning. Nature 405: 243–253.

Marshall, D. R., & A. H. D. Brown. 1975. Optimum sampling strategies in genetic conservation (pp. 3–80). In O. H. Frankel & J. H. Hawkes (eds.). Crop genetic resources for today and tomorrow. Cambridge University Press, Cambridge, UK.

Marshall, I., & F. Fenner. 1958. Studies in the epidemiology of infectious myxomatosis of rabbits. V. Changes in the innate resistance of Australian wild rabbits exposed to myxomatosis. J Hygiene 56: 288–302.

Marshall, I., & F. Fenner. 1960. Studies in the epidemiology of infectious myxomatosis of rabbits. VII. The virulence of strains of myxoma virus recovered from Australian wild rabbits between 1951 and 1959. J Hygiene 58: 485–488.

Martin, R. D. (ed.). 1975. Breeding endangered species in captivity. Academic Press, London, UK.

Martin, P. S. 1984. Prehistoric overkill: The global model (pp. 354–403). In P. S. Martin & R. G. Klein (eds.). Quaternary extinctions. University Arizona Press, Tucson, Ariz.

Marvier, M. 2004. Risk assessment of GM crops warrants higher rigor and reduced risk tolerance than traditional agrichemicals. Nat schutz Biol Vielfalt. 1: 119–129.

Marvier, M., & R. VanAcker. 2005. Can crop transgenes be kept on a leash? Front Ecol Environ. 3: 99–106.

Maschinski, J., J. E. Baggs, P. E. Quintana-Ascencio, & E. S. Menges. 2006. Using population viability analysis to predict the effects of climate change on the extinction risk of an endangered limestone endemic shrub, Arizona cliffrose. Conserv Biol. 20: 218–228.

Mattson, W., H. Vanhanen, T. Veteli, et al. 2007. Few immigrant phytophagous insects on woody plants in Europe: Legacy of the European crucible? Biol Invasions 9: 957–974 DOI 10.1007/s10530-007-9096-y.

Mayr, E., & R. J. O'Hara. 1986. The biogeographical evidence supporting the Pleistocene forest refuge hypothesis. Evolution 40: 55–67.

McCarty, J. P. 2001. Ecological consequences of recent climate change. Conserv Biol. 15: 320–331.

McColl, G., A. A. Hoffmann, & S. W. McKechnie. 1996. Response of two heat shock genes to selection for knockdown heat resistance in *Drosophila melanogaster*. Genetics 143: 1615–1627.

McColl, G., & S. McKechnie. 1999. The *Drosophila* heat shock *hsr-omega* gene: An allele frequency cline detected by quantitative PCR. Mol Biol Evol. 16: 1568–1574.

McCracken, K. G., W. P. Johnson, & F. H. Sheldon. 2001. Molecular population genetics, phylogeography, and conservation biology of the mottled duck (*Anas fulvigula*). Conserv Gen. 2: 87–102.

McDade, T. W. 2003. Life history theory and the immune system: Steps toward a human ecological immunology. Yearb Phys Anthrop. 46: 100–125.

McGowan, A. J. 2004a. Ammonoid taxonomic and morphologic recovery patterns after the Permian–Triassic. Geology 32: 665–668.

McGowan, A. J. 2004b. The effect of the Permo-Triassic bottleneck on Triassic ammonoid morphological evolution. Paleobiology 30: 369–395.

McGraw, J. B. 2001. Evidence for decline in stature of American ginseng plants from herbarium specimens. Biol Conserv. 98: 25–32.

McGraw, J. B., & N. Fetcher. 1992. Response to tundra plant populations to climatic change (pp. 359–376). In F. S. Chapin, III, R. L. Jeffreis, J. F. Reynolds, GR Shaver & J Svoboda (eds.). Arctic ecosystems in a changing climate. Academic Press, San Diego, Calif.

McKay, J. K., & R. G. Latta. 2002. Adaptive population divergence: Markers, QTL and traits. Trends Ecol Evol. 17: 285–291.

McKechnie, S. W., M. M. Halford, G. McColl, & A. A. Hoffmann. 1998. Both allelic variation and expression of nuclear and cytoplasmic transcripts of *Hsr-omega* are closely associated with thermal phenotype in *Drosophila*. Proc Natl Acad Sci USA 95: 2423–2428.

McKinney, M. L., & J. L. Lockwood. 2005. Community composition and homogenization: Eveness and abundance of native and exotic species (pp. 365–381). In D. F. Sax, J. J. Stachowicz, & S. D. Gaines (eds.). Species invasions: Insights into ecology, evolution and biogeography. Sinauer Associates, Sunderland, Mass.

McLachlan, J. S., J. S. Clark, & P. S. Manos. 2005. Molecular indicators of tree migration capacity under rapid climate change. Ecology 86: 2088–2098.

McRae, B. H. 2006. Isolation by resistance. Evolution 60: 1551–1561.

Mealor, B. A., & A. L. Hild. 2006. Potential selection in native grass populations by exotic invasion. Mol Ecol. 15: 2291–2300.

Meekins, J. F., & B. C. McCarthy. 2001. Effects of environmental variation on the invasive success of a nonindigenous forest herb. Ecol Appl. 11: 1336–1348.

Meffe, G. K., & C. R. Carroll (eds.). 1994. Principles of conservation biology. Sinauer Associates, Sunderland, Mass.

Meffe, G. K., & C. R. Carroll (eds.). 1997. Principles of conservation biology. 2nd ed. Sinauer Associates, Sunderland, Mass.

Melbourne, B. A., H. V. Cornell, K. F. Davies, C. J. Dugaw, S. Elmendorf, A. L. Freestone, R. J. Hall, S. Harrison, A. Hastings, M. Holland, M. Holyoak, J. Lambrinos, K. Moore & H. Yokomizo. 2007. Invasion in a heterogeneous world: Resistance, coexistence or hostile takeover? Ecol Lett. 10: 77–94.

Mendoza-Cuenca, L., & R. Macias-Ordonez. 2005. Foraging polymorphism in *Heliconius charitonia* (Leptidoptera: Nypmphalidae): Morphological constraints and behavioural compensation. J Trop Ecol. 21: 407–415.

Merilä, J., & P. Crnokrak. 2001. Comparison of genetic differentiation at marker loci and quantitative traits. J Evol Biol. 14: 892–903.

Metz, M., & J. Futterer. 2002. Suspect evidence of transgenic contamination. Nature 416: 600–601.

Mezquida, E. T., & C. W. Benkman. 2005. The geographic selection mosaic for squirrels, crossbills, and Aleppo pine. J Evol Biol. 18: 348–357.

Millennium Ecosystem Assessment. 2005a. Ecosystems and human well-being: Biodiversity synthesis. World Resources Institute, Washington, D.C.

Millennium Ecosystem Assessment. 2005b. Responses assessment: Biodiversity. Island Press, Washington, D.C.

Miller, K. M., K. H. Kaukinen, T. D. Beacham, & R. E. Withler. 2001. Geographic heterogeneity in natural selection on an MHC locus in sockeye salmon. Genetica 111: 237–257.

Millington, A. C., P. J. Styles, & R. W. Critchley. 1992. Mapping forests and savannas in sub-Saharaan Africa from advanced very high resolution radiometer (AVHRR) imagery (pp. 37–62). In P. A. Furley, J. Proctor, & J. A. Ratter (eds.). Nature and dynamics of forest–savanna boundaries. Chapman & Hall, New York.

Mills, L. S., & P. E. Smouse. 1994. Demographic consequences of inbreeding in remnant populations. Am Nat. 144: 412–431.

Minh, B. Q., S. Klaere, & A. Von Haeseler. 2006. Phylogenetic diversity within seconds. Syst Biol. 55: 769–773.

Misenhelter, M. D., & J. T. Rotenberry. 2000. Choices and consequences of habitat occupancy and nest site selection in sage sparrows. Ecology 81: 2892–2901.

Mitchell, C. E., A. A. Agrawal, J. D. Bever, G. S. Gilbert, R. A. Hufbauer, J. N. Klironomos, J. L. Maron, W. F. Morris, I. M. Parker, A. G. Power, E. W. Seabloom, M. E. Torchin & D. P. Vazquez. 2006. Biotic interactions and plant invasions. Ecol Lett. 9: 726–740.

Mitchell, C. E., & A. G. Power. 2003. Release of invasive plants from fungal and viral pathogens. Nature 421: 625–627.

Mitchell-Olds, T., & D. Bradley. 1996. Genetics of *Brassica rapa*. 3. Costs of disease resistance to three fungal pathogens. Evolution 50: 1859–1865.

Mitchell-Olds T. & R. G. Shaw. 1987. Regression analysis of natural selection: statistical inference and biological interpretation. Evolution 41:1149–1161.

Mittelbach, G. G., A. M. Turner, D. J. Hall, & J. E. Rettig. 1995. Perturbation and resilience: A long-term whole-lake study of predator extinction and reintroduction. Ecology 76: 2347–2360.

Mittermeier, R. A., N. Myers, P. R. Gil, & C. G. Mittermeier. 1999. Hotspots: Earth's biologicaly richnest and most endangered terrestrial ecoregions. Cemex, Conservation International and Agrupacion Sierra Madre, Monterrey, Mexico.

Mitton, J. B. 1997. Selection in natural populations. Oxford University Press, Oxford, UK.

Mitton, J. B., B. R. Kreiser, & R. G. Latta. 2000. Glacial refugia of limber pine (*Pinus flexilis* James) inferred from the population structure of mitochondrial DNA. Mol Ecol. 9: 91–97.

Moczek, A. P., & H. F. Nijhout. 2003. Rapid evolution of a polyphenic threshold. Evol Devel. 5: 259–268.

Moeller, D. A. 2004. Facilitative interactions among plants via shared pollinators. Ecology 85: 3289–3301.

Moilanen, A., & M. Cabeza. 2002. Single-species dynamic site selection. Ecol Appl. 12: 913–926.

Mollet, F. M., S. B. M. Kraak, & A. D. Rijnsdorp. 2007. Fisheries-induced evolutionary changes in maturation reaction norms in North Sea sole (*Solea solea*). Mar Ecol Prog Ser. 351: 189–199.

Mooers, A. Ø., & R. A. Atkins. 2003. Indonesia's threatened birds: Over 500 million years of evolutionary heritage at risk. Anim Conserv. 6: 183–188.

Mooney, E. H., & J. B. McGraw. 2007. Alteration of selection regime resulting from harvest of American ginseng, *Panax quinquefolius*. Conserv Genet. 8: 57–67.

Mooney, H. A., R. N. Mack, J. A. McNeely, L. E. Neville, P. J. Schei, & J. K. Waage. 2005. Invasive alien species: a new synthesis. Island Press, Washington, D.C.

Morgan, M. J. 2000. Estimating spawning stock biomass in 2J3KL cod using a cohort maturation model and variable sex ratios.

Morgan, J.M. 2000. Estimating spawning stock biomass in 2J3KL cod using a cohort maturation model and variable sex ratio. Canadian Stock Assessment Secretariat Research Document 2000/110. Department of Fisheries and Oceans, Ottawa.

Morgan, T. J., & T. F. C. Mackay. 2006. Quantitative trait loci for thermotolerance phenotypes in *Drosophila melanogaster*. Heredity 96: 232–242.

Morin, P. A., G. Luikart, R. K. Wayne, & the SNP Workshop Group. 2004. SNPs in ecology, evolution and conservation. Trends Ecol Evol. 19: 208–216.

Moritz, C. 2002. Strategies to protect biological diversity and the evolutionary processes that sustain it. Syst Biol. 51: 238–254.

Moritz, C., & C. Cicero. 2004. DNA barcoding: Promise and pitfalls. PLoS Biol. 2: e354.

Moritz, C., & D. P. Faith. 1998. Comparative phylogeography and the identification of genetically divergent areas for conservation. Mol Ecol. 7: 419–430.

Morris, W. F., & D. F. Doak. 2002. Quantitative conservation biology: Theory and practice of population viability analysis. Sinauer Associates, Sunderland, Mass.

Munte, A., J. Rozas, M. Aguade, & C. Segarra. 2005. Chromosomal inversion polymorphism leads to extensive genetic structure: A multilocus survey in *Drosophila subobscura*. Genetics 169: 1573–1581.

Murren, C. J. 2002. Effects of habitat fragmentation on pollination: Pollinators, pollinia viability and reproductive success. J Ecol. 90: 100–107.

Murren, C. J. 2003. Spatial and demographic population genetic structure in *Catasetum viridiflavum* across a human-disturbed habitat. J Evol Biol. 16: 333–342.

Murren, C. J., L. Douglass, A. Gibson, & M. R. Dudash. 2006. Individual and combined effects of Ca/Mg ratio and water on trait expression of *Mimulus guttatus*. Ecology 87: 2591–2602.

Myers, N. 2002. Biodiversity hotspots for conservation priorities. Nature 403: 853–858.

Myers, N., R. A. Mittermeier, C. G. Mittermeier, G. de Fonesca & J. Kent. 2000a. Biodiversity hotspots for conservation priorities. Nature 403: 853–858.

Myers, J. H., D. Simberloff, A. M. Kuris, & J. R. Carey. 2000b. Eradication revisited: Dealing with exotic species. Trends Ecol Evol. 15: 316–320.

Mylius, S. D., & O. Dieckmann. 1995. On evolutionarily stable life histories, optimization and the need to be specific about density dependence. Oikos 74: 218–224.

Nadis, S. 2005. The lands where species are born. Natl Wildl Mag. 43: 34T–34X.

National Research Council. 2004. Biological confinement of genetically engineered crops. The National Academies Press, Washington, D.C.

Natural Environment Research Council. 1999. The global population dynamics database. NERC Centre for Population Biology, Imperial College. Online. http://www3.imperial.ac.uk/cpb/research/patternsandprocesses/gpdd. Accessed on March 2, 2008.

NatureServe (2007). NatureServe conservation status. NatureServe, Arlington, Virginia. Online. http://www.natureserve.org/explorer/ranking.htm. Accessed 1 May 2007.

Navarro, C., S. Cavers, A. Pappinen, P. Tigerstedt, A. Lowe & J. Merila. 2005. Contrasting quantitative traits and neutral genetic markers for genetic resource assessment of Mesoamerican *Cedrela odorata*. Silvae Genet. 54: 281–292.

Neel, M. C. 2008. Patch connectivity and genetic diversity conservation in the federally endangered and narrowly endemic plant species *Astragalus albens* (Fabaceae) 141: 938–955.

Neel, M. C., & M. P. Cummings. 2003a. Effectiveness of conservation targets in capturing genetic diversity. Conserv Biol. 17: 219–229.

Neel, M. C., & M. P. Cummings. 2003b. Genetic consequences of ecological reserve design guidelines: An empirical investigation. Conserv Genet. 4: 427–439.

Neel, M. C., K. McGarigal, & S. A. Cushman. 2004. Behavior of class-level landscape metrics across gradients of class aggregation and area. Landsc Ecol. 19: 435–455.

Nei, M., & A. L. Hughes. 1991. Polymorphism and evolution of the major histocompatibility compex loci in mammals (pp. 222–247). In R. K. Selander, A. Clark, & S. Thomas (eds.). Evolution at the molecular level. Sinauer Associates, Sunderland, Mass.

Nelson, K., & M. Soulé. 1987. Genetical conservation of exploited fishes (pp. 345–368). In N. Ryman & F. Utter (eds.). Population genetics and fishery management. Washington Sea Grant Program, Seattle, Wash.

Nentwig, W. (ed.). 2007. Biological invasions. Springer Verlag, Berlin.

Newman, D., & D. Pilson. 1997. Increased probability of extinction due to decreased genetic effective population size: Experimental populations of *Clarkia pulchella*. Evolution 51: 354–362.

Nicholls, A. O. 1998. Integrating population abundance, dynamics and distribution into broad-scale priority-setting (pp. 251–272). In G. M. Mace, A. Balmford, & J. R. Ginsberg (eds.). Conservation in a changing world. Cambridge University Press, Cambridge, UK.

Nicholson, E., M. I. Westphal, K. Frank, W. A. Rochester, R. L. Pressey, D. B. Lindenmayer & H. P. Possingham. 2006. A new method for conservation planning for the persistence of multiple species. Ecol Lett. 9: 1049–1060.

Nolte, A. W., J. Freyhof, K. C. Stemshorn, & D. Tautz. 2005. An invasive lineage of sculpins, *Cottus* sp (Pisces, Teleostei) in the Rhine with new habitat adaptations has originated from hybridization between old phylogeographic groups. Proc R Soc Lond B. 272: 2379–2387.

Norris, R. D. 1996. Symbiosis as an evolutionary innovation in the radiation of Paleoecene planktic Foraminifera. Paleobiology 22: 461–480.

Norris, S. 2006. Evolutionary tinkering. Conserv Pract. 7: 28–34.

Norris, K., & M. R. Evans. 2000. Ecological immunology: Life history trade-offs and immune defense in birds. Behav Ecol. 11: 19–26.

Norry, F. M., J. Dahlgaard, & V. Loeschcke. 2004. Quantitative trait loci affecting knockdown resistance to high temperature in *Drosophila melanogaster*. Mol Ecol. 13: 3585–3594.

Nosil, P., B. J. Crespi, & C. P. Sandoval. 2002. Host–plant adaptation drives the parallel evolution of reproductive isolation. Nature 417: 440.

Noss, R.F., 1996. Protected areas: how much is enough. In: R. G. Wright (ed.), National parks and protected areas: their role in environmental protection. Blackwell Science, Cambridge, USA, pp. 91–119.

Nuismer, S. L. 2006. Parasite local adaptation in a geographic mosaic. Evolution 60: 24–30.

Nuismer, S. L., & M. Kirkpatrick. 2003. Gene flow and the evolution of parasite range. Evolution 57: 746–754.

Nuismer, S. L., J. N. Thompson, & R. Gomulkiewicz. 1999. Gene flow and geographically structured coevolution. Proc R Soc Lond B. 266: 605–609.

Nunn, C. L., & S. Altizer. 2005. The global mammal parasite database: An online resource for infectious disease records in wild primates. Evol Anthrop 14: 1–2.

Nunn, C. L., S. Altizer, W. Sechrest, K. E. Jones, R. A. Barton & J. L. Gittleman. 2004. Parasites and the evolutionary diversification of primate clades. Am Nat. 164: S90–S103.

Nunney, L. 1993. The influence of mating system and overlapping generations on effective population size. Evolution 47: 1329–1341.

Nunney, L. 2000. The limits to knowledge in conservation genetics: The value of effective population size. Evol Biol. 32: 179–194.

Nunney, L. 2002. The effective size of annual plant populations: The interaction of a seed bank with fluctuating plant numbers. Am Nat. 160: 195–204.

Nunney, L. 2003. The cost of natural selection revisited. Ann Zool Fenn. 40: 185–194.

Obedzinski, M., & B. H. Letcher. 2004. Variation in freshwater growth and development among five New England Atlantic salmon (*Salmo salar*) populations reared in a common environment. Can J Fish Aq Sci. 61: 2314–2328.

O'Brien, S. J., & J. F. Evermann. 1988. Interactive influence of infectious disease and genetic diversity in natural populations. Trends Ecol Evol. 3: 254–259.

O'Brien, S. J., et al. 1996. Conservation genetics of the felidae (pp. 50–74). In J. C. Avise & J. L. Hamrick (eds.). Conservation genetics: Case histories from nature. Chapman & Hall, New York.

O'Brien, S.J., D. E. Wildt, D. Goldman, C. R. Merril, & M. Bush. 1983. The cheetah is

depauperate of genetic variation. Science 221:459–462.
Ogden, R., & R. S. Thorpe. 2002. Molecular evidence for ecological speciation in tropical habitats. Proc Natl Acad Sci USA 99: 13612–13615.
O'Grady, J. J., B. W. Brook, D. H. Reed, J. D. Ballou, D. W. Tonkyn, & R. Frankham. 2006. Realistic levels of inbreeding depression strongly affect extinction risk in wild populations. Biol Conserv. 133: 42–51.
O'Grady, J. J., D. H. Reed, B. W. Brook, & R. Frankham. 2004. What are the best correlates of extinction risk? Biol Conserv. 118: 513–520.
O'Hanley, J. R., R. L. Church, & J. K. Gilless. 2007. Locating and protecting critical reserve sites to minimize expected and worst-case losses. Biol Conserv. 134: 130–141.
O'Hara, R. B., & J. Merilä. 2003. Bias and precision in Qst estimates: Problems and some solutions. Genetics 171: 1331–1339.
O'Keefe, K. J., & J. Antonovics. 2002. Playing by different rules: The evolution of virulence in sterilizing pathogens. Am Nat. 159: 597–605.
Oldroyd, B. P. 1999. Coevolution while you wait: *Varroa jacobsoni*, a new parasite of western honeybees. Trends Ecol Evol. 14: 312–315.
Oliver, J. C. 2006. Population genetic effects of human mediated plant range expansions on native phytophagous insects. Oikos 112: 456–463.
Olivieri, I., Y. Michalakis, & P.- H. Gouyon. 1995. Metapopulation genetics and the evolution of dispersal. Am Nat. 146: 202–228.
Olsen, E. M., M. Heino, G. R. Lilly, MJ Morgan, J Brattey, B Ernande & U Dieckmann 20 2004. Maturation trends indicative of rapid evolution preceded the collapse of northern cod. Nature 428: 932–935.
Olsen, E. M., G. R. Lilly, M. Heino, M. J. Morgan, J. Brattey, & U. Dieckmann. 2005. Assessing changes in age and size at maturation in collapsing populations of Atlantic cod (*Gadus morhua*). Can. J. Fish. Aquat. Sci. 62:811–823.
Oostermeijer, J. G. B., S. H. Luijten, & J. C. M. den Nijs. 2003. Integrating demographic and genetic approaches in plant conservation. Biol Conserv. 113: 389–398.
Orme, C. D. L., R. G. Davies, M. Burgess, F. Eigenbrod, N. Pickup, V. A. Olson, A. J. Webster, T. S. Ding, P. C. Rasmussen, R. S. Ridgely, A. J. Stattersfield, P. M. Bennett, T. M. Blackburn, K. J. Gaston. & I. P. F. Owens. 2005. Global hotspots of species richness are not congruent with endemism or threat. Nature 436: 1016.
Orr, M. R., & T. B. Smith. 1998. Ecology and speciation. Trends Ecol Evol. 13: 502–506.
Ortiz-Garcia, S., E. Ezcurra, B. Schoel, F. Acevedo, J. Soberón, & A. A. Snow. 2005. Absence of detectable transgenes in local landraces of maize in Oaxaca, Mexico (2003–2004). Proc Natl Acad Sci USA 102: 12338–12343.
Osterhaus, A. D. M. E., & E. J. Vedder. 1988 Identification of a virus causing recent seal deaths. Nature 335: 20.
Ottersen, G., D. Oe Hjermann, & N. C. Stenseth. 2006. Changes in spawning stock structure strengthen the link between climate and recruitment in a heavily fished cod (*Gadus morhua*) stock. Fish Oceanogr. 15: 230–243.
Ouberg, N. G., A. Biere, & C. L. Muddle. 2000. Inbreeding effects on resistance and transmission-related traits in the *Silene–Microbotryum* pathosystem. Ecology 81: 520–531.
Ouborg, N. J., & R. Van Treuren. 1994. The significance of genetic erosion in the process of extinction. 4. Inbreeding load and heterosis in relation to population-size in the mint *Salvia pratensis*. Evolution 48: 996–1008.
Ouborg, N. J., & R. Van Treuren. 1995. Variation in fitness-related characters among small and large populations of *Salvia pratensis*. J Ecology 83: 369–380.
Ouborg, N. J., P. Vergeer, & C. Mix. 2006. The rough edges of the conservation genetics paradigm for plants. J Ecol. 94: 1233–1248.
Ovaskainen, O. 2002. The effective size of a metapopulation living in a heterogeneous patch network. Am Nat. 160: 612–628.
Ovaskainen, O., & I. Hanski. 2002. Transient dynamics in metapopulation response to perturbation. Theor Pop Biol. 61: 285–295.
Ovaskainen, O., & I. Hanski. 2003. How much does an individual habitat fragment contribute to metapopulation dynamics and persistence? Theor Pop Biol. 64: 481–495.
Overgaard, J., J. G. Sorensen, S. O. Petersen, V Loeschcke & M Holmstrup 2006. Reorganization of membrane lipids during fast and slow cold hardening in *Drosophila melanogaster*. Physiol Entomol. 31: 328–335.
Page, R. D. M. 2003. Tangled trees: Phylogenies, cospeciation and coevolution. University of Chicago Press, Chicago, Ill.
Pagel, M. D., R. M. May, & A. R. Collie. 1991. Ecological aspects of the geographical distribution and diversity of mammalian species. Am Nat. 137: 791–815.
Palo, J. U., R. B. O'Hara, A. T. Laugen, A Laurila, CR Primmer & J Merilä 2 2003. Latitudinal

divergence of common frog (*Rana temporaria*) life history traits by natural selection: Evidence from a comparison of molecular and quantitative genetic data. Mol Ecol 12: 1963–1978.

Palsbøll, P. J. 1999. Genetic tagging: Contemporary molecular ecology. Biol J Linn Soc. 68: 3–22.

Palumbi, S. 2001a. Humans as the world's greatest evolutionary force. Science 293: 1786–1790.

Palumbi, S. R. 2001b. The evolution explosion: How humans cause rapid evolution change. WW Morton, New York.

Pannell, J. R. 2002. The evolution and maintenance of androdioecy. Annu Rev Ecol Syst. 33: 397–425.

Pannell, J. R. 2003. Coalescence in a metapopulation with recurrent local extinction and recolonization. Evolution 57: 949–961.

Pannell, J. R., & S.C. H. Barrett. 1998. Baker's law revisited: Reproductive assurance in a metapopulation. Evolution 52: 657–668.

Pannell, J. R., & S. C. H. Barrett. 2001. Effects of population size and metapopulation dynamics on a mating-system polymorphism. Theor Pop Biol. 59: 145–155.

Pannell, J. R., & B. Charlesworth. 1999. Neutral genetic diversity in a metapopulation with recurrent local extinction and recolonization. Evolution 53: 664–676.

Pannell, J. R., & B. Charlesworth. 2000. Effects of metapopulation processes on measures of genetic diversity. Phil Trans R Soc Lond. 355: 1851–1864.

Paoletti, M., K. W. Buck, & C. M. Brasier. 2006. Selective acquisition of novel mating type and vegetative incompatibility genes via interspecies gene transfer in the globally invading eukaryote *Ophiostoma novo-ulmi*. Mol Ecol. 15: 249–262.

Parchman, T. L., & C. W. Benkman. 2002. Diversifying coevolution between crossbills and black spruce on Newfoundland. Evolution 56: 1663–1672.

Parchman, T. L., C. W. Benkman, & S. C. Britch. 2006. Patterns of genetic variation in the adaptive radiation of North American crossbills (AVES: *Loxia*). Mol Ecol. 15: 1873–1887.

Parchman, T. L., C. W. Benkman, & E. T. Mezquida. 2007. Coevolution between Hispaniolan crossbills and pine: Does more time allow for greater phenotypic escalation at lower latitude? Evolution 61: 2142–2153.

Parker, M. A. 1992. Disease and plant population genetic structure (pp. 345–362). In R. S. Fritz & E. L. Simms (eds.). Plant resistance to herbivores and pathogens. University of Chicago Press, Chicago, Ill.

Parker, J. D., D. E. Burkepile, & M. E. Hay. 2006a. Opposing effects of native and exotic herbivores on plant invasions. Science 311: 1459–1461.

Parker, J. D., D. E. Burkepile, & M. E. Hay. 2006b. Response to comment on "Opposing effects of native and exotic herbivores on plant invasions." Science 313: 298.

Parmesan, C. 2006. Ecological and evolutionary responses to recent climate change. Annu Rev Ecol Evol Syst. 37: 637–669.

Parmesan, C., & G. Yohe. 2003. A globally coherent fingerprint of climate change impacts across natural systems. Nature 42: 37–42.

Parrish, C. 1990. Emergence, history, and variation of canine, mink, and feline parvoviruses. Adv Virus Res. 38: 403–450.

Parsell, D. A., & S. Lindquist. 1993. The function of heat-shock proteins in stress tolerance: Degradation and reactivation of damaged proteins. Annu Rev Genet. 27: 437–496.

Parvinen, K. 2005. Evolutionary suicide. Acta Biotheoretica 53: 241–264.

Paschke, M., G. Bernasconi, & B. Schmid. 2003. Population size and identity influence the reaction norm of the rare, endemic plant *Cochlearia bavarica* across a gradient of environmental stress. Evolution 57: 496–508.

Paterson, A. H., K. F. Schertz, Y. Lin, S. Liu, & Y. Chang. 1995. The weediness of wild plants: Molecular analysis of genes influencing dispersal and persistence of johnsongrass, *Sorghum halepense* (L.) Pers. Proc Natl Acad Sci USA 92: 6127–6131.

Paterson, S., K. Wilson, & J. M. Pemberton. 1998. Major histocompatibility complex (MHC) variation associated with juvenile survival and parasite resistance in a large ungulate population (*Ovis aries* L). Proc Natl Acad Sci USA 95: 3714–3719.

Pauly, D. 1995. Anecdotes and the shifting base-line syndrome of fisheries. Trends Ecol Evol. 10: 430.

Payne, J. L. 2005. Evolutionary dynamics of gastropod size across the end-Permian extinction and through the Triassic recovery interval. Paleobiology 31: 269–290.

Payne, J. L., D. J. Lehrmann, J. Wei, M. J. Orchard, D. P. Schrag, & A. H. Knoll. 2004. Large perturbations in the carbon cycle during recovery from the end-Permian extinction. Science 305: 506–509.

Payne, R. B., L. L. Payne, J. L. Woods, & M. D. Sorenson. 2000. Imprinting and the origin of parasite–host species associations in brood-parasitic indigobirds, *Vidua chalybeata*. Anim Behav. 59: 69–81.

Pearman, P. B., & T. W. J. Garner. 2005. Susceptibility of Italian agile frog populations to an emerging strain of Ranavirus parallels population genetic diversity. Ecol Lett. 8: 401.

Pearson, D.E. & R. M. Callaway. 2006. Biological control agents elevate hantavirus by subsidizing deer mouse populations. Ecol Lett 9:443–450.

Pearson, R. G., & T. P. Dawson. 2003. Predicting the impacts of climate change on the distribution of species: Are bioclimate envelope models useful? Glob Ecol Biogeogr. 12: 361–371.

Pease, C. M., R. Lande, & J. J. Bull. 1989. A model of population growth, dispersal, and evolution in a changing environment. Ecology 70: 1657–1664.

Pedersen, A. B., & A. Fenton. 2007. Putting the ecology in parasite community ecology. Trends Ecol Evol. 22: 133–139.

Pedersen, A. B., K. Jones, C. L. Nunn, & S. Altizer. 2007. Infectious diseases and extinction risk in wild mammals. Conserv Biol. 21: 1269–1279.

Pedersen, A. B., M. Poss, S. Altizer, A. Cunningham, & C. Nunn. 2005. Patterns of host specificity and transmission among parasites of wild primates. Inter J Parasit. 35: 647–657.

Penn, D. J., K. Damjanovich, & W. K. Potts. 2002. MHC heterozygosity confers a selective advantage against multiple-strain infections. Proc Natl Acad Sci USA 99: 11260–11264.

Peperkorn, R., C. Werner, & W. Beyschlag. 2005. Phenotypic plasticity of an invasive *acacia* versus two native Mediterranean species. Funct Plant Biol. 32: 933–944.

Percival, S. P., Jr., & A. G. Fischer. 1977. Changes in calcareous nanoplankton in the Cretaceous–Tertiary biotic crisis at Zumaya, Spain. Evol Theory 2: 1–35.

Percy, D., R. Page, & Q. Cronk. 2004. Plant–insect interactions: Double-dating associated insect and plant lineages reveals asynchronous radiations. Syst Biol. 53: 120–127.

Perry, W. L., D. M. Lodge, & J. L. Feder. 2002. Importance of hybridization between indigenous and nonindigenous freshwater species: An overlooked threat to North American biodiversity. Syst Biol. 51: 255–275.

Peterson, K. J. 2005. Macroevolutionary interplay between planktic larvae and benthic predators. Geology 33: 929–932.

Peterson, R. K. D., S. J. Meyer, A. T. Wolf, J. D. Wolt & P. M. Davis. 2006. Genetically engineered plants, endangered species, and risk: A temporal and spatial exposure assessment for Karner blue butterfly larvae and *Bt* maize pollen. Risk Anal. 26: 845–858.

Petit, R. J., A. El Mousadik, & O. Pons. 1998. Identifying populations for conservation on the basis of genetic markers. Conserv Biol. 12: 844–855.

Pew Trust. 2001. Harvest on the horizon: Future uses of agricultural biotechnology. The Pew Initiative on Food and Biotechnology. Online. pewagbiotech.org/research/harvest/harvest.pdf. Accessed March 1, 2008.

Pfrender, M. E., K. Spitze, J. Hicks, K Morgan, L Latta & M Lynch 2 2000. Lack of concordance between genetic diversity estimates at the molecular and quantitative-trait levels. Conserv Genet. 1: 263–269.

Phillips, S. J., R. P. Anderson, & R. E. Schapire. 2006. Maximum entropy modeling of species geographic distributions. Ecol Mod. 190: 231–259.

Phillips, B., & R. Shine. 2004. Adapting to an invasive species: Toxic cane toads induce morphological change in Australian snakes. Proc Natl Acad Sci USA 101: 17150–17155.

Pico, F. X., N. J. Ouborg, & J. M. Van Greonendael. 2004. Evaluation of the extent of among-family variation in inbreeding depression in the perennial herb *Scabiosa columbaria* (Dipsaceae). Am J Bot. 91: 1183–1189.

Pike, N., T. Tully, P. Haccou, & R. Ferrière. 2004. The effect of autocorrelation in environmental variability on the persistence of populations: An experimental test. Proc R Soc Lond B. 271: 2143–2148.

Pillon, Y., M. F. Faya, A. B. Shipunova, & M. W. Chase. 2006. Species diversity versus phylogenetic diversity: A practical study in the taxonomically difficult genus *Dactylorhiza* (Orchidaceae). Biol Conserv. 129: 4–13.

Pilson, D., & H. R. Prendeville. 2004. Ecological effects of transgenic crops and the escape of transgenes into wild populations. Annu Rev Ecol Evol Syst. 35: 149–174.

Pimentel, D., L Lach, R. Zuniga & D. Morrison. 2000. Environmental and economic costs of invasive species in the United States. BioScience 50: 53–65.

Pimm, S. L. 1991. Balance of nature? Ecological issues in the conservation of species and communities. University Chicago Press, Chicago, Ill.

Pimm, S. L., L. Dollar, & O. L. Bass, Jr. 2006. The genetic rescue of the Florida panther. Anim Conserv. 9: 115–122.

Pinto, M. A., W. L. Rubink, J. C. Patton, RN Coulson & JS Johnston 2005. Africanization in the United States: Replacement of feral European honeybees (*Apis mellifera* L) by an African hybrid swarm. Genetics 170: 1653–1665.

Podos, J. 2001. Correlated evolution of morphology and vocal signal structure in Darwin's finches. Nature 409: 185–188.

Poiani, K. A., B. D. Richter, M. G. Anderson, & H. E. Richter. 2000. Biodiversity conservation at multiple scales: Functional sites, landscapes and networks. Bioscience 50: 133–146.

Pollack, A. 2006. Lax oversight found in tests of gene-altered crops. New York Times. Jan 3, F2.

Pollock, M. M., R. J. Naiman, H. E. Erickson, C. A. Johnston, J. Pastor & G. Pinay. 1995. Beaver as engineers: Influences on biotic and abiotic characteristics of drainage basins (pp. 117–126). In C. G. Jones & J. H. Lawton (eds.). Linking species and ecosystems. Chapman & Hall, New York.

Pörtner, H. O., & R. Knust. 2007. Climate change affects marine fishes through the oxygen limitation of thermal tolerance. Science 315: 95–97.

Possingham, H. P., I. R. Ball, & S. Andelman. 2000. Mathematical methods for identifying representative reserve networks (pp. 291–305). In S. Ferson & M. Burgman (eds.). Quantitative methods for conservation biology. Springer-Verlag, New York.

Poulin, R., L. J. Marshall, & H. G. Spencer. 2000. Metazoan parasite species richness and genetic variation among freshwater fish species: Cause or consequence? Int J Parasit. 30: 697–703.

Pounds, J. A., M. P. L. Fogden, & J. H. Campbell. 1999. Biological response to climate change on a tropical mountain. Nature. 398: 611–615.

Pressey, R. L., R. M. Cowling. & M. Rouget. 2003. Formulating conservation targets for biodiversity pattern and process in the Cape Floristic Region, South Africa. Biol Conserv. 112: 99–127.

Pressey, R. L., C. J. Humphries, C. R. Margules, RI Vanewright & PH Williams 1993. Beyond opportunism: Key principles for systematic reserve selection. Trends Ecol Evol. 8: 124–128.

Pressey, R. L., I. R. Johnson, & P. D. Wilson. 1994. Shade of irreplaceability: Towards a measure of the contribution of sites to a reservation goal. Biodiv Conserv. 3: 242–262.

Pressey, R. L., & A. O. Nicholls. 1989. Efficiency in conservation evaluation: Scoring versus iterative approaches. Biol Conserv. 50: 199–218.

Price, G. R. 1970. Selection and covariance. Nature 227: 520–521.

Price, P. W. 1980. Evolutionary biology of parasites. Princeton University Press, Princeton, N.J.

Price, T. D., A. Qvarnstrom, & D. E. Irwin. 2003. The role of phenotypic plasticity in driving genetic evolution. Proc R Soc Lond B. 270: 1433–1440.

Pritchard, J. K., M. Stephens, & P. Donnelly. 2000. Inference of population structure using multilocus genotype data. Genetics 155: 945–959.

Proctor, H. C. 1991. Courtship in the water mite neumania–papillator: Males capitalize on female adaptations for predation. Anim Behav. 42: 589–598.

Proctor, H. C. 1992. Sensory exploitation and the evolution of male mating behavior: A cladistic test using water mites (Acari, Parasitengona). Anim Behav. 44: 745–752.

Prodöhl, P. A., A. F. Walker, R. Hynes, J. B. Taggart, & A. Ferguson. 1997. Genetically monomorphic brown trout (*Salmo trutta* L) populations, as revealed by mitochondrial DNA, multilocus and single-locus minisatellite (VNTR) analyses. Heredity 79: 208–213.

Pruss, S., M. Fraiser, & D. J. Bottjer. 2004. Proliferation of Early Triassic wrinkled structures: Implications for environmental stress following the end-Permian mass extinction. Geology 32: 461–464.

Pulliam, H. R. 1988. Sources, sinks, and population regulation. Am Nat. 132: 652–661.

Purvis, A., T. Brooks, & J. Gittleman (eds.). 2005. Phylogeny and conservation. Cambridge University Press, Cambridge, UK.

Puurtinen, M., K. E. Knott, S. Suonpää, T. van Ooik, & V. Kaitala. 2004. Genetic variability and drift load in populations of an aquatic snail. Evolution 58: 749–756.

Qian, H., & R. E. Ricklefs. 2006. The role of exotic species in homogenizing the North American flora. Ecol Lett. 9: 1293–1298.

Quinn, T. P. 2004. The behaviour and ecology of Pacific salmon and trout. University of Washington Press, Seattle, Wash.

Quinn, T. P., M. T. Kinnison, & M. J. Unwin. 2001. Evolution of chinook salmon (*Oncorhynchus tshawytscha*) populations in New Zealand: Pattern, rate, and process. Genetica 112–113: 493–513.

Quinn, J. A., & J. D. Wetherington. 2002. Genetic variability and phenotypic plasticity in flowering phenology in populations of two grasses. J Torr Bot Soc. 129: 96–106.

Quist, D., & I. H. Chapela. 2001. Transgenic DNA introgressed into traditional maize

landraces in Oaxaca, Mexico. Nature 414: 541–543.
Rabinowitz, D. 1981. Seven forms of rarity (pp. 205–217). In H. Synge (ed.). The biological aspects of rare plant conservation. John Wiley, New York.
Rahel, F. J. 2002. Homogenization of freshwater fauna. Ann Rev Ecol Syst. 33: 291–315.
Rahel, F. J. 2007. Biogeographic barriers, connectivity and homogenization of freshwater faunas: Iit's a small world after all. Freshw Biol. 52: 696–710.
Ralls, K., J. D. Ballou, & A. Templeton. 1988. Estimates of lethal equivalents and the cost of inbreeding in mammals. Conserv Biol. 2: 185–193.
Ralls, K., K. Brugger, & J. Ballou. 1979. Inbreeding and juvenile mortality in small populations of ungulates. Science 206: 1101–1103.
Ratnasingham, S. & P. D. Hebert. 2007. BOLD: The Barcode of Life Data System (www.barcodinglife.org). Molecular Ecology Notes (2007) doi: 10.1111/j.1471-8286.2006.01678.x.
Rauch, E. M., & Y. Bar-Yam. 2004. Theory predicts the uneven distribution of genetic diversity within species. Nature 431: 449–452.
Raup, D., & J. Sepkoski. 1982. Mass extinctions in the marine fossil record. Science 215: 1501–1503.
Real, L. A., J. C. Henderson, R. Biek, J. Snaman, T. L. Jack, J. E. Childs, E. Stahl, L. Waller, R. Tinline, & S. Nadin-Davis. 2005. Unifying the spatial population dynamics and molecular evolution of epidemic rabies virus. Proc Natl Acad Sci USA 102: 12107–12111.
Réale, D., A. G. McAdam, S. Boutin, & D. Berteaux. 2003. Genetic and plastic responses of a northern mammal to climate change. Proc R Soc Lond B. 270: 591–596.
Redding, D. W., & A. Ø. Mooers. 2006. Incorporating evolutionary measures into conservation prioritization. Conserv Biol. 20: 1670–1678.
Reed, D. H. 2004. Extinction risk in fragmented habitats. Anim Conserv. 7: 181–191.
Reed, D. H. 2005. Relationship between population size and fitness. Conserv Biol. 19: 563–568.
Reed, D. H., D. A. Briscoe, & R. Frankham. 2002. Inbreeding and extinction: The effect of environmental stress and lineage. Conserv Genet. 3: 301–307.
Reed, D. H., & E. H. Bryant. 2000. Experimental tests of minimum viable population size. Anim Conserv. 3: 7–14.
Reed, D. H., & R. Fankham. 2001. How closely correlated are molecular and quantitative measures of genetic variation? A meta-analysis. Evolution 55: 1095–1103.
Reed, D. H., & R. Frankham. 2003. Correlation between fitness and genetic diversity. Conserv Biol. 17: 230–237.
Reed, D. H., & G. R. Hobbs. 2004. The relationship between population size and temporal variability in population size. Anim Conserv. 7: 1–8.
Reed, D. H., E. H. Lowe, D. A. Brisoce, & R. Frankham. 2003a. Fitness and adaptation in a novel environment: Effect of inbreeding, prior environment, and lineage. Evolution 57: 1822–1828.
Reed, D. H., A. C. Nicholas, & G. E. Stratton. 2007a. Inbreeding levels and prey abundance interact to determine fecundity in natural populations of two species of wolf spider. Conserv Genet. 8: 1061–1071.DOI: 10.1007/s10592-006-9260-4.
Reed, D. H., A. C. Nicholas, & G. E. Stratton. 2007b. The genetic quality of individuals directly impacts population dynamics. Anim Conserv. 10: 275–283.
Reed, D. H., J. J. O'Grady, J. D. Ballou, & R. Frankham. 2003b. Frequency and severity of catastrophic die-offs in vertebrates. Anim Conserv. 6: 109–114.
Reed, D. H., J. J. O'Grady, B. W. Brook, J. D. Ballou, & R. Frankham. 2003c. Estimates of minimum viable population sizes for vertebrates and factors influencing those estimates. Biol Conserv. 113: 23–34.
Regan, J. L., L. M. Meffert, & E. H. Bryant. 2003. A direct experimental test of founder–flush effects on the evolutionary potential for assortative mating. J Evol Biol. 16: 302–312.
Rehfeldt, G. E., C. C. Ying, D. L. Spittlehouse, & D. A. Hamilton, Jr. 1999. Genetic responses to climate in *Pinus contorta*: Niche breadth, climate change, and reforestation. Ecol Monogr. 69: 375–407.
Reichman, J. R., L. S. Watrud, E. H. Lee, C. A. Burdick, M. A. Bollman, M. J. Storm, G. A. King, & C. Mallory-Smith. 2006. Establishment of transgenic herbicide-resistant creeping bentgrass (*Agrostis stolonifera* L) in nonagronomic habitats. Mol Ecol. 15: 4243–4255.
Reid, J. M., P. Arcese, & L. F. Keller. 2003. Inbreeding depresses immune response in song sparrows (*Melopsiza melodia*): Direct and inter-generational effects. Proc R Soc Lond B. 270: 2151–2157.
Rejmanek, M., & D. M. Richardson. 1996. What attributes make some plant species more invasive? Ecology 77: 1655–1661.

Retallack, G. J., J. J. Veevers, & R. Morante. 1996. Global coal gap between Permian–Triassic extinction and Middle Triassic recovery of peat-forming plants. Geol Soc Am Bull. 108: 195–207.

Reznick, D. N., & C. K. Ghalambor. 2005. Can commercial fishing cause evolution? Answers from guppies (*Poecilia reticulata*). Can J Fish Aquat Sci. 62: 791–801.

Reznick, D., H. Rodd, & L. Nunney. 2004. Empirical evidence for rapid evolution (pp. 244–264). In R. Ferrière, U. Dieckmann, & D. Couvet (eds.). Evolutionary conservation biology. Cambridge University Press, Cambridge, UK.

Rhymer, J. M. 2001. Evolutionary relationships and conservation of the Hawaiian anatids. Stud Avian Biol. 22: 61–67.

Rhymer, J. M. 2006. Extinction by hybridization and introgression in Anatinae. Acta Zool Sinica. 52(Suppl.): 583–585.

Rhymer, J. M., & D. Simberloff. 1996. Extinction by hybridization and introgression. Annu Rev Ecol Syst. 27: 83–109.

Ricciardi, A., & S. K. Atkinson. 2004. Distinctiveness magnifies the impact of biological invaders in aquatic ecosystems. Ecol Lett. 7: 781–784.

Ricciardi, A., & H. J. MacIsaac. 2000. Recent mass invasion of the North American Great Lakes by Ponto-Caspian species. Trends Ecol Evol. 15: 62–65.

Ricciardi, A., & J. M. Ward. 2006. Comment on "Opposing effects of native and exotic herbivores on plant invasions." Science 313: 298.

Rice, K. J., & N. C. Emery. 2003. Managing microevolution: Restoration in the face of global change. Front Ecol Environ. 1: 469–478.

Rice, R. R., & E. E. Hostert. 1993. Laboratory experiments on speciation: What have we learned in 40 years? Evolution 47: 1637–1653.

Ricker, W. E. 1981. Changes in the average size and average age of Pacific salmon. Can J Fish Aquat Sci. 38: 1636–1656.

Ricker, W. E. 1995. Trends in the average size of Pacific salmon in Canadian catches. Can Spec Pub Fish Aqu Sci. 121: 593–602.

Rideout, S. G., & L. W. Stolte. 1989. Restoration of Atlantic salmon to the Connecticut and Merrimack rivers (pp. 67–81). In R. H. Stroud (ed.). Present and future Atlantic salmon management. Marine Recreational Fisheries, Savannah, Ga.

Rieseberg, L. H., O. Raymond, D. M. Rosenthal, Z. Lai, K. Livingstone, T. Nakazato, J. Durphy, A. E. Schwarzbach, L. A. Donovan & C. Lexer 2. 2003. Major ecological transitions in wild sunflowers facilitated by hybridization. Science 301: 1211–1216.

Rijnsdorp, A. D. 1993. Fisheries as a large-scale experiment on life-history evolution: Disentangling phenotypic and genetic effects in changes in maturation and reproduction of North Sea plaice, *Pleuronectes platessa* L. Oecologia 96: 391–401.

Ritland, K. 2002. Extensions of models for the estimation of mating systems using n independent loci. Heredity 88: 221–228.

Rizzo, D. M., & M. Garbelotto. 2003. Sudden oak death: Endangering California and Oregon forest ecosystems. Front Ecology Environment 1: 197–204.

Roberts, S. P., J. H. Marden, & M. E. Feder. 2003. Dropping like flies: Environmentally induced impairment and protection of locomotor performance in adult *Drosophila melanogaster*. Phys Biochem Zool. 76: 615–621.

Rodriguez-Trelles, F., G. Alvarez, & C. Zapata. 1996. Time-series analysis of seasonal changes in the O inversion polymorphism of *Drosophila subobscura*. Genetics 142: 179–187.

Rodriguez-Trelles, F., & M. A. Rodriguez. 1998. Rapid micro-evolution and loss of chromosomal diversity in *Drosophila* in response to global warming. Evol Ecol. 12: 829–838.

Roelke, M. E., J. S. Martensen, & S. J. O'Brien. 1993. The consequences of demographic reduction and genetic depletion in the endangered Florida panther. Curr Biol. 3: 340–350.

Roff, D. A. 1994. Habitat persistence and the evolution of wing dimorphsim in insects. Am Nat. 144: 772–798.

Rolff, J., & M. Siva-Jothy. 2003. Invertebrate ecological immunology. Science 301: 472–475.

Roman, J., & B. W. Bowen. 2000. The mock turtle syndrome: Genetic identification of turtle meat purchased in the southeastern United States of America. Anim Conserv. 3: 61–65.

Ronce, O., & I. Olivieri. 1997. Evolution of reproductive effort in a metapopulation with local extinctions and ecological succession. Am Nat. 150: 220–249.

Rose, R., S. McCammon, & S. Lively. 2006. Proceedings: Workshop on Confinement of Genetically Engineered Crops During Field Testing. September 13–14, 2004. Biotechnology Regulatory Services, USDA. Online. www.aphis.usda.gov/brs/pdf/conf_ws_proc2.pdf. Accessed February 29, 2008.

Rose, M. R., L. N. Vu, S. U. Park, & J. L. Graves. 1992. Selection on stress resistance increases

longevity in *Drosophila melanogaster*. Exp Gerontol. 27: 241–250.
Rosenfield, J. A., S. Nolasco, S. Lindauer, C. Sandoval, & A. Kodric-Brown. 2004. The role of hybrid vigor in the replacement of Pecos pupfish by its hybrids with sheepshead minnow. Conserv Biol. 18: 1589–1598.
Rosenzweig, M. L. 1995. Species diversity in space and time. Cambridge University Press, Cambridge, UK.
Rosenzweig, M., K. M. Brennan, T. D. Tayler, P. O. Phelps, A. Patapoutian, & P. A. Garrity. 2005. The *Drosophila* ortholog of vertebrate *TRPA1* regulates thermotaxis. Genes Dev. 19: 419–424.
Rosenzweig, M. L., & R. D. McCord. 1991. Incumbent replacement: Evidence of long-term evolutionary progress. Paleobiology 17: 202–213.
Rouget, M., R. M. Cowling, A. T. Lombard, AT Knight & IHK Graham 2 2006. Designing large-scale conservation corridors for pattern and process. Conserv Biol. 20: 549–561.
Rouget, M., R. M. Cowling, R. L. Pressey, & D. M. Richardson. 2003. Identifying spatial components of ecological and evolutionary processes for regional conservation planning in the Cape Floristic Region, South Africa. Divers Distrib. 9: 191–210.
Rowe, G., & T. J. Beebee. 2003. Population on the verge of a mutational meltdown? Fitness costs of genetic load for an amphibian in the wild. Evolution 57: 177–181.
Roy, B. A. 2001. Patterns of association between crucifers and their flower-mimic pathogens: Host jumps are more common than coevolution or cospeciation. Evolution 55: 41–53.
Rudgers, J. A., & S. Y. Strauss. 2004. A selection mosaic in the facultative mutualism between ants and wild cotton. Proc R Soc Lond B. 271: 2481–2488.
Ruesink, J. L., I. M. Parker, M. J. Groom, & P. M. Kareiva. 1995. Reducing the risks of nonindigenous species introductions: Guilty until proven innocent. Bioscience 45: 465–477.
Rundle, H. D., L. Nagel, J. W. Boughman, & D. Schluter. 2000. Natural selection and parallel speciation in sympatric sticklebacks. Science 287: 306–308.
Rutter, C. 1902. Natural history of the quinnat salmon. A report of investigations in the Sacramento River, 1886–1901. Bull US Fish Comm. 22: 65–141.
Ryder, O. A. 1986. Species conservation and the dilemma of subspecies. Trends Ecol Evol. 1: 9–10.
Ryder, O. A., A. McLaren, S. Brenner, YP Zhang & K Benirschke 2000. DNA banks for endangered animal species. Science 288: 275–277.
Ryman, N., & F. Utter (eds.). 1987. Population genetics and fishery management. University of Washington Press, Seattle, Wash.
Saccheri, I., & I. Hanski. 2006. Natural selection and population dynamics. Trends Ecol Evol. 21: 341–347.
Saccheri, I. J., M. Kuussaari, M. Kankare, P Vikman, W Fortelius & I Hanski 1998. Inbreeding and extinction in a butterfly metapopulation. Nature 392: 491–494.
Sacchi, P., D. Soglia, S. Maione, G Meneguz, M Campora & R Rasero 2004. A non-invasive test for sex identification in short-toed eagle (*Circaetus gallicus*). Mol Cell Probes 18: 193–196.
Sakai, A. K., F. W. Allendorf, J. S. Holt, D. M. Lodge et al. 2001. The population biology of invasive species. Annu Rev Ecol Syst. 32: 305–332.
Sambatti, J. B., & K. J. Rice. 2006. Local adaptation, patterns of selection, and gene flow in the Californian serpentine sunflower (*Helianthus exilis*). Evolution 60: 696–710.
Sambatti, J. B., & B. Sickler. 2006. Perl and Python codes to estimate Hardy-Weinberg and linkage disequilibrium related parameters. Mol Ecol Notes 6: 594–596.
Sanjuan, R., & S. F. Elena. 2006. Epistasis correlates to genomic complexity. Proc Natl Acad Sci USA 103: 14402–14405.
Sargent, R. D., & S. P. Otto. 2006. The role of local species abundance in the evolution of pollinator attraction in flowering plants. Am Nat 167: 67–80.
Sato, K., H. Matsuda, & A. Sasaki. 1994. Pathogen invasion and host extinction in lattice structured populations. J Math Biol. 32: 251–268.
Saunders, W. B., D. M. Work, & S. V. Nikolaeva. 2004. The evolutionary history of shell geometry in Paleozoic ammonoids. Paleobiology 30: 19–43.
Sax, D. F., J. J. Stachowicz, & S. D. Gaines (eds.). 2005. Species invasions: Insights into ecology, evolution and biogeography. Sinauer Associates, Sunderland, Mass.
Schaeffer, S. W., M. P. Goetting-Minesky, M. Kovacevic, J. R. Peoples, J. L. Graybill, J. M. Miller, K. Kim, J. G. Nelson, & W. W. Anderson. 2003. Evolutionary genomics of inversions in *Drosophila pseudoobscura*: Evidence for epistasis. Proc Natl Acad Sci USA 100: 8319–8324.

Scheiner, S. M., & L. Y. Yampolsky. 1998. The evolution of *Daphnia pulex* in a temporally varying environment. Genet Res 72: 25–37.

Schemske, D. W., B. C. Husband, M. H. Ruckelshaus, C Goodwillie, IM Parker & JG Bishop 1994. Evaluating approaches to the conservation of rare & endangered plants. Ecology 75: 584–606.

Schemske, D. W., & R. Lande. 1985. The evolution of self-fertilization and inbreeding depression in plants. II. Empirical observations. Evolution 39: 41–52.

Schlaepfer, M. A., P. W. Sherman, B. Blossey, & M. C. Runge. 2005. Introduced species as evolutionary traps. Ecol Lett. 8: 241–246.

Schluter, D. 2000. The ecology of adaptive radiation. Oxford University Press, New York.

Schmid-Hempel, P. 2004. Evolutionary ecology of insect immune defenses. Annu Rev Entomol. 50: 529–551.

Schmid-Hempel, P., & D. Ebert. 2002. On the evolutionary ecology of specific immune defence. Trends Ecol Evol. 18: 27–32.

Schmidt, P. S., L. Matzkin, M. Ippolito, & W. F. Eanes. 2005a. Geographic variation in diapause incidence, life-history traits, and climatic adaptation in *Drosophila melanogaster*. Evolution 59: 1721–1732.

Schmidt, P. S., A. B. Paaby, & A. S. Heschel. 2005b. Genetic variance for diapause expression and associated life histories in *Drosophila melanogaster*. Evolution 59: 2616–2625.

Schneider, C. 2000. Natural selection and speciation. Proc Natl Acad Sci USA 97: 12398–12399.

Schneider, C., T. B. Smith, B. Larison, & C. Moritz. 1999. A test of alternative models of diversification in tropical rainforests: Ecological gradients vs rainforest refugia. Proc Natl Acad Sci USA 94: 13869–13873.

Schonewald-Cox, C. M., S. M. Chambers, B. MacBryde, & L. Thomas (eds.). 1983. Genetics and conservation. Benjamin-Cummings, Menlo Park, Calif.

Schrag, S. J., & P. Wiener. 1995. Emerging infectious disease: What are the relative roles of ecology and evolution? Trends Ecol Evol. 10: 319–324.

Schtickzelle, N., & M. Baguette. 2004. Metapopulation viability analysis of the bog fritillary butterfly using RAMAS/GIS. Oikos 104: 277–290.

Schwarz, D., B. M. Matta, N. L. Shakir-Botteri, & B. A. McPheron. 2005. Host shift to an invasive plant triggers rapid animal hybrid speciation. Nature 436: 546–549.

Scott, J. M., F. Davis, B. Csuti, R. Noss, B. Butterfield, C. Groves, H. Anderson, S. Caicco, F. D'Erchia, T. C. Edwards, Jr., J. Ulliman, & R. G. Wright. 1993. GAP analysis: A geographic approach to protection of biological diversity. Wildl Monogr 123: 1–41.

Sechrest, W., T. M. Brooks, G. A. B. da Fonseca, WR Konstant, RA Mittermeier, A Purvis, AB Rylands & JL Gittleman 2002. Hotspots and the conservation of evolutionary history. Proc Natl Acad Sci USA 99: 2067–2071.

Seddon, N. 2005. Ecological adaptation and species recognition drives vocal evolution in neotropical suboscine birds. Evolution 59: 200–215.

Seehausen, O. 2006. Conservation: Losing biodiversity by reverse speciation. Curr Biol. 16: R334–R337.

Seehausen, O., J. J. M. Van Alphen, & F. Witte. 1997. Cichlid fish diversity threatened by eutrophication that curbs sexual selection. Science 277: 1808–1811.

Seger, J., & W. D. Hamilton. 1988. Parasites and sex (pp. 176–193). In R. E. Michod & B. R. Levin (eds.). Evolution of sex: an Examination of current ideas. Sinauer Associates, Sunderland, Mass.

Sehgal, R. N. M., H. I. Jones, & T. B. Smith. 2001. Host specificity and incidence of trypanosoma in some African rainforest birds: A molecular approach. Mol Ecol. 10:2319–2327.

Sgrò, C. M., & A. A. Hoffmann. 1998. Effects of stress combinations on the expression of additive genetic variation for fecundity in *Drosophila melanogaster*. Genet Res. 72: 13–18.

Shackelton, L., C. Parrish, U. Truyen, & E. Holmes. 2005. High rate of viral evolution associated with the emergence of carnivore parvovirus. Proc Natl Acad Sci USA 102: 379–384.

Shaffer, M. L. 1981. Minimum viable population sizes for species conservation. Bioscience 31: 131–134.

Shaw, A. J., C. J. Cox, & S. B. Boles. 2003. Global patterns in peatmoss biodiversity. Mol Ecol. 12: 2553–2570.

Shaw, R. G., C. J. Geyer, S. Wagenius, H. H. Hangelbroek, & J. R. Etterson. 2008. Unifying life history analyses for inference of fitness and population growth. American Naturalist, in press.

Shaw, R. G., & F. H. Shaw. 1994. Quercus: programs for quantitative genetic analysis using maximum likelihood. Online. http://www.cbs.umn.edu/eeb/events/quercus.shtml. Accessed September 21, 2007.

Sheehan, P. M. 1975. Brachiopod synecology in a time of crisis (Late Ordovician–Early Silurian). Paleobiology 1: 205–212.
Sheehan, P. M., & M. T. Harris. 2004. Microbialite resurgence after the Late Ordovician extinction. Nature 430: 75–78.
Sheldon, B. C., & S. Verhulst. 1996. Ecological immunology: Costly parasite defences and trade-offs in evolutionary ecology. Trends Ecol Evol. 11: 317–321.
Sherald, J. L., T. M. Stidham, J. M. Hadidian, & J. E. Hoeldtke. 1996. Progression of the dogwood anthracnose epidemic and the status of flowering dogwood in Catoctin Mountain Park. Plant Dis. 80: 310–312.
Sheridan, P., & D. Karowe. 2000. Inbreeding, outbreeding, and heterosis in the yellow pitcher plant, *Sarracenia flava* (Sarraceniaceae), in Virginia. Am J Bot. 87: 1628–1633.
Shigesada, N., & K. Kawasaki. 1997. Biological invasions: Theory and practice. Oxford University Press, New York.
Siemann, E., & W. E. Rogers. 2001. Genetic differences in growth of an invasive tree species. Ecol Lett. 4: 514–518.
Siemann, E., & W. E. Rogers. 2003. Reduced resistance of invasive varieties of the alien tree *Sapium sebiferum* to a generalist herbivore. Oecologia 135: 451–457.
Siepielski, A. M., & C. W. Benkman. 2005. A role for habitat area in the geographic mosaic of coevolution between red crossbills and lodgepole pine. J Evol Biol. 18: 1042–1049.
Signor, P. W., & G. J. Vermeij. 1994. The plankton and the benthos: origins and early history of an evolving relationship. Paleobiology 20: 297–319.
Sih, A., A. M. Bell, J. C. Johnson, & R. E. Ziemba. 2004. Behavioral syndromes: An integrative overview. Q Rev Biol. 79: 241–277.
Silbermann, R., & M. Tatar. 2000. Reproductive costs of heat shock protein in transgenic *Drosophila melanogaster*. Evolution 54: 2038–2045.
Silvertown, J., P. Poulton, E. Johnston, G. Edwards, M Heard & PM Biss 2006. The Park Grass Experiment 1856–2006: Its contribution to ecology. J Ecol. 94: 801–814.
Simberloff, D. 2006. Invasional meltdown 6 years later: Important phenomenon, unfortunate metaphor, or both? Ecol Lett. 9: 912–919.
Sinclair, B. J., & S. P. Roberts. 2005. Acclimation, shock and hardening in the cold. J Therm Biol. 30: 557–562.
Singer, F. J., W. T. Swank, & E. E. C. Clebsch. 1984. Effects of wild pig rooting in a deciduous forest. J Wild Manag. 48: 464–473.
Singer, F. J., L. C. Zeigenfuss, & L. Spicer. 2001. Role of patch size, disease, and movement in rapid extinction of bighorn sheep. Conserv Biol. 15: 1347–1354.
Slabbekoorn, H., & T. B. Smith. 2002. Habitat-dependent song divergence in the little greenbul: An analysis of environmental selection pressures on acoustic signals. Evolution 56: 1849–1858.
Slatkin, M. 1977. Gene flow and genetic drift in a species subject to frequent local extinction. Theor Popul Biol. 12: 253–262.
Smith, P. J. 1994. Genetic diversity of marine fisheries resources: Possible impacts of fishing. Food and Agriculture Organization (FAO) fish technical paper no. 344. FAO, Rome.
Smith, J. W., & C. W. Benkman. 2007. A coevolutionary arms race causes ecological speciation in crossbills. Am Nat. 169: 455–465.
Smith, T. B., M. W. Bruford, & R. K. Wayne. 1993. The preservation of process: The missing element of conservation programs. Biodive Lett. 1: 164–167.
Smith, T. B., R. Calsbeek, R. K. Wayne, K. H. Holder, D. Pires, & C. Bardeleben. 2005a. Testing alternative mechanisms of evolutionary divergence in an African rain forest passerine bird. J Evol Biol. 18: 257–268.
Smith, T. B., L. A. Freed, J. K. Lepson, & J. H. Carothers. 1995. Evolutionary consequences of extinctions in populations of a Hawaiian honeycreeper. Conserv Biol. 9: 107–113.
Smith, B. R., C. M. Herbinger, & H. R. Merry. 2001. Accurate partition of individuals into full-sib families from genetic data without parental information. Genetics 158: 1329–1338.
Smith, T. B., S. Saatchi, C. H. Graham, H. Slabbekoorn, & G. Spicer. 2005b. Putting process on the map: Why ecotones are important for preserving biodiversity (pp. 166–197). In A. Purvis, J. Gittleman, & T. Brooks (eds.). Phylogeny and conservation. Cambridge University Press, Cambridge, UK.
Smith, K. F., D. F. Sax, & K. D. Lafferty. 2006. Evidence for the role of infectious disease in species extinction and endangerment. Conserv Biol. 20: 1349–1357.
Smith, T. B., & R. K. Wayne (eds.). 1996. Molecular genetic approaches in conservation. Oxford University Press, New York.
Smith, T. B., R. K. Wayne, D. J. Girman, & M. W. Bruford. 1997. A role for ecotones in generating rainforest biodiversity. Science 276: 1855–1857.

Smouse, P. E., R. J. Dyer, R. D. Westfall, & V. L. Sork. 2001. Two-generation analysis of pollen flow across a landscape. I. Male gamete heterogeneity among females. Evolution 55: 260–271.

Snaydon, R. W. 1970. Rapid population differentiation in a mosaic environment. I. The response of *Anthoxanthum odoratum* populations to soils. Evolution 24: 257–269.

Snow, A. A., & P. M. Palma. 1997. Commercialization of transgenic plants: Potential ecological risks. Bioscience 47: 86–96.

Sokal, R. R., & F. J. Rohlf. 1985. Biometry: The principles and practice of statistics in biological research. 3rd ed. WH Freeman, New York.

Soleri, D., & D. A. Cleveland. 2006. Transgenic maize and Mexican maize diversity: Risk synergy? Agr Human Values 23: 27–31.

Soltis, D. E., C. H. Haufler, D. C, Darrow, & G. J. Gastony. 1983. Starch gel electrophoresis of ferns: A compilation of grinding buffers, gel and electrode buffers and staining schedules. Am Fern J. 73: 9–27.

Sorensen, J. G., T. N. Kristensen, & V. Loeschcke. 2003. The evolutionary and ecological role of heat shock proteins. Ecol Lett. 6: 1025–1037.

Sorenson, M. D., K. M. Sefc, & R. B. Payne. 2003. Speciation by host switch in brood parasitic indigobirds. Nature 424: 928–931.

Sork, V. L., J. Nason, D. R. Campbell, & J. F. Fernandez. 1999. Landscape approaches to historical and contemporary gene flow in plants. Trends Ecol Evol. 14: 219–223.

Sotka, E. E. 2005. Local adaptation in host use in marine invertebrates. Ecol Lett. 8: 448–459.

Soulé, M. E. 1987. Introduction (pp. 1–10). In M. E. Soulé (ed.). Viable populations for conservation. Cambridge University Press, Cambridge, UK.

Soulé, M. E., J. A. Estes, J. Berger, & C. Martinez del Rio. 2003. Ecological effectiveness: Conservation goals for interactive species. Conserv Biol. 17: 1238–1250.

Soulé, M. E., & M. A. Sanjayan. 1998. Conservation targets: Do they help? Science 279: 2060–2061.

Soulé, M. E., & D. Simberloff. 1986. What do genetics & ecology tell us about the design of nature reserves? Biol Conserv. 35: 19–40.

Soulé, M. E., & B. A. Wilcox. 1980. Conservation biology: An ecological–evolutionary perspective. Sinauer Associates, Sunderland, Mass.

Spielman, D., B. W. Brook, D. A. Briscoe, & R. Frankham. 2004a. Does inbreeding and loss of genetic diversity decrease disease resistance? Conserv Genet. 5: 439–448.

Spielman, D., B. W. Brook, & R. Frankham. 2004b. Most species are not driven to extinction before genetic factors impact them. Proc Natl Acad Sci USA 101: 15261–15264.

Spitze, K. 1993. Population structure in *Daphnia obtusa*: Quantitative genetic and allozymic variation. Genetics 135: 367–374.

Spong, G., M. Johansson, & M. Björkland. 2000. High genetic variation in leopards indicates large and long-term stable effective population size. Mol Ecol. 9: 1773–1782.

Squires, R. L. 2003. Turnovers in marine gastropod faunas during the Eocene–Oligocene transition, west coast of the United States (pp. 14–35). In D. R. Prothero, L. C. Ivany, & E. A. Nesbitt (eds.). From greenhouse to icehouse: The marine Eocene–Oligocene transition. Columbia University Press, New York.

Stanley, G. D., Jr., & P. K. Swart. 1995. Evolution of the coralzooxanthellae symbiosis during the Triassic: A geochemical approach. Paleobiology 21: 179–199.

Stearns, S. C., & J. C. Koella. 1986. The evolution of phenotypic plasticity in life-history traits: Prediction of reaction norms for age and size at maturity. Evolution 40: 893–913.

Stebbins, G. L. 1969. The significance of hybridization for plant taxonomy and evolution. Taxon 18: 26–35.

Steel, M. 2005. Phylogenetic diversity and the greedy algorithm. Syst Biol. 54: 527–529.

Steets, J. A., J. L. Hamrick, & T.- L. Ashman. 2006. Consequences of vegetative herbivory for maintenance of intermediate outcrossing in an annual plant. Ecology 87: 2717–2727.

Stewart, N. C. Jr. 2006. Go with the glow: Fluorescent proteins to light transgenic organisms. Trends Biotech. 24: 155–162.

Stewart, C. N., Jr., M. D. Halfhill, & S. I. Warwick. 2003. Transgene introgression from genetically modified crops to their wild relatives. Nat Rev Genet. 4: 806–817.

Stockwell, C. A., A. P. Hendry, & M. T. Kinnison. 2003. Contemporary evolution meets conservation biology. Trends Ecol Evol 18: 94–101.

Stokes, T. K., J. M. McGlade, & R. Law (eds.). 1993. The exploitation of evolving resources. Lecture notes in biomathematics 99. Springer-Verlag, Berlin, Germany.

Storfer, A. 1996. Quantitative genetics: A promising approach for the assessment of genetic variation in endangered species. Trends Ecol Evol. 11: 343–348.

Storfer, A., & A. Sih. 1998. Gene flow and ineffective antipredator behavior in a stream breeding salamander. Evolution 52: 558–565.
Strauss, S. H., & H. D. Bradshaw. 2004. The bioengineered forest: Challenges for science and society. Resources for the Future, Washington, D.C.
Strauss, S. Y., J. A. Lau, & S. P. Carroll. 2006a. Evolutionary responses of natives to introduced species: What do introductions tell us about natural communities? Ecol Lett. 9: 357–374.
Strauss, S. Y., C. O. Webb, & N. Salamin. 2006b. Exotic taxa less related to native species are more invasive. Proc Natl Acad Sci USA 103: 5841–5845.
Strayer, D. L., V. T. Eviner, J. M. Jeschke, & M. L. Pace. 2006. Understanding the long-term effects of species invasions. Trends Ecol Evol. 21: 645–651.
Street, G. T., G. R. Lotufo, P. Q. Montagna, & J. W. Fleeger. 1998. Reduced genetic diversity in a meiobenthic copepod exposed to a xenobiotic. J Exptl Marine Biol Ecol. 222: 93–111.
Strickland, B. K., S. Demerais, L. E. Castle, J. W. Lipe, W. H. Lunceford, H. A. Jacobson, D. Frels, & K. V. Miller. 2001. Effects of selective-harvest strategies on white-tailed deer antler size. Wildlife Soc Bull. 29: 509–520.
Su, C., D. Evans, R. Cole, J. Kissinger, J. Ajiola, & L. Sibley. 2003. Recent expansion of *Toxoplasma* through enhanced oral transmission. Science 299: 414–416.
Swain, D. P., A. F. Sinclair, & J. M. Hanson. 2007. Evolutionary response to size-selective mortality in an exploited fish population. Proc R Soc Lond B. 274: 1015–1022.
Swindell, W. R. 2006. The association among gene expression responses to nine abiotic stress treatments in *Arabidopsis thaliana*. Genetics 174: 1811–1824.
Swindell, W. R., & J. L. Bouzat. 2005. Modeling the adaptive potential of isolated populations: Experimental simulations using *Drosophila*. Evolution 59: 2159–2169.
Swindell, W. R., & J. L. Bouzat. 2006. Gene flow and adaptive potential in *Drosophila melanogaster*. Conserv Genet. 7: 79–89.
Switzer, W. M., M. Salemi, V. Shanmugam, F. Gao, M. Cong, C. Kuiken, V. Bhullar, B. E. Beer B. E., D. Vallet, A. Gautier-Hion, Z. Tooze, F. Villinger, E. C. Holmes, W. Heneine 2005. Ancient co-speciation of simian foamy viruses and primates. Nature 434: 376–380.
Taberlet, P., J.- J. Camarra, S. Griffin, E. Uhrès, O. Hanotte, L. P. Waits, C. Dubois-Paganon, T. Burke, & J. Bouvet. 1997. Noninvasive genetic tracking of the endangered Pyrenean brown bear population. Mol Ecol. 6: 869–876.
Takacs, D. 1996. The idea of biodiversity: Philosophies of paradise. Johns Hopkins University Press, Baltimore, Md.
Tallmon, D., G. Luikart, & R. S. Waples. 2004. The alluring simplicity and complex reality of genetic rescue. Trends Ecol Evol 19: 489–496.
Tanaka, Y. 1997. Extinction of populations due to inbreeding depression with demographic disturbances. Res Popul Ecol. 39: 57–66.
Tao, W., & E. G. Boulding. 2003. Associations between single nucleotide polymorphisms in candidate genes and growth rate in Arctic charr (*Salvelinus alpinus* L.). Heredity 91: 60–69.
Tautz, D. 1989. Hypervariability of simple sequences as a general source for polymorphic DNA markers. Nucl Acids Res. 17: 6463–6471.
Taylor, E. B. 1991. A review of local adaptation in Salmonidae, with particular reference to Pacific and Atlantic salmon. Aquaculture 98: 185–207.
Taylor, E. B., W. Boughman, M. Groenenboom, M. Sniatynski, D. Schluter, & J. L. Gow. 2006. Speciation in reverse: Morphological and genetic evidence of the collapse of a three-spined stickleback (*Gasterosteus aculeatus*) species pair. Mol Ecol. 15: 343–355.
Tear, T. H., P. Kareiva, P. L. Angermeier, P. Comer, B. Czech, R. Kautz, L. Landon, D. Mehlman, K. Murphy, M. Ruckelshaus, J. M. Scott, & G. Wilhere. 2005. How much is enough? The recurrent problem of setting measurable objectives in conservation. Bioscience 55: 835–849.
Templeton, A. R., K. Shaw, E. Routman, & S. K. Davis. 1990. The genetic consequences of habitat fragmentation. Ann Mo Bot Gard. 77: 13–27.
Terborgh, J., J. A. Estes, P. Paquet, K Ralls, D Boyd-Heigher, BJ Miller & RF Noss 1999. The role of top carnivores in regulating terrestrial ecosystems (pp. 39–54). In M. Soulé & J. Terborgh (eds.). Continental conservation: Scientific foundations of regional reserve networks. Island Press, Washington, D.C.
Tewksbury, L., R. Casagrande, & A. Gassmann. 2002. Swallow-warts (pp. 38–47). In R. van Driesche, B. Blossey, M Hoddle, S Loyon & R Reardon (eds.). Biological control of invasive plants in eastern United States. Forest Health Enterprise Team, Morgantown, W.V.

Thévenon, S., & D. Couvet. 2002. The impact of inbreeding depression on population survival depending on demographic parameters. Anim Conserv. 5: 53–60.

Thomas, C. D. 1990. What do real population dynamics tell us about minimum viable population sizes? Conserv Biol. 4: 324–327.

Thomas, C. D., E. J. Bodsworth, R. J. Wilson, A. D. Simmons, Z. G. Davies, M. Musche & L. Conradt. 2001. Ecological and evolutionary processes at expanding range margins. Nature 411: 577–588.

Thomas, C. D., A. Cameron, R. E. Green, M. Bakkenes, L. J. Beaumont, Y. C. Collingham, B. F. N Erasmus, M. F. de Siqueira, A. Grainger, L. Hannah, L. Hughes, B. Huntley, A. S van Jaarsveld, G. F. Midgley, L. Milles, M. A. Ortega-Huerta, A. T. Peterson, O. L. Phillips, & S. E. Williams. 2004. Extinction risk from climate change. Nature 427: 145–148.

Thomas, C. D., R. J. Wilson, & O. T. Lewis. 2002. Short-term studies underestimate 30-generation changes in a butterfly metapopulation. Proc R Soc Lond B. 269: 563–569.

Thompson, J. N. 1994. The coevolutionary process. University of Chicago Press, Chicago, Ill.

Thompson, J. N. 1996. Evolutionary ecology and the conservation of biodiversity. Trends Ecol Evol. 11: 300–303.

Thompson, J. N. 1998. Rapid evolution as an ecological process. Trends Ecol Evol. 13: 329–332.

Thompson, J. N. 1999. Specific hypotheses on the geographic mosaic of coevolution. Am Nat. 153: S1–S14.

Thompson, J. N. 2005. The geographic mosaic of coevolution. University Chicago Press, Chicago, Ill.

Thompson, R. S., K. H. Anderson, & P. J. Bartlein. 2000. Atlas of relations between climatic parameters and distributions of important trees and shrubs in North America. US Geological Survey professional paper, 1650 A-B. US Geological Survey, Wahsington, D.C.

Thompson, J. N., & B. M. Cunningham. 2002. Geographic structure and dynamics of coevolutionary selection. Nature 417: 735–738.

Thorne, E. T., & E. S. Williams. 1988. Disease and endangered species: The black-footed ferret as a recent example. Conserv Biol. 2: 66–74.

Thornhill, N. W. (ed.). 1993. The natural history of inbreeding and outbreeding. University of Chicago Press, Chicago, Ill.

Thrall P. H., J. J. Burdon, & J. D. Bever. 2002. Local adaptation in the *Linum marginale*—*Melampsora lini* host-pathogen interaction. Evolution 56:1340-1351 [Section 4]

Thrall, P. H., & J. J. Burdon. 2003. Evolution of virulence in a plant host–pathogen metapopulation. Science 299: 1735–1737.

Thuillet, A. C., D. Bru, J. David, P. Roumet, S. Santoni, P. Sourdille, & T. Bataillon. 2002. Direct estimation of mutation rate for 10 microsatellite loci in durum wheat, *Triticum turgidum* (L.) Thell. ssp *durum* desf. Mol Biol Evol. 19: 122–125.

Tilman, D., R. M. May, C. L. Lehman, & M. A. Nowak. 1994. Habitat destruction and the extinction debt. Nature 371: 65–66.

Toju, H., & T. Sota. 2006. Imbalance of predator and prey armament: Geographic clines in phenotypic interface and natural selection. Am Nat. 167: 105–117.

Tomimatsu, H., & M. Ohara. 2006. Evaluating the consequences of habitat fragmentation: A case study in the common forest herb *Trillium camschatcense*. Popul Ecol. 48: 189–198.

Torchin, M. E., K. D. Lafferty, A. P. Dobson, VJ McKenzie & AM Kuris 2003. Introduced species and their missing parasites. Nature 421: 628–630.

Tordoff, H. B., & P. T. Redig. 2001. Role of genetic background in the success of reintroduced peregrine falcons. Conserv Biol. 15: 528–532.

Totland, Ø. 1999. Effects of temperature on performance and phenotypic selection on plant traits in alpine *Ranunculus acris*. Oecologia 120: 242–251.

Traverset, A., & D. M. Richardson. 2006. Biological invasions as disruptors of plant reproductive mutualisms. Trends Ecol Evol. 21: 208–216.

Travis, J. M. J. 2003. Climate change and habitat destruction: A deadly anthropogenic cocktail. Proc R Soc Lond B. 270: 467–473.

Trippel, E. A. 1995. Age at maturity as a stress indicator in fisheries. Bioscience 45: 759–771.

Trouve, S., L. Degen, F. Renaud, & J. Goudet. 2003. Evolutionary implications of a high selfing rate in the freshwater snail *Lymnaea truncatula*. Evolution 57: 2303–2314.

Tsutsui, N. D., & A. V. Suarez. 2003. The colony structure and population biology of invasive ants. Conserv Biol. 17: 48–58.

Tufto, J. 2001. Effects of releasing maladapted individuals: A demographic–evolutionary model. Am Nat. 158: 331–340.

Turchin, P. 2003. Complex population dynamics. Princeton University Press, Princeton, N.J.

Twiddy, S. S., E. C. Holmes, & A. Rambaut. 2003. Inferring the rate and time-scale of

dengue virus evolution. Mol Biol Evol. 20: 1822–1829.
Twitchett, R. J., L. Krystyn, A. Baud, J. R. Wheeley & S. Richoz. 2004. Marine recovery after the end-Permian mass extinction event in the absence of marine anoxia. Geology 32: 805–808.
Umina, P. A., A. R. Weeks, M. R. Kearney, S. W. McKechnie & A. A. Hoffmann. 2005. A rapid shift in a classic clinal pattern in *Drosophila* reflecting climate change. Science 308: 691–693.
UNEP [United Nations Environment Programme]. 1992. Convention on biological diversity.?United Nations Environment Programme, Nairobi, Kenya.
Urban, D., & T. Keitt. 2001. Landscape connectivity: A graph–theoretic perspective. Ecology 82: 1205–1218.
Urban, M. C. 2006. Maladpatatoin and mass effects in a metacommunity: consequences for species coexistence. Amer. Natur. 168: 28–40.
Urban, M. C., B. L. Phillips, D. K. Skelly, & R. Shine. 2007. The cane toad's (*Chaunus* [*Bufo*] *marinus*) increasing ability to invade Australia is revealed by a dynamically updated range model. Proc R Soc Lond B. 274: 1413–1419.
U.S. Fish and Wildlife Service. 2007 Draft recovery plan for the northern spotted owl, *Strix occidentalis caurina*: Merged options 1 and 2. U.S. Fish and Wildlife Service. Portland, Ore.
USC [United States Congress] 1973. Endangered species act. Public Law 93–205, 87 Stat. 884, 16 United States Congress. §1531–§1544, Washington, DC.
Vacher, C., A. E. Weiss, D. Hermann, Kossler, C. Young & M. E. Hochberg 20. 2004. Impact of ecological factors on the initial invasion of *Bt* transgenes into wild populations of birdseed rape (*Brassica rapa*). Theor Appl Gen. 109: 806–814.
Vähä, J.- P., & C. R. Primmer. 2006. Efficiency of model-based Bayesian methods for detecting hybrid individuals under different hybridization scenarios and with different numbers of loci. Mol Ecol. 15: 63–72.
Vamosi, J. C., T. M. Knight, J. A. Steets, S. J. Mazer, M. Burd & T. L. Ashman. 2006. Pollination decays in biodiversity hotspots. Proc Natl Acad Sci USA 103: 956–961.
Van Buskirk, J., & Y. Willi. 2006. The change in quantitative genetic variation with inbreeding. Evolution 60: 2428–2434.
Van der Meijden, R., B. Odé, C. L. G. Groen, J.-P. M. Witte, & D. Bal 2 2000. Bedreigde en kwetsbare vaatplanten in Nederland. Basisrapport met voorstel voor de Rode Lijst. Gorteria 26: 85–208.
van Frankenhuyzen, K., & T. Beardmore. 2004. Current status and environmental impact of transgenic forest trees. Can J For Res. 34: 1163–1180.
van Oers, K., M. Klunder, & P. J. Drent. 2005. Context dependence of personalities: Risk-taking behavior in a social and a nonsocial situation. Behav Ecol. 16: 716–723.
van Oosterhout, C., W. G. Zijlstra, M. K. van Heuven, & P. M. Brakefield. 2000. Inbreeding depression and genetic load in laboratory metapopulations of the butterfly *Bicyclus anynana*. Evolution 54: 218–225.
Van Riper, C., S. G. Van Riper, M. L. Goff, & M. Laird. 1986. The epizootiology and ecological significance of malaria in Hawaiian land birds. Ecol Monogr. 56: 327–344.
Van Treuren, R., R. Bijlsma, N. J. Ouborg, & W. Van Delden. 1993. The significance of genetic erosion in the process of extinction IV. Inbreeding depression and heterosis effects caused by selfing and outcrossing in *Scabiosa columbaria*. Evolution 47: 1669–1680.
Van Valkenburgh, B. 1999. Major patterns in the history of carnivorous mammals. Earth Plan Sci Rev. 26: 463–493.
Vane-Wright, R. I., C. J. Humphries, & P. H. Williams. 1991. What to protect? Systematics and the agony of choice. Biol Conserv. 55: 235–254.
Vasemägi, A., J. Nilsson, & C. R. Primmer. 2005. Expressed sequence tag-linked microsatellites as a source of gene-associated polymorphisms for detecting signatures of divergent selection in Atlantic Salmon (*Salmo salar* L.). Mol Biol Evol. 22: 1067–1076.
Vellend, M., K. Verheyen, H. Jacuemyn, A. Kolb, H. Van Calster, G. Peterken, & M. Hermy. 2006. Extinction debt persists for more than a century following habitat fragmentation. Ecology 87: 304–311.
Verbeek, M. E. M., A. Boon, & P. J. Drent. 1996. Exploration, aggressive behavior and dominance in pair-wise confrontations of juvenile male great tits. Behaviour 133: 945–963.
Vergeer, P., R. Rengelink, A. Copal, & N. J. Ouborg. 2003. The interacting effects of genetic variation, habitat quality and population size on performance of *Succisa pratensis*. J Ecol. 91: 18–26.
Vergeer, P., E. Sonderen, & N. J. Ouborg. 2004. Introduction strategies put to the test: Local adaptation versus heterosis. Conserv Biol. 18: 812–821.

Vermeij, G. J. 1977. Biogeography and adaptation. Princeton University Press, Princeton, N.J.
Vermeij, G. J. 1987. Evolution and escalation: An ecological history of life. Princeton University Press, Princeton, N.J.
Vermeij, G. J. 1996. An agenda for invasion biology. Biol Conserv. 78: 3–9.
Vermeij, G. J. 2001. Community assembly in the sea: Geologic history of the living shore biota (pp. 39–60). In M. D. Bertness, S. D. Gaines, & M. E. Hay (eds.). Marine community ecology. Sinauer Associates, Sunderland, Mass.
Vermeij, G. J. 2004. Nature: An economic history. Princeton University Press, Princeton. N.J.
Vermeij, G. J. 2005a. From Europe to America: Pliocene to recent trans-Atlantic expansion of cold-water North Atlantic molluscs. Proc R Soc Lond B 272: 2545–2550.
Vermeij, G. J. 2005b. Invasion as expectation: A historical fact of life (pp. 315–339). In D. F. Sax, J. J. Stachowicz, & S. D. Gaines (eds.). Species invasions: Insights into ecology, evolution, and biogeography. Sinauer Associates, Sunderland, Mass.
Vermeij, G. J. 2005c. One-way traffic in the western Atlantic: Causes and consequences of Miocene to early Pleistocene molluscan invasions in Florida and the Caribbean. Paleobiology 31: 624.
Vermeij, G. J., & D. Dorritie. 1996. Late Permian extinctions. Science 274: 1550.
Vermeij, G. J., & G. Rosenberg. 1993. Giving and receiving: The tropical Atlantic as donor and recipient region for invading species. Am Malacol Bull. 10: 181–194.
Vigouroux, Y., J. S., Jaqueth, Y. Matsuoka, O. S. Smith, W. D. Beavis, J. S. Smith, & J. Doebley. 2002. Rate and pattern of mutation at microsatellite loci in maize. Mol Biol Evol. 19: 1251–1260.
Vilà. C., A.-K. Sundqvist, Ø. Flagstad, J. Seddon, S. Björnerfeldt, I. Kojola, A. Casulli, H. Sand, P. Wabakken, & H. Ellegren. 2003. Rescue of a severely bottlenecked wolf (*Canis lupus*) population by a single immigrant. Proc R Acad Lond B 270: 91–97.
Villier, L., & D. Korn. 2004. Morphological disparity of ammonoids and the mark of Permian mass extinctions. Science 306: 264–266.
Viranta, S. 2003. Geographic and temporal ranges of middle and late Miocene carnivores. J Mamm. 84: 1267–1278.
Visser, M. E., & L. J. M. Holleman. 2001. Warmer springs disrupt the synchrony of oak and winter moth phenology. Proc R Soc Lond B 268: 289–294.
Visser, M. E., A. J. van Noordwijk, J. M. Tinbergen, & C. M. Lessells. 1998. Warmer springs lead to mistimed reproduction in great tits (*Parus major*). Proc R Soc Lond B 265: 1867–1870.
Vitalis, R., & D. Couvet. 2001. Estimation of effective population size and migration rate from one- and two-locus identity measures. Genetics 157: 911–925.
Vitalis, R., S. Glemin, & I. Olivieri. 2004. When genes go to sleep: The population genetic consequences of seed dormancy and monocarpic perenniality. Am Nat. 163: 295–311.
Vitousek, P. M., C. M. D'Antonio, L. L. Loope, M Rejmanek, & R Westbrooks 1997. Introduced species: A significant component of human-caused global change. N Zeal J Ecol. 21: 1–16.
Vogler, D. W., & S. Kalisz. 2001. Sex among the flowers: The distribution of plant mating systems. Evolution 55: 202–204.
Vucetich, J. A., & R. O. Peterson. 2004. The influence of prey consumption and demographic stochasticity on population growth rate of Isle Royale wolves *Canis lupus*. Oikos 107: 309–320.
Vucetich, J. A., & T. A. Waite. 1999. Erosion of heterozygosity in fluctuating populations. Conserv Biol. 13: 860–868.
Wakeley, J. 1996. Distinguishing migration from isolation using the variance of pairwise differences. Theor Popul Biol. 49: 369–386.
Wakeley, J., & N. Aliacar. 2001. Gene genealogies in a metapopulation. Genetics 159: 893–905.
Wakeley, J., & S. Lessard. 2003. Theory of the effects of population structure and sampling on patterns of linkage disequilibrium applied to genomic data from humans. Genetics 164: 1043–1053.
Walker, P., & D. P. Faith. 1994. DIVERSITY-PD: Procedures for conservation evaluation based on phylogenetic diversity. Biodivers Lett. 2: 132–139.
Wallace, B. 1968. Polymorphism, population size & genetic load (pp. 87–108). In R. C. Lewontin (ed.). Population biology and evolution. Syracuse University Press, Syracuse, N.Y.
Walsh, B. 2001. Quantitative genetics in the age of genomics. Theor Popul Biol. 59: 175–184.
Walsh, P. D., K. A. Abernethy, M. Bermejo, R. Beyersk, P. De Wachter, M. E. Akou, B. Huljbregis, D. I. Mambounga, A. K. Toham, A. M. Kilbourn, S. A. Lahm, S. Latour, F. Maisels, C. Mbina, Y. Mihindou, S. N. Obiang, E. N. Effa, M. P. Starkey, P. Telfer, M. Thibault, C. E. Tutin,

L. J. White, D. S. Wilkie. 2003. Catastrophic ape decline in western equatorial. Afr Nat. 422: 611–614.
Walsh, P. D., R. Biek, & L. A. Real. 2005. Wave-like spread of Ebola Zaire. PLoS Biol. 3: E371.
Wang, J., & M. C. Whitlock. 2003. Estimating effective population size and migration rates from genetic samples over space and time. Genetics 163: 429–446.
Waples, R. S. 1989. A generalized approach for estimating effective population size from temporal changes in allele frequency. Genetics 121: 379–391.
Waples, R. S. 1995. Evolutionary significant units and the conservation of biological diversity under the endangered species act (pp. 8–27). In J. L. Nielsen (ed.) Evolution and the aquatic ecosystem: defining unique units in population conservation. Symposium 17. American Fisheries Society, Bethesda, Md.
Ward, J. K., J. Antonovics, R. B. Thomas, & B. R. Strain. 2000. Is atmospheric CO2 a selective agent on model C-3 annuals? Oecologia 123: 330–341.
Ward, J. R., K. Kim, & C. D. Harvell. 2007. Temperature drives coral disease resistance and pathogen growth. Mar Ecol Prog Ser. 329: 115–121.
Wares, J. P., A. R. Hughes, & R. K. Grosberg. 2005. Mechanisms that drive evolutionary change: Insights from species introductions and invasions (pp. 229–257). In D. F. Sax, J. J. Stachowicz, & S. D. Gaines (eds.). Species invasions: Insights into ecology, evolution, and biogeography. Sinauer Associates, Sunderland, Mass.
Waser, N. M., & M. V. Price. 1994. Crossing-distance effects in *Delphinium nelsonii*: Outbreeding and inbreeding depression in progeny fitness. Evolution 48: 842–852.
Waser, N. M., M. V. Price, & R. G. Shaw. 2000. Outbreeding depression varies among cohorts of *Ipomopsis aggregata* planted in nature. Evolution 54: 485–491.
Watrud, L. S., E. H. Lee, A. Fairbrother, C. Burdick, J. R. Reichman, M. Bollmman, M. Storm, G. King & P. K. Van de Water. 2004. Evidence for landscape-level, pollen-mediated gene flow from genetically modified creeping bentgrass with CP4 EPSPS as a marker. Proc Natl Acad Sci USA 101: 14533–14538.
Watters, J. V., S. C. Lema, & G. A. Nevitt. 2003. Phenotype management: A new approach to habitat restoration. Biol Conserv. 112: 435–445.
Watters, J. V., & C. L. Meehan. 2007. Different strokes: Can managing behavioral types increase post-release success? Appl Anim Behav Sci. 102: 364–379.
Webb, C. J., & D. G. Lloyd. 1986. The avoidance of interference between the presentation of pollen and stigmas in angiosperms II. Herkogamy. New Zealand J Bot. 24: 163–178.
Weeks, A. R., S. W. McKechnie, & A. A. Hoffmann. 2002. Dissecting adaptive clinal variation: Markers, inversions and size/stress associations in *Drosophila melanogaster* from a central field population. Ecol Lett. 5: 756–763.
Weill. M., F. Chandra, C. Brengues, S Manguin, M Akogbeto, N Pasteur, P Guillet & M Raymond 2000. The *kdr* mutation occurs in the Mopti form of *Anopheles gambiae s.s.* through introgression. Insect Mol Biol. 9: 451–455.
Weiner, J. 1990. Asymmetric competition in plant populations. Trends Ecol Evol. 5: 360–364.
Weinig, C., M. T. Brock, J. A. Dechaine, & S. M. Welch. 2007. Resolving the genetic basis of invasiveness and predicting invasions. Genetica 129: 205–216.
Weinig, C., M. C. Ungerer, L. A. Dorn, N. C. Kane, Y. Toyonaga, S. S. Halldorsdottir, T. F. C. Mackay, M. D. Purugganan & J. Schmitt. 2002. Novel loci control variation in reproductive timing in *Arabidopsis thaliana* in natural environments. Genetics 162: 1875–1884.
Weitzman, M. L. 1992. On diversity. Q J Econ. 107: 363–405.
Weitzman, M. L. 1998. The Noah's Ark problem. Econometrica 66: 1279–1298.
Weller, S. G. 1979. Variation in heterostylous reproductive systems among populations of *Oxalis alpine* in southeastern Arizona. Syst Bot. 4: 57–71.
Weller, S. G., A. K. Sakai, & W. L. Wagner. 2001. Artificial and natural hybridization in *Schiedea* and *Alsinidendron* (Caryophyllaceae: Alsinoideae): The importance of phylogeny, genetic divergence, breeding system, and population size. Syst Bot. 26: 571–584.
Westemeier, R. L., J. D. Brawn, S. A. Simpson, T. L. Esker, R. W. Jansen, J. W. Walk, E. L. Kershner, J. L. Bouzat, & K. N. Paige. 1998. Tracking the long-term decline and recovery of an isolated population. Science 282: 1695–1698.
Westerling, A. L., H. G. Hidalgo, D. R. Cayan, & T. W. Swetnam. 2006. Warming and earlier spring increase western US forest wildfire activity. Science 313: 940–943.

Whitlock, M. C., & N. H. Barton. 1997. The effective size of a subdivided population. Genetics 146: 427–441.

Whitlock, M. C., & D. E. McCauley. 1990. Some population genetic consequences of colony formation and extinction: Genetic correlations within founding groups. Evolution 44: 1717–1724.

Whittaker, J. C., R. M. Harbord, N. Boxall, I. Mackay, G. Dawson, & R. M. Sibly. 2003. Likelihood-based estimation of microsatellite mutation rates. Genetics 164: 781–787.

Wignall, P. B., & R. J. Twitchett. 1996. Oceanic anoxia and the end Permian mass extinction. Science 272: 155–158.

Wilcove, D. S., J. Rothstein, J. Dubow, A Phillips & E Losos 1998. Quantifying threats to impreiled species in the United States. Bioscience 48: 607–615.

Wilding, C. S., R. K. Butlin, & J. Grahame. 2001. Differential gene exchange between parapatric morphs of *Littorina saxatilis* detected using AFLP makers. J Evol Biol. 14: 611–619.

Wildt, D. E., W. F. Rall, J. K. Critser, S. L. Monfort & U. S. Seal. 1997. Genome resource banks. Bioscience 47: 689–698.

Wildt, D. E., & C. Wemmer. 1999. Sex and wildlife: The role of reproductive science in conservation. Biodivers Conserv. 8: 965–976.

Wilf, P., & K. R. Johnson. 2004. Land plant extinction at the end of the Cretaceous: A quantitative analysis of the North Dakota megafloral record. Paleobiology 30: 347–368.

Wilgers, D. J., & E. A. Horne. 2006. Effects of different burn regimes on tallgrass prairie herpetofaunal species diversity and community composition in the Flint Hills, Kansas. J Herpetol. 40: 73–84.

Wilkinson, M. J., L. J. Elliott, J. Allainguillaume, M. W. Shaw, C. Norris, R. Welters, M. Alexander, J. Sweet, & D. C. Mason. 2003. Hybridization between *Brassica napus* and *B. rapa* on a national scale in the United Kingdom. Science 302: 457–459.

Will, K. W., & D. Rubinoff. 2004. Myth of the molecule: DNA barcodes for species cannot replace morphology for identification and classification. Cladistics 20: 47–55.

Willi, Y., & M. Fischer. 2005. Genetic rescue in interconnected populations of small and large size of the self-incompatible *Ranunculus reptans*. Heredity 95: 437–443.

Willi, Y., J. Van Buskirk, & A. A. Hoffmann. 2006. Limits to the adaptive potential of small populations. Annu Rev Ecol Evol Syst. 37: 433–458.

Williams, C. G. 2006a. Landscapes, genomics and transgenic conifers. Springer Press, Dordrecht, the Netherlands.

Williams, C. 2006b. Opening Pandora's box: Governance for genetically modified forests. ISB News Report, January: 1–5.

Williams, C. L., R. C. Brust, T. T. Fendley, G. R. Tiller Jr, & O. E. Rhodes Jr. 2005a. A comparison of hybridization between mottled ducks (*Anas fulvigula*) and mallards (*A. platyrhynchos*) in Florida and South Carolina using microsatellite DNA analysis. Conserv Genet. 6: 445–453.

Williams, P., D. Faith, L. Manne, W. Sechrest, & C. Prestone. 2005b. Complementarity analysis: Mapping the performance of surrogates for biodiversity. Biol Conserv. 128: 253–264.

Williams, D. A., E. Muchugu, W. A. Overholt, & J. P. Cuda. 2007. Colonization patterns of the invasive Brazilian peppertree, *Schinus terebinthifolius*, in Florida. Heredity 98: 284–293.

Williams, J. W., B. N. Shuman, T. Webb, III, P. J. Bartlein & P. L. Leduc. 2004. Late-Quaternary vegetation dynamics in North America: Scaling from taxa to biomes. Ecol Monogr. 74: 309–334.

Williamson, M., & A. Fitter. 1996. The invaders. Ecology 77: 1661–1666.

Williamson, E. G., & M. Slatkin. 1999. Using maximum likelihood to estimate population size from temporal changes in allele frequencies. Genetics 152: 755–761.

Willis, J. H., & H. A. Orr. 1993. Increased heritable variation following population bottlenecks: The role of dominance. Evolution 47: 949–957.

Wills, R. T. 1993. The ecological impact of *Phytophthora cinnamomi* in the Stirling Range National Park, Western-Australia. Aust J Ecol. 18: 145–159.

Wilson, E. O. 1992. The diversity of life. Harvard University Press, Cambridge, Mass.

Wilson, E. O. 1997. Forward (pp.ix–x). In D. Simberloff, D. C. Schmitz, & T. C. Brown (eds.). Strangers in paradise. Island Press, Washington, D.C.

Wilson, A. C., R. L. Cann, S. M. Carr, M. George, U. B. Gyllensten, K. M. Helm-Bychowski, R. G. Higuchi, S. R. Palumbi, E. M. Prager, R. D. Sage, & M. Stoneking. 1985. Mitochondrial DNA and two perspectives on evolutionary genetics. Biol J Linn Soc. 26: 375–400.

Wilson, A. J., & M. M. Ferguson. 2002. Molecular pedigree analysis in natural populations of fishes: Approaches, applications and practical considerations. Can J Fish Aquat Sci. 59: 1696–1707.

Wilson, P. J., S. K. Grewal, I. D. Lawford, J. N. M. Heal, A. G. Granacki, D. Pennock, J. B.

Theberge, M. T. Theberge, D. R. Voigt, W. Waddell, R. E. Chambers, P. C. Paquet, G. Goulet, D. Cluff & B. N. White. 2000. DNA profiles of the eastern Canadian wolf and the red wolf provide evidence for a common evolutionary history independent of the gray wolf. Can J Zool. 78: 2156–2166.

Wilson, P. J., S. Grewal, T. McFadden, R. C. Chambers, & B. N. White. 2003a. Mitochondrial DNA extracted from eastern North American wolves from the 1800s is not of gray wolf origin. Can J Zool. 81: 936–940.

Wilson, A. J., G. McDonald, H. K. Moghadam, et al. C. M. Herbinger, & M. M. Ferguson. 2003b. Marker-assisted estimation of quantitative genetic parameters in rainbow trout, *Oncorhynchus mykiss*. Genet Res. 81: 145–156.

Wing, S. L., & G. J. Harrington. 2001. Global response to rapid warming in the earliest Eocene and implications for concurrent faunal change. Paleobiology 27: 539–563.

Wing, S. L., G. J. Harrington, F. A. Smith, J. I. Bloch, D. M. Boyer, & K. H. Freeman 2. 2005. Transient global change and rapid global warming at the Paleocene–Eocene boundary. Science 310: 993–996.

Witting, L., & V. Loescke. 1995. The optimization of biodiversity conservation. Biol Conserv. 71: 205–207.

Wolf, D. E., N. Takebayashi, & L. H. Rieseberg. 2001. Predicting the risk through hybridization. Conserv Biol. 15: 1039–1053.

Wolfe, L. M., J. A. Elzinga, & A. Biere. 2004. Increased susceptibility to enemies following introduction in the invasive plant *Silene latifolia*. Ecol Lett. 7: 813–820.

Wolfenbarger, L. L., & P. R. Phifer. 2000. The ecological risks and benefits of genetically engineered plants. Science 290: 2088–2093.

Woodruff, R. C., J. N. Thompson, Jr., & S. Gu. 2004. Premeiotic clusters of mutation and the cost of natural selection. J Hered. 95: 277–283.

Woods, A. D., D. J. Bottjer, M. Mutti, & J. Morrison. 1999. Lower Triassic large sea-floor carbonate cements: Their origin and a mechanism for the prolonged biotic recovery from the end-Permian mass extinction. Geology 27: 645–648.

Woodworth, B. L., C. T. Atkinson, D. A. LaPointe, P. J. Hart, C. S. Spiegel, E. J. Tweed, C. Henneman, J. LeBrun, T. Denette, R. DeMots, K. L. Kozar, D. Triglia, D. Lease, A. Gregor, T. Smith, D. Duffy. 2005. Host population persistence in the face of introduced vector-borne diseases: Hawaii amakihi and avian malaria. Proc Natl Acad Sci USA 102: 1531–1536.

Woolhouse, M. E. J., D. T. Haydon, & R. Antia. 2005. Emerging pathogens: The epidemiology and evolution of species jumps. Trends Ecol Evol. 20: 238–244.

Wright, S. 1931. Evolution in mendelian populations. Genetics 16: 97–159.

Wright, S. 1965. The interpretation of population structure by F-statistics with special regard to systems of mating. Evolution 19: 395–420.

Yazdanbakhsh, M., P. G. Kremsner, & R. van Ree. 2002. Allergy, parasites, and the hygiene hypothesis. Science 296: 490–494.

Yesson, C., & A. Culham. 2006. A phyloclimatic study of Cyclamen. BMC Evol Biol. 6: 72.

Yoshida, T., L. E. Jones, S. P. Ellner, G. F. Fussman. & N. G. Hairston. 2003. Rapid evolution drives predator–prey dynamics in a predator-prey system. Nature 424: 303–306.

Young, T. P. 1991. Diversity overrated. Nature 352: 10.

Young, A., T. Boyle, & A. H. D. Brown. 1996. The population genetic consequences of habitat fragmentation for plants. Trends Ecol Evol. 11: 413–418.

Young, A. G., & G. M. Clarke. 2000. Genetics, demography and viability of fragmented populations. Cambridge University Press, New York.

Yuhki, N., & S. O'Brien. 1990. DNA variation of the mammalian major histocompatibility complex reflects genomic diversity and population history. Proc Natl Acad Sci USA 87: 836–840.

Zachos, J., M. Pagani, L. Sloan, E. Thomas & K. Billups. 2001. Trends, rhythms, and aberrations in the global climate 65 Ma to present. Science 292: 686–693.

Zangerl, A. R., & M. R. Berenbaum. 2003. Phenotype matching in wild parsnip and parsnip webworms: Causes and consequences. Evolution 57: 806–815.

Zangerl, A. R., & M. R. Berenbaum. 2005. Increase in toxicity of an invasive weed after reassociation with its coevolved herbivore. Proc Natl Acad Sci USA 102: 15529–15532.

Zani, P. A., S. E. T. Swanson, D. Corbin, L. W. Cohnstaedt, M. D. Agotsch, W. E. Bradshaw, & C. M. Holzapfel. 2005. Geographic variation in tolerance of transient thermal stress in the mosquito *Wyeomyia smithii*. Ecology 86: 1206–1211.

Zartman, C. E., S. F. McDaniel, & J. Shaw. 2006. Experimental habitat fragmentation increases linkage disequilibrium but does not affect genetic diversity or population structure in the Amazonian liverwort *Radula flaccida*. Mol Ecol. 15: 2305–2315.

Zatsepina, O. G., V. V. Velikodvorskaia, V. B. Molodtsov, D. Garbuz, D. N. Lerman, B. R. Bettencourt, M. E. Feder, & M. B. Evgenev. 2001. A *Drosophila melanogaster* strain from sub-equatorial Africa has exceptional thermotolerance but decreased Hsp70 expression. J Exp Biol. 204: 1869–1881.

Zeyl, C., M. Mizesko, & J. A. G. M. De Visser. 2001. Mutational meltdown in laboratory yeast populations. Evolution 55: 909–917.

Zhong, D., A. Pai, & G. Yan. 2005. Costly resistance to parasitism: Evidence from simultaneous quantitative trait loci mapping for resistance and fitness in *Tribolium castaneum*. Genetics 169: 2127–2135.

Zhong, D., E. A. Temu, T. Guda, L. Gouagna, D. Menge, A. Pai, J. Githure & G. Yan 2. 2006. Dynamics of gene introgression in the African malaria vector, *Anopheles gambiae*. Genetics 172: 2359–2365.

Zoological Society of San Diego. 2008. Genome Studies for Conservation of the California Condor. http://cres.sandiegozoo.org/projects/gr_condor_genome.html. Accessed March 4, 2008.

Index